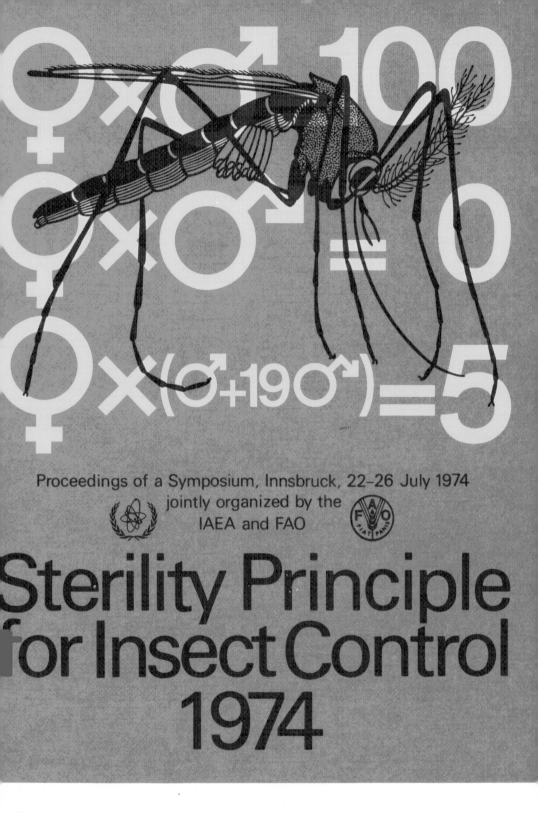

Proceedings of a Symposium, Innsbruck, 22–26 July 1974
jointly organized by the
IAEA and FAO

Sterility Principle for Insect Control 1974

INTERNATIONAL ATOMIC ENERGY AGENCY, VIENNA, 1975

STERILITY PRINCIPLE
FOR INSECT CONTROL 1974

PROCEEDINGS SERIES

STERILITY PRINCIPLE
FOR INSECT CONTROL 1974

PROCEEDINGS OF THE SYMPOSIUM ON
THE STERILITY PRINCIPLE FOR INSECT CONTROL
JOINTLY ORGANIZED BY THE
INTERNATIONAL ATOMIC ENERGY AGENCY
AND THE FOOD AND AGRICULTURE ORGANIZATION
OF THE UNITED NATIONS,
AND HELD IN INNSBRUCK,
22 - 26 JULY 1974

INTERNATIONAL ATOMIC ENERGY AGENCY
VIENNA, 1975

STERILITY PRINCIPLE FOR INSECT CONTROL 1974
IAEA, VIENNA, 1975
STI/PUB/377
ISBN 92–0–010275–1

Printed by the IAEA in Austria
February 1975

FOREWORD

The Symposium, held in Innsbruck, Austria, from 22 to 26 July 1974, jointly organized by the International Atomic Energy Agency and the Food and Agriculture Organization of the United Nations, was devoted to the sterility principle of insect control and related entomological studies. A total of 91 participants from 37 countries and five international organizations attended the meeting. The 53 papers presented and the discussions resulting from these papers are published in this volume.

In 1970 a similar symposium was held in Athens, Greece. At that time, most of the developed countries in the world had surpluses of many agricultural commodities. Now, however, this is no longer true. There are at the present time, and probably will be for the next few years, severe shortages of agricultural products in many parts of the world. In addition, the rapidly increasing cost and the shortage of insecticides has resulted in many developing countries actively searching for alternative methods of insect control. It was against this background that the Innsbruck Symposium was held.

The sterile insect technique is a species-specific method of insect control or, under certain conditions, eradication. The requirements for effectively implementing insect control by the sterility principle are not particularly difficult to meet. They include information on the basic biology, ecology, rearing, sterilization and releasing of the target insect. Perhaps the most difficult aspect of implementing insect control by this method is organizational, since for effective results it must be applied over a large area where the total population of the noxious insect is to be controlled. Individual producers cannot use the method successfully unless all the producers in the area use it.

The numerous successful field experiments to test the effectiveness of the sterile insect technique are encouraging. The next step, practical commercial usage, should be accomplished with a few insect species within the next five years.

The organizers of the Symposium are grateful to the Government of Austria for acting as host. The close co-operation of the staff of the Innsbruck Kongresshaus was indispensable in making the Symposium a success.

EDITORIAL NOTE

CONTENTS

COMPUTER MODELS AND APPLICATIONS OF THE
STERILE-MALE TECHNIQUE (Session 3)

CONTROL OF FRUIT FLIES BY THE STERILE-MALE
TECHNIQUE (Session 4)

EFFECTS OF STERILIZATION BY RADIATION AND CHEMICALS
(Session 5)

GENETIC MECHANISMS OF INSECT CONTROL (Session 6)

STERILE-MALE TECHNIQUE
FIELD PROGRAMMES
(Session 1; Session 2, Part 1)

USE OF THE STERILE-MALE TECHNIQUE IN PLANT PROTECTION PROGRAMMES

G. E. Cavin
US Department of Agriculture,
Animal and Plant Health Inspection Service,
Plant Protection and Quarantine Programs,
Hyattsville, Md.,
United States of America

Abstract

USE OF THE STERILE-MALE TECHNIQUE IN PLANT PROTECTION PROGRAMMES.
The world food crisis emphasizes the need for new approaches to pest control, such as offered by the sterile-male technique. Various plant protection programmes in which the sterile-male technique is being used or is proposed are discussed, and other possibilities for its use are suggested.

In recent months in the newspapers and magazines, we have read almost daily the views of world leaders on the current crisis in agriculture. There are many varied opinions as to the future of the world's food supply.

Recently, J. Lawrence Apple, an eminent plant pathologist from the North Carolina State University, stated that the possibility of a world food famine in the near future is real. Two things have happened: US surpluses and surplus lands have disappeared, and the affluent nations' appetite for meat is pricing basic food supplies beyond the reach of poor people. When these grim statistics are considered in the light of the ever-present possibility of a catastrophic epidemic of a disease and/or insect pest which could cause staggering losses in food production in developing countries, the picture becomes even more bleak. The spectre of famine is further intensified by the serious shortage of petroleum products which is being felt by the developing countries in the form of pesticide and fertilizer shortages. Those essential supplies which are available are far more costly than before. This combination of circumstances presents a threat of famine greater than the world has ever known.[1]

According to Quentin West, administrator of the US Department of Agriculture's Economic Research Service, the prospects are possibly not quite so bad. He believes there are prospects for improvement, but we must make them happen. The shortage of pesticides is one of the potentials for disaster in food and health. The building blocks for pesticides are petroleum derivatives which have become critically short and much more costly. These include intermediates such as ethylene, butylene, propylene, benzene, xylene, phenol, chlorine and other related materials. Pesticide containers, solvents and emulsifiers are petrochemically based also. These shortages will affect all countries, whether they are exporters or importers. A number of developing countries, including the Philippines, India and Nicaragua, have reported pesticide shortages for the current growing season.

---

[1] APPLE, J.L., Intensified pest management needs of developing nations, Bioscience 22 (1972) 461.

At the Food and Agriculture Organization (FAO) Conference on Ecology in Relation to Plant Pest Control held in Rome on 11 December 1972, Ray F. Smith, Chairman of the Department of Entomological Sciences, University of California, pointed out that there is great concern in the world today about better management of our limited resources. Involved in this concern are the conservation of a limited, often meagre, supply of natural resources; the avoidance of pollution; the preservation of endangered populations of plants and animals; the maintenance of genetic variability in our biotas to meet future requirements; the preservation of primitive areas, as well as of large uncultivated areas and of stable agro-ecosystems for a better human environment. The system of traditional agriculture characteristic of many areas in developing nations is beginning to give way to modern agricultural technology.

The changed agro-ecosystems resulting from the introduction of new methodologies will provide shifts in and probably an intensification of pest problems.

The present drought in the Sahel of Africa vividly depicts the need for new methods of pest control. Vast areas are at present uninhabitable for man and his animals. Those peoples displaced by the drought must find new lands to cultivate or forage if they are to survive. Teams of entomologists, veterinarians, range managers, climatologists and agricultural economists have spent months studying the Sahel in an attempt to predict its future and the future of its remaining inhabitants. One of the primary conclusions reached has been the urgent need for tsetse fly control to open new lands to agriculture.

The possibility of using the sterilization technique in an integrated approach to tsetse control was seriously considered but then discarded because of the unfortunately slow progress being achieved in the mass rearing of this serious pest. The alternatives are tree removal or the maintenance of pesticide barriers to re-infestation once the pest is eliminated from an area. Both methods have inherent drawbacks. Although it is unfortunate that so little can be done immediately, perhaps this crisis will awaken people everywhere to the urgent need for expanded research on some of the world's most debilitating diseases of man, animals and plants for which insects are vectors. Conquering the tsetse fly alone would open nearly one-fifth of the land area of the continent of Africa to human habitation and agricultural production.

Although the tsetse fly problem has not been fully resolved, there are many other uses of the sterile-male technique in plant protection programmes that are being advocated, and there are numerous instances where it is being used at present. A substantial number of these programmes involve the fruit flies. The International Organization for Biological Control is supporting action towards the eradication of Mediterranean fruit fly from Cyprus. Peru has interest in a programme of progressive eradication of Mediterranean fruit fly from its infested mountain valleys.

A fruit fly proposal that has aroused interest and stirred controversy for a number of years is the one to eradicate the Mediterranean fruit fly from Central America utilizing a combination of methods in which the sterile technique would be a major component. A recent review conducted by the US Agency for International Development, after it had been approached to assist in financing the project, questioned the values to be gained in relation to the expected cost of eradication. The report also questioned the

concept of eradication and its feasibility in a tropical environment. The pros
and cons of this determination will not be discussed here. Most of the
arguments for and against the proposal have been aired in the past.

The programme that really started it all was the Screwworm Eradication
Program. With the signing of an agreement with Mexico on 28 August 1972,
the United States-Mexico Joint Commission for the Eradication of Screwworms
was formed and offices established in Mexico City, Mexico.

Production facilities have been designed and the sterile fly plant site
selected at Tuxtla Gutierrez, Chiapas. Negotiations with contractors are
underway for construction of the main building of a facility to produce more
than 300 million sexually sterile screwworm flies each week. This production
plant is scheduled for completion in 1975.

Livestock producers of the Southwestern United States of America and
Mexico had formally requested that both countries conduct a joint screwworm
eradication programme in Mexico. They pointed out that such a programme
would better protect screwworm-free areas of the United States of America,
and cut costs of operating the barrier, which could then be established across
a narrower section of southern Mexico. It is expected that several years
will be required to eliminate this costly pest from Mexico. When eradication
is completed north of Tuxtla Gutierrez, a permanent barrier will be formed
against northward migrations of the insect by continuously releasing sterile
flies in a wide path across Mexico's narrow Isthmus of Tehuantepec. Periodically, the United States of America is now re-infested from Mexico because
highly favourable climatic conditions allow screwworm buildups in Mexico
which overwhelm the present sterile fly barrier. Even with these occasional
re-infestations, the livestock industry and the American public have received
tremendous benefit. Annual savings are estimated at more than $100 000 000.

Despite these successes, some scientists are questioning the validity
of the sterile fly technique in the eradication of screwworms. An outside
study has been authorized and a task force is conducting an extensive
evaluation.

The use of the sterile-male technique in an integrated control approach
is being widely recognized. The boll weevil is often referred to as the most
important agricultural insect pest in the United States of America. Since its
discovery in west Texas in 1892, it is estimated to have cost the cotton
industry in excess of 12 thousand million dollars. At present it costs
200-300 million dollars annually through direct destruction of the cotton
crop and cost of control.

Elimination of this pest from the United States of Ameria has been the
aim of the cotton industry for many years. Until just recently, research has
been directed primarily at development of effective control measures based
almost exclusively upon chemical pesticides. In the early 1960s, through
the efforts of the National Cotton Council and other cotton organizations, the
research programme was intensified in an attempt to develop a more acceptable solution to the boll weevil problem. Important new developments in
boll weevil control were accrued from this research effort.

These developments prompted leaders in the cotton industry to consider
the possibility that the necessary technology was available to eliminate this
pest. In 1969, the National Cotton Council adopted a resolution to look into
the possibility of eradication. The Council appointed a committee to evaluate
the state of technology from the standpoint of eradication.

This committee concluded that possibly the necessary technology was available to achieve this objective, but it would require a large-scale pilot test to determine if the necessary operational capability was on hand. An eradication trial was conducted in southern Mississippi and adjacent parts of Louisiana and Alabama.

This trial was completed in August 1973. The committee concluded that the experiment proved that the technology and operational capability were available for elimination of the boll weevil as an economic pest from the United States of America by methods which were ecologically acceptable. One of the key components in this successful trial was the use of sterile insects integrated with various cultural and chemical control techniques.

At present, a proposed plan for implementation is under study in the Department; but no commitment has been made to begin the programme. The Department will make additional studies and evaluation of this highly complex and costly proposal before a decision can be reached.

So far only programmes in which sterilization is accomplished through the use of irradiation or chemosterilization have been mentioned. Sterilization by means of genetic manipulation has also been receiving considerable attention lately. In most cases, progress has not developed enough to provide widespread use of these techniques in plant protection programmes. However research is progressing rapidly in this direction. This is a particularly intriguing approach to insect control. At the Metabolism and Radiation Research Laboratory of the Agricultural Research Service, US Department of Agriculture, Fargo, North Dakota, scientists have developed strains of house flies in which the laboratory-reared males, when mated with wild females, produce only male progeny.

The United States Forest Service is investigating the possible presence of a lethal gene in a strain of the gypsy moth found in Japan. If this strain is introduced into the gypsy moth population of the northeastern United States, this lethal factor could spell doom to a destructive pest of our hardwood forests. Though the existence of this lethal factor has been known for quite some time, regulatory officials until only recently have been reluctant to allow this particular strain of gypsy moth to be introduced into the United States for research purposes. Their concern was valid. The gypsy moth was originally established in the United States when a research scientist brought it in to attempt to cross it with the silkworm moth. His colony of gypsy moths escaped. We in Federal Plant Protection Programs are watching with interest and enthusiasm the progress which is being made in research in this new concept of pest control. Regulatory entomologists have a natural reluctance to allow indiscriminant movement of exotic strains. Controls on the movement of insects, plants and animal products which they enforce are necessary to ensure that adequate security is maintained to prevent pests escaping. Regulatory entomologists are trying to prevent incidents such as occurred in the United States of America with the gypsy moth.

The use of sterilization as a tool in regulatory entomology is also receiving increased attention. There are at least two ways in which the sterile technique is being used or considered in regulatory entomology.

First it can be used to exclude a pest by maintaining a barrier of sterile flies. An example is the Mexican fruit fly release programme in northern Baja California, Republic of Mexico. In this programme, flies are released each year between June und November when infested contraband fruit is liable to reach the markets. Since initiation of these releases in 1964, the Mexican fruit fly has been excluded from northwestern Mexico and adjacent California and Arizona in the United States. Before initiation of these releases, insecticide treatments were required nearly every year. The programme has been so successful that it has been proposed that the technique be used in the lower Rio Grande Valley of Texas. This would eliminate the need for fumigation when the Mexican fruit fly invades the Rio Grande Valley each year from Mexico.

A second regulatory use of the sterile technique is insect control through the irradiation of commodities. Widespread application of this technique in the near future for preserving freshness and destroying insects is unlikely. If irradiation of commodities ever becomes widely used, it would have a significant effect on quarantine activities. Present quarantines and treatments would in large part become obsolete. Large volumes of commodities could enter with no inspection beyond proof of appropriate radiation.

Losses from pests and diseases vary greatly but are generally considered to take regularly from 25 to 50% of the crop in tropical areas and often result in total loss. Moreover, these pests may at times threaten an economy or prevent essential food crops from being grown.

All too often in the past approaches to control of agricultural pests and diseases have, for the most part, involved application of chemicals with little regard for ecological complexities of individual crops and fields, danger of environmental contamination, the deleterious modifications of the biosphere or the socio-economic implications of the control measures. Chemical controls have important limitations, yet insects must be brought under ecologically sound management in order to sustain increases of food and other crops and provide for orderly economic development. Despite the scepticism of the use of the sterile technique in some large-scale plant protection programmes that has been voiced recently, I am confident the technique will continue to be a major contributor to modern pest control.

DISCUSSION

J. L. MONTY: Are chemicals still used to sterilize the boll weevil in the United States of America?

G. E. CAVIN: Yes.

J. L. MONTY: Has official clearance been given for field application?

G. E. CAVIN: No. Since the technique is still in the developmental stage, there has been no urgency to expend the time and effort involved in registration.

D. W. WALKER: Could you give us some idea of how one goes about soliciting funds for these very large programmes?

G. E. CAVIN: Support normally comes from the producer or general public. The value of the crop loss or degree of nuisance to the public by the pest greatly influences the degree of public reaction. The environmental aspects are also being more widely recognized by the public. Projects such as the proposed boll weevil eradication programme gain added support from the realization that success could reduce the use of pesticides in the United States of America by up to 30%.

A.-F. M. WAKID: You said that the irradiation sterilization for the boll weevil is more successful than the chemical sterilization. However, I understand that the boll weevil midgut is affected by radiation. Have you solved this problem?

L. E. LaCHANCE (Chairman): Perhaps I could answer that one. Recent studies using dose-fractionation for boll weevil sterilization look most promising. However, this does not rule out chemicals, growth regulators, etc. When a programme for the boll weevil is started, the best method available will be used.

R. PAL: Since this is a general session, could we hear more about the scepticism expressed recently concerning the use of the sterile-male technique?

G. E. CAVIN: Much of the scepticism seems to revolve around the question of genetic variability in laboratory strains compared with wild insects. The disease problem, especially in Lepidopterous insects with long life cycles, also discourages some people. Better quality control would go a long way towards eliminating this scepticism.

R. PAL: Fears have been expressed that the release of large numbers of chemosterilized insects could pose serious health hazards. How do you feel about this?

G. E. CAVIN: Some chemosterilants such as Tepa and Apholate are highly toxic to most life-forms and can be used only under rigidly controlled conditions, thus limiting the ways they can be applied in the field. However, many chemicals that show chemosterilant tendencies have relatively low mammalian toxicities, and for these there is a wide range of possible applications.

H. LEVINSON: Could you tell us more about the lethal factor in Lymantria dispar mentioned in your oral presentation, particularly its physiological manifestations? Also, can this strain be readily reared and how do you intend to use it?

G. E. CAVIN: I suggest you contact the Northeastern Forest Experiment Station, Forest Service, US Department of Agriculture, New Haven, Connecticut.

L. C. MADUBUNYI: During a study tour of the United States of America in 1972 I got the impression that the Department of Agriculture was limiting the use of chemosterilized insects in large-scale pilot eradication programmes. If my impression is correct, could you tell me if this policy is still in force and, if so, why?

G. E. CAVIN: The new Federal Insecticide, Fungicide and Rodenticide Act (FIFRA) requires more rigid and time-consuming tests than in the past. These are costly for the chemical industry. In addition it is considered by many scientists that chemosterilants such as Tepa and Apholate are too toxic to use in the field as spray applications or associated with attractants. This limits their use to laboratory sterilization. Therefore, unless these chemosterilants provide more effective results than radiation, the use of radiation is the method of choice, as it is more easily handled and generally less expensive.

M. M. HOSSAIN: It is known that the success of the sterile-male technique depends to a great extent on ecological isolation of the pest. If natural isolation does not exist, is it possible to create a barrier by artificial means?

G. E. CAVIN: Yes, in some instances. Pesticide barriers are one solution. This has been considered for the tsetse fly and will be used in the experiment in Tanzania in 1975. It was used in Tunisia for the medfly. Herbiciding or cultural practices to provide a host-free barrier are also possible, especially in the case of insects which are not highly mobile.

M. FITZ-EARLE: We have heard rumours that native screwworm females are able to select against irradiated males that are released. Could you please comment on this problem?

G.E. CAVIN: There are many rumours being spread as to why a greater number of screwworm cases have recently appeared in the southwestern United States of America. Some of these rumours may have a degree of substance while others are completely erroneous. To clear up these mis-understandings, a task force has been formed with W. Eden of the University of Florida as Chairman, to evaluate the use of the sterile technique in screwworm eradication. The task force has just begun its work and has not had time to report any conclusions.

T.P. BOGYO: There seems to be evidence of very rapid genetic selection in the codling moth laboratory material, i.e. a situation in which the insects prefer to mate among themselves rather than with the females of the wild species. This, I think, requires very urgent attention.

G.E. CAVIN: We have not observed this tendency in either our pink bollworm or Mexican fruit-fly rearing facilities. We have been rearing the fruit fly since mid-1960 and wild native flies are added periodically to maintain vigour. We have greater difficulty rearing Lepidopterous insects than the flies, but this is generally on account of disease rather than genetic change.

F. OGAH: Milan Topis (unpublished work) of the University of Notre Dame observed no change in the house entry behaviour of silvan A. aegypti after maintaining them in the laboratory for over a year. I hope this will help to allay Mr. Bogyo's fears of rapid evolutionary change in behaviour of laboratory-maintained cultures.

L.E. LaCHANCE (Chairman): I have come across no evidence of selection occurring in the screwworm programme. We recognize that the possibility does exist, however, and are conducting investigations. If necessary, material from the release area can be added to the laboratory strain to improve the situation.

THE STERILE-MALE TECHNIQUE
FOR CONTROL OF THE MEDITERRANEAN
FRUIT FLY, Ceratitis capitata Wied.,
IN THE MEDITERRANEAN BASIN

C. SERGHIOU
Agricultural Research Institute,
Ministry of Agriculture and Natural Resources,
Nicosia, Cyprus

Presented by D.A. Lindquist

Abstract

THE STERILE-MALE TECHNIQUE FOR CONTROL OF THE MEDITERRANEAN FRUIT FLY, Ceratitis capitata Wied.,
IN THE MEDITERRANEAN BASIN.

Certain problems caused by the use of insecticides in the management of agricultural pests, such as
environmental pollution, insecticide resistance and disturbance of biological balance, has led to the development of selective pest control methods. A prominent place among these for the control of the Mediterranean
fruit fly has been attained by the species specific sterile-insect technique (SIT). This study reviews the status of
field programmes in countries of the Mediterranean basin, and of related mass rearing, irradiation and field
release methodology. The SIT has been successfully tested in Spain, Italy, Israel and Cyprus. In all these
cases, however, tests were conducted in small semi-isolated areas where at best a high degree of suppression but
not eradication of the fly could be obtained since immigration of gravid females was always possible. The SIT
programme in Cyprus and data on medfly ecology in the island is here reviewed in more detail. A proposal
is made for the eradication of medfly from Cyprus by the use of an integration of methods, namely bait spraying,
cultural practices and sterile-insect releases.

INTRODUCTION

The Mediterranean fruit fly or medfly, Ceratitis capitata Wied., a
cosmopolitan species, is an important pest in the Mediterranean basin where
it attacks citrus, deciduous, stone fruits and other cultivated hosts. From
the point of view of area cultivated and export value medfly is particularly
important as a pest of citrus (Table I).

When no control measures are taken losses sustained can be extremely
heavy, and may be as much as 80% fruit infestation [1]. The conventional
·method of control in citrus consists of poisoned bait sprays, an organophosphorus such as malathion, trichlorfon or fenthion-protein hydrolysate
mixture, applied from the ground or air. In some countries such as Spain
and Israel control measures are compulsory and are taken collectively by
the state; in others such as Greece and Cyprus they are left to the discretion
of individual growers. In the 1960 to 1965 seasons when no compulsory
control measures were taken, the average citrus fruit infestation in Spain
was 1 - 2.5%, this figure subsequently dropping to 0.2 - 0.5% [2]. The cost of
control depends on the mode of application, i.e. ground or air, combination
of insecticide and attractant used, cost of labour and number of applications
per season which in turn relate to fruit susceptibility and climatic conditions.

11

TABLE I.   AREA, PRODUCTION, QUANTITY EXPORTED AND
VALUE OF CITRUS IN COUNTRIES IN THE MEDITERRANEAN BASIN

Country	Area[a] (ha)	Total production[b] (1000 metric tons)	Quantity[c] exported (metric tons)	Export value[c] (1000 US Dollars)
Spain	197 000	2337	236 086	181 972
Portugal	12 000	116	8	11
France	1 000[d]	6	7 873	5 928
Greece	40 000	620	72 948	21 428
Italy	160 000	2550	334 159	82 737
Tunisia	14 000	88	3 918	2 813
Israel	45 000	1240	259 672	82 990
Turkey	45 000[d]	655	26 807	8 128
Lebanon	12 120	265	9 319	6 742
Algeria	35 000[d]	471		
Morocco	55 000[d]	901	41 406	39 781
Cyprus	17 000	178	62 461	17 245

a Ciba-Geigy data.
b FAO 1971 Production Yearbook.
c FAO 1971 Trade Yearbook.
d Estimated.

In Israel, where accurate data are available, the total cost of control in the
1973 season for a citrus area of 48 000 ha was about $600 000, i.e. about
$12 per hectare; four treatments were applied on the average and spraying
was mostly carried out from the air (S. S. Kamburov, personal communication).[1]
     Bait spraying for the control of medfly is generally efficient and econo-
mical, but serious losses have been reported despite control measures [3].

THE STATUS OF THE STERILE-INSECT TECHNIQUE FOR MEDFLY
CONTROL IN THE MEDITERRANEAN BASIN

     Certain problems caused by insecticide applications in the management
of agricultural pests, such as undesirable chemical residues, environmental
pollution, insecticide resistance and disturbance of biological balance, has
led to the development of selective pest control methods.   A prominent place
among these for the suppression or eradication of medfly has been attained
by the species specific sterile-insect technique (SIT).
     Eleven laboratories in five countries around the Mediterranean basin are
currently involved in research activities related to medfly control by the

---

[1] S.S. KAMBUROV, Biological Control Institute, Citrus Marketing Board of Israel, 27 Keren Kayemet Street
Rehovot, Israel.

sterile-insect technique [4], and successful field programmes have been conducted in four of these countries, Spain, Italy, Israel and Cyprus.

In Spain, the National Institute for Agricultural Research (INIA) has been carrying out a programme for the control of medfly with the sterile-insect technique since 1965. Preliminary field trials were conducted in 1966-68 in the island of Tenerife, and in 1969 in the province of Murcia on the Spanish mainland. These tests showed that the method is effective when applied to small and relatively isolated areas [5-11]. In 1972 and 1973 successful tests were conducted in the province of Granada employing a new concept, i.e. releases of sterile insects as a peripheral barrier to protect a semi-isolated area against invasion by wild flies [12,13].

In Italy tests using the SIT for the control of medfly have been success-fully conducted in the Parthenopean islands by the Laboratory for Application of Nuclear Techniques to Agriculture (CNEN) in collaboration with the Italian Ministry of Agriculture and Forestry (MAF), Euratom and the IAEA [14, 15].

This success prompted CNEN to initiate in 1971 on the island of Procida a four-year pilot test to obtain definitive, practical and economic data on the application of SIT. During this test, considerable data on medfly ecology were collected, improvement in methodology effected, costs lowered and the effectiveness of the method conclusively demonstrated by repeatedly eradicating the fly for short periods from the island [16,17]. The fly nevertheless re-invaded Procida through import of infested fruits and immigration of gravid females from Monte di Procida and Ischia. In fact recaptures of marked adults released in Monte di Procida and Ischia confirmed the thesis of a continuous immigration of medfly from these areas to Procida from August onwards [16, 17].

In Israel success with SIT has been limited, primarily because of poor isolation of release sites. In 1972 a preliminary test was conducted in a village area of about 360 ha. Suppression of fly population through sterile releases could be demonstrated for about one month only, because of the dense wild fly population in groves surrounding the release site, and the resulting vast immigration of fertile females in the release area in the absence of any bait spray barrier.

In 1973 a larger area was used, about 1000 ha containing commercial citrus groves and surrounded by a 500-metre wide low volume bait spray barrier. In this instance successful control of the wild fly population was obtained for several months, and a clear suppression until July; thereafter, wild fertile females penetrated through the barrier into the release area, a fact which was subsequently confirmed by releases of marked flies [18].

The status and prospects of the SIT programme in Cyprus will be discussed below.

Mass rearing

Mass rearing of medfly is now at a very advanced and refined stage of development. The cost of materials for rearing the flies has been reduced to $10 per million flies, the combined cost of labour and materials to $20, and 15 000 - 18 000 pupae are produced per litre of larval medium, which approaches the maximum possible without reduction of pupal size [19]. The space requirements for the weekly production of 1 million flies are estimated at 15 m^2 total rearing facility surface [20].

The main nutrient components in the larval diet are dried brewer's yeast and sugar, though some additional nutrients are likely to be supplied by bran which makes up most of the bulk. Fungi and bacteria are controlled by adjusting the pH with HCl to 4.5, and by including sodium benzoate and Nipagin ® (methyl-p-hydroxybenzoate). Adult diet is composed of a mixture of yeast hydrolysate enzymatic and sugar.

Studies are currently being conducted to substitute expensive nutrients such as the yeast hydrolysate enzymatic in the adult diet with more inexpensive and readily available materials [21]. Nipagin has been substituted by correspondingly increasing the more inexpensive sodium benzoate in the larval diet in Seibersdorf and other laboratories (D. J. Nadel, personal communication).[2] Insecticide contamination in the bran is often a problem, and in Italy bran has been recently substituted by crushed straw [22].

There is a growing awareness and concern regarding fitness, competitiveness and prevention of genetic and physiological deterioration of the mass-produced flies, and the need for developmental research in quality control and quality control techniques. This will be essential for determining the most important biological parameters for sterile-fly releases. Such techniques will allow the early screening of the flies best suited for a field programme as well as the continuous monitoring of the effectiveness of the released flies [20]. An improvement in fly quality will reduce costs by lowering the sterile-to-wild medfly ratio necessary to achieve eradication.

Irradiation

Flies are sterilized by irradiation. Male sterility is assessed by recording the egg hatch resulting from mating of irradiated males with untreated females. Little difference in sterility was found between males irradiated 1-3 days before eclosion. Under these conditions, egg hatch of less than 0.5% was obstained with a 10 krad dosage [23].

Male competitiveness has been similarly evaluated from egg hatch resulting when irradiated males were confined in various ratios with untreated pairs of insects. Competitiveness decreased with increased dose. This factor counteracted the increased degree of sterility induced by increased dose, so that doses of 5 - 11 krad led to similar reductions in egg hatch [24].

In practice the dose used is a compromise between acceptable levels of sterility and competitiveness. Thus, with pupal irradiation, an 8 or 9 krad dose has been commonly used in field programmes around the Mediterranean basin [12, 18, 25]. With the more difficult to sterilize adults, irradiation dose is increased to 10 krad or more [17].

Field release methods

Field releases in these small scale programmes are made from the ground. The method usually employed is to allow flies to emerge in paper bags in the laboratory, transfer the bags at a predetermined release point in the field and slit them open to allow the flies to escape. Recently aerial releases with chilled adults have been successfully tested in Israel [18].

---

[2] D.J. NADEL, Division of Research and Laboratories, International Atomic Energy Agency, P.O. Box 590, 1011 Vienna, Austria.

The effects of chilling on medfly are still being investigated but it seems that 4° - 5°C can be safely used for knocking down and holding the inactivated flies for several hours (S. S. Kamburov, personal communication; E. J. Harris, personal communication).[3] Under these conditions most of the flies recover when exposed to ambient temperatures of 20° - 25°C within 5 - 10 minutes. It is very likely that aerial releases with chilled adults will be the method to be used in large-scale programmes.

Evaluation of results

Evaluation of results is based on comparative egg hatch and fruit infestation data obtained from the release and a control site. Since marked sterile flies are usually released, trapping data provide additional information on the suppression of the wild fly population.

Whereas the effectiveness of the SIT has been repeatedly demonstrated in tests conducted in semi-isolated areas, at best only a high degree of suppression but not eradication was achieved since immigration of gravid females was always possible.

It is believed that the time is ripe to embark on a large-scale programme in an appropriate area with the aim to eradicate the fly. The island of Cyprus is proposed for such a programme.

MEDFLY STATUS AND PROSPECTS OF ERADICATION IN CYPRUS

The annual losses and cost of medfly control in Cyprus are currently estimated at $750 000 with a tendency to increase because of contributing factors: (a) increased acreages of citrus, deciduous and stone fruits coming into production; (b) extension of the picking season of valencia oranges from February-March when they virtually escaped infestation, to March-June, thus making them liable to medfly attack; (c) increased cost of control due to more expensive labour and materials; and (d) cost of fumigation of citrus fruits against medfly intended for markets in continental Europe, a prerequisite not thus far necessary in the case of the UK market.

Field tests for the suppression of medfly using the sterile-insect technique were initiated in 1972 with the co-operation and assistance of the International Atomic Energy Agency, and were continued in 1973 and 1974. In the 1974 programme, which is still in progress, in addition to sterile releases, the effectiveness of methods which could be integrated with sterile releases in a programme for the eradication of medfly from Cyprus, i. e. removal and destruction of sour oranges and early season bait spraying, are being investigated. Collection of data on medfly ecology preceded the sterile-insect programme and continued during its implementation. These include data on medfly population fluctuation through the year, succession of susceptible hosts and the most important overwintering hosts, and medfly incidence at different elevations, as well as data on dispersal and survival of the released flies.

---

[3] E.J. HARRIS, US Dept. of Agriculture Agricultural Research Service, Western Region, Southern California—Hawaii Area, Hawaiian Fruit Flies Investigations, P.O. Box 2280, Honolulu, Hawaii 96804.

Medfly population dynamics

Trap surveys conducted over a number of years in different localities of
the island permitted the general pattern of the adult medfly population fluctuation through the year to be determined. Thus it was found that medfly
population was very low in the winter and spring, increased in the summer
and reached a peak in the autumn [26].

Succession of medfly hosts

The succession of the most important medfly hosts that perpetuate
infestations is well established. Medflies overwinter as larvae in unpicked
citrus fruits, especially sour oranges. There are nevertheless trap data
showing that adult fly activity occurs throughout the year at least for certain
years and localities. Apricots and the early maturing varieties of peaches
become susceptible in late spring and early summer, figs contribute to the
build-up of medfly populations in high numbers in late summer and autumn,
with citrus becoming susceptible in the autumn.

Medfly incidence at different elevations

Information on medfly incidence at different elevations was obtained from
an admittedly limited survey conducted in 1972 and 1973. Data were obtained
from ten single trap sites established from sealevel at Morphou up to 6000 ft
on Mount Olympus, the island's highest point Flies were trapped in both
years at all elevations from sealevel to 3500 ft, but none at higher elevations.

Dispersal of sterile medflies

Data on dispersal of sterile flies were obtained during the sterile release
programme at Rizokarpaso and in specific tests conducted at Deftera, a
village near Nicosia.

During the 1972 and 1973 programme, marked sterile flies were trapped
up to 12 km from the nearest release point, and in the current year's
programme up to 16 km. Nevertheless, most flies did not seem to move
more than a few hundred metres, and in order to ensure a uniform distribution we had to use close release points about 100 - 150 m apart. In fact, we
found that good distribution was as important as the number of the released
flies.

Dispersal of flies is, of course, influenced by a number of factors such
as temperature, host density and distribution of prevailing winds.

Both in the 1973 and the current year's release programme, severalfold
fewer marked flies were caught in the cooler spring months than later in
the season even though we were releasing a constant number of flies per
week, were using the same number of traps and trapping sites and a fixed
release pattern. These data suggest a lower mobility of the flies in the
spring, but in addition it is likely that the attractiveness of traps and
especially the attractiveness of the hosts on which traps were placed varied
with the season.

In two tests at Deftera, which were conducted in warm weather in early
autumn of 1972, flies were released from a single point and traps were
placed on the periphery of concentric circles around the release point. Flies

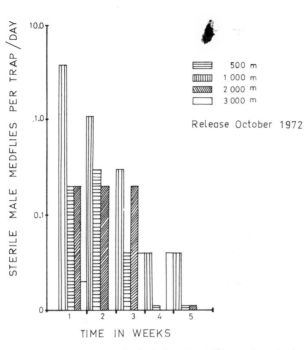

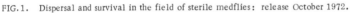

FIG.1.   Dispersal and survival in the field of sterile medflies: release October 1972.

were caught up to 3 km from the release point, which was the maximum
distance tested (Fig. 1). Most of the flies were caught in traps at 250 m
(1023 flies per trap, one week after release) which were nearest to the
release point (Fig. 2). Most traps had been placed on figs which at that
time had susceptible fruits.

These tests were repeated using the same set-up in April and May 1974.
In April only a single fly was trapped at 250 m and none at greater distances.
In May few flies were caught at 250 m, two flies at 500 m and one fly at
1 km (Fig. 3).

Survival of sterile flies under field conditions

Data on medfly survival were similarly obtained at Rizokarpaso and
Deftera.

At Rizokarpaso marked flies were recaptured six weeks after suspension
of the releases in 1972 and five weeks in 1973, but most of the flies nevertheless probably did not live more than one or two weeks as suggested by the
sharp decline in trap catches (Fig. 4).

At Deftera a maximum survival of nine weeks was obtained (Fig. 1).

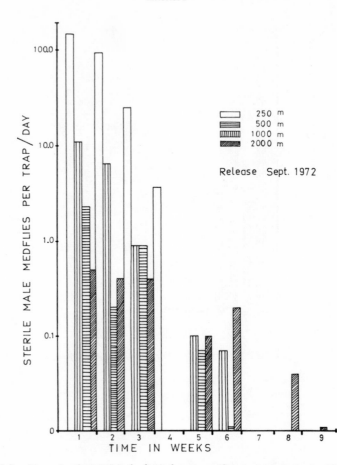

FIG.2.   Dispersal and survival in the field of sterile medflies: release September 1972.

STERILE-RELEASE PROGRAMME

Preliminary field releases with sterile medflies were initiated in the late summer of 1972 when the wild medfly population was already high and it was apparent that it could not be suppressed.   That programme, during which approximately 26 million flies were released, consequently aimed at obtaining information and acquiring experience on logistics of release, estimates of wild fly population and sterile fly movement and dispersal, which proved very useful in the 1973 test.

The aim of the 1973 programme was to test the effectiveness of sterile releases alone, unaided by any other means such as bait sprays or cultural practices in suppressing the wild medfly population in a restricted semi-isolated environment.   The village of Rizokarpaso in the Karpass peninsula was used as the release site;   it comprises an area of approximately 6 km²

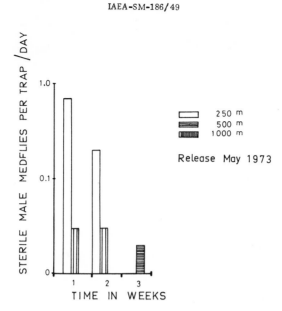

FIG.3.   Dispersal and survival in the field of sterile medflies: release May 1973.

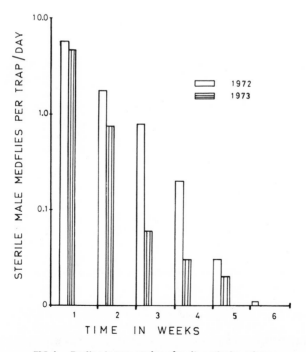

FIG.4.   Decline in trap catches after discontinuing releases.

in which a variety of medfly hosts are grown mainly in home gardens; several small commercial citrus (jaffa orange) groves are also present. Yialousa, a village similar in size and host distribution to Rizokarpaso, was used as the control site. Between the two villages there is an 18 km stretch of low forest area in which very few medfly hosts are grown. Ground releases were initiated in early March to take advantage of the natural low level of the wild medfly population, and were continued until early October. During this period approximately 73 million sterile flies were released. Evaluation of results was based on comparative fruit infestation data obtained from the two villages. Sterile releases reduced fruit infestation at Rizokarpaso compared with Yialousa by 95% in sour oranges, by 100% in jaffa oranges, by 95% in apricots and by 98% in figs [26].

THE 1974 PROGRAMME

Since this programme is still in progress results are preliminary and conclusions reached are tentative.

Sterile releases at Rizokarpaso

Sterile releases were resumed in mid-March and are being continued at the rate of about 3 million flies per week. No larvae were obtained from 10 sour orange samples, averaging 200 fruits each, collected from January through late April. From that time onwards a low level infestation in the sour oranges was detected with some tendency to increase. Available data through May nevertheless show a considerable reduction in both fruit infestation and adult population during the 1974 compared with the 1973 sterile release programme (Tables II and III).

Destruction of sour orange fruits at Yialousa

The importance of sour orange fruits as a medfly overwintering host had been established in previous years. It was decided consequently to test the effect of destroying sour oranges on the overwintering medfly population. A detailed survey at Yialousa and two neighbouring villages, Ayia Trias and Melanarga, showed 274 sour orange trees. Fruit samples were collected in order to provide an estimate of the larval population to be destroyed. A 500 ground fruit sample revealed an infestation of 1.36 larvae per fruit and a similar sample picked from the tops showed 0.27 larvae. An estimated 15 000 ground fruits were removed and destroyed in early February and an estimated 75 000 were picked and destroyed from the tops in early March.

The adult medfly population was monitored in Yialousa in 1973 and 1974 by the use of 30 Nadel traps baited with trimedlure and dibrom. The total number of flies trapped in the months February through May in 1973 was 779 whereas the corresponding number for 1974 was 6, suggesting a reduction in adult population of more than 99%.

Early season bait spraying

The effectiveness of early season bait spraying was tested in Koma tou Yialou, a village in the Karpass peninsula, with a second village in the

TABLE II.   COMPARATIVE INFESTATION IN SOUR ORANGES IN
EARLY PART OF THE 1973 AND THE 1974 STERILE RELEASE
PROGRAMME AT RIZOKARPASO

Month	1973 No. of larvae		1974 No. of larvae		Percentage reduction of infestation	
	per fruit	per kg	per fruit	per kg	per fruit	per kg
March	0.11	0.97	0.00	0.00	100	100
April	0.31	3.1	0.013	0.13	96	96
May	0.12	1.07	0.037	0.32	69	70
	0.16	1.43	0.022	0.21	86.3	85.3

TABLE III.   COMPARATIVE NUMBER OF WILD MALE MEDFLIES
CAUGHT IN 30 NADEL TRAPS BAITED WITH TRIMEDLURE IN THE
EARLY PART OF THE 1973 AND THE 1974 STERILE
RELEASE PROGRAMME AT RIZOKARPASO

Month	1973	1974	Percentage reduction of adult medfly population
March	16	0	100
April	26	1	96
May	175	3	98
	217	4	98

peninsula, Leonarisso, serving as control. Bait spraying was applied only
on fruit-bearing citrus. All citrus in dooryards were sprayed. In the case
of regular plantations bait spraying was applied on alternate trees on all
rows. The bait spray solution consisted of 1% malathion + 5% Naziman 73
(a protein hydrolysate commercial product), and was applied with a knapsack
sprayer on a foliage area of 2-3 ft^2. Five sprays were applied at 2-to-3 week
intervals between 19 March and 27 May. Evaluation of results was based
on trap catches from 10 one-foot square metal sticky traps baited with
trimedlure which were placed in each village. All bait sprays except the
last did not seem to have an appreciable effect in suppressing the adult
medfly population (Fig. 5). The poor results obtained by bait spraying in the
cool spring months are probably due to one or a combination of the following
factors: lower mobility of medflies, reduced attractiveness of the bait
and/or lower insecticidal efficiency of malathion.

TOTAL MEDFLY SUPPRESSION ON THE KARPASS PENINSULA

This programme, which is planned for 1975, aims at total suppression
or even eradication of medfly from the Karpass peninsula by an integration of

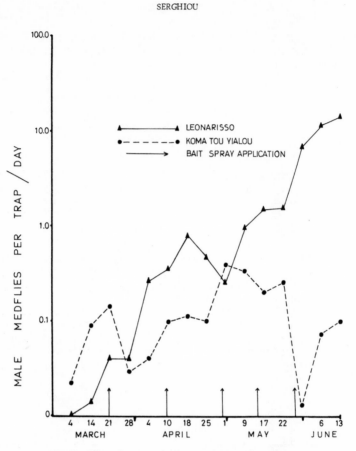

FIG. 5.   Effect of early season bait-spray on medfly population.

methods: sterile-insect releases, cultural practices and bait spraying.
This programme is considered to be an exercise to develop logistics, acquire
experience in aerial releases and improve integrated methodology. It will
also serve in identifying and delineating the problems likely to occur in a
country-wide programme. These problems could then be solved before the
initiation of a survey programme. The total medfly suppression programme
on the Karpass peninsula provides for a barrier zone and a release area.
The aim of the barrier zone will be to prevent as far as possible the immigration of gravid female medflies into the release area. The barrier zone
will be 5-7 miles wide, and will include the area inside a line drawn from the
coast at Yioti, to Ephtakomi, to Galatia, to the southern coast, along the
southern coast to near Gastria, to Gastria, to Kridhia, and to the northern
coast at Galounia (Fig. 6). The main medfly overwintering hosts such as
sour oranges and other fallen citrus fruits will be removed and destroyed
in January-February. In addition to fruit destruction, all medfly hosts in
season in all 11 villages in the barrier zone will be treated with bait sprays
on a 7-10 day schedule, as needed, from mid-March until the project is

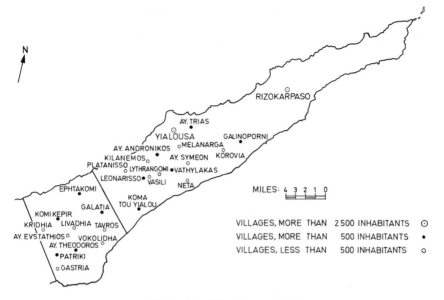

FIG.6.   Map of the Karpass peninsula.

terminated in early winter 1975.   From 10 - 15 sticky traps will be placed
in each village and serviced weekly, and trapping results will guide bait
spraying.

The release area will extend east of the barrier zone to the tip of the
Karpass peninsula and includes 16 villages.

The medfly population in the release area will be similarly reduced by
the destruction of overwintering hosts and bait spraying, followed by sterile-
insect releases.   At the time that the total suppression programme for the
Karpass peninsula was formulated, results of the 1974 programme were not
available, and early season bait spraying was planned.   It now seems that a
better scheme would be for the bait spraying to be applied in the autumn preceding
the releases, when it is known to be effective.

Sterile releases with chilled, marked medflies will be initiated in April,
at the rate of 15 - 18 million flies per week, and will continue until the
autumn.   Unless alternate hosts are located before the initiation of the
release programme, all releases will be made in villages and around isolated
houses with susceptible hosts.   All releases will be made by air, three to
four times per week.   Flight lines will be 250 m apart, but only every other
line will be used on each release.

Evaluation of results will be based on trap data (20 Nadel traps in each
village) and on fruit infestation data.   Follow-up extensive surveys will be
conducted in 1976.

A detailed survey and mapping of cultivated medfly hosts in all villages
in the barrier zone and release area will be completed by the late summer
of 1974 in order to delineate and have precise knowledge of the areas to be
treated.

The estimated expenditure for the total medfly suppression programme
in the peninsula is estimated at about $120 000.

PLAN FOR MEDFLY ERADICATION FROM CYPRUS

I. Pre-eradication phase

A. Objectives:

   (1) to obtain technical data necessary for the final decision to proceed
        with eradication;
   (2) to prepare a comprehensive project document
   (3) to establish administrative mechanism for management and funding
        of the eradication project.

B. Technical data necessary:

   (1) to determine the possibility of medfly immigration from Turkey.
            This information is expected to be obtained from the following:

        (a) survey of medfly hosts along southern coast of Turkey;
        (b) survey of medfly population along southern coast of Turkey;
        (c) a separate study of the series of events leading to medfly
             migration and complementary studies on flight behaviour;
        (d) massive releases of sterile marked flies along the southern
             coast of Turkey and an attempt to recapture the flies in traps
             which will be set up along northern coast of Cyprus.

   (2) Determination of numbers, stage and sites of medfly overwintering
        in Cyprus.

            Overwintering of medfly on Cyprus must be delineated to
        determine the total area which will require treatment by insecticides
        and/or sterile releases. The stage (adult, larvae, pupae) of over-
        wintering medflies in citrus (particularly sour orange and untreated
        dooryard citrus) will be detailed. An extensive search for other
        hosts, both cultivated and wild, that serve as overwintering sites,
        will be made, and a complete map of Cyprus showing all over-
        wintering sites will be prepared.

   (3) Host distribution survey.

            In connection with (2) above, an island-wide survey and mapping
        of all known hosts will be undertaken, and a parallel search for other
        hosts, both cultivated and wild, will be made.
            It is assumed at present that all villages will receive 2-3 bait
        sprays either from the ground or air, that all citrus trees at isolated
        houses will be treated from the ground, and that all potential over-
        wintering hosts will receive 2-3 bait sprays.

   (4) Economic survey.

            A detailed economic survey of losses caused by medfly on
        Cyprus and the cost of medfly control will be undertaken. The
        advantages of the sale of insecticide-free fruit and the ease of

expansion of exports to countries imposing strict quarantine against medfly will be weighed.

Estimates of medfly eradication costs making use of a wide range of assumptions is also planned.

C. Administrative and financial:

(1) Meeting of interested parties.

A meeting should be called as soon as a management committee has been formed and an outline of an overall eradication programme has been prepared. The meeting should include and concern both technical and financial matters. Organizations such as the IAEA, FAO, COPR and CIDA, as well as possible financial donors such as the EEC, UNDP and the World Bank, will be invited.

(2) Management committee.

The establishment of a management committee to co-ordinate the activities of the various organizations involved in or interested in medfly eradication from Cyprus is considered essential. This committee would be responsible for policy, finances and management of the project.

(3) Technical committee.

A technical committee consisting of scientists with knowledge of the sterile-insect technique and the medfly should be established to advise the Government of Cyprus on the technical aspects of the medfly eradication programme.

(4) Design of a rearing building.

Preliminary design and specifications for a medfly escape-proof mass-rearing factory is expected to be prepared as soon as possible. The building must be completely medfly escape-proof and have a rearing capacity of 200 million flies weekly. An alternate scheme would be for the Cyprus Government to contract the supply of flies on a regular or emergency basis from supporting laboratories in the Mediterranean basin.

(5) Detailed technical and financial plans for eradication

A detailed plan, including quarantine procedures, is expected to be developed by the end of 1975.

II. Eradication plan

This will be essentially a scaling up the Karpass peninsula project.

A. Objective.

To eradicate the medfly from Cyprus using a combination of cultural practices, insecticide bait sprays and the sterile-insect technique.

B.   General plan of eradication campaign.

(1)   Timing.

It is expected that the necessary research data will be completed
by early 1976.   Therefore the eradication campaign could be started
in January 1977 and be completed (including the follow-up survey) by
July 1979.

(2)   Population reduction.

The medfly population will be reduced as in the Karpass
peninsula project, i. e. by cultural practices such as destruction of
sour oranges and fallen citrus in January-February 1977, and also
bait spraying.   The bait sprays will be applied by ground and air
equipment to all citrus, both commercial, semi-commercial and
dooryard.   From two to four sprays will be applied, based on
trapping data.

(3)   Sterile-insect technique.

After the medfly population has been reduced by cultural
practices and insecticides, the aerial release of marked, chilled
sterile medflies will be initiated.   The rate of release will vary
from 200 000/km^2 to 1 000 000/km^2 weekly, depending on the density
of medfly hosts.   Large areas where no hosts are present will not
be treated.   Scattered dooryard hosts and wild hosts will be treated
weekly with bait sprays and no sterile flies.   From 1500 to 3000 traps
will be placed in medfly hosts around the island.   These traps will
be inspected weekly.   Intensive fruit sampling will be conducted on an
island-wide basis to determine fruit infestation.

After the end of July there should be no wild medflies caught in
traps.   If unmarked (and presumably wild) medflies are caught, an
intensive fruit inspection will immediately be conducted in the
immediate area (2-4 km^2);   if either infested fruit or additional
unmarked wild flies are captured, the area will receive 2-3 aerial
bait sprays at 5-day intervals.

The released flies will be sterilized with 7 ± 10% krad for the
first two months of release (mid-March to mid-May).   The dosage
will be increased to 8 ± 10% krad for the next 6 weeks (mid-May to
1 July) and increased to 9 ± 10% krad thereafter.   This procedure
ensures the most competitive flies during the critical early part of
the programme.   Releases will continue until the autumn or early
winter.

(4)   Follow-up survey.

In January and February 1978 all sour oranges on Cyprus will
be picked, and representative samples will be placed in emergence
cages.   All fallen citrus will be handled in the same way.   These
emergence cages will be checked twice a week for pupae and adults.
Areas of the island where infested fruit is found will be bait-sprayed

four times on a 5-7 day schedule. Sterile flies will be released for 3 months in large numbers immediately after the last bait spray.

The follow-up survey to ensure eradication will include extensive trapping with sticky traps in the preferred medfly hosts and fruit infestation sampling. Eradication will be considered accomplished if no flies or infested fruit are found by December 1979.

REFERENCES

[1] CRAMER, H.H., Plant protection and world crop production, PflSchNachr.Bay. 20 (1967) 1.
[2] KOPPELBERG, B., CRAMER, H.H., Control of Mediterranean fruit fly Ceratitis capitata Wied. in Spain: Methods and economic importance, PflSchNachr.Bay. 22 (1969) 1.
[3] PICECE, P., "Report on problems dealing with the presence of Ceratitis capitata Wied. in Italy", Joint EPPO/OILB Conference on Ceratitis capitata, Madrid, 19-21 May, 1970.
[4] Report of the IIIrd meeting of the OILB-SROP (IOBC/WPPRS) Working Group on Genetic Control of the Mediterranean Fruit Fly (IAEA, Vienna, 17 November 1973) (unpublished).
[5] MELLADO, L., CABALLERO, F., ARROYO, M., JIMENEZ, A., Ensayos sobre erradicacion de Ceratitis Capitata Wied. por el metodo de los "machos estériles" en la isla de Tenerife, Boletin de la Estación de Fitopatología Agrícola, No.399, I.N.I.A., Madrid (1966).
[6] MELLADO, L., ARROYO, M., JIMENEZ, A., CABALLERO, F., Ensayos sobre erradicación de Ceratitis capitata Wied. por el método de los "machos estériles" en la isla de Tenerife: Progresos realizados durante el ano 1967, Boletin de la Estación de Fitopatología Agrícola, XXX, I.N.I.A., Madrid (1967-1968).
[7] ARROYO, M., MELLADO, L., JIMENEZ, A., ROS, P., Lucha biológica contra Ceratitis capitata por el método de "machos estériles", Progresos realizados en 1971, Progress Report 1971, I.N.I.A., Madrid.
[8] MELLADO, L., ARROYO, M., JIMENEZ, A., CABALLERO, F., Ensayos sobre erradicación de Ceratitis capitata Wied. por el método de "machos estériles" en la isla de Tenerife, Progresos realizados en 1968, Boletin de la Estación de Patologia Vegetal y Entomologia Agrícola XXX, I.N.I.A., Madrid (1969).
[9] MELLADO, L., NADEL, D.J., ARROYO, M., JIMENEZ, A., "Mediterranean fruit fly suppression experiment on the Spanish mainland in 1969", Sterile-male Technique for Control of Fruit Flies (Proc.Panel Vienna, 1969), IAEA, Vienna (1970) 91 (short contribution).
[10] MELLADO, L., ARROYO, M., JIMENEZ, A., CASTILLO, E., Ensayos de lucha autocida contra Ceratitis capitata Wied., Progresos realizados en 1969, Anales Instituto Nacional Inr.Agrarias Serie Protección Vegetal No.2, I.N.I.A., Madrid (1972).
[11] ARROYO, M., MELLADO, L., JIMENEZ, A., CASTILLO, E., Ensayos de lucha autocida contra Ceratitis capitata Wied., Progresos realizados en 1970, Anales Instituto Nacional Inr.Agrarias, Serie Protección Vegetal No.2, I.N.I.A., Madrid (1972).
[12] MELLADO, L., ARROYO, M., ROS, P., "Control of Ceratitis capitata Wied. by the sterile-male technique in Spain", The Sterile-Insect Technique and its Field Applications (Proc.Panel Vienna, 1972), IAEA, Vienna (1974) 63.
[13] MELLADO, L., ROS, P., ARROYO, M., CASTILLO, E., "Genetic control of Ceratitis capitata; Practical application of the sterile-male technique", Panel on the Sterile-Male Technique for Control of Fruit Flies, IAEA, Vienna, 12-16 November 1973 (unpublished).
[14] de MURTAS, I.D., CIRIO, V., GUERRIERI, G., ENKERLIN, D., "An experiment to control the Mediterranean fruit fly on the island of Procida by sterile-insect technique", Sterile-Male Technique for Control of Fruit Flies (Proc.Panel Vienna, 1969), IAEA, Vienna (1970) 59.
[15] NADEL, D.J., GUERRIERI, G., "Experiments on Mediterranean fruit fly control with the sterile-male technique", Sterile-Male Technique for Eradication or Control of Harmful Insects (Proc.Panel Vienna, 1968), IAEA, Vienna (1969) 97.
[16] CIRIO, U., de MURTAS, I.", Status of Mediterranean fruit fly control by the sterile-male technique on the island of Procida", The Sterile-Insect Technique and its Field Applications (Proc.Panel Vienna, 1972), IAEA, Vienna (1974) 5.
[17] CIRIO, U., "The Procida medfly pilot experiment: Present status of the medfly control after two years of sterile insect releases", Panel on the Sterile-Male Technique for Control of Fruit Flies, IAEA, Vienna, 12-16 November 1973 (unpublished).

[18] KAMBUROV, S.S., YAWETZ, A., NADEL, D.J., "Application of the sterile-male technique for control of
      Mediterranean fruit flies in Israel under field conditions", Panel on the Sterile-Male Technique for
      Control of Fruit Flies, IAEA, Vienna, 12-16 November 1973   (unpublished).
[19] STEINER, L.F., "Mediterranean fruit fly research in Hawaii for the sterile fly release program", Insect
      Ecology and the Sterile-Male Technique (Proc.Panel Vienna, 1967), IAEA, Vienna (1969) 73 .
[20] INTERNATIONAL ATOMIC ENERGY AGENCY, Specific statements and recommendations, Panel on the
      Sterile-Male Technique for Control of Fruit Flies, IAEA, Vienna, 12-16 November 1973 (unpublished).
[21] MOURAD, A., "Preliminary experiments in reducing the cost of medfly diets", these Proceedings,
      IAEA-SM-186/15.
[22] KAMBUROV, S.S., Report on the IAEA fellowship to study sterile insect techniques and mass rearing
      methods (unpublished)
[23] HOOPER, G.H.S., "Sterilization of the Mediterranean fruit fly, a review of laboratory data", Sterile-
      Male Technique for Control of Fruit Flies (Proc.Panel Vienna, 1969), IAEA, Vienna (1970) 3.
[24] HOOPER, G.H.S., Sterilization of the Mediterranean fruit fly with gamma radiation: effect on male
      competitiveness and change in fertility of females alternately mated with irradiated and untreated males,
      J.Econ.Entomol. 65 (1972) 1.
[25] SERGHIOU, C., BALOCK, J.W., "Suppression of the Mediterranean fruit fly or medfly, Ceratitis
      capitata Wied., in a semi-isolated area in Cyprus by the use of the sterile-insect release method",
      Panel on the Sterile-Male Technique for Control of Fruit Flies, IAEA, Vienna, 12-16 November 1973
      (unpublished).
[26] SERGHIOU, C., ZYNGAS, J., KRAMBIAS, A., "Preliminary studies of Mediterranean fruit fly, Ceratitis
      capitata Wied., populations in Cyprus", Computer Models and Application of the Sterile-Male
      Technique (Proc.Panel Vienna, 1971), IAEA, Vienna (1973) 165.

USE OF TRAPS AND FRUIT INFESTATION IN THE EVALUATION OF MEDFLY FIELD POPULATIONS

J.E. SIMON F.
Estación Experimental Agraria,
La Molina, Lima

O. VELARDE R.
Estación Experimental Agraria,
Tacna

J. ZAVALA P.
Oficina Agraria,
Moquegua,
Peru

Abstract

USE OF TRAPS AND FRUIT INFESTATION IN THE EVALUATION OF MEDFLY FIELD POPULATIONS.
The medfly, Ceratitis capitata Wied., was reported in Peru for the first time in 1956. Estimates of the size of the medfly population have been made by either McPhail or Steiner traps, the data being used to estimate the efficiency of systems for controlling the pest. The results of a study in the Moquegua valley ey in Peru are presented, with an additional form of evaluation: percentage of damage.

INTRODUCTION

The mediterranean fruit fly, Ceratitis capitata (Wied.), was reported in Peru for the first time in 1956. From that time onwards estimates of the size of the fly population have been made either with McPhail or Steiner traps [1, 2], the data being used to demonstrate the efficiency of systems for controlling the pest.

The authors, through their work in the Moquegua valley in Peru, felt the need for an additional form of evaluation since the data from the traps was not always in accordance with the results of the yield obtained by the farmers. Some work was done taking fruit samples, but economic and time factors made them unable to compare both sets of numbers until they obtained from the joint FAO/IAEA Division of Atomic Energy in Food and Agriculture (IAEA) Research Contract No. 1202/RB and its renewal Research Contract No. 1202/RI/RB. The results are presented here.

METHODS AND MATERIALS

Weekly samples of ripening fruits were collected from various orchards situated in the Moquegua valley, and retained for 7 days in special cabinets that enable larvae and pupae of medfly to develop, if they were present in the sample.

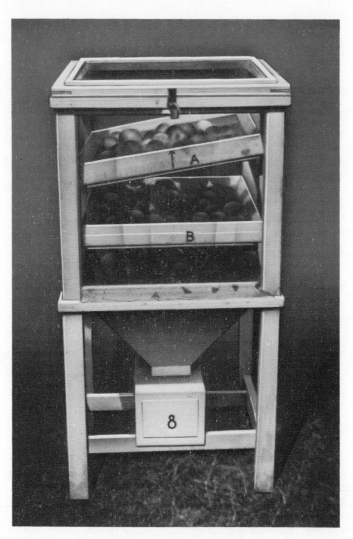

FIG.1. Fruit sampling cabinet.

The Moquegua valley from east to west is about 65 km long. The upper and middle zones receive sun all the year round which is why several species of fruits ripen throughout the whole year and thus give the medfly a good chance for continuous reproduction.

The valley was divided into three sections: the upper of about 3.89 square miles (996 ha), in which only insecticides are used for controlling the flies, was supplied with 15 fruit sampling cabinets. In the middle zone, where an integrated system of sterile insects with insecticides and parasites is used, 37 cabinets were distributed over its 9 square miles area (2304 cultivated hectares). Finally, the lowest part of the valley, the delta called Ilo, with a surface of 1.8 square miles, or 468 cultivated hectares, received only 8 fruit sampling cabinets, mainly because most of the area is planted with olive tree and only 8000 host trees of medfly are present.

The fruit sampling cabinets (Fig.1) are made of wood and iron mesh. Their dimensions are 60 × 60 × 60 cm with four legs 0.60 m high. The upper part is a cube with the roof like a door which is used to fill with the fruit the three 55 × 55 × 10 cm inside screens or frames. The lower part is a funnel made of triply ending in a box of 30 × 20 × 20 cm, with a door in front through which it is possible to introduce a tray with sand.

After each cabinet is filled in the presence of the farmer (the owner of the orchard) it is closed and locked for a week. Seven days later the project officer and the owner open the box at the lower part of the cabinet, take out the tray and sieve the sand. If some larvae and pupae are found, the fruit is examined in order to find out how many of them have been damaged by the medfly.

The number of damaged fruit multiplied by 100 divided by the total number of fruit in the cabinet gives the "percentage of damage", and the total of immature stages divided by the number of damaged fruit gives the "intensity". Both data — intensity and percentage of damage — are registered, and the fruit sampling cabinet is refilled or taken to another orchard when the harvest season has finished.

Once a year, or when the cabinets are in need of repair, they are brought back to the laboratory for service or maintenance.

The data from the cabinets are compared with that obtained from the Steiner traps installed in the orchard or nearby. The Steiner traps are examined also weekly, and normally the difference in timing is no more than 24 hours.

Table I shows a list of crops among which cabinets had been installed.

Text continued on p. 47

TABLE I. LIST OF CROPS AMONG WHICH FRUIT SAMPLER CABINETS WERE INSTALLED IN THE MOQUEGUA VALLEY

Crop	Periods of sampling
Apple	Apr. - Dec. 73 & May 74
Apricot	Oct. - Dec. 73
Cherimolia	May - Oct. 73 & May 74
Fig	Jan. - Jul. 73 & Mar. - May 74
Guava	Apr. - Nov. 73 & Apr. - May 74
Inga	Jan. - Feb. 73
Locuat	Apr. - Dec. 73 & Jan. - May 74
Mandarin	May - July 73 & May - Jun. 74
Mango	Jan. - Feb. 73 & Mar. 74
Olive	Apr. - Jun. 73
Orange	Apr. - Dec. 73 & Jan. - May 74
Peach	Nov. 72 - Apr. 73 & Dec. - Apr. 74
Pear	Jan. 73 & Jan. - Feb. 74
Quince	Mar. - May 73 & Mar. - May 74

TABLE II. RESULTS OF THE MEDFLY POPULATION EVALUATION THROUGHOUT THE YEAR WITH STEINER TRAPS (F) AND FRUIT SAMPLER CABINETS (D) AT MOQUEGUA, PERU: INSECTICIDE ZONES

Month	Week	TUMILACA (Insecticide zone)											TORATA (Insecticide zone)			
		Apple		Guava		Loquat		Peach		Pear			Loquat		Quince	
		D	F	D	F	D	F	D	F	D	F		D	F	D	F
Feb.	6															
	7									0	0					
	8									0	0					
	9									0	0					
Mar.	10							8	0							
	11							18	0							
	12							-	-							
	13							0	0							
Apr.	14	0	0			28	45	8	3							
	15	4	30			37	16	-	-							
	16	4	0			34	11	4	4				14		0	0
	17	0	21			29	22	4	6				8	13		
May	18	5	88			34	48						21	36		
	19	1	19			32	37						15	37		
	20	2	46			-	-						20	41		
	21	0	16			23	14						11	6		
	22	0	4			11	14						15	6		
Jun.	23	0	6			2	61						15	7		
	24	0	19			0	66						5	19		
	25	0	0			1							2			
	26	0	5			0	76						0	25		
Jul.	27	0	6			0	26						1	19		
	28	0	3			0	16						0	14		
	29	0	1			0	2						2	6		
	30	0	1	24	1								0	2		
Aug.	31	0	1	0	3								0	1		
	32	0	1	0	0								0	1		
	33	0	0	0	1								0	0		
	34	-	-	0	1								0	0		
	35	0	0	0	0								-	-		
Sep.	36	0	0													
	37	0	0													
	38	0	0													
	39	-	-													
	40	-	-													

TABLE III. RESULTS OF THE MEDFLY POPULATION EVALUATION THROUGH THE YEAR WITH STEINER TRAPS (F) AND FRUIT SAMPLERS (D) AT MOQUEGUA, PERU: ESTUAQUIÑA (STERILE-INSECT RELEASE ZONE)

Month	Week	Apple D	Apple F	Cherimolia D	Cherimolia F	Fig D	Fig F	Guava D	Guava F	Loquat D	Loquat F	Inga D	Inga F	Mango D	Mango F	Orange D	Orange F	Peach D	Peach F	Quince D	Quince F	Month	Week	Apricot D	Apricot F
Jan.	4	0	1									56	0									Sep.	39		
	5	0	5			0	0					42	3	0	5							Sep.	40		
Feb.	6	0	0	54	1							68	0	0	0							Oct.	41		
	7	0	2	42	0									-	-							Oct.	42		
	8	0	2	42	128									36	18							Oct.	43		
	9	0	2	-	-									78	0			76	9			Oct.	44	0	0
Mar.	10	0	0	-	-													69				Nov.	45	-	-
	11	0	0	-	-													66	81			Nov.	46	0	0
	12	0	4	8	0													77	8			Nov.	47	10	0
	13	0	2	16	0													87	3			Nov.	48	6	1
Apr.	14	0	8	0	0													40	0			Dec.	49	7	0
	15	0	9	2	0																	Dec.	50	28	1
	16	3	5	-	-															7	5	Dec.	51	26	2
	17	0	5	4	0											0	1			18	7	Dec.	52	-	-
May	18	1	5	1	2	0										2	2			6	5		1		
	19	87	12	1	5	-	-									14	1						2		
	20	0	1	7	5	-	-									-	-						3		
	21	0	1	5	2	-	-			26						-	-					X	X		
	22	0	15	2	5			13	8	22	1					-	-					X	X		
Jun.	23	0	5	6	3			4	0	-	-					18	2								
	24	0	5	0	6			24	1	27	10					42	0								
	25	0	0	4	2			5	0	-	-					12	0								
	26	0	0	0	0			47	0	-	-														
Jul.	27	0	39	10	1			21	2	42															
	28	0	20	2	2			23	0	-	-														
	29	0	0	3	3			30	1	-	-														
	30	0	8	6	3			35	0	42															
Aug.	31	0	0	2	3			28	3																
	32	0	0	2	2			15	3																
	33	0	1	0	4			11	8																
	34	0	3	7	8			10	1																
	35	0	7	0	2			6	0																
Sep.	36	0	0	2	11			19	2																
	37	0	5	9	7			52	4																
	38	0	5	3	5			-	-																

TABLE IV. RESULTS OF THE MEDFLY POPULATION EVALUATION THROUGH THE YEAR WITH STEINER TRAPS (F) AND FRUIT SAMPLERS (D) AT MOQUEGUA, PERU: SAMEGUA (STERILE-INSECT RELEASE ZONE)

Month	Week	Apple		Cheri-molia		Fig		Guava		Loquat		Orange		Peach		Quince	
		D	F	D	F	D	F	D	F	D	F	D	F	D	F	D	F
Jan.	1											70					
	2											30					
	3											0					
	4											-	-	36	1		
	5	0	0									0	4	24	2		
Feb.	6	0	0									30	1	56	3		
	7	0	3									18	0	68	12		
	8	0	4									36	12	78	10		
	9	0	10									-	-	44	10		
Mar.	10	1	0									-	-	-	-	20	0
	11	0	0									-	-	87	1	88	0
	12	3	10									-	-	30	17	88	10
	13	1	15			6	4					-	-			-	-
Apr.	14	-	-									-	-			-	-
	15	8	16									-	-			29	16
	16	5	16									0	0			18	16
	17	1	10									3	110				
May	18	2	10					84	2			1	3				
	19	0	2	11	2			-	-	14	2	6	3				
	20	1	3	10	2			-	-	-	-	17	3				
	21	0	3	-	-			-	-	-	-	41	2				
	22	0	6	-	-			6	2	-	-	22	4				
Jun.	23	2	0	-	-					-	-	16	2				
	24	0	3	-	-					27	2	38	7				
	25	0	0	20	0					13		16	6				
	26	0	0	7	2					26							
Jul.	27	0	0	13	2					0							
	28	0	0	0	2					12							
	29	0	0	-	-					0							
	30	0	0	10	0					-	-						
Aug.	31	0	0	-	-					-	-						
	32	0	0	0	1					-	-						
	33	0	0	6	0					13							
	34	0	0	-	-					21							
	35	0	0	-	-					-	-						
Sep.	36	0	0							0							
	37	0	0							-	-						
	38	0	0							-	-						
	39	-	-							-	-						

TABLE IV. (cont.)

Month	Week	Apple		Cheri-molia		Fig		Guava		Loquat		Orange		Peach		Quince	
		D	F	D	F	D	F	D	F	D	F	D	F	D	F	D	F
Oct.	40	-	-							-	-						
	41	-	-							-	-						
	42	0	7							0	0	8	4				
	43	0	6							0	0	-	-				
	44	0	0							0	-	10	3				
Nov.	45	0	1							5	1	-	-				
	46	1	1							0	0	10	3				
	47	0	1							-	-	30	1				
	48	0	2							-	-	4	2				
Dec.	49	0	1							-	-	2	3				
	50	0	1							-	-	6	3				
	51	0	1							0	0	15	1				
	52	0	1							0	3	60	2				

TABLE V. RESULTS OF THE MEDFLY POPULATION EVALUATION THROUGHOUT THE YEAR WITH STEINER TRAPS (F) AND FRUIT SAMPLERS (D) AT MOQUEGUA, PERU: VALLE (STERILE-INSECT RELEASE ZONE)

Month	Week	Apple		Fig		Guava		Mandarin		Orange		Peach		Pear		Quince	
		D	F	D	F	D	F	D	F	D	F	D	F	D	F	D	F
Jan.	1											0					
	2											2	0	12	0		
	3											36	0	1	0		
	4											17	1	20	0		
	5											11	2	-	-		
Feb.	6											31	1	17	1		
	7											16	1	39	2		
	8											15	0	47	0		
	9											30	2	-	-		
Mar.	10											-	-				
	11											47	3				
	12											-	-				
	13	0	7									63	1				
Apr.	14	4	31	9	0							35	1				
	15	4	34	-	-											1	0
	16	5	16	3	0					0	2					0	0
	17	3	16	4	0					1	1					0	1
May	18	4	19	12	1	81	8			0	1						
	19	4	14	0	1	63	5			4	0						
	20	6	33			86	18			2	2						
	21	1	10			68				2	1						
	22	1	4			-	-			4	2						
Jun.	23	3	87							12	0						
	24	1	2							29	1						
	25	2								17	1						
	26	1						0	0								
Jul.	27	1						0	0								
	28	1						0	0								
	29	2						0	0								
	30	0						0	0								
Aug.	31	0						0	0								
	32	0															
	33	0															
	34	0															
	35	0															
Dec.	49											72	0				
	50											14	0				
	51											36					
	52											6					

TABLE VI. RESULTS OF THE MEDFLY POPULATION EVALUATION
THROUGHOUT THE YEAR WITH STEINER TRAPS (F) AND FRUIT
SAMPLERS (D) AT MOQUEGUA, PERU: ILO (CONTROL ZONE)

Month	Week	Fig		Guava		Mandarin		Olive		Orange	
		D	F	D	F	D	F	D	F	D	F
Mar.	13	100	26								
Apr.	14										
	15			62	9			0	12		
	16			47	12			0	30		
	17			29	38			0	38		
May	18			18	62			0	36		
	19			24	29			-	-		
	20			8	19	20	12	0	12		
	21			18	38	19	36	0	37		
	22			14	6	22	26	0	57		
Jun.	23			14	37	28	60	0	12	58	145
	24			38	73	-	-	-	-	6	123
	25			9	-	-	-	-	-	10	122
	26			11	15	3	55	0	14		
Jul.	27			6	40						
	28			5	42						
	29			5	60						
	30			12	57						
Aug.	31			-	-						
	32			3	19						
	33			7	48						
	34			19	51						
	35			3	12						
Sep.	36			0	12						
	37			7	19						
	38			5	18						
	39			5	10						
Oct.	40			-	-						
	41			-	-						
	42			10							
	43			11							
	44			15							
Nov.	45			23							
	46			27							
	47			30							
	48			54	84						

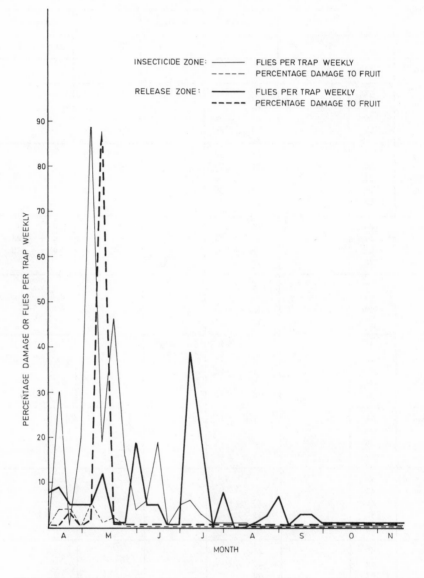

FIG.2. Percentage damage and number of flies in traps: apple, 1973; insecticide and release zones.

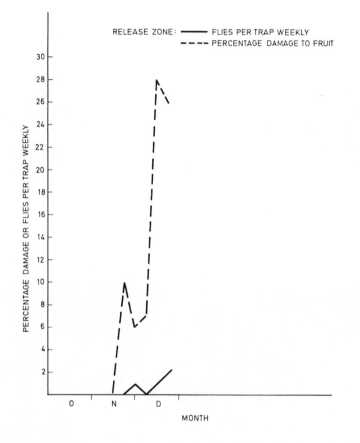

FIG.3. Percentage damage and number of flies in traps: apricot, 1973; release zone.

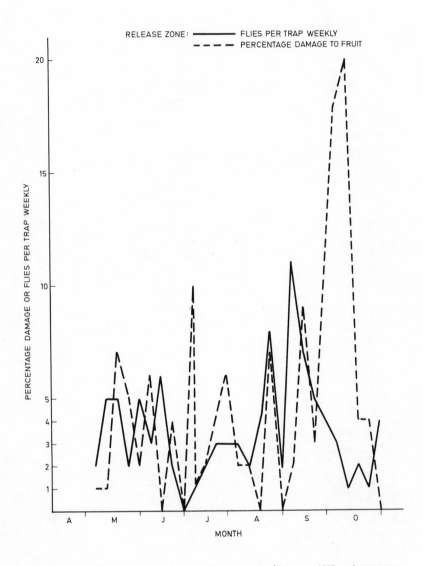

FIG.4. Percentage damage and number of flies in traps: chirimoya, 1973; release zone.

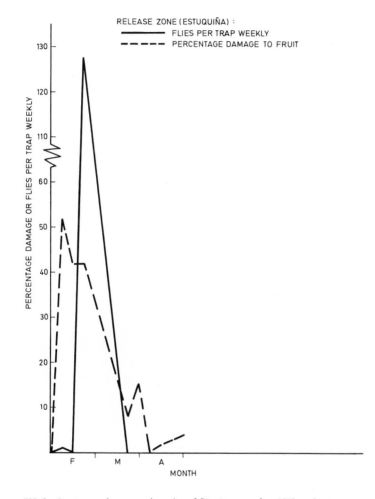

FIG.5. Percentage damage and number of flies in traps: fig, 1973; release zone.

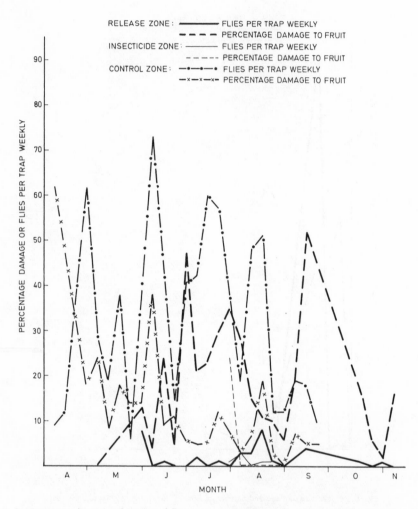

FIG.6. Percentage damage and number of flies in traps: guayaba, 1973; release, insecticide and control zones.

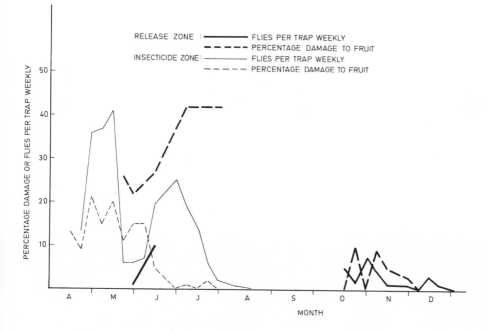

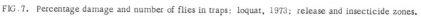

FIG.7. Percentage damage and number of flies in traps: loquat, 1973; release and insecticide zones.

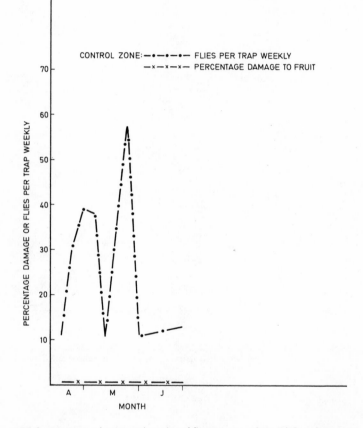

FIG.8. Percentage damage and number of flies in traps: olive, 1973; release zone.

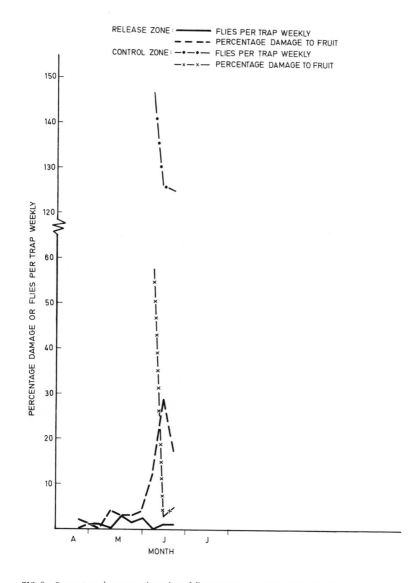

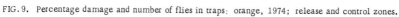

FIG. 9. Percentage damage and number of flies in traps: orange, 1974; release and control zones.

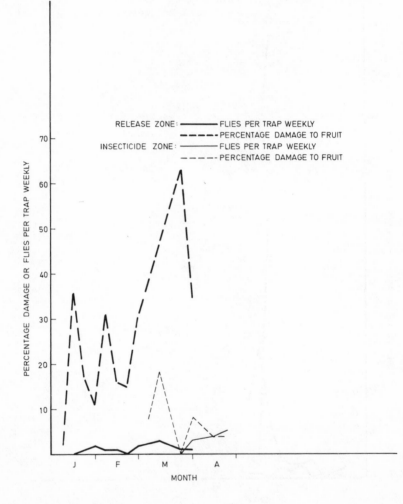

FIG.10. Percentage damage and number of flies in traps: peach, 1974; insecticide and release zones.

RESULTS AND DISCUSSION

Tables II - VI show the weekly averages of all the cabinets and the averages of all the cabinets and their nearest traps filled with each of 14 different fruit species throughout the year. It is evident that the percentage of damage (D) and the number of flies per trap per week (F) do not coincide either in numbers, or in trends. In some cases absence of flies at a trap occurs at the same time as severe damage of fruit, and also conversely in the case of olives no damage was sustained although hundreds of flies were present. It might be assumed that the olive is not a host of medfly, but we also have almost the same figure in the case of apples at Estuquiña (one of the sterile-insect release zones) where only in 3 out of 50 weeks did we detect damaged fruit and in one case 87% damage with 12 flies in contrast to no damage during Week 27 with 39 flies present.

The problem becomes serious when the whole emphasis of a project is concentrated on the number of flies captured or the absence of damage to fruit or animals. No single set of data shows what is happening in the field, as Figs 2-10 show. The evaluation of project performances is in need of revision, whether it applies to chemistry control, biological control or sterile-insect control programmes. It is possible in this particular case that other factors must be taken into consideration in addition to the number of flies captured in Steiner or other traps.

ACKNOWLEDGEMENTS

The authors thank the workers who assisted in the project and the farmers in the valley, the technician P.A. Duilio Cohaila, and the head of the Agriculture Office in Moquegua, R. Lema, for providing facilities. The carpenter, G. Quispe Ñaca, is also thanked for constructing the cabinets, and the IAEA Joint FAO/IAEA Division of Atomic Energy in Food and Agriculture for providing financial assistance through research contracts.

REFERENCES

[1] WILLE, J.E., La mosca Mediterránea Ceratitis capitata (Wied.), en el Peru, Revta Peru. Entomol. Agríc. 1 (1958) 59.

[2] GAMERO de la TORRE, O., Trabajos de control de las moscas de la fruta Ceratitis capitata (Wied.) y Anastrepha striata Shin (Trypetidae), Revta Peru. Entomol. Agríc. 1 (1958) 60.

BIBLIOGRAPHY

SIMON F., J.E., Insect control with gamma rays, Basic Life Science 2 (1973) 337.

SIMON F., J.E., The status and prospects of Mediterranean fruit fly control by the sterile-male technique in Peru, Panel organized by the Joint FAO/IAEA Division of Atomic Energy in Food and Agriculture, IAEA, Vienna, November 1973 (unpublished).

STEINER, L.F., Low cost plastic fruit fly trap, J. Econ. Entomol. 50 (1957) 508.

STEINER, L.F., "Methods of estimating the size of populations of sterile pest tephritidae in release programs", Insect Ecology and the Sterile-Male Technique (Proc. Panel Vienna, 1967), IAEA, Vienna (1969) 63.

DISCUSSION

G. E. CAVIN: What is the distance between traps? I ask this because there is evidence of competition between traps if placed too close together and, although the total catch is high, a breaking point exists where the total catch per trap is reduced. A recent test in Hawaii indicates that this breaking point may be about 1000 feet.

J. E. SIMON F.: We have 275 traps in the Moquegua Valley. There is one trap to each 16 ha and the traps are approximately 400 m apart.

L. C. MADUBUNYI: How do you explain the high level of damage with low medfly populations and, conversely, the low level of damage with very high medfly populations in some of the fruit orchards?

J. E. SIMON F.: Unfortunately we do not know the reason. It could be that because the Moquegua Valley has sunshine all the year round and various fruits are ripening at all times, the flies that we captured in a trap located in one orchard came from another orchard, in which a different kind of fruit is grown, and were thus not attracted by the fruit being sampled.

J. P. ROS: How reliable is the evaluation of damage from Ceratitis attacks by this new system? Have you compared the results with data derived from the fruit count?

J. E. SIMON F.: The sample is taken by the owner of the orchard together with the Project Officer at the time of harvest, so it is representative of the fruit that the farmer sells.

J. TICHELER: As you found little correlation between numbers of flies trapped and percentage of fruits infested, perhaps the results could be better explained on the basis of the intensity of fly attack, that is to say the number of immature stages per fruit?

J. E. SIMON F.: I presented the intensity figures at the panel[1] on fruit flies held in November 1973 in Vienna and, here again, there was no relation between those figures and the number of flies caught in each trap.

---

[1] The Sterile-Male Technique for Control of Fruit Flies, Panel held in Vienna by the IAEA, November 197̃ (unpublished).

LA STERILISATION DE LA MOUCHE MEDITERRANEENNE DES FRUITS Ceratitis capitata Wied.:
Méthode industrielle étudiée par le CNEN dans le cadre de la lutte autocide

I.D. DE MURTAS, U. CIRIO
Laboratorio di Agricoltura del Comitato
Nazionale per l'Energia Nucleare,
Santa Maria di Galeria,
Rome, Italie

Abstract—Résumé

STERILIZATION OF THE MEDITERRANEAN FRUIT FLY Ceratitis capitata Wied.: AN INDUSTRIAL METHOD STUDIED BY CNEN FOR PEST AUTOCIDAL CONTROL BY THE STERILE INSECT TECHNIQUE.
A new method is suggested for industrial irradiation of the Mediterranean fruit fly, Ceratitis capitata Wied. The gamma irradiation pilot plant belonging to the CNEN Agricultural Laboratory, with which unlimited numbers of fruit flies can be treated, is described. The flies are irradiated at the adult stage, since research has shown that their sexual competitivity is then greater than that of insects irradiated at the pupal stage.

LA STERILISATION DE LA MOUCHE MEDITERRANEENNE DES FRUITS Ceratitis capitata Wied.: METHODE INDUSTRIELLE ETUDIEE PAR LE CNEN DANS LE CADRE DE LA LUTTE AUTOCIDE.
Les auteurs présentent une nouvelle méthode industrielle d'irradiation de la mouche méditerranéenne des fruits, Ceratitis capitata Wied. Ils décrivent l'installation pilote d'irradiation gamma du Laboratoire d'agriculture du CNEN, qui permet de traiter des quantités illimitées de Ceratitis. Celles-ci sont irradiées au stade adulte, les recherches effectuées ayant montré que leur compétitivité sexuelle était supérieure à celle des insectes irradiés au stade pupal.

INTRODUCTION

En Italie, le Laboratoire d'agriculture du CNEN, en collaboration avec le Ministère de l'Agriculture et des Forêts, la Division mixte FAO/AIEA de l'énergie atomique dans l'alimentation et l'agriculture (Vienne) et la Division de biologie de l'Euratom, conduit depuis 1967, dans les îles Parthénopées, diverses expériences de lutte autocide contre Ceratitis capitata Wied., dont les résultats sont toujours plus prometteurs.

Cependant, au cours des premiers essais (Capri, 1967; Procida, 1969; Procida, 1972), nous avons pu constater que la stérilisation des insectes irradiés au stade pupal, 1 ou 2 jours avant l'émergence, s'accompagnait d'une série d'inconvénients indésirables concernant la longévité, la mobilité de vol et la vigueur sexuelle.

Ainsi le point crucial à étudier, dans ces programmes de lutte, était la réduction, dans certains cas, de la compétitivité des individus irradiés.

En outre les opérations techniques et manuelles d'irradiation des pupes étaient quelque peu laborieuses.

Une nouvelle méthode d'irradiation des adultes de Ceratitis est exposée dans ce travail.

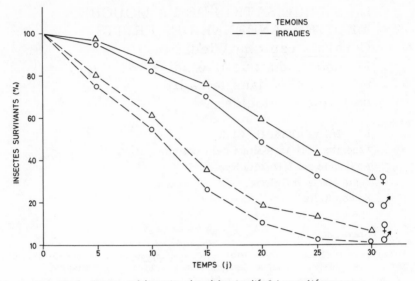

FIG.1. Survie, en laboratoire, des adultes irradiés 3 j avant l'émergence.

MATERIEL ET METHODE

Les études de stérilisation de la mouche méditerranéenne des fruits au moyen des rayonnements ionisants ont été très nombreuses ces dix dernières années. En particulier Katiyar [1] en Amérique-centrale; Féron [2] en France; Keiser et Schneider [3], Anwar et coll. [4], Schroeder et coll. [5] aux Etats-Unis d'Amérique; Arroyo et coll. [6] en Espagne; Hooper [7] à Vienne; Cavalloro et Delrio [8], De Murtas et coll. [9] en Italie, ont décrit les effets des rayonnements ionisants sur les aspects les plus importants de la bio-éthologie de Ceratitis.

Dans la quasi-totalité de ces expériences les irradiations ont été effectuées à l'aide d'une source «Gamma cell».

La «Gamma cell» est une unité d'irradiation complète, munie de sa propre source de cobalt et d'une gaine de protection. Elle est employée normalement aujourd'hui pour stériliser Ceratitis dans les programmes de lutte autocide (Chypre, Tunisie, Maroc, Vienne, etc.) même si les modestes capacités de sa chambre de réception du matériel ne permet pas, dans le cas de la mouche des fruits, de stériliser une quantité appréciable d'individus, surtout au stade adulte.

Il est donc clair que, travaillant dans un centre nucléaire où il est possible d'accéder à des sources de rayonnements plus puissantes, nos essais de radiostérilisation de la mouche des fruits se sont orientés principalement vers le problème de la mise au point d'une méthode d'irradiation industrielle la plus simple possible, et qui permette naturelle-ment de maintenir et si possible d'améliorer les résultats obtenus jusqu'à présent sur l'induction de la stérilité de ce trypétide.

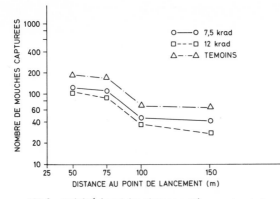

FIG.2. Mobilité de vol des adultes irradiés au stade pupal.

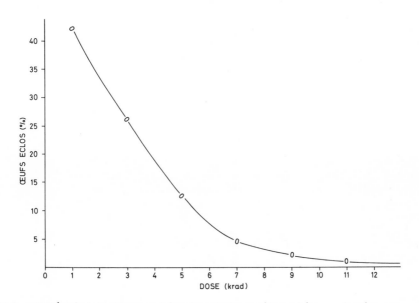

FIG.3. Fertilité après accouplement entre femelles normales et mâles irradiés peu après l'émergence, en fonction de la dose.

Nos recherches ont commencé avec les expériences Capri 1967, Procida 1968 et 1969, où les pupes irradiées nous étaient expédiées, par avion, d'Israël et de Vienne. Nous nous étions aperçu, par exemple, que la survivance chez les <u>Ceratitis</u> irradiées au stade pupal 3 jours avant l'émergence diminuait fortement la première semaine (fig. 1). En outre, en ce qui concerne la mobilité de vol et la dispersion, la différence de la recapture était nettement en faveur des individus non irradiés (fig. 2).

TABLEAU I. FREQUENCE ET DUREE DES ACCOUPLEMENTS DE Ceratitis capitata Wied. EN FONCTION DU
STADE ET DE LA DOSE D'IRRADIATION

Critères considérés	Témoins	Pupes		Adultes	
		7,5 krad	9,0 krad	7,5 krad	9,0 krad
Nombre moyen d'accouplements	3,1	1,4	1,2	2,9	1,8
Durée totale des accouplements (h)	12,0	3,8	1,1	5,7	3,9

Cependant, concernant le problème de la compétitivité, deux paramètres fondamentaux devaient encore être étudiés:
- la fertilité de l'insecte irradié exprimée en pourcentage d'œufs éclos après accouplement entre femelles normales et mâles irradiés (fig. 3)
- l'activité sexuelle comparée des Ceratitis irradiées au stade de pupes et d'adultes (tableau I).

Les résultats ont confirmé l'intérêt des objectifs de nos recherches; par conséquent nous avons pu adopter pour notre projet de lutte Procida 1973 la méthode d'irradiation suivante:

On a utilisé l'installation pilote d'irradiation gamma du Laboratoire d'agriculture; la source radioactive, constituée de 48 barres de cobalt-60, avait à l'origine une activité spécifique de 15 Ci/g et une activité totale de 60 000 Ci.

Le matériel à irradier a été placé dans des cages spéciales (70 × 30 × 60 cm), à raison de 100 000 pupes par cage. Jusqu'à l'émergence les cages étaient rangées dans des cellules bien climatisées. Puis les cages contenant les jeunes imagos étaient installées, une par une, sur un système transporteur qui permettait l'introduction continue du matériel à irradier dans la chambre d'irradiation, et la sortie automatique après traitement. La géométrie de la source était exploitée différemment selon les doses à administrer, en faisant varier de manière adéquate la vitesse du système transporteur et, par conséquent, la durée de séjour du matériel dans la chambre d'irradiation. En se servant du circuit le plus rapide, il a été possible d'irradier à la dose de 10 krad, en un peu plus d'une heure, 21 wagonnets contenant chacun 100 000 individus. Le débit de dose a été fixé à 3000 rad/min, compte tenu de la décroissance naturelle du cobalt-60. La dosimétrie était effectuée tous les 30 jours par la méthode de Fricke.

CONCLUSION

On peut affirmer que, dans le domaine de la stérilisation de la mouche méditerranéenne des fruits, des progrès intéressants ont été accomplis, surtout en Italie. Cependant, même si la situation actuelle peut être considérée comme une phase réellement avancée qui permettrait de lancer dans le futur immédiat de vastes opérations de radiostérilisation, nous devons déplorer un certain retard quant à l'application de cette méthode de lutte.

Il est donc nécessaire de parvenir à une coordination plus efficace en matière de projets de lutte à grande échelle, en communiquant, par l'entremise des services appropriés, les résultats des recherches aux organes responsables, aux instituts qui effectuent les expérimentations, aux divers opérateurs, et même au public. En Italie, la Sardaigne, immense laboratoire naturel, semble vouloir mettre en application les résultats des recherches traditionnellement consacrées à la lutte biologique et autocide.

REFERENCES

[1] KATIYAR, K.P., «Possibilities of eradication of the Mediterranean fruit fly, Ceratitis capitata Wied., from Central America by gamma-irradiated males», Proc. Fourth Inter-American Symposium on the Peaceful Applications of Nuclear Energy, Mexico City, 9-13 April 1962 (1962) 211-17.

[2] FERON, M., Stérilisation de la mouche méditerranéenne des fruits, Ceratitis capitata Wied., par irradiation des pupes aux rayons gamma, Ann. Epiphyt. 17 (1966) 229-39.

[3] KEISER, I., SCHNEIDER, E.L., Need for immediate sugar and ability to withstand thirst by newly emerged oriental fruit fly, melon flies and Mediterranean fruit flies sexually sterilized with tepa or radiation, J. Econ. Entomol. 62 (1969) 682-85.

[4] ANWAR, M., CHAMBERS, D.L., OHINATA, K., KOBAYASHI, R.M., Radiation-sterilisation of the Mediterranean Fruit-Fly, Ceratitis capitata Wied. (Diptera Tephritidae), Comparison of spermatogenesis in flies treated as pupae or adults, Ann. Entomol. Soc. Am. 64 (1971) 627-33.

[5] SCHROEDER, W.J., CHAMBERS, D.L., MIYBARA, R.Y., Mediterranean fruit fly: Propensity to flight of sterilized flies, J. Econ. Entomol. 66 6 (1973) 1261-62.

[6] ARROYO, M., JIMENEZ, A., MELLADO, L., CABALLERO, F., Aplicación de isótopos radiactivos a la investigación de métodos sobre la lucha biológica contra las plagas, III. Obtención de «machos estériles» mediante la irradiación de sus pupas con rayos gamma, Trabajos (Serie fitopatológica) núm. 390, Estación de Fitopatología Agrícola de Madrid (1965).

[7] HOOPER, G.H.S., «Sterilization of the Mediterranean fruit fly», Sterile-Male Technique for Control of Fruit Flies (C.R. Groupe d'étude Vienne, 1969), AIEA, Vienne (1970) 3.

[8] CAVALLORO, R., DELRIO, G., Studi sulla radiosterilizzazione di Ceratitis capitata Wied. e sul comportamento dell'insetto normale e sterile, REDIA 52 (1971) 511-47.

[9] DE MURTAS, I.D., CIRIO, U., SIMBOLOTTI, P., «Induzione della sterilità nella mosca mediterranea della frutta Ceratitis capitata Wied. con le radiazioni ionizzanti», Accademia Italiana di Entomologia, IX congresso Nazionale di Entomologia, Sienne, juin 1972 (1972) 161-73.

DISCUSSION

J.P. ROS: How do you collect the adults in order to irradiate them?

I.D. DE MURTAS: We irradiate the adults directly in the container in which they emerge.

H. LEVINSON: Are the physiological and biochemical causes of the inferior mating competitivity and reduced longevity of irradiated insects known?

I.D. DE MURTAS: Not definitely as yet but perhaps Mr. LaChance could shed some light on this.

L.E. LaCHANCE: Not very much. There are too many unknown variables involved in various species. It would take far too long to discuss all of them. In the case of the cotton boll weevil, Anthonomus grandis, we do know that reduced longevity is related to cell damage in the midgut. Reduced mating ability has a variety of causes in different insects.

F. OGAH: What is the effect of your standard dosage on the early and late stages of the pupae of the flies, in terms of competitiveness, longevity, etc.?

I.D. DE MURTAS: As is well known, the effects of radiation vary considerably with the stage of the insect and the age at the stage treated. Radiation of young pupae results in inferior insects. Only adult or very late stage pupae should be radiated.

G.W. RAHALKAR: What is the variation in the dose with which the insects are irradiated?

I.D. DE MURTAS: With our dynamic irradiation system it is possible to keep the dose variation to within ± 10%.

L. WIESNER: What are the dimensions of the wagons you use to carry the cages through the facility?

I.D. DE MURTAS: They measure approximately 100 × 40 × 70 cm.

D. ENKERLIN S.: Is there any difference in the dose delivered by a static cobalt bomb system and your moving system?

I. D. DE MURTAS: In the static system the dose variation may exceed ± 15% but in the dynamic system described it is not more than ± 10%.

H. J. HAMANN: Did you observe any effect of dose rate, dose splitting, radiation quality and absence of oxygen on sterility?

I. D. DE MURTAS: In our experiments dose splitting did not have any appreciable effect on the results. For the rest we have not carried out any detailed investigations.

G. DELRIO: Our laboratory experiments have shown that there is very little difference between the number of matings of insects irradiated as adults and those irradiated as pupae. Moreover, experiments on the sexual competivity of caged insects have shown that adults derived from irradiated pupae are more effective. Do you not feel that the question of the best stage at which to sterilize Ceratitis capitata Wied. ought to be examined more closely?

I. D. DE MURTAS: I am very familiar with your work but we came to quite the opposite conclusions and we prefer to irradiate the adults directly.

CONTROL GENETICO CONTRA C. capitata Wied. POR EL METODO DE INSECTOS ESTERILES
Trabajos realizados en España (1969-73)

J.P. ROS
Instituto Nacional de Investigaciones Agrarias,
Madrid, España

Abstract—Resumen

GENETIC CONTROL OF C. capitata Wied. BY THE STERILE-MALE TECHNIQUE: ACTIVITIES CARRIED OUT IN SPAIN (1969-73).

Progress made between 1969 and the last field season in 1973 is described. The first part (A) deals with studies on the dispersion, longevity and population dynamics of sterile males released in the field. The final conclusion is that these tests are subject to so many variables (wind direction, humidity, orientation, etc.) that they are valid only for the experimental site, and the data obtained cannot be applied to other sites having different characteristics. The second part (B) studies the ecology of Ceratitis in Granada province. Studies carried out during two successive years show displacement of the pest from the coast (south), where it lives in the adult stage throughout the year, to the interior (north) as the temperatures continue to rise and fruits ripen. This aspect is being investigated in order to verify whether the displacement reflects migration of the insect or in situ wintering with eclosion when the optimum temperatures are reached. The third part (C) gives the details of three experiments on the control of C. capitata by the sterile-male technique in Murcia (1969) and Granada (1972 and 1973). The results, as well as the progress made in the transport of refrigerated adults and methods of release in the field, were highly satisfactory.

CONTROL GENETICO CONTRA C. capitata Wied. POR EL METODO DE INSECTOS ESTERILES: TRABAJOS REALIZADOS EN ESPAÑA (1969-73).

En el presente trabajo se da cuenta de los progresos realizados desde 1969 hasta la última campaña de 1973. La primera parte (A) corresponde a los estudios realizados sobre dispersión, longevidad y dinámica de las poblaciones de insectos estériles liberados en el campo. La consecuencia final a que se llega es que estos ensayos están sometidos a tal número de variables (dirección del viento, humedad, orientación, etc), que únicamente tienen validez para el lugar de la experiencia, no sirviendo los datos obtenidos para otro con características distintas. La segunda parte (B) corresponde al estudio de la ecología de Ceratitis en la provincia de Granada. En los estudios efectuados durante dos años consecutivos se ha visto un desplazamiento de la plaga desde la costa (sur), en la que vive en estado adulto todo el año, hacia el interior (norte) a medida que las temperaturas iban en aumento y los frutos madurando. Se continúa trabajando en este aspecto para averiguar si es migración del insecto o es invernación in situ con eclosión cuando se alcanzan temperaturas óptimas. La tercera parte (C) detalla las tres experiencias para el control de C. capitata por el método de insectos estériles llevadas a cabo en Murcia (1969) y Granada (1972 y 1973). Los resultados obtenidos han sido altamente satisfactorios, así como los progresos realizados en el transporte de insectos adultos refrigerados y métodos de suelta en el campo.

A. DISPERSION, LONGEVIDAD Y DINAMICA DE LAS POBLACIONES DE INSECTOS ESTERILES LIBERADOS EN EL CAMPO

1. MATERIAL Y METODOS

1.1. Experiencia N° 1 (Provincia de Murcia, 1969)

En una zona experimental de 25 ha, se efectuaron sueltas de insectos estériles (pupas irradiadas a 9 krad). Dicha zona comprendía plantaciones de albaricoques y melocotones, con un relativo aislamiento por áreas colindantes de monte y terreno cultivado y sin vegetación espontánea que pudiera servir de soporte a la mosca (figura 1).

Todos los insectos marcados (Day-glo fluorescent pigment) se soltaron en el punto 0 en el interior del círculo A (figura 2) en las fechas y cantidades especificadas en el cuadro I.

Los mosqueros (tipo Nadel) se dispusieron por toda la zona experimental, según se detalla en la figura 2.

1.2. Experiencia N° 2 (Provincia de Granada, 1971)

En la provincia de Granada se efectuó un ensayo análogo en una finca experimental situada en la Vega de Granada (850 m sobre el nivel del mar), toda ella de regadío, con una plantación de melocotoneros de 50 ha y un aislamiento casi perfecto.

Las liberaciones de insectos estériles (pupas irradiadas a 9 krad) se efectuaron desde 6 jaulas de suelta colgadas en dos árboles (X e Y separados por una distancia de 10 m) en la parte central de la plantación. Los mosqueros de captura se colocaron en dos filas perpendiculares (letras A, B, C, D, E, F, G, H, I, J, K, L, M, N, O, R, S) según el esquema siguiente:

```
                    S
                    R
                    M
                    L
                    K
                    X
          G F E     Y    A B C D
                    H
                    I
                    J
                    N
                    O
```

La distancia entre dos mosqueros de una fila fue de 25 m exceptuando los N y R que se distanciaron 50 m y los O y S, 55 m.

El experimento se inició el 16 de junio. En dicha fecha no existía en la plantación fruta receptiva a la picadura de la mosca. Para la tinción de adultos se utilizaron colorantes (pigmentos fluorescentes Day-glo) mezclados con las pupas a razón de 1 g de colorante por cada 20 000 pupas. En cada jaula se dispusieron las pupas teñidas de la siguiente forma:

Arbol	Jaula	N° de pupas	Color
	1	50 000	Rojo
X	2	50 000	Azul
	3	50 000	Amarillo
	1	50 000	Rojo
Y	2	50 000	Azul
	3	50 000	Amarillo

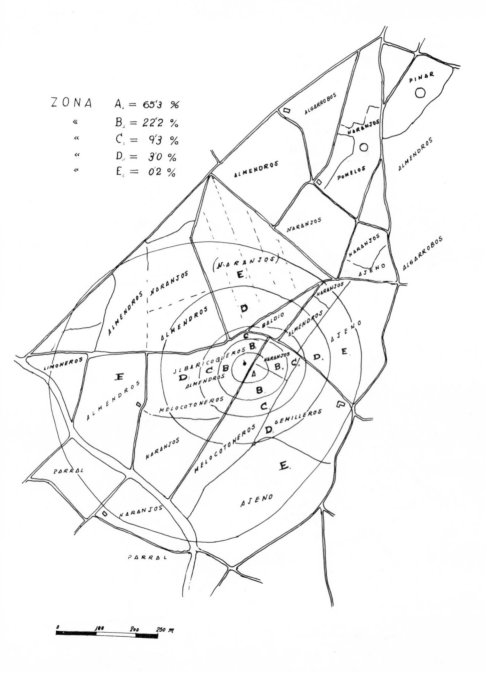

ZONA   A. = 65'3 %
  «   B. = 22'2 %
  «   C. = 9'3 %
  «   D. = 3'0 %
  «   E. = 0'2 %

FIG.1. Dispersión de insectos marcados.

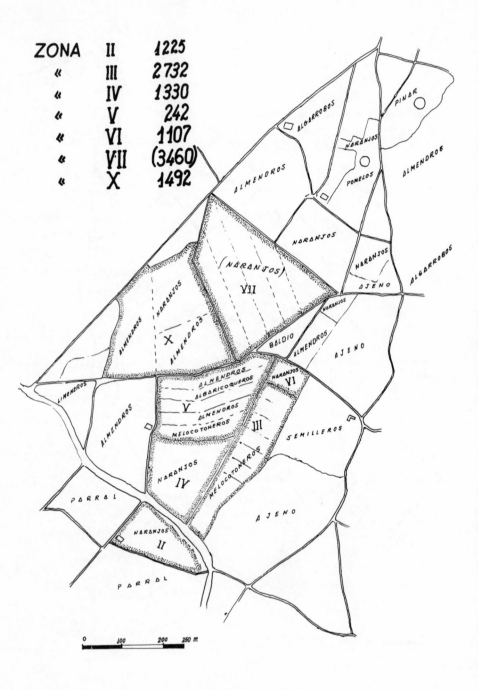

FIG. 2. Areas de cultivo.

CUADRO I. INSECTOS MARCADOS LIBERADOS EN PUNTO 0

Fecha	Cantidad	Colorante
28/IV	100 000	Azul
5/V	300 000	Rojo
5/V	300 000	Amarillo
24/V	310 000	Rojo
13/VI	100 000	Amarillo
7/VII	100 000	Rojo

Se ensayaron tres tipos de alimentación de los adultos emergidos, introduciendo en las jaulas unos saquitos de malla conteniendo:

a) pupas rojas: azúcar + proteína (3:1) sólido.
b) pupas azules: azúcar (sólido).
c) pupas amarillas: sin alimentar.

Se retuvo a los adultos 2 días después de la eclosión en las jaulas, a fin de que se alimentaran suficientemente. Pasado este período se abrieron las jaulas, liberándose los adultos que habían emergido hasta ese momento. Seguidamente se retiraron las jaulas del campo, no liberándose más adultos teñidos para evitar interferencias en la interpretación de los resultados. Por este método y analizando el contenido de las jaulas se pudo determinar con toda exactitud el N° de adultos liberados de cada color (cuadros II y IIa).

A las 48 h se dispusieron los mosqueros en los lugares marcados, recogiéndose al cabo de 24 h. Con los mismos intervalos de tiempo se repitió la operación hasta que se dejaron de obtener capturas.

Se recogieron las moscas capturadas en los mosqueros, determinándose su coloración por su examen con luz ultravioleta.

2.   RESULTADOS

2.1.   Experiencia N° 1

2.1.1.   Dispersión

El cuadro III recoge los datos de capturas de esta experiencia. Con estos datos se calcularon unos índices de densidad relativa de insectos capturados (figura 3).

2.1.2.   Longevidad

Los resultados de las capturas de moscas marcadas parecen indicar que la población liberada disminuye de forma que su número total se reduce a la mitad (aproximadamente) cada siete días. En efecto:

CUADRO II. CONTENIDO DE LAS JAULAS Y CALCULO DE ADULTOS LIBERADOS (GRANADA, 1971)

Arbol	Jaula	Cantidad de pupas	Color	Pupas sin eclosionar (%)	Eclosión (%)	a) Moscas muertas (%) b) Moscas liberadas (%)		N° moscas liberadas
X	1	50 000	Rojo	48,4	51,6	a)	19,8	15 750
						b)	31,8	
	2	50 000	Azul	83,4	16,6	a)	6,8	4 861
						b)	9,8	
	3	50 000	Amarillo	33,4	66,6	a)	38,8	13 900
						b)	27,8	
Y	1	50 000	Rojo	48,0	52,0	a)	13,9	19 050
						b)	38,1	
	2	50 000	Azul	49,6	50,4	a)	28,7	10 850
						b)	21,7	
	3	50 000	Amarillo	59,2	40,8	a)	20,3	10 250
						b)	20,5	

CUADRO IIa.   COMPOSICION DE LA POBLACION LIBERADA

Color		N° total de moscas	% del total
Rojo	(R)	34 800	46,6
Azul	(B)	15 711	21,0
Amarillo	(A)	24 150	32,4
Total		74 661	100,0

Moscas teñidas de amarillo:
   capturas del 13 de mayo, total 773 moscas
   capturas del 20 de mayo, total 384 moscas

Moscas teñidas de rojo:
   capturas del 13 de mayo, total 482 moscas
   capturas del 20 de mayo, total 149 moscas

2.2. Experiencia N° 2

De los resultados de las capturas expresados en el cuadro IV se pueden deducir los siguientes resultados.

2.2.1. Dispersión

Para una primera estimación de la dispersión se han expresado gráficamente (figura 3) las capturas totales de los mosqueros (cuadro IV), suponiendo que cada uno es representativo del área trapezoidal en que está situado.
La dispersión de insectos total (suma de los tres colores) viene representada en la figura 4.

2.2.2. Longevidad

Un primer examen de los resultados parece indicar una mortalidad excesivamente elevada. Si sólo se tiene en cuenta los datos de las capturas la vida media parece ser inferior a tres días. Sin embargo, es preciso tener en consideración que una gran proporción de las moscas liberadas escapa de la zona controlada por los mosqueros, según se desprende del estudio de la dispersión.

3.   CONCLUSIONES

3.1. Dispersión

Los datos de la experiencia N° 1 (provincia de Murcia) muestran que la dispersión es suficiente para los objetivos de aplicación del método de los «insectos estériles». Así a una distancia de 120 m del punto de suelta

CUADRO III. CAPTURAS DE MOSCAS MARCADAS (MURCIA, 1969)

Zona	Mosquero	Fecha	Moscas capturadas	
			Rojas	Amarillas
II	1	13/V	1	2
II	6	20/V	1	13
II	6	17/VI	0	1
III	1	13/V	15	67
III	2	13/V	2	22
III	1	20/V	6	35
III	2	20/V	1	2
III	3	20/V	2	4
III	1	17/VI	2	2
IV	1	13/V	1	12
IV	2	13/V	7	24
IV	3	13/V	2	2
IV	4	13/V	0	1
IV	5	13/V	1	5
IV	3	20/V	0	4
V	1	13/V	5	11
V	2	13/V	1	1
V	3	13/V	3	1
V	4	13/V	5	2
V	5	13/V	175	196
V	1	20/V	3	5
V	3	20/V	4	8
V	4	20/V	0	3
V	5	20/V	64	92
V	1	17/VI	1	1
V	2	17/VI	2	2
V	3	17/VI	0	40
V	4	17/VI	0	53
V	5	17/VI	90	115
VI	1	13/V	24	9
VI	2	13/V	47	103
VI	3	13/V	64	96
VI	4	13/V	16	8

CUADRO III (Continuación)

Zona	Mosquero	Fecha	Moscas capturadas	
			Rojas	Amarillas
VI	5	13/V	103	204
VI	1	20/V	13	19
VI	3	20/V	61	146
VI	5	20/V	34	42
VI	1	17/VI	3	3
VI	2	17/VI	0	1
VI	3	17/VI	0	1
VI	5	17/VI	0	6
X	1	20/V	2	0
X	1	17/VI	0	1
X	2	17/VI	2	0
P (Periferia)	III	13/V	2	0
P	V	13/V	1	1
P	VI	13/V	1	0
P	VII	13/V	5	4
P	VIII	13/V	0	1
P	X	13/V	1	1
P	III	20/V	0	2
P	V	20/V	0	3
P	VI	20/V	1	0
P	VII	20/V	2	5
P	X	20/V	0	1
P	II	17/VI	0	2
P	III	17/VI	1	0
P	IV	17/VI	1	0
P	XX	17/VI	0	2

(zona C) la densidad relativa es de aproximadamente un 30% de la densidad de la zona A. Esto significa que en condiciones favorables sería posible una cobertura adecuada de una zona determinada, escalonando los puntos de suelta cada 100 m. A los 200 m (zona D) la densidad relativa es ya 6 veces menor que en la zona A. A partir de los 350 m (zona F) es de 500 veces menor que en A.

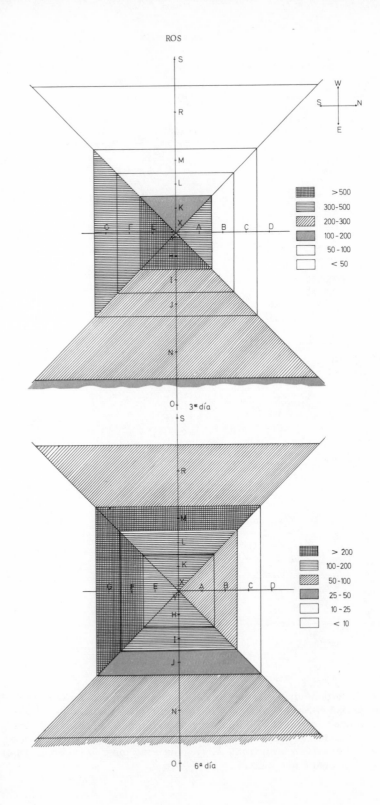

3ᵉ día

6ª día

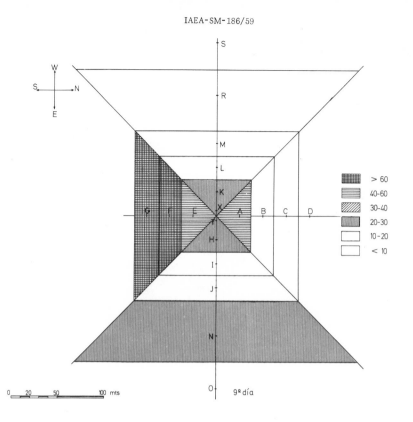

FIG.3. Densidad de población de los insectos marcados.

Sin embargo estos resultados no excluyen la posibilidad de que algunos insectos no sean capaces de volar (o ser arrastrados por el viento) a distancias muy superiores. (En dicho ensayo se hallaron tres adultos en mosqueros situados a 2 km.)

De las gráficas de la experiencia N° 2 (provincia de Granada) se deduce claramente un desplazamiento de las moscas en dirección sur. A partir del 6° día, la máxima densidad de moscas corresponde a los mosqueros más alejados hacia el sur. Ello parece indicar, por otra parte, que un elevado número de moscas están escapando a la zona controlada por los mosqueros.

Las curvas que limitan dichas zonas no pretenden ofrecer un cuadro exacto de la distribución de los insectos, pues se carece de datos para ello. Sin embargo, consideramos que ofrece una representación suficientemente aproximada para fines de aplicación práctica.

La distribución de moscas de todos los colores es muy homogénea, no apreciándose diferencias significativas entre las moscas rojas, azules y amarillas en cuanto a su dispersión.

CUADRO IV.   CAPTURAS (GRANADA, 1971)

Día	Mosquero	Distancia media a los puntos de suelta (m)	Rojas	Azules	Amarillas	Total
3°	A	25	124	200	95	419
	B	50	27	16	12	55
	C	75	1	1	3	5
	D	100	3	2	3	8
	E	25	336	230	282	848
	F	50	192	101	80	373
	G	75	177	190	101	468
	H	30	455	258	233	946
	I	55	128	89	77	294
	J	80	115	75	47	237
	K	30	44	81	43	168
	L	55	21	9	12	42
	M	80	13	11	2	26
	N	130	74	71	62	207
	O	185	49	32	25	106
	R	130	-	1	2	3
	S	185	1	-	2	3
	Total		1760	1367	1081	4208
6°	A	25	19	18	21	58
	B	50	15	39	37	91
	C	175	2	3	2	7
	D	100	2	4	9	15
	E	25	49	65	41	155
	F	50	107	89	68	264
	G	75	71	67	130	268
	H	30	72	65	45	182
	I	55	67	43	35	145
	J	80	8	23	2	33
	K	30	41	57	41	139
	L	55	52	38	76	166
	M	80	68	116	54	238
	N	130	43	25	31	99
	O	185	37	18	26	81

CUADRO IV (Continuación)

Día	Mosquero	Distancia media a los puntos de suelta (m)	Rojas	Azules	Amarillas	Total
	R	130	7	12	43	62
	S	185	-	6	6	12
	Total		660	688	667	2015
9°	A	25	18	27	7	52
	B	50	2	7	-	9
	C	75	1	2	-	3
	D	100	2	3	1	6
	E	25	10	51	-	61
	F	50	31	61	10	102
	G	75	35	48	16	99
	H	30	13	10	1	24
	I	55	6	2	1	9
	J	80	2	9	-	11
	K	30	1	20	1	22
	L	55	4	9	3	16
	M	80	1	4	1	6
	N	130	9	15	3	27
	O	180	3	15	1	19
	R	130	1	11	-	12
	S	180	1	4	-	5
	Total		140	298	45	483
12°	A	25	-	-	-	-
	B	50	-	-	1	1
	C	75	-	-	-	-
	D	100	-	-	-	-
	E	25	-	1	1	2
	F	50	-	1	-	1
	G	75	-	-	-	-
	H	30	-	-	-	-
	I	55	-	-	1	1
	J	80	-	-	-	-
	K	30	-	-	-	-

CUADRO IV (Continuación)

Día	Mosquero	Distancia media a los puntos de suelta (m)	Rojas	Azules	Amarillas	Total
	L	55	-	-	-	-
	M	80	1	-	1	2
	N	130	-	-	-	-
	O	180	-	1	-	1
	R	130	1	2	-	3
	S	180	-	-	-	-
	Total		2	5	4	11
15°	N° de capturas = 0					

3.2. Longevidad

Con los datos de la experiencia N° 1 (provincia de Murcia) se formularon varias hipótesis sobre la supervivencia de la población estéril liberada en el campo. La hipótesis que mejor encaja con los datos experimentales es la siguiente: la mortalidad de la población liberada sigue una función logarítmica, con una «vida media» de unos seis días. Es decir, que cada seis días aproximadamente la población se reduce a la mitad.

En la experiencia N° 2 (provincia de Granada) no se pueden dar cifras de supervivencia ya que parte de las moscas escapaban a la acción de los mosqueros. Unicamente se pueden dar cifras comparativas entre las moscas de distintos colores.

Se observa que en el conjunto de la población aumenta cada vez más el porcentaje de moscas azules con relación a las amarillas y rojas. Ello puede indicar una mayor supervivencia de dichos adultos en relación con los otros o una menor dispersión de las moscas azules (figura 5).

Es de destacar que los ensayos sobre dispersión de insectos estériles en el campo, según nuestras experiencias, están sometidos a tal número de variables, que dichos ensayos tienen validez únicamente para el lugar de la experiencia, ya que la influencia de la dirección del viento dominante, humedad, orientación, vegetación circundante, estado de maduración de los frutos, etc., hacen que los datos obtenidos en un lugar no sean válidos para otros.

B. ESTUDIOS SOBRE LA ECOLOGIA DE Ceratitis capitata EN LA PROVINCIA DE GRANADA

1. ANTECEDENTES

En la costa de la provincia de Granada, C. capitata constituye una plaga endémica y dadas las condiciones climáticas el insecto existe en forma

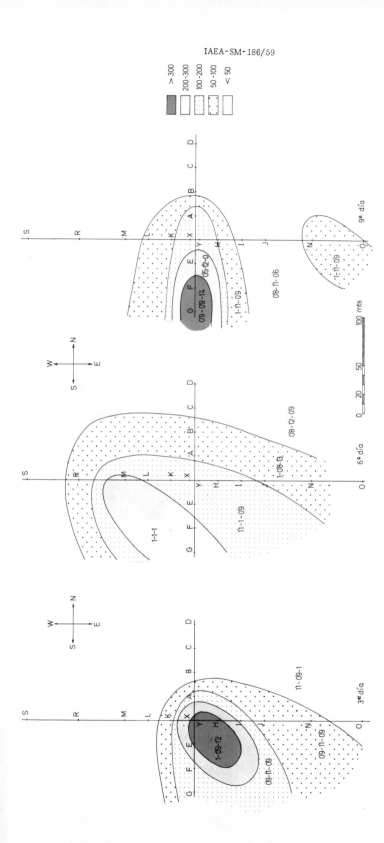

FIG.4. Densidad relativa de insectos marcados.

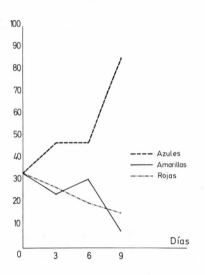

FIG.5. Comparación de la publación en porciento.

adulta durante todo el año. En dicha zona son plantas hospedantes importantes: melocotón, albaricoque, naranjo e higuera. Otras especies
características de la zona, que no se encuentran en el resto de la península,
son: chirimoyo, guayabo, mango y aguacate. Como huésped silvestre es
de gran importancia la Opuntia ficus índica.

En el interior de la provincia, las especies de plantas huéspedes se
reducen básicamente a frutales de hueso en plantaciones regulares. Los
frutos más tempranos escapan normalmente al ataque de C. capitata.
Dicho ataque se produce más avanzada la estación, suponiéndose que es
debido a la migración del insecto desde las zonas litorales más cálidas.

Uno de los objetivos de los ensayos de 1972, 1973 y 1974 consistió en
estudiar la aparición sucesiva de adultos en toda la zona comprendida entre
la costa y el interior (figura 6). Dicho estudio viene facilitado por las
características geográficas de la zona. Si efectivamente se trata de una
migración de insectos desde la costa, las únicas vías de penetración natu-
rales serían el valle del Lecrín y el valle de Almuñécar. Este último
por sus características topográficas es más improbable pese a lo cual se
ha incluido en el estudio.

2. MATERIAL Y METODOS

Se instalaron mosqueros de plástico conteniendo atrayente (Trimedlure)
e insecticida (DDVP) en las localidades de Jete, Motril, Almuñécar, Vélez
de Benaudalla, Guajar, Fraguit, Padul, Santa Fe, Benalúa, Guadix y
Villanueva de las Torres (figura 1).

Se instalaron cuatro mosqueros en cada una de las localidades citadas
en plantaciones regulares de melocotón (en algunas existían además otros
frutales: naranjos, nísperos y albaricoques). Se inspeccionaron todos los
mosqueros dos veces por semana, prosiguiéndose dicha inspección hasta
comprobar la implantación del insecto.

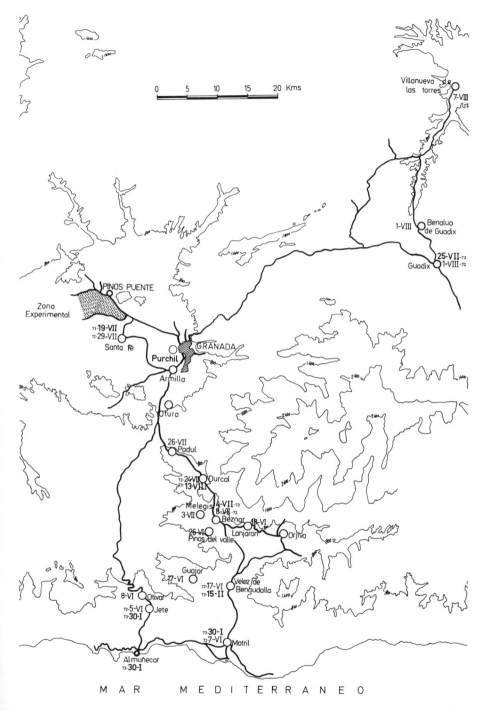

FIG.6. Area geográfica de la primera aparición de adultos de <u>Ceratitis</u>.

3. RESULTADOS

En el cuadro V se recogen las fechas de la primera aparición de adultos de C. capitata en las diferentes localidades (de Sur a Norte).

En 1972 los primeros mosqueros se pusieron el 5 de junio, careciéndose de datos para los meses anteriores, aunque es conocido que durante todo el invierno hubo ataque leve en frutos tropicales. Todo ello demuestra que en la zona costera (Almuñécar, Motril, Jete) la mosca existe en estado adulto durante todo el año.

En la localidad de Beznar, cabecera del valle de Lecrín, es donde se puede asegurar que la mosca no existe en forma adulta en los meses invernales ya que en los mosqueros instalados desde el mes de enero (73) no capturaron adultos hasta el mes de julio y en los cultivos de cítricos no existen ataques en esos meses.

Desde esta localidad hacia el interior de la provincia y según las capturas de los mosqueros se observa un desplazamiento progresivo de la plaga (figura 6) de sur a norte y en aumento de altura/nivel del mar.

Comparando las capturas de 1972 con las de 1973 en las localidades de Beznar, Dúrcal, Santa Fe y Guadix, se observa que en el año 1973 la mosca se adelantó con respecto al año 1972 en un promedio de siete días. Climatológicamente, el año 1973 fue más benigno que el 72 (figura 7).

Por observaciones en todo el valle del Lecrín y por los muestreos llevados a cabo en la finca experimental y parcelas testigo (año 1972) en 1973 la fruta (melocotón) maduró 10-15 días antes que en el año anterior.

CUADRO V. FECHAS DE PRIMERA APARICION DE ADULTOS DE C. capitata

Localidad	1972	1973
Almuñécar		30/I
Motril	7/VI	30/I
Jete	5/VI	30/I
Otivar	8/VI	
Vélez	10/VI	15/II
Guajar	17/VI	
Lanjarón	18/VI	
Pinos Valle	26/VI	
Melegis	3/VII	
Beznar	8/VII	4/VII
Dúrcal	24/VII	13/VII
Padul	26/VII	
Santa Fe	29/VII	19/VII
Benalúa	1/VIII	
Guadix	1/VIII	25/VIII
Villanueva	7/VIII	

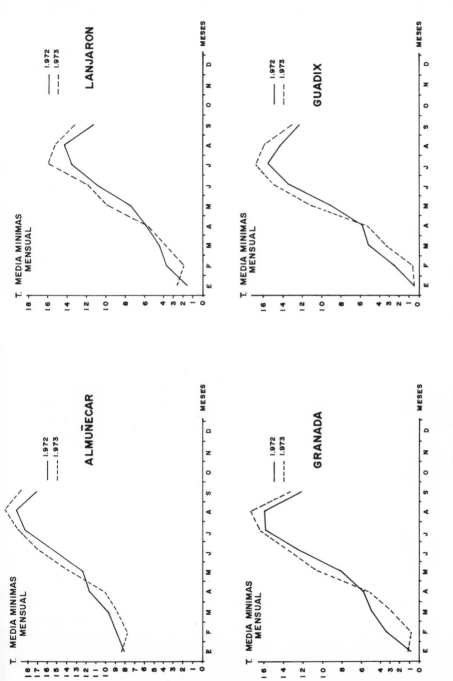

FIG.7. Curvas de temperaturas en diferentes lugares.

CUADRO VI. SUELTAS MASIVAS DE INSECTOS ESTERILES (CANTIDAD DE PUPAS IRRADIADAS; MURCIA, 1969)

Zona	Procedencia de las pupas M = Madrid V = Viena	Marzo	Abril	Mayo	Junio	Julio	Agosto	Totales
II	M	-	-	1 685 000	845 000	1 330 000	-	3 860 000
	V	-	-	220 000	-	-	-	220 000
III	M	-	80 000	510 000	2 335 000	3 400 000	1 260 000	7 585 000
	V	-	350 000	650 000	-	-	-	1 000 000
IV	M	1 420 000	1 230 000	720 000	810 000	1 610 000	-	5 790 000
	V	-	500 000	530 000	150 000	-	-	1 180 000
V	M	-	100 000	830 000	880 000	1 820 000	660 000	4 290 000
	V	-	600 000	800 000	660 000	-	-	2 060 000
VI	M	-	-	250 000	595 000	200 000	-	1 045 000
	V	-	350 000	440 000	-	-	-	790 000
VII	M	-	425 000	-	-	-	-	425 000
	V	-	450 000	-	-	-	-	450 000
X	M	-	-	1 065 000	1 230 000	750 000	-	3 045 000
	V	-	-	-	500 000	-	-	500 000
Totales	M	1 420 000	1 835 000	5 060 000	6 695 000	9 110 000	1 920 000	26 040 000
	V	-	2 250 000	2 640 000	1 310 000	-	-	6 200 000

4.   CONCLUSIONES

A la vista de lo expuesto anteriormente los datos pueden ser inter-
pretados de la siguiente manera:

1.   C. capitata vive en forma de adulto durante todo el año en la zona
     costera.
2.   Se observa una progresión de la plaga de Sur a Norte conforme
     van ascendiendo, en relación directa, las temperaturas.
3.   Se precisará estudiar a fondo esta cuestón en los próximos años
     a fin de determinar si esta progresión de la plaga obedece a una
     migración del insecto desde la costa al interior o bien a una
     invernación in situ con eclosión posterior cuando las temperaturas
     sean óptimas.

C.   CONTROL GENETICO DE C. capitata POR EL METODO
     DE MACHOS ESTERILES

1.   MATERIAL Y METODOS

1.1.   Experiencia N° 1 (Provincia de Murcia, 1969)

Se utilizaron pupas procedentes de la unidad de cría masiva de Madrid
y del laboratorio de Seibersdorf (OIEA).   Las pupas irradiadas (9 krad) se
transportaron en bolsas de tela, estratificadas en bandejas dispuestas en
cajas de madera especialmente diseñadas con objeto de facilitar su aireación
y evitar una elevación excesiva de la temperatura.

Su envío a Murcia se efectuó por carretera (viaje nocturno) para evitar
el posible daño a causa de las temperaturas elevadas en los meses de verano.

En la zona experimental (figura 1), las pupas se sacaron de sus
recipientes, distribuyéndose en bolsas de papel (con orificios para su
aireación) a razón de 3000 pupas/bolsa.   No se suministró alimento a los
adultos eclosionados en las bolsas, ni se utilizaron colorantes para su
tinción.

Cuando la eclosión era la adecuada, las bolsas se iban transportando
al campo, colgándose de los árboles y rasgándose para permitir la salida
de los adultos estériles.

El resumen de las sueltas de insectos estériles en cada una de las
zonas del área experimental se recoge en el cuadro VI especificándose la
procedencia de las pupas M = Madrid;   V = Viena, y la zona de suelta.

1.2.   Experiencia N° 2 (Provincia de Granada, 1972)

En la finca anteriormente descrita (véase A.1.2.) la liberación de
insectos estériles se efectuó en siete puntos fijos, alejados 100-200 m de
la plantación con el fin de proteger ésta con una barrera de esterilidad
(figura 8).

Cada punto de suelta consistía en tres jaulas iguales a las usadas para
los ensayos de dispersión.   Las pupas irradiadas (9 krad) procedentes de
la unidad de cría de Madrid se transportaron por ferrocarril (viaje nocturno).

En el total de la campaña se liberaron 25 millones de insectos útiles
en el campo (cuadro VII).   No se les suministró a los insectos ningún tipo
de alimento en las jaulas ni se marcaron con colorantes.

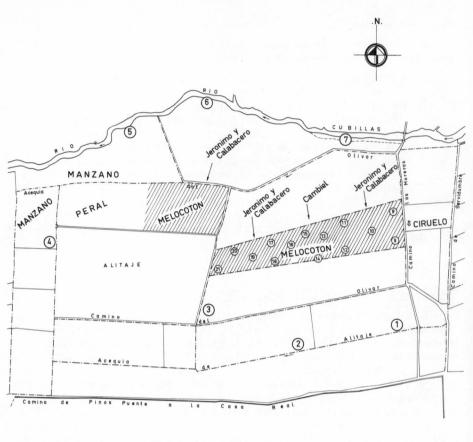

FIG.8. Area experimental con puntos de suelta fijos.

1.3. Experiencia N° 3 (Provincia de Granada, 1973)

1.3.1. Suelta de insectos estériles

La campaña de sueltas de insectos estériles se inició a últimos de Marzo, en las localidades de Almuñécar, Jete, Motril y Vélez, por estimar estos puntos como reductos invernales de la mosca.

A partir del 11 de Julio, todas las sueltas se efectuaron en la zona experimental de Pinos Puente.

Todos los insectos procedían del laboratorio de cría artificial masiva de Madrid (El Encín) enviados en estado de pupa o en estado adulto.

Las sueltas en forma de pupa se realizaron por el mismo método que en 1972 y los adultos se liberaron en lugares adecuados en pleno campo, donde el sol los reavivara rápidamente, después de haber estado sometidos varias horas a baja temperatura.

CUADRO VII. SUELTA DE INSECTOS ESTERILES (GRANADA, 1972)

Fecha		Cantidad de pupas	Eclosión (%)	Equivalencia real de adultos útiles liberados	Observaciones
Junio	6	300 000	69,4	208 200	-
	9	300 000	72,3	216 900	-
	14	300 000	65,7	197 100	-
	18	100 000	32,1	32 100	Descenso brusco temperatura
	24	200 000	28,3	56 600	Tormenta de aire y agua
	28	250 000	31,4	78 500	Temperaturas bajas
Julio	1	300 000	45,7	137 100	id.
	5	300 000	52,3	156 900	id.
	7	375 000	64,1	240 375	-
	12	300 000	60,3	180 900	-
	20	300 000	62,3	186 900	-
	22	300 000	0,0	0	Retraso en el transporte
	27	700 000	65,1	455 700	-
Agosto	1	1 500 000	45,3	679 500	Baja eclosión debido a irra-
	5	1 800 000	56,2	1 011 600	diación desfasada en rela-
	9	1 850 000	56,8	1 050 800	ción al ciclo
	12	1 000 000	76,4	764 000	-
	17	1 250 000	75,9	948 750	-
	19	1 700 000	72,3	1 229 100	-
	22	1 500 000	78,0	1 140 000	-
	24	1 300 000	75,4	980 200	-
	26	2 200 000	68,3	1 502 600	-
	29	1 600 000	71,4	1 142 400	-
	31	2 100 000	69,3	1 455 300	-
Sept.	2	1 600 000	25,0	400 000	Temporal (descenso de la
	5	1 600 000	45,0	720 000	temperatura y lluvia)
	7	900 000	64,3	578 700	Mejora temperatura
	9	900 000	78,3	704 700	-
	12	900 000	73,4	660 600	Buen tiempo
	14	600 000	76,3	457 800	-
	16	900 000	68,7	618 300	-
	19	900 000	79,6	716 400	-
	21	900 000	59,2	532 800	Lluvia

CUADRO VII (Continuación)

Fecha		Cantidad de pupas	Eclosión (%)	Equivalencia real de adultos útiles liberados	Observaciones
	23	2 200 000	69,7	1 533 400	-
	26	1 500 000	71,3	1 069 500	-
	28	1 500 000	65,7	985 500	-
	30	1 500 000	53,4	801 000	Temporal
Oct.	3	1 000 000	25,3	253 000	Lluvia intensa
	5	1 000 000	32,5	325 000	Lluvia
	7	1 000 000	57,1	571 000	-
Totales		40 725 000	61,3	24 979 225	

Todas las sueltas efectuadas en la zona experimental se localizaron en los puntos 1 al 7 (figura 8), excepto en la defensa de la variedad tardía Campiel en que las sueltas se realizaron en el interior de la parcela.

El calendario de sueltas y cantidad de insectos liberados se recogen en el cuadro VIII.

1.3.2. Nuevo método de envío de pupas

El envío de pupas de C. capitata desde el centro de cría masiva a la zona experimental se efectuó en jaulas formadas por un bastidor paralelepipédico de madera, forrado por todas sus caras de malla metálica muy fina.

En dicho recipiente se introducen a manera de cajoncillos las unidades que contienen las pupas (100 000) cerradas en su parte superior por una malla de plástico que permite salir de los cajones a los adultos que emergen por el camino y distribuirse por todo el volumen del recipiente.

De este forma, a la recepción del envío en el campo, un 45% de los insectos llegan a estado adulto y en perfectas condiciones de vitalidad.

1.3.3. Técnica de laboratorio para la obtención de adultos irradiados

Para almacenar el mayor número de adultos irradiados, empleando el menor espacio posible, se montaron unos cilindros de malla de tela de 2,50 m de altura y 45 cm de diámetro. En la parte inferior se les adaptó unas cubetas circulares de plástico y en la parte superior un embudo cuyo diámetro mayor fuera igual al de la cubeta inferior.

En el fondo de la cubeta se coloca una capa de serrín y encima otra de pupas (50 000). El conjunto se cubre con un círculo de malla metálica, por cuyos orificios pueden pasar los adultos recién eclosionados, pero no así las pupas.

Estos cilindros se cuelgan del techo. A medida que las pupas van eclosionando, los adultos atraviesan la malla, vuelan hacia arriba y se reparten por toda la superficie del cilindro.

Cuando hay suficiente cantidad de adultos, las jaulas se pliegan y se introducen en una cámara frigorífica durante unos momentos hasta que los adultos pierden actividad. Seguidamente se vuelven a colgar del techo en sentido inverso. Los adultos van resbalando por la tela y caen al embudo recogiéndose en recipientes adecuados. Las pupas no pueden caer, pues quedan retenidas por la malla metálica que cubre la cubeta.

El sistema descrito es sencillo, de escaso coste y de fácil operación. Puede utilizarse indistintamente con pupas previamente irradiadas o con pupas no estériles, si se desea efectuar la irradiación en estado adulto.

En el primer caso (método utilizado para las sueltas de 1973) los adultos se recogieron en recipientes cilíndricos, con bases de tela metálica muy fina, con capacidad para 30 000 insectos.

1.3.4. Técnicas de envíos de adultos refrigerados

Una vez cargados los recipientes con los adultos son introducidos en neveras portátiles con varias bolsas de hielo.

De esta manera hacen el viaje nocturno (por tren) hasta la zona experimental. Una vez allí, los adultos se liberan en lugares adecuados al sol para que la recuperación de temperatura sea rápida y puedan iniciar su actividad normal en el menor tiempo posible.

En la práctica este sistema de envíos ha sido muy eficaz y salvo algunos accidentes (rotura de bolsas de hielo) el porcentaje de mortandad ha oscilado siempre entre límites muy bajos (cuadro VIII).

Según la temperatura ambiente durante el viaje a la zona experimental, se incluye mayor o menor número de bolsas de hielo para que la temperatura interior de la nevera oscile de 5°C a 8°C.

1.3.5. Ensayos preliminares de sueltas aéreas

En la presente campaña se han efectuado varios ensayos de sueltas aéreas con avionetas tipo Piper. El fin primordial que se persigue es hallar un método simple, económico y aplicable a gran escala por personal no especializado.

En estos ensayos se utilizó como material soporte de los adultos de Ceratitis viruta de corocho y bolitas de poliestireno, comportándose este último material mejor que el primero. Se empleó en estos experimentos un total de 2 500 000 adultos.

Los resultados son prometedores, pero el sistema requiere perfeccionamiento pues se observó una mortandad muy elevada.

2. RESULTADOS

2.1. Experiencia N° 1

Durante la recolección de la fruta se apartó en la totalidad de la cosecha de la zona experimental toda aquella fruta que presentaba algún síntoma de ataque o deterioro siendo examinada posteriormente para determinar si el ataque se debe a C. capitata. Simultáneamente se realizaron muestreos de fruta (en árbol y caída) en todas las zonas de control para determinar el grado de ataque de Ceratitis. Los resultados se resumen en los cuadros IX (zona experimental) y X (resumen comparativo zona experimental y zonas de control).

CUADRO VIII. SUELTA DE INSECTOS ESTERILES (GRANADA, 1972)

Fecha	Cantidad		Eclosión pupas (%)	Mortandad adultos (%)	Equivalencia real de adultos útiles liberados	Lugar de suelta
	Pupas	Adultos				
Marzo 29	1 000 000	600 000	58	4	1 156 000	Jete
Abril 3	1 000 000	400 000	60	2	992 000	Vélez
26	1 000 000	500 000	48	3	965 000	Jete
Mayo 3	1 000 000	300 000	51		510 000	Motril
8	1 000 000		45	5	735 000	Almuñécar
17		600 000		8	552 000	Jete
29		1 000 000		12	880 000	Motril
Junio 12		250 000		8	230 000	Vélez
14		400 000		50	200 000	Jete
15	1 000 000	600 000	67	10	1 210 000	Jete
16	1 000 000	800 000	70	20	1 340 000	Vélez
20	700 000	350 000	68	25	738 500	Jete
23	750 000		65		487 500	Vélez
27	700 000	120 000	68	40	548 000	Vélez
28		450 000		10	405 000	Melegis
29		450 000		40	270 000	Melegis
Julio 3	750 000	300 000	68	3	801 000	Pinos del V.
4		600 000	50		300 000	Beznar
6	800 000		55		440 000	Melegis
7	1 000 000	200 000	61	8	794 000	Beznar
11		500 000		20	400 000	Parcela exper.
12		450 000		40	270 000	id.
13	800 000	200 000	58	15	634 000	id.
14	800 000	200 000	56	10	628 000	id.
17		500 000		10	450 000	id.
20	750 000		55		412 500	id.

Fecha							Parcela exp.
Julio	21	1 000 000	450 000	53	20	890 000	
	24	800 000	300 000	65	10	790 000	id.
	28	1 000 000		55		550 000	id.
	31	1 000 000	500 000	61	8	1 070 000	id.
Agosto	1	1 100 000	700 000	68	10	630 000	id.
	2	1 000 000	900 000	62	10	1 558 000	id.
	4	1 000 000	150 000	68	8	758 000	id.
	7		500 000		40	980 000	id.
	8		200 000		8	184 000	id.
	9		550 000		10	500 000	id.
	10		300 000		10	270 000	id.
	11	800 000		64		512 000	id.
	14	1 250 000	1 000 000	52	35	650 000	id.
	17	2 500 000		68		2 350 000	id.
	18	1 500 000	1 500 000	65	50	975 000	id.
	21	2 400 000	500 000	55	10	2 070 000	id.
	23	1 100 000		50		1 000 000	id.
	24	1 100 000		62		680 000	id.
	25	900 000		55		495 000	id.
	28	1 000 000		60		600 000	id.
	29	2 000 000	600 000	68	8	1 912 000	id.
	30		800 000		10	720 000	id.
	31	1 000 000		68		680 000	id.
Sept.	1	2 700 000	300 000	50		1 350 000	id.
	4	1 650 000		60	8	1 266 000	id.
	6	1 000 000	800 000	55	10	1 270 000	id.
	8	2 200 000		62		1 364 000	id.
	11	1 100 000	600 000	65	15	1 225 000	id.
	12	1 000 000	200 000	62	4	812 000	id.
	13		600 000		12	528 000	id.
	15	600 000		59		354 000	id.
	16	600 000		60		360 000	id.
	19	800 000	200 000	68	5	734 000	id.
	20	1 000 000		60		600 000	id.

CUADRO VIII (Continuación)

Fecha	Cantidad		Eclosión pupas (%)	Mortandad adultos (%)	Equivalencia real de adultos útiles liberados	Lugar de suelta
	Pupas	Adultos				
Sept. 22	1 500 000		65		975 000	Parcela exp.
25		600 000		60	240 000	id.
26	1 100 000		68		878 000	id.
28	2 350 000		50	35	1 175 000	id.
Oct. 3		800 000		60	320 000	id.
Total	54 100 000	23 020 000			50 625 500	

CUADRO IX. CONTROL DE FRUTA EN ZONA EXPERIMENTAL (MURCIA, 1969)

Fecha	Zona 3 Melocotón	Zona 5 Melocotón	Zona 5 Albaricoque	Zona 7 Naranjo	Zona 8 Mandarina	Zona 10 Naranjo
29 Abril				Sin ataque	Ataque: 6%	Ataque: 11%
10 Mayo	Fruto no receptivo	No receptivo		Sin ataque		
22 Mayo	Probablemente receptivo	Receptivo				
2 Junio			Sin ataque			
4 Junio			1 fruto con 12 larvas			
9 Junio			Finalizó la recolección sin rechazar ningún fruto			
10 Junio			291 picaduras sin huevo			
25 Junio	Comienza la recolección		67 picaduras sin huevo			
3 Julio	Ataque: 0,001%		Sin fruta			
7 Julio	id.: 0,01%					
11 Julio	id.: 0,06%					
17 Julio	id.: 0,2%					
19 Julio	id.: 0,3%					
21 Julio	id.: 0,44%					
22 Julio	id.: 0,75%					
23 Julio	id.: 1,6%					
29 Julio	id.: 13%					
30 Julio	id.: 13%					

CUADRO X.   PORCENTAJE DE FRUTA ATACADA (EN ARBOL;
MURCIA, 1969)

Zona experimental				Julio			
Fruta	Abril	Mayo	Junio	1-10	10-20	20-25	25-31
Agrios	10	-	-	-	-	-	-
Albaricoque	-	0	0	-	-	-	-
Melocotón	-	-	0	0,1	0,2	1,0	10
Zona de control							
A. Melocotón (tratado)	-	-	-	0	0	0	0
B. Albaricoque	-	0	0	50	90	-	-
C. Albaricoque	-	0	0	40	90	-	-
C. Melocotón	-	-	-	40	90	100	-
E. Melocotón	-	-	-	-	-	-	60
F. Albaricoque	-	-	-	-	90	-	-
G. Paraguayo	-	-	-	-	-	100	-
H. Melocotón	-	-	-	-	25	100	-

2.2.   Experiencia N° 2

En el cuadro XI se recogen los datos sobre el ataque de C. capitata
al melocotón en las zonas testigo (Armillar y Santa Fe) y en la zona
experimental (Pinos Puente).

2.3.   Experiencia N° 3

En la zona experimental y zonas testigo se efectuaron muestreos de
fruta según se iba recolectando (incluyendo la fruta caída en el suelo).   El
muestreo se realizó tomando fruta de cajas al azar en distintos días de
la recolección.   El detalle del número de frutos examinados en cada
muestreo y de la cosecha total en cada caso viene recogido en los
cuadros XII y XIII.

En la variedad tardía Campiel el muestreo se realizó en la mesa de
recepción de la fruta y a medida que se fue envasando en cajas (cuadro XIV).

En los cuadros XII y XIII se recogen los datos sobre el ataque de
C. capitata en melocotón temprano en las zonas testigo (Armillar, Santa Fe
y Purchil) y en la zona experimental (Pinos Puente).

En el cuadro XIV se recogen los datos de ataque de C. capitata a la
variedad de melocotón tardía Campiel, en la zona testigo (Purchil) y en
la zona experimental (Pinos Puente).

CUADRO XI. CONTROL DE FRUTA (MELOCOTON; GRANADA, 1972)

Localidades	Fecha	Cantidad recogida (kg)	N° de frutos examinados[a]	Estado de los frutos					
				Sanos	(%)	Ataque de Ceratitis capitata	(%)	Ataque otras causas (insectos, hongos, pájaros, etc.)	(%)
Armilla Total cosecha	21/8	11 430 11 430	1 350	1 202	(89,0)	67	(4,9)	81	(6,0)
Santa Fe	22/8	12 700	1 500	1 211	(80,7)	151	(10,1)	138	(9,2)
	23/8	10 260	1 500	1 326	(88,4)	72	(4,8)	102	(6,8)
	25/8	12 080	1 500	1 141	(76,0)	219	(14,6)	140	(9,3)
	28/8	11 660	1 500	1 286	(85,7)	122	(8,1)	92	(6,1)
Total cosecha		86 458							
Pinos Puente Variedades: Jerónimo y Calabacero	21/8	14 380	1 500	1 428	(95,2)	0		72	(4,8)
	26/8	15 720	1 500	1 409	(93,9)	0		91	(6,0)
	28/8	13 860	1 500	1 394	(92,9)	0		106	(7,0)
	29/8	18 700	1 500	1 419	(94,6)	0		81	(5,4)
Total cosecha		220 000							
Variedad Campiel (Fruta del suelo antes de la recolección)	25/9	640	899	271	(30,1)	15	(1,6)	613	(68,1)
(Recolección)	28/9	4 150	2 850	2 586	(90,7)	15	(0,52)	249	(8,7)
	30/9	3 560	2 425	2 189	(90,2)	21	(0,86)	215	(8,8)
	4/10	6 200	4 340	3 897	(89,7)	45	(1,0)	398	(9,1)
	7/10	5 800	3 950	3 584	(90,7)	54	(1,3)	312	(7,8)
Total cosecha		26 000							

[a] Promedio: 1 kg = 7 melocotones.

CUADRO XII.   PORCENTAJE DE ATAQUE DE Ceratitis capitata EN
ZONA EXPERIMENTAL (VARIEDADES JERONIMO Y CALABACERO;
GRANADA, 1973)

Lugar	Fecha	Fruta recogida (kg)	Fruta atacada (kg)	(%)
Pinos Puente	4,5,6/8	5000	0,5	0,01
	7/8	1500	20	1,30
	8/8	5000	0,5	1,01
	14/8	4000	1	0,02
	15/8	3500	1,5	0,04
	16/8	2000	1	0,05
	17/8	2000	2	0,10
	21/8	1000	-	0,00
	29/8	1500	5	0,30

3.   CONCLUSIONES

3.1.   Experiencia N° 1

a)   En la zona experimental protegida únicamente por la suelta de insectos
     estériles, el ataque de C. capitata fue inferior al 1% durante toda la
     experiencia, excepto durante la última semana, cuando quedaba muy
     poca fruta por recoger.
b)   En la zona experimental se halló un único albaricoque atacado, de una
     cosecha total de unos 7500 kg.
c)   En la zona de control tratada con insecticidas organofosforados cada
     7 días (a partir del comienzo de cambio de coloración del fruto), no se
     registró ataque.
d)   En las zonas de control no tratadas con insecticidas, el ataque osciló
     entre el 60% y el 100%.

3.2.   Experiencia N° 2

a)   El ataque de C. capitata a las variedades más tempranas de melocotón
     (Jerónimo y Calabacero) ha sido nulo en la zona experimental, protegida
     exclusivamente por la suelta de insectos estériles (en años anteriores
     se venía registrando, en estas variedades, un ataque del 5% a pesar de
     los tratamientos químicos convencionales).
b)   En las zonas testigo colindantes, el ataque de C. capitata a las mismas
     variedades y durante el mismo periodo ha oscilado entre un 5% y un
     14% a pesar de los tratamientos químicos correspondientes.
c)   En la variedad tardía de melocotón (Campiel) en la zona experimental,
     el ataque de C. capitata ha oscilado entre el 0,5 y el 1,6% (en años
     anteriores se registraron ataques del orden del 15%).

CUADRO XIII.   PORCENTAJE DE ATAQUE DE Ceratitis capitata EN
PARCELAS TESTIGO (VARIEDADES JERONIMO Y CALABACERO;
GRANADA, 1973)

Lugar	Fecha	Fruta recogida (kg)	Fruta atacada (kg)	(%)	Observaciones
Santa Fe	30/7	1100	28	2,50	2 tratamientos
	31/7	360	6	1,70	(Lebaycid)
	7/8	3000	49	1,60	1/8 y 11/8
	8/8	1500	18	1,20	
	9/8	500	30	6,00	
	10/8	1700	46	2,70	
	11/8	2000	39	1,85	
	13/8	1700	41	2,40	
	15/8	500	20	4,00	
Purchil	3/8	1250	44	3,50	2 tratamientos
	4/8	2480	97	3,90	(Lebaycid)
	5/8	2230	101	4,50	26/7 y 10/8
	6/8	3450	210	6,10	
	7/8	1860	54	2,90	
	8/8	2660	98	3,70	
	15/8	3845	357	9,30	
	16/8	2300	200	8,70	
	17/8	3640	375	10,30	
	20/8	5435	717	13,20	
	24/8	1824	312	17,10	
Armilla	31/7-5/8	12600	542	4,30	2 tratamientos
	7/8-10/8	6348	387	6,10	(Lebaycid)
	11/8-17/8	8300	482	5,80	26/7 y 10/8
	17/8-24/8	6840	875	12,80	
Purchil	10/8	3224	2960	92,00	Sin tratamiento químico
Armilla	6/8	2840	2840	100,00	Sin tratamiento químico

d)   Las conclusiones anteriores parecen confirmar la hipótesis de que es
posible proteger una zona, relativamente aislada mediante un cordón
sanitario periférico, utilizando exclusivamente el método de insectos
estériles.

3.3.   Experiencia N° 3

De los cuadros se deduce:

a)   En todos los muestreos, el ataque de C. capitata a las variedades más
tempranas de melocotón (Jerónimo y Calabacero) ha sido inferior al
1,3% en la zona experimental, protegida exclusivamente por la suelta
de insectos estériles.

CUADRO XIV.   PORCENTAJE DE ATAQUE DE Ceratitis capitata
(VARIEDAD TARDIA CAMPIEL; GRANADA, 1973)

Lugar	Fecha	N° de frutos recogidos	N° de frutos atacados	(%)	Observaciones
Purchil	26/9	8400	53	0,63	Zona testigo
	26/9	300	18	6,00	Tratamientos
	27/9	5280	110	2,08	29/8 y 13/9
	2/10	7580	93	1,22	
	3/10	10 370	122	1,17	
	4/10	7160	83	1,18	
	7/10	7020	89	1,26	
	8/10	7100	98	1,38	
	14/10	2820	219	7,70	
	15/10	2960	339	11,50	
Pinos Puente	25/9	7954	4	0,05	Zona experi-
	26/9	7250	21	0,28	mental
	27/9	1920	4	0,20	Sin tratamiento
	30/9	4520	18	0,39	
	2/10	7680	21	0,27	
	3/10	2340	25	1,06	
	7/10	7548	28	0,37	

En los muestreos de las zonas testigo (tratadas químicamente)
el ataque ha oscilado entre un 1,2% y un 17%.   En las zonas no tratadas,
el ataque ha sido del 92% al 100%.

b)   Los promedios de ataque, han sido los siguientes:
Zona experimental   0,12%
Zonas testigo
Santa Fe   2,20%

Purchil   8,30%
Armilla   6,70%

Variedades tardías
Zona experimental   0,30%
Zona testigo (Purchil)   2,10%

c)   Del cuadro VIII se deduce que es más eficaz el envío de insectos en
estado adulto (refrigerados) que en estado de pupa.   Estos resultados
pueden sintetizarse en las siguientes cifras (se incluyen los datos de
1972 a efectos comparativos).

	1972	1973
Envío total de pupas (millones)	40.7	54.1
Total de insectos útiles (millones)	25.0	32.4
Eficacia del método (%)	61,3	60,0
Envío total de adultos (millones)	-	23,0
Total de insectos útiles (millones)	-	18,1
Eficacia del método (%)		79,0

Estos datos indican que de los envíos de pupas resultan un 60% de insectos útiles, frente a un 79% en el caso de envío de adultos.

d) Los resultados de 1973 confirman los de 1972 y 1969 en el sentido de que el método de insectos estériles es plenamente eficaz en la defensa contra C. capitata. La reducción que se ha logrado en el coste de producción y la mayor eficacia del método de suelta de adultos refrigerados constituyen en progreso importante para la aplicación práctica del método a gran escala.

BIBLIOGRAFIA

MELLADO, L., CABALLERO, F., ARROYO, M., JIMENEZ, A., Ensayos sobre erradicación de Ceratitis capitata Wied. por el método de los «machos estériles» en la isla de Tenerife, Bol. Estac. Fitopatol. Agríc. N° 399, INIA, Madrid (1966).

ARROYO, M., MELLADO, L., JIMENEZ, A., ROS, J.P., Lucha biológica contra Ceratitis capitata por el método de «machos estériles». Progresos realizados en 1971, Progress Report 1971, INIA, Madrid (1971).

ARROYO, M., JIMENEZ, A., MELLADO, L. CABALLERO, F., Aplicación de isótopos radiactivos a la investigación de métodos sobre lucha biológica contra plagas, Bol. Estac. Fitopatol. Agríc., Vol. XXVIII N° 388-392, INIA, Madrid (1965).

MELLADO, L., ARROYO, M., JIMENEZ, A., CABALLERO, F., Ensayos sobre erradicación de Ceratitis capitata Wied. por el método de los «machos estériles» en la isla de Tenerife. Progresos realizados durante el año 1967, Bol. Estac. Fitopatol. Agríc., Vol. XXX, INIA, Madrid (1967-68).

ARROYO, M., MELLADO, L., JIMENEZ, A., CABALLERO, F., Influencia en la puesta de Ceratitis capitata Wied. de distintas dietas alimenticias, Bol. Estac. Fitopatol. Agríc., Vol. XXX, INIA, Madrid (1967-68).

MELLADO, L., ARROYO, M., JIMENEZ, A., CABALLERO, F., Ensayos sobre erradicación de Ceratitis capitata Wied. por el método de «machos estériles» en la isla de Tenerife. Progresos realizados en 1968, Bol. Patol. Veg. y Entomol. Agríc., Vol. XXXI, INIA, Madrid (1969).

MELLADO, L., ARROYO, M., JIMENEZ, A., «Sterile-male technique for control of Ceratitis capitata. Work at the Instituto Nacional de Investigaciones Agronómicas, Madrid», Sterile-Male Technique for Control of Fruit Flies (Actas Panel Viena, 1969), OIEA, Viena (1970) 90.

MELLADO, L., NADEL, D.J., ARROYO, M., JIMENEZ, A., «Mediterranean fruit fly suppression experiment on the Spanish mainland in 1969», Sterile-Male Technique for Control of Fruit Flies (Actas Panel Viena, 1969), OIEA, Viena (1970) 91.

MELLADO, L., «La técnica de machos estériles en el control de la mosca del Mediterráneo: programas realizados en España» (Survey Paper), Sterility Principle for Insect Control or Eradication (Actas Simp. Atenas, 1970), OIEA, Viena (1971) 49.

MELLADO, L., ARROYO, M., JIMENEZ, A., CASTILLO, E., Ensayos de lucha autocida contra Ceratitis capitata Wied. Progresos realizados en 1969, An. Inst. Nac. Invest. Agrar., Serie Protec. Veg. N° 2, INIA, Madrid (1972).

VARGAS, C., MELLADO, L., Efectos del tratamiento combinado de refrigeración e irradiación gamma en adultos de Ceratitis capitata Wied., An. Inst. Nac. Invest. Agrar., Serie Protec. Veg. N° 2, INIA, Madrid (1972).

ARROYO, M., MELLADO, L., JIMENEZ, A., CASTILLO, E., Ensayos de lucha autocida contra Ceratitis capitata Wied. Progresos realizados en 1970, An. Inst. Nac. Invest. Agrar., Serie Protec. Veg. N° 2, INIA, Madrid (1972).

MELLADO, L., ARROYO, M., ROS, P., «Control of Ceratitis capitata Wied. by the sterile-male technique in Spain», The Sterile-Insect Technique and its Field Application (Actas Panel Viena, 1972), OIEA, Viena (1974) 63.

DISCUSSION

K. SYED: Did you sample natural populations of the fly, when you made releases, in order to establish the ratio of sterile to wild flies?

J. P. ROS: No.

J. E. SIMON F.: Have you had any problems with birds at your release stations? I am asking this because we have encountered them both in Panama and in Moquegua, Peru.

J. P. ROS: Yes, we have. This is just a small problem but it is still a nuisance. In particular we find that swallows fly over low and snap up our insects just as they are gaining height after release. However, it should not be too difficult to find a solution to this.

J. L. MONTY: What is the extent of damage to fruit caused by the sting of sterile females?

J. P. ROS: In the experiments in Murcia in 1969 we had a problem with sterile female stings which was caused, I think, by the release of a large number of flies in a small plantation and, as a result, each fruit had up to 20 sterile stings. We overcame this by protecting the plantations with sterility barriers. We still get some stings but these are not extensive and a fruit can have a sterile sting without suffering loss of quality.

J. L. MONTY: What is the reaction of farmers to the release of sterile flies?

J. P. ROS: At first they are alarmed at seeing so many flies in their orchards, even if they are sterile, but later on when they see the results, they come and ask us to protect their crops with the sterile-insect technique.

M. S. H. AHMED: What kind of insecticides do you use for control purposes?

J. P. ROS: In Spain we generally use Lebaycid (0.1%) in total spray treatments or a mixture of Lebaycid and Buminal (hydrolized protein) in quantities of 0.75%, and 0.75% in bait sprays.

APPLICATION OF STERILIZATION TECHNIQUES
TO Anastrepha suspensa Loew
IN FLORIDA, UNITED STATES OF AMERICA

A.K. BURDITT, Jr., F. LOPEZ-D.,
L.F. STEINER, D.L. von WINDEGUTH
Subtropical Horticulture Research Unit,
Florida/Antilles Area, Southern Region,
Agricultural Research Service,
US Department of Agriculture, Miami, Fla.

R. BARANOWSKI
Agricultural Research and Education Center,
University of Florida, Homestead, Fla.,
United States of America

M. ANWAR
Atomic Energy Agricultural Research Centre,
Tandojam, Pakistan

Abstract

APPLICATION OF STERILIZATION TECHNIQUES TO Anastrepha suspensa Loew IN FLORIDA, UNITED STATES OF AMERICA.

Research at the USDA Subtropical Horticulture Research Station in Miami, Florida, and at the University of Florida Agricultural Research and Education Center at Homestead, Florida, demonstrated that the Caribbean fruit fly could be reared on several different rearing media containing sugar-cane bagasse-citrus pulp, rice hulls or ground corn cobs as bulking agents or on a diet based on agar. These diets were used to rear large numbers of larvae for irradiation studies and sterilization research. Irradiation studies have shown that adults as well as pupae can be sterilized with five to eight krad of gamma irradiation from a ^{60}Co source. The higher dose of irradiation caused some injury to the flies. Studies of Caribbean fruit fly populations on Key West showed that release of up to 115 000 sterile flies per square kilometre weekly would suppress the fruit fly population. Research to determine the effects of gamma irradiation on Caribbean fruit fly is continuing in Miami. Our research has demonstrated that suppression would be possible. However, since most of the host fruits for this fly are not commercially produced in Florida, there is no economic justification for suppression at this time. Procedures have been developed for mass rearing, sterilization, release and monitoring populations of the Caribbean fruit fly, and a suppression campaign could be undertaken if economically justified.

INTRODUCTION

The Caribbean fruit fly, Anastrepha suspensa Loew, has been introduced into Florida on at least three occasions, the last being in April 1965. Since then it has spread throughout the southern and central part of the state until over 30 counties are at present infested. Lopez [1] reviewed the research programmes on the Caribbean fruit fly at a panel organized by FAO/IAEA in September 1969. The present report reviews the fruit fly sterilization research programme and related problems in Florida since Lopez's report.

In 1968, a United States Department of Agriculture (USDA) laboratory was established at Miami, Florida, to conduct research on biology and control of the Caribbean fruit fly. Research previously conducted at the USDA laboratory in Orlando, Florida, was transferred to the Miami location where it could be closely co-ordinated with related co-operative research being carried out at the University of Florida Agricultural Research and Education Center in Homestead, Florida. This location has proved ideal for such research since many of the fruit fly hosts are found here, and survey personnel conducting detection and control programmes for the USDA and the Florida Department of Agriculture and Consumer Services are quartered here.

Caribbean fruit fly research initially concentrated on chemical control and developing feasible methods of mass rearing. Once these studies were underway, research was expanded to find better fruit fly attractants and to determine the effects of gamma irradiation on this species. Later, an experimental programme was undertaken to determine the feasibility of fruit fly suppression by release of flies irradiated as adults. Research also is being conducted to improve techniques of survey and detection of the Caribbean fruit fly, and, recently, studies of commodity treatments were initiated.

MASS REARING OF THE CARIBBEAN FRUIT FLY

Kamasaki and co-workers [2] found that the dehydrated carrot medium used for the Mexican fruit fly, A. ludens Loew, could also be used for the Caribbean fruit fly. However, the medium used in Hawaii for the oriental fruit fly, Dacus dorsalis Hendel, and the Mediterranean fruit fly, Ceratitis capitata Wiedemann, was not satisfactory.

During the past five years, four diets have been developed for rearing Caribbean fruit fly larvae, two at the USDA laboratory in Miami and two at the Homestead station. Procedures for collecting and handling eggs, pupae and adults are essentially similar at both locations.

At Homestead, Baranowski and his co-workers at the University of Florida Agricultural Research and Education Center have reared Caribbean fruit fly larvae on two rearing media under a co-operative agreement support by the USDA, and have produced in excess of 115 million pupae for use in the joint fly sterilization suppression programme conducted in Key West. Larvae initially were reared on a dehydrated sugar cane bagasse - citrus pulp substrate that was later replaced by whole and ground rice hulls; nutrients were supplied by sucrose, torula yeast and wheat germ. Larvae mature on these diets in seven days at 26°C. Thereafter the trays of medium are inverted on wire mesh panels that are then stacked over trays containing moistened vermiculite (40% moisture by weight). Larvae crawl out of the medium and pupate in the vermiculite. Twenty-four hours later the pupae and vermiculite are removed, mixed uniformly, and held at 26°C and 80% r.h. Adults emerge in 14 days.

At the USDA Miami laboratory, one of the two larval diets has corn cob grits as the bulking agent; the other uses agar as a jelling agent. The corn cob diet consists of 64.5% water, 0.15% tego, 0.10% sodium benzoate, 6% sugar, 3% wheat germ, 4% torula yeast, 0.7% hydrochloric acid and 21.55% ground corn cobs (20-40 mesh). All the ingredients except the corn cobs are mixed for 20 minutes; then corn cob is added, and the mixing is continued for 15 more minutes. Three litres of medium are placed in each 23 × 64-cm tray. The semi-solid agar diet was developed for research studies that require a medium from which larvae can be separated easily. This medium consists of 80.95% water, 0.3% agar, 0.1% tego, 0.1% sodium benzoate, 0.05% cholesterol, 6% sugar, 6% embo (defatted wheat germ), 6% torula and 0.5% HCl. The medium is prepared by boiling half the water and adding agar while stirring continuously. Once the agar is dissolved, heating is discontinued; however stirring should continue during cooling. The balance of the water and remaining ingredients are mixed thoroughly and added to the agar solution when the solution has cooled to 42°-45°C. Mixing is continued for 2 or 3 minutes, and the medium is poured into 23×64-cm stainless-steel trays to a depth of 1 cm. Eggs can be placed on the medium when it has cooled to 25°C.

With the corn cob diet, larvae crawl off the trays of medium and drop into trays of vermiculite for pupation. With the agar diet the larvae must be washed from the medium by using a 32-mesh screen; then excess water is allowed to drain from the larvae before they are placed on vermiculite for pupation. Pupae are held at 26°C in vermiculite containing 40% water by weight. After 13 days, adults begin to emerge. A 10-mesh screen can be used to separate pupae from the vermiculite, if desired.

Adult flies are maintained in 0.2 m^3 cages in a room at 25°-27°C and 70-80% r.h. Approximately 28 000 flies are held in each cage. The cages are made with wood frames and 18-mesh aluminium screen. Fluorescent lights are placed above each cage. Lighting is on a 12 : 12 light-dark cycle. Four blocks of 1% agar, to supply water, and 454 g of cubed sugar are placed on top of each cage. A 16-mesh screen panel approximately 20×60 cm is hung lengthwise in the cage. A thick mixture of hydrolysed yeast and water is applied to the screen panel in the cage. Fresh yeast may be placed on top of the cage at weekly intervals. Female flies oviposit through panels of cheese-cloth impregnated with a hot mixture of paraffin/petrolatum (2:1 by weight) to which a small amount of red candle dye has been added. Two panels form one side of the cage. The panels are covered externally with moistened cellulose sponges and wood to prevent egg desiccation.

A fine mist spray of water is used to wash the eggs from the panels. After the eggs are collected, the water is decanted off, and the eggs are washed with 0.025% sodium hypochlorite for 20-30 seconds, rinsed several times in water, and then mixed with 0.03% solution of sodium benzoate in water. The eggs are measured volumetrically (22 500/mlitre) and spread uniformly on paper towelling at the rate of approximately 1 mlitre/100 cm^2 of towelling.

IRRADIATION STUDIES

Lopez tested the effects of gamma irradiation dosages in the range of 3-10 krad on mature pupae and on chilled 0 - 6-day-old adults (unpublished data). He found that when adults were exposed to 3, 4, 5 and 6 krad, mortality of males and females was about 100% and 80% greater respectively than that of normal (N) flies during the first four weeks after treatment. There was little difference among dosages. The fecundity of treated (T) females paired (750/test) with N males was reduced 96% by 3, 4 and 5 krad and 99% by 6 krad; hatch was 6.2, 1.2, 0.3 and 0% respectively, compared with 87.5% in the controls. In the same tests, pairing T males with N females produced hatches of 5.3, 2.1, 0.8 and 0.1% respectively. When

TABLE I. HATCH OF EGGS LAID BY NORMAL FEMALE CARIBBEAN
FRUIT FLIES CROSSED WITH IRRADIATED MALES [a]

Stage treated	Irradiation dose (krad)		
	4	6	8
10-day-old pupae	0.00	0.00	0.00
12-day-old pupae	0.20	0.03	0.00
1-day-old adults	0.30	0.09	0.00

[a] Hatch of eggs laid by flies crossed with untreated males was 83.84%.

1 - 6-day-old adults were treated at a rate of 6 krad and 7500 pairs were
combined in 0.2-m^3 cages in ratios of N to T of 750:6750 (1:9), 300:7200
(1:24), and 150:7350 (1:49), the mean hatches during four weeks of egg
collecting were reduced 80, 88 and 94% from the 88% hatch in control cages.

M. Anwar, working with us while on an IAEA training assignment at
Miami, studied the effects of irradiation on the longevity and egg production of the Caribbean fruit flies. In these studies, flies were irradiated
as 10 and 12-day-old pupae and as 1 to 2-day-old adults. Each group was
irradiated at 0, 4, 6 or 8 krad. Both the eclosed and young adults were
chilled and separated according to sex; then 75 males or 75 females were
placed in 0.014-m^3 cages with 75 N females or males. Each test was
replicated three times. Seventeen 24-hour egg collections were made at
intervals from 10 to 40 days after adult emergence using 85-mm-diameter
containers covered with cheese-cloth impregnated with wax.

The female Caribbean fruit flies irradiated at 4, 6 or 8 krads and placed
in cages with N males failed to produce eggs in any of the collections (Table I).
The N females placed with T males continued to produce eggs through the
40-day test period. However, none of the eggs produced by N females caged
with T males irradiated as 10-day-old pupae hatched. Some of the eggs
produced by N females caged with T males irradiated at 4 or 6 krad as
12-day-old pupae or 1-day-old adults did hatch.

The effect of irradiation on mortatility and longevity of fruit flies
was determined in the same study. We found that females were more apt to
be injured and that the result was increased initial mortality (Table II).
Thereafter, the increase in mortality of T and N females remained constant
over the six weeks of the experiment. The increase in mortality of males
was greater than that of females. Males irradiated at 8 krad were more
susceptible to injury than N males.

At present scientists at the Miami laboratory are investigating the
effects of ionizing radiation on the immature forms of A. suspensa. The
age/dose/mortality curves for this fly are similar to the curves determined
for Dacus dorsalis, Ceratitis capitata, and D. cucurbitae Coquillett by
Burditt and Seo in Hawaii [3]. For this study eggs are irradiated bare,
and larvae are irradiated in the semi-solid agar medium. Mean time for
hatch of A. suspensa eggs at 25°C is about 60 hours so the oldest eggs were
irradiated at 55 hours. At this age a dose of >40 krad is required to stop

TABLE II.   EFFECT OF GAMMA IRRADIATION ON MORTALITY
OF CARIBBEAN FRUIT FLIES

Age of flies (days)	Percentage mortality at irradiation dose (krad)			
	0	4	6	8
	Males treated			
8	11	13	15	26
15	17	19	27	38
22	41	53	50	60
29	59	70	64	74
36	76	76	73	84
43	90	83	81	87
	Females treated			
8	2	16	12	16
15	13	28	31	35
22	23	43	44	51
29	34	52	54	61
36	53	58	61	68
43	65	66	69	75

egg hatch although 10 krad will prevent larvae from pupating. Doses in
excess of 40 krad are required to stop 6-day-old larvae from forming puparia
although most (95%) of the puparia are malformed. A dose of 5 krad will
stop emergence of adults from these larvae.

SUPPRESSION OF FRUIT FLY POPULATION ON KEY WEST

Beginning in 1970, a sterile-fly release test was conducted co-operatively
by research and regulatory personnel of the US Department of Agriculture,
the University of Florida Agricultural Research and Education Center,
Homestead, Florida, and the Florida State Department of Agriculture and
Consumer Services, Division of Plant Industry, to determine the feasibility
of suppressing the Caribbean fruit fly. Key West, the southernmost of a
chain of islands at the southern tip of Florida, was selected as the site for
the programme. The test was continued over a two-year period and ended
June 1972.

Larvae reared at the Homestead Center on a sugar cane bagasse-citrus
pulp or ground rice hulls substrate were used in the programme. The pupae
were brought to the Miami station where they were placed in vermiculite
coated with red, green, or yellow Dayglo (Dayglo Color Division, Switzer

Bros. Inc., 4732 St. Clair Ave., Cleveland, Ohio 44103). As the adults
emerged, they picked up the dye on their ptilinum and elsewhere on their
bodies. The adults were held in 0.5 m^3 cages. In general, flies were
1 - 4-days-old at the time of irradiation.

Immediately before irradiation, the flies were inactivated by exposure
to 4.5°C. They were treated at the rate of approximately 3.5 krad min at
dosages of from 5 to 7.5 krad. After treatment, the flies were kept either
in a portable refrigerator or in styrofoam cartons and transported to Key
West. Flies were released 3 - 24 hours after irradiation, either from ground
release stations or from aircraft.

Approximately 0.5 million flies per week were released, primarily by
aircraft, for 8 months, beginning in January 1971. The native population
declined gradually until May, but then increased sharply during June and
July (Fig.1). Therefore, four applications of ULV malathion were made at
weekly intervals beginning in August to suppress partially the fly population.
The aerial releases were continued at an average rate of 1.2 million flies
per week for the following five months, at which time ground releases of
1.3 million per week were made for two months. Releases were terminated
in April 1972. The fly population reached a low level at this time and continued thus for the two months following the ending of releases.

Estimates of the relative fly population density, sex and ratio of sterile
to native flies, the effects of the sprays and the ultimate effects of the programme were obtained from about 208 glass invaginated traps rebaited
each week with 3% torula yeast hydrolysate and 4% borax [4], and from fruit
samples collected weekly to determine the effect of releases on larval
population.

During the 16 months of the suppression experiment, 61 million adult
irradiated flies were released. The release rates gradually increased from
160 000 per week in January 1971 to over 1.5 million per week in February
1972. Recovery of irradiated flies from baited glass traps increased from
a mean of 3.2 flies weekly per trap in January to 20.6 in August 1971 (Fig.1).
During the application of ULV malathion, recaptures of irradiated flies
dropped to 2.4, even though the releases were continued, but recovered as
soon as spraying was discontinued.

Peak recapture (31.9 irradiated flies weekly per trap) occurred in
December and the rate continued high until distribution of flies was discontinued in April 1972.

Capture of native flies in the traps gradually declined during the first
five months of sterile-fly releases in 1971 (Fig.1), but in June, when the
release rate dropped slightly, there was an increase in recovery of native
flies. As noted, the application of ULV malathion caused a sharp drop in
the capture of native flies that was followed by a rapid increase. However,
this increase was short-lived, and continued releases of sterile flies brought
about rapid suppression of the native fly population. The population then
remained at a low level for two months after sterile fly releases ceased.
The week ending 3 June 1972, six weeks after the last release, was the first
and only week without any catch of native flies. Catches remained below 0.1
flies weekly per trap for the next four weeks, and then increased rapidly;
by September the catch was approaching the level existing before releases.

Weekly fruit samples were collected whenever available beginning in
August 1970 from 38 species of trees or hedges known or suspected to be
hosts in the Key West area. Fruit were counted, weighed, and held over

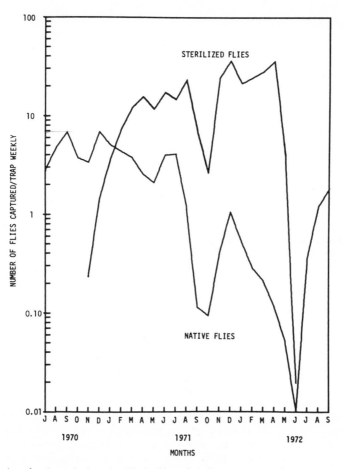

FIG. 1.   Catches of native and released sterile Caribbean fruit flies on Key West from July 1970 to September 1972, in 208 glass invaginated traps baited with torula yeast hydrolysate-borax.

sand; the pupae or larvae were screened from the sand at weekly intervals [5]. Fly suppression was measured in the seven hosts (Malpighia punicifolia L. (Barbados cherry), Citris mitis Blanco (Calamondin), Psidium guajava L. (guava), Eriobotrya japonica Thumb., Lindl. (loquat), Manilkara zapota L., van Royen (sapodilla), Eugenia uniflora L. (Surinam cherry) and Terminalia catappa L. (tropical almond)) that formed more than 95% of the local fly population.

The larval recovery varied widely depending upon the season of year and availability of fruit. During the final months of the experiment, the larval population reached a low point in tropical almond, and larvae could not be found in the other common species of hosts. By late June 1972, the population had begun to recover because of migration of native flies from adjacent islands and emergence of adults from larvae surviving in the tropical almonds. Like the trap catches, the fruit samples showed that population of flies recovered rapidly and returned to normal by late August 1972.

RELATED RESEARCH

The emergence of adult Caribbean fruit flies from lots of pupae ranged from 36 to 80%. Dissection revealed that most mortality of pupae occurred soon after pupation. Tests of loading rates in the emergence cages showed that adult mortality was related to fly density. Mortality in cages loaded with 46 000 - 69 000, 70 000 - 92 000, and 105 000 - 140 000 pupae averaged 12, 19 and 32% respectively.

Exposure of flies to chilling at 4.5°C for 2, 12 or 24 hours did not affect longevity over a 38-day period, but the chilled flies produced 25% less eggs than the controls, though fertility was unaffected. When flies were chilled for 24-48 hours or partially warmed and then rechilled, mortalities above 50% were sometimes observed at the time of release. Also, the slow recovery of flies inactivated by chilling led to substantial losses when they fell on hot surfaces or into water puddles.

Comparative tests made in a Miami mango grove showed no differences in effectiveness of aerial and ground releases of chilled flies on to low vegetation. Therefore, aerial releases that expose flies to the lethal effects of hot surfaces or water would cause higher mortality than ground releases.

CONCLUSION

Investigation of the effects of gamma irradiation on the Caribbean fruit fly is continuing in Miami. Control of the native population is difficult because of the wide range of hosts available in southern Florida as well as the wide area of the state through which the fly has dispersed. Our research has demonstrated that suppression would be possible; however, most of the host fruits for this fly are not commercially produced in Florida, so there is no economic justification for suppression of the fly at the present time. The primary importance of our investigation of the Caribbean fruit fly is therefore the demonstration that the same techniques can be used against Anastrepha species of fruit flies as against the Ceratitis and Dacus species. We have developed procedures for mass rearing, sterilization, release and monitoring populations of the Caribbean fruit fly.

REFERENCES

[1] LOPEZ-D., F., "Sterile-male technique for eradication of the Mexican and Caribbean fruit flies: Review of current status", Sterile-male Technique for Control of Fruit Flies (Proc. Panel, Vienna, 1969), IAEA, Vienna (1970) 111.

[2] KAMASAKI, H., SUTTON, R., LOPEZ-D., F., SELHIME, A., Laboratory culture of the Caribbean fruit fly, Anastrepha suspensa, in Florida, Ann. Entomol. Soc. Am. 63 (1970) 639.

[3] BURDITT, A.K., Jr., SEO, S.T., "Dose requirements for quarantine treatment of fruit flies with gamma irradiation", Disinfestation of Fruit by Irradiation (Proc. Panel, Vienna, 1970), IAEA, Vienna (1970) 33.

[4] LOPEZ-D., F., STEINER, L.F., HOLBROOK, F.R., A new yeast hydrolysate-borax bait for trapping the Caribbean fruit fly, J. Econ. Entomol. 64 (1971) 1541.

[5] VON WINDEGUTH, D.L., PIERCE, W.H., STEINER, L.F., Infestations of Anastrepha suspensa in fruit on Key West, Florida, and adjacent islands, Fla. Entomol. 56 2 (1973) 127.

DISCUSSION

B. BUTT: Will these flies move north into the peach-growing areas of Georgia and is climate a limiting factor?

R.M. BARANOWSKI: Movement of A. suspensa into the peach-growing areas of Georgia appears to be limited by the lack of other kinds of host fruits rather than climate. The flies can tolerate much lower temperatures than those occurring in the peach-growing areas.

J.E. SIMON F.: You say that chilling has a certain effect on the females. I do not think that this is very important for the sterile-insect technique. However, may I ask what effect it has on the male insects?

R.M. BARANOWSKI: Chilling did increase mortality, but we felt this was not significant and could be compensated for.

РАЗВИТИЕ ИССЛЕДОВАНИЙ ПО МЕТОДУ ЛУЧЕВОЙ СТЕРИЛИЗАЦИИ ВРЕДНЫХ НАСЕКОМЫХ В СССР

С.В.АНДРЕЕВ, Б.К.МАРТЕНС
Всесоюзный научно-исследовательский институт
защиты растений,
Ленинград,
Союз Советских Социалистических Республик

Abstract–Аннотация

RESEARCH ON THE TECHNIQUE OF RADIATION STERILIZATION OF INSECT PESTS IN THE USSR.
Great importance is attached in the USSR to the radiation sterilization technique, which is considered to be an effective means of controlling insect pests. Attention is devoted especially to the study of the biological action of ionizing radiations on agricultural pests (grains, pulses, dried fruits, etc.). These studies have established the basic parameters and conditions of radiation disinfestation of foodstuffs during storage. The data obtained have been used for designing facilities for disinfestation of grains and pulses in continuous operation. Large-scale research is being conducted on the application of the radiation sterilization technique for controlling agricultural pests. The research material used includes the turnip moth (Argotis sagetum Schiff), cotton bollworm (Chlorides obsoleta F.), cabbage fly (Chortophila brassicae L.), melon fly (Muiopardalis pardalina Big.), mallow moth (Lepidoptera, Gelechidae), apple-tree ermine moth (Hyponomenta maliness L.), peach moth (Grapholitha molesta), beet fly (Pegomyia hyoscyami Pz.), codling moth (Laspeyresia pomonella), bean weevil (Acanthoscelides obtectus Say), pea beetle (Bruchus pisorum L.) and a number of other pests. The studies have served as a basis for determining the optimum values of sterilizing doses, the most suitable sterile-fertile insect ratios and other characteristics needed for practical application of the radiation sterilization technique (insect sterilization facilities, automated facilities for large-scale rearing, nutrient media for growing insects under artificial conditions, etc.).

РАЗВИТИЕ ИССЛЕДОВАНИЙ ПО МЕТОДУ ЛУЧЕВОЙ СТЕРИЛИЗАЦИИ ВРЕДНЫХ НАСЕКОМЫХ В СССР.
В СССР методу лучевой стерилизации придается большое значение. Этот метод рассматривается как эффективное средство борьбы с вредными насекомыми. Особое внимание уделяется исследованиям биологического действия ионизирующих излучений на насекомых − вредителей сельскохозяйственных продуктов (зерна, зернобобовых, сухофруктов и т.д.). Этими исследованиями установлены основные параметры и режимы лучевой дезинсекции продуктов при их хранении на складах. Полученные данные использованы при разработке установок для дезинсекции зерна и зернобобовых в потоке. В широком плане проводятся исследования, связанные с применением метода лучевой стерилизации в борьбе с насекомыми − вредителями сельскохозяйственных растений. В качестве объектов исследования взяты: озимая совка (Agrotis sagetum Schiff), хлопковая совка (Chlorides obsoleta F.), капустная муха (Chortophila brassicae L.), дынная муха (Muiopardalis pardalina Big.), мальвовая моль (Lepidoptera, Gelechidae), яблонная моль (Hyponomenta maliness L.), восточная плодожорка (Grapholitha molesta), свекловичная муха (Pegomyia hyoscyami Pz.), яблонная плодожорка (Laspeyresia pomonella), фасолевая зерновка (Acantoscelides obtectus Say), гороховая зерновка (Bruchus pisorum L.) и ряд других вредителей. Проводящиеся исследования позволили уточнить оптимальные значения стерилизующих доз, наивыгоднейшие соотношения стерилизованных насекомых к фертильным и другие показатели, необходимые для практического использования метода лучевой стерилизации (установки для стерилизации насекомых, автоматизированные устройства для массового разведения насекомых, питательные среды для воспитания насекомых в искусственных условиях и т.п.).

В СССР большое значение придается исследованиям, связанным с разработкой автоцидного метода лучевой стерилизации вредных насекомых − вредителей продуктов хранения и сельскохозяйственных растений.

В целях решения этой сложной проблемы исследуются различные
аспекты, имеющие значение для практического применения этого метода в производственных условиях.

Изучается действие ионизирующих излучений на: амбарного долгоносика (Calandra granaria L.), рисового долгоносика (Calandra oryzae L.),
хлебного точильщика (Stegotium paniceum), суринамского мукоеда (Oryzaephilus surinamensis), зернового точильщика (Rhizopertha dominica),
малого хрущака (Tribolium confusum) и других вредителей продуктов
запаса.

Вакар А.Б. и др. [1] исследовали динамику радиационного пора-
жения амбарного долгоносика (Calandra granaria) в зависимости от
возраста и стадии развития с учетом температуры и мощности дозы.

Изучена биохимия зерна, обработанного различными дозами радиации, в результате чего было установлено, что стерилизующие насекомых дозы не оказывают заметных изменений в пищевых качествах зерна в пределах 10-20 крад.

Аналогичные исследования были проведены Щеголевой Г.И. [2] в
отношении вредителей сухофруктов и их качества.

Юсифов Н.И. и др. [3] провели теоретические исследования по изучению механизма действия радиации на различные стадии развития мель-
ничной огневки (Ephestia kuhniella Zell). Исследования подтвердили
данные, полученные ранее многими авторами, что радиация оказывает
тормозящее действие на развитие насекомых, т.е. удлиняет сроки пе-
рехода из одной стадии в другую.

Опыты, проведенные с гусеницами (Ephestia kuhniella Zell), показали, что в результате воздействия радиации на гусениц после вве-
дения им препарата экдиссона последний значительно снижает тормозящее действие радиации на процессы линьки у насекомых. Это ука-
зывает на то, что в организме насекомых под влиянием радиации про-
исходят глубокие нарушения механизмов, управляющих процессами линьки.

Широкое развитие получили исследования, связанные с разработкой метода лучевой стерилизации вредителей сельскохозяйственных
растений: озимой совки (Agrotis segetum Schiff.), хлопковой совки
(Chloridea armigere A.), капустной мухи (Chortophila floralis Fill),
дынной мухи (Muipardalis pardalina Big), мальвовой моли (Pectinophora malvella Hb.), яблонной моли (Hyponomenta malinellus Zell) и
других вредителей.

Исследования Бедного В.Д. [4] позволили уточнить оптимальные
дозы для американской белой бабочки (Hyphatria cunea Br) – 20 крад.

Кванчантирадзе М.С. и Кипиани Р.Н. [5] установили, что для восточной плодожорки (Graphalitha molesta B) стерилизующая доза составляет 30-35 крад. Исследования показали, что дальнейшее повыше-
ние дозы снижает половую активность бабочек, а при облучении куколок
дозой в 60 крад бабочки выходят уродливыми и вскоре погибают после
вылета. Климпиня А.Е. [6] провела исследования по изучению действия радиации на свекловичную муху (Pegomyia hyosciami Pz.) различных возрастов и на разные стадии развития. Было установлено, что
гамма-радиация 5-10 крад вызывает морфологические изменения в по-
ловых железах мух, задерживает или полностью останавливает процесс
сперматогенеза и оогенеза.

Доманский В.Н. [7] в целях повышения конкурентоспособности
стерилизованных самцов яблонной плодожорки (Laspeyresia pomonella)

облучал их более низкими дозами. В результате проведенных исследований показана возможность получения стерилизующего эффекта субстерилизующими дозами, величина которых составляет 15-20 крад. Такие дозы обеспечивают высокую половую активность стерилизованных насекомых. Дозы 30-35 крад, хотя и вызывают полную стерилизацию насекомых, однако значительно снижают конкурентоспособность стерилизованных особей по отношению к фертильным.

Андреевым С.В. и др. [8] на основе сравнительного анализа особенностей биологии экономически значимых вредителей было установлено, что метод лучевой стерилизации с наименьшими затратами средств и капиталовложений может быть применен против вредителей зернобобовых культур: фасолевой зерновки (Acanthoscelides obtectus Say) и гороховой зерновки (Bruchus pisorum L.). При применении метода лучевой стерилизации в борьбе с гороховой зерновкой (Bruchus pisorum) исключается необходимость массового разведения этого вредителя. В этом случае весь собранный урожай подвергается лучевой обработке на току или в амбарах. При применении метода стерилизации против фасолевой зерновки (Acanthoscelides obtectus), наряду с облучением вредителя в зерне на току или в амбарах, может быть использован метод выпуска стерилизованных жуков, полученных в искусственных условиях. С целью разработки метода борьбы с фасолевой зерновкой (Acanthoscelides obtectus) были проведены детальные исследования действия радиации на этого вредителя. Показателем стерилизующего действия радиации являлось снижение отрождения нового поколения по сравнению с контролем. В результате проведенных исследований было установлено, что оптимальная стерилизующая доза находится в пределах 10-12 крад (рис.1).

Другим вопросом, имеющим практическое значение для применения метода, является выбор оптимального соотношения между стерилизованными и фертильными насекомыми. Данные исследований, проведенных в этом направлении, представлены на рис.2. На этом рисунке по оси X отложены отношения фертильных самцов и самок к стерилизованным самцам. По оси Y отложены десятичные логарифмы от процента отрождения ($LgF_1\%$). В целях определения влияния расселения в популяции стерилизованных самок на эффективность применения метода лучевой стерилизации был проведен ряд экспериментов, в которых были применены следующие соотношения: 1:1:20, 1:1:20:20, т.е. в первом случае выпускались только стерилизованные самцы, а во втором вместе со стерилизованными самцами выпускались также стерилизованные самки.

На рис.3а показаны результаты указанных экспериментов — зависимость отрождения нового поколения от величины стерилизующей дозы. По оси X отложены дозы облучения, а по оси Y — десятичные логарифмы процента отрождения ($LgF_1\%$). На рисунке изображены две кривые, соответствующие случаям, когда в природу выпускались только стерилизованные самцы при соотношении 1:1:20, а также вместе со стерилизованными самцами выпускались стерилизованные самки при соотношении 1:1:20:20. Из рис.3б видно, что расселение стерилизованных самок несколько снижает эффективность метода.

При разработке метода лучевой стерилизации гороховой зерновки (Bruchus pisorum) были установлены стерилизующие дозы в пределах 6-8 крад. При указанных дозах у гороховой зерновки (Bruchus pisorum)

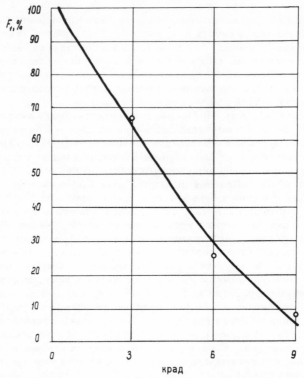

Рис. 1. Отрождение первого поколения F_1 в зависимости от величины стерилизующей дозы в крад.

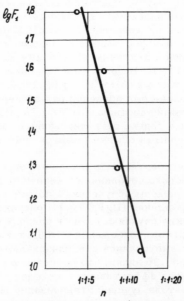

Рис. 2. Зависимость отрождения первого поколения (десятичный логарифм от процента отрождения первого поколения $Lg F_1$) от соотношения фертильных самцов к стерилизованным.

резких физиолого-биохимических изменений не отмечалось. Наблюдалось некоторое изменение в активности каталазы и интенсивности дыхания. При более высоких дозах (30 крад) наблюдались сильные изменения в генеративных органах облученных самцов.

Для практического использования метода лучевой стерилизации в хозяйствах в ряде научных центров СССР ведутся работы по созданию различных технических средств. Из большого числа выполненных инженерных работ практический интерес представляет передвижной гамма-дезинсектор оригинальной конструкции, разработанный доктором Бибергалем А.В. и др. [9]. Дезинсектор рассчитан на обработку небольших партий хранящегося в хозяйствах зерна, зараженного вредными насекомыми.

На рис.4 показана принципиальная схема этой установки. Облучатель 2 собран из стандартных источников гамма-излучения изотопа цезия (^{137}Cs) общей активностью 21 кг-экв.радия. Облучатель находится между двумя роторными полыми кольцами, расположенными одно над другим. Кольца разделены на 24 камеры, представляющие собой открытые сверху ящики с откидывающимся дном, которое открывается при вращении колец в определенном месте, минуя облучатель. Зараженное вредителями зерно транспортером загружается поочередно в камеры верхнего роторного кольца во время его движения. Совершив один оборот над облучателем, днище у камер поочередно откидываются и зерно пересыпается в камеры нижнего роторного кольца. Затем, совершив второй круг под облучателем, зерно пересыпается в приемный бункер для облученного зерна.

За время двукратного прохождения зерна в зоне облучения насекомые получают стерилизующую дозу, которая может быть установлена путем регулирования скорости вращения роторных колец.

Производительность гамма-дезинсектора при дозе 10 крад составляет 4 тонны в час.

Андреев С.В., Мартенс Б.К. и Каушанский Д.А. [10,11] разработали для стерилизации насекомых, получаемых в искусственных условиях с целью расселения в естественной популяции вредителей, специализированную гамма-установку типа "Генетик" (рис.5).

В целях получения в массовых количествах стерилизованных насекомых интенсивно проводятся исследования по их массовому размножению и подбору питательных сред.

В результате проведенных исследований по испытанию большого количества питательных сред была подобрана рецептура полусинтетической среды для воспитания яблонной плодожорки (Laspeyresia pomonella), состоящая из 13 компонентов. Борисова А.Е. [12] на этой среде воспитала 28 поколений яблонной плодожорки без признаков заболевания или вырождения. Имаго, выращенные на среде, по физиологической активности не отличались от особей природных популяций. В результате дальнейших исследований была получена еще более простая питательная среда, состоящая из 5 компонентов: зародышей пшеницы — 12г, хмеля — 3г, ячменных проростков — 2г, аскорбиновой кислоты — 0,6г, агара — 2г. На этой среде было получено несколько поколений, по физиологическим показателям не отличавшихся от особей нормальной популяции и от особей, воспитанных на более сложной среде.

Следует отметить, что введение в качестве компонентов хмеля, ячменных проростков и некоторого увеличения количества зародышей

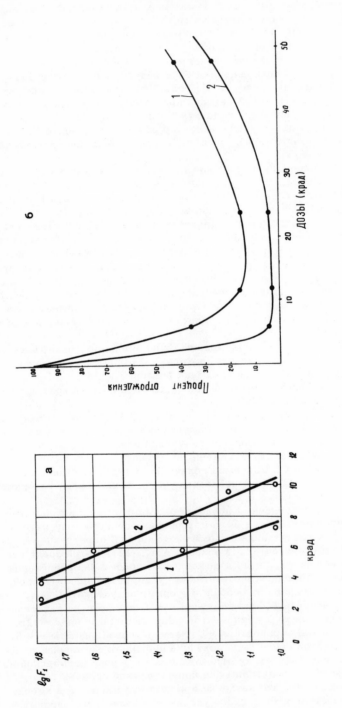

Рис. 3. Зависимость отрождения первого поколения (десятичный логарифм от процента отрождения первого поколения $\lg F_1$) от величины стерилизующей дозы. Кривая 1-для соотношения 1 : 1 : 20. Кривая 2- для соотношения 1 : 1 20 : 20.

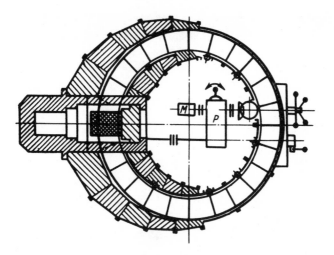

Рис.4. Схема роторного гамма-дезинсектора.

пшеницы позволило исключить казеин, сахарозу, витамины группы B, сухие пивные дрожжи и целлюлозу. Эта среда имеет и то преимущество, что входящие в ее состав компоненты являются отходами пивоваренного производства. Их использование в 3 раза удешевляет стоимость среды [12].

Сравнительные испытания различных сред для воспитания капустной совки (Barathra brassicae L.), проведенные Кондратовым Е.С. и Пономаревым И.А., позволили использовать известную среду Оуи с казеиновой основой. Путем добавления в нее пептона, семян и листьев капусты была получена полусинтетическая среда, положительно влияющая на биологические показатели развития капустной совки: общий выход бабочек, их плодовитость не ниже, чем при выращивании гусениц на свежих капустных листьях.

Исследования Успенской Н.В. и Хлистовского Е.Д. [13] по подбору питательных сред для воспитания совок: хлопковой (Chloridae obsoleta H.), малой наземной (Laphygma exigna Hb.) и озимой (Agrotis segetum Schiff) позволили подобрать полусинтетическую среду, представляющую собой видоизмененную среду Шори. Состав этой среды следующий: набухшие семена маша (Phaseolus aureus), автолизат пивных дрожжей, агар, аскорбиновая кислота и т.д.

Для предохранения среды от микозного и бактериального заражения использовалась в качестве антисептика смесь формалина 0,05% и метабена (0,2% от веса среды). Эта противомикробная смесь практически не оказывала токсического действия на гусениц. Данный состав питательной среды обеспечивает возможность разведения указанных насекомых без снижения жизнеспособности последующих поколений. В данном случае были приведены примеры применения полусинтетических питательных сред для воспитания некоторых насекомых. Более детальное изложение вопроса можно найти в монографии Эдельмана Н.М. [14].

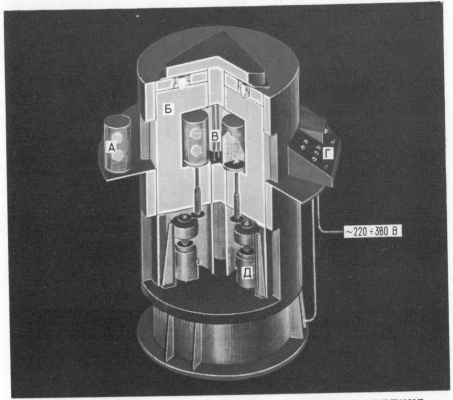

A – КОНТЕЙНЕР   Г – ПУЛЬТ УПРАВЛЕНИЯ

Б – БИОЛОГИЧЕСКАЯ ЗАЩИТА   Д – ЭЛЕКТРОПРИВОД

В – ИСТОЧНИКИ
    ГАММА-РАДИАЦИИ

Рис.5. Гамма-стерилизатор "Генетик" (показан в разрезе)

Техническая характеристика:

1. Максимальный объем рабочей камеры не более 5000 см^3
2. Число рабочих камер – 3 шт.
3. Мощность дозы в центре рабочей камеры не менее 1500 Р/мин
4. Степень неравномерности облучения в пределах 10%
5. Источники излучения: цезий-137 тип XII

6. Мощность экспозиционной дозы (МЭД) на расстоянии 1 м – $2,1 \cdot 10^{-2}$ Р/с
7. Защита обеспечивает мощность дозы на поверхности контейнера-облучателя не более 2,8 мР/ч
8. Габаритные размеры: длина – 1460 мм, ширина – 1460 мм, высота – 1600 мм
9. Общий вес не более 6000 кг
10. Потребляемая мощность электрооборудования не более 1 кВт

Рис.6. Пульт централизованного управления технологическими процессами биофабрики.

Развитие исследований и дальнейший успех по практическому применению автоцидных методов, основанных на лучевой и химической стерилизации вредных насекомых, и биологических методов, в которых используются полезные хищные и паразитические насекомые, а также методов получения энтомопатогенных вирусов на вредных насекомых зависит от уровня разработки способов и средств массового разведения насекомых.

В связи с этим особое значение приобретают работы по совершенствованию и автоматизации технологических процессов массового разведения насекомых. На первом этапе этих работ в качестве модельных объектов для проверки принципов автоматизации были взяты яйцепаразиты рода трихограмма (Trichogramma), широко применяемые на полях в биологическом методе борьбы, и зерновая моль (Sitotzoga cerealella Oliv), на яйцах которой разводят трихограмму.

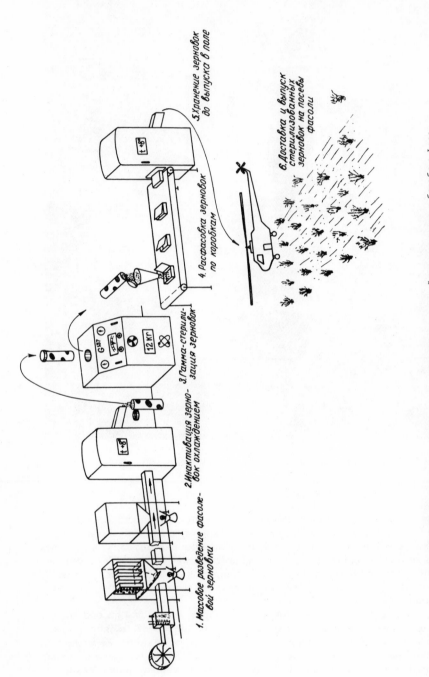

1. Массовое разведение фасоле-вой зерновки

2. Инактивация зерно-вок охлаждением

3. Гамма-стерили-зация зерновок

4. Расфасовка зерновок по коробкам

5. Хранение зерновок до выпуска в поле

6. Доставка и выпуск стерилизованных зерновок на посевы фасоли

Рис. 7. Принципиальная схема применения метода лучевой стерилизации в борьбе с фасо-левой зерновкой.

В целях замены трудоемких операций и повышения производительности труда при массовом разведении этих насекомых были разработаны принципы автоматизации технологических процессов. При разработке отдельных узлов и агрегатов предварительно были изучены специфические особенности биологии этих насекомых, их поведенческие реакции (таксисы) на физические факторы (свет, температура, влажность и т.п.). На основе этих исследований была разработана принципиальная схема автоматизированной линии производительностью 3 млн. яйцепаразитов в сутки при затрате труда одного человека в течение 2-3 часов в день. Для проверки правильности решения ряда вопросов биологического и технического плана была построена экспериментальная автоматизированная линия массового разведения указанных насекомых. Пульт централизованного управления технологическим процессом показан на рис.6. После проверки всех показателей автоматизированной линии в настоящее время на юге страны построены автоматизированные биофабрики промышленного типа производительностью 12-24 и 36 млн. яйцепаразитов в сутки.

Разработан типовой проект биофабрики производительностью 50 млн. яйцепаразитов в сутки, на которой в технологическом процессе будет занято всего 4 человека. Оформляются патенты в ряде стран (США, Канада, Франция, Швейцария и др.).

Опыт по созданию биофабрики по разведению трихограммы позволил разработать аналогичные автоматизированные устройства для разведения других видов насекомых, борьба с которыми может проводиться методом лучевой стерилизации. Так, например, разработана автоматизированная линия для разведения вредителей фасоли — фасолевой зерновки (Acomtoscelides obtectus Say), рис.7. Принципиальная схема автоматизированной линии заключается в следующем: предварительно зараженная жуками некондиционная фасоль засыпается в кассеты, которые устанавливаются в светозащищенные боксы, для массового разведения жуков (1). В дальнейшем размножившиеся и вышедшие через перфорацию в кассетах вредители под влиянием источников света, установленных под выходным отверстием боксов, концентрируются в полости инсектопровода. Сконцентрированные насекомые периодически пневматически из инсектопровода переносятся в камеры с пониженной температурой в целях снижения их физиологической активности (2), из камеры собранные насекомые помещаются в специальные контейнеры, которые поступают в активную зону гамма-стерилизатора, где насекомые подвергаются действию стерилизующих доз радиации в пределах 10-12 крад(3). После этого стерилизованные насекомые расфасовываются по коробкам (4) и сохраняются до выпуска в поле при пониженной температуре (4-8°С) в течение нескольких недель. По мере поступления сигналов службы прогнозов о необходимости выпуска стерилизованных насекомых на посевы фасоли, коробки с насекомыми транспортируются вертолетами в места массового размножения естественной популяции вредителей зернобобовых с целью выпуска стерилизованных жуков (6). После спарирования выпущенных стерилизованных жуков с жуками естественной популяции самки откладывают нежизнеспособные яйца, из которых новое поколение не отрождается, и, таким образом, популяция вредителей резко снижается. Повторными выпусками стерилизованных насекомых популяция вредителей может быть ликвидирована.

Работы, проведенные в Краснодарском крае, подтверждают эффективность этого метода борьбы с фасолевой зерновкой на посевах фасоли.

В дальнейшем предусматривается создание крупных предприятий — биофабрик промышленного типа с целью получения различных видов насекомых в количествах, необходимых для использования их в борьбе с вредными насекомыми того же вида в зонах их вредоносности. Создание подобных биофабрик предполагается в Крыму для борьбы с яблонной плодожоркой (Laspeyresia pomonella), в районах возделывания хлопчатника с целью подавления хлопковой совки (Chloridea obsoleta), карадрины (Laphygma exigua) и других вредителей.

ЛИТЕРАТУРА

[1] ВАКАР, А.Б., ЗАКЛАДНОЙ, Г.А., ПЕРЦОВСКИЙ, Е.С., РАТАНОВА, В.Ф.,
СОСЕДОВ, Н.И.,"Итоги и перспективы применения ионизирующих излучений в
обеспечении сохранности зерна", Радиационная Обработка Пищевых Продуктов.
Атомиздат, М., 1971.

[2] ЩЕГОЛЕВА, Г.И., "Радиационная дезинсекция сушеных фруктов, овощей и пищевых концентратов", Радиационная Обработка Пищевых Продуктов, Атомиздат,М.,
1971.

[3] ЮСИФОВ, Н.И., КОЛОМИЙЦЕВА, И.Н., КУЗИН, А.М., Исследование причин радиационного нарушения окукливания гусениц, Радиобиология, 8 4 (1968) 510.

[4] БЕДНЫЙ, В.Д., Влияние гамма-излучения на половую активность и продолжительность жизни американской белой бабочки, Материалы совещания по прогрессивным
методам борьбы с вредителями, М., 1973.

[5] КВАНЧАНТИРАДЗЕ, М.С., КИПИАНИ, Р.Н., Половая стерилизация восточной
плодожорки, Материалы совещания по прогрессивным методам борьбы с вредителями сельскохозяйственных растений, М., 1973.

[6] КЛИМПИНЯ, А.Е., Ионизирующие излучения в борьбе с вредными насекомыми,
Изд. "Знание", Рига, 1971.

[7] ДОМАНСКИЙ, В.Н., Влияние гамма-облучения самцов яблонной плодожорки
(Laspeyresia pomonella) на плодовитость потомства, Материалы совещания по
прогрессивным методам борьбы с вредителями сельскохозяйственных культур, М.,
1973.

[8] АНДРЕЕВ, С.В., МАРТЕНС, Б.К., КАБЛОВ, В.В., Метод лучевой стерилизации в
борьбе с вредителями сельскохозяйственных растений, Изотопы в СССР, № 21, 1971.

[9] БИБЕРГАЛЬ, А.В., ПРИМАК-МИРОЛЮБОВ, В.Н., АРАКЕЛОВ, О.Г., ГАЛИЧ-
СКИЙ, А.К., Радиационная Обработка Пищевых Продуктов, Атомиздат, М., 1971.

[10] АНДРЕЕВ, С.В., МАРТЕНС, Б.К., КАУШАНСКИЙ, Д.А., Метод лучевой стерилизации насекомых и создание типовой установки "Генетик". Тезисы докладов Всесоюзной конференции по использованию радиационной техники в сельском хозяйстве,
Кишинев, 1972.

[11] АНДРЕЕВ, С.В., Мартенс, Б.К., КАУШАНСКИЙ, Д.А., Метод гамма-дезинсекции
зерна и зернобобовых и создание передвижной установки "Дезинсектор", там же.

[12] БОРИСОВА,А.Е., Поиски сред для массового разведения яблонной плодожорки,
Материалы совещания по прогрессивным методам борьбы с вредителями сельскохозяйственных культур, М., 1973.

[13] УСПЕНСКАЯ, Н.В., ХЛИСТОВСКИЙ, Е.Д., Приемы массовой выкормки на полусинтетической среде гусениц вредных совок, Труды XIII Международного энтомологического конгресса, Изд. "Наука", 1971.

[14] ЭДЕЛЬМАН, Н.М., Массовое разведение насекомых-фитофагов, Биологические методы борьбы с вредными насекомыми, Изд. ВИНТИ, 1972.
КОЛМАКОВ,П.Г., Влияние Ионизирующих Излучений на Насекомых, М.,
Атомиздат, 1970.
КОНДРАТОВ, Е.С., ПОНОМАРЕВА, И.А., Искусственные питательные среды для
разведения капустной совки, Сельскохозяйственная Биология, 8 2(1973) 284.
ПРИСТАВКО, В.П., ОРГЕЛЬ, Г.С., Влияние рентгеновского и гамма-излучения на
репродуктивные способности яблонной плодожорки (Laspeyresia pomonella),
Зоологический Журнал, 1971.
ПТИЦЫНА, Н.В., ГРЕСС, П.Я., Питательные среды для массового разведения яблонной плодожорки, Материалы совещания по прогрессивным методам борьбы с
вредителями сельскохозяйственных культур, М., 1973.

DISCUSSION

B. AMOAKO-ATTA: Since you do not irradiate to kill in your radiation disinfestation studies on grain, I should be interested to know whether any serious feeding damage was caused to the stored grain by the sterilized insects.

S.V. ANDREEV: Treating grain pests with sterilizing rather than lethal doses has the advantage that a much lower dose is required and the productivity of the sterilizers is therefore much higher. Moreover, there is no risk of impairing the biochemical quality of the grain itself. The damage caused by sterilized insects is very limited because the irradiation treatment is applied to all the grain in the store and 95% of the insects in it are affected. With such a high incidence of sterile insects the pest is very soon eliminated.

J.P. ROS: Is the quality of the irradiated grain not affected in any way?

S.V. ANDREEV: After applying gamma doses of 10-20 krad, to sterilize beetles, we did not observe any appreciable changes in the palatability of the grain, nor did biochemical analyses reveal any changes.

D.W. WALKER: What species of Trichogramma are you rearing automatically?

S. V. ANDREEV: Several different species of Trichogramma are being bred in millions in our automated biofactories, for example, Trichogramma ivanescens, Trichogramma minutum, Trichogramma altaiscaia and other species which are found and removed from their natural habitat for rearing and subsequent release in the field.

B. BUTT: What is the status of your Laspeyresia pomonella programme in the USSR?

S.V. ANDREEV: Intensive laboratory and field investigations are currently being carried out on this pest. Methods of mass-rearing on synthetic media are being developed and the most effective and cheapest media have been established. Positive results have been obtained in field trials of the sterilization method in orchards in the Crimea.

Á. SZENTESI: We doubt whether the use of the sterile-male technique will be very effective in the case of Acanthoscelides obtectus. Females originating from stores will probably already have mated with normal males. The first mating has a decisive effect on the viability of eggs laid by the female, so further mating(s) will have no influence on it. Therefore the release of sterile insects will not have much effect on the natural population.

S.V. ANDREEV: Successful application of the sterile-male technique to Acanthoscelides obtectus depends on the correct timing of the release, and we release our insects on the basis of information received from our forecasting service on the development of the pest population in areas sown with beans. It is a well-known fact that the females of this species do not mate right away and so the sterilized insects are released a short time in advance of the forecast appearance of the natural population in the bean plantation. Our aim is not to eradicate the population but to reduce it to economically acceptable levels and we employ the technique largely to control populations which have already suffered losses due to hibernation in the field. As far as beetles in stores are concerned, on large socialist farms it is the usual practice to destroy them by fumigation with methyl bromide, which naturally excludes any possiblity of their mating in the store and escaping into the open. The number of pests entering the field from the much smaller stores of private farms is insignificant and can be discounted.

I.A. KANSU: Do you have any integrated control programme or research project which includes the use of the sterile-male technique against any insect in the Soviet Union?

S.V. ANDREEV: We are actually on the point of introducing the sterile-male technique as a regular agricultural practice to control Acanthoscelides obtectus and Laspeyresia pomonella.

H. LEVINSON: Have you considered or tried combining sex pheromones with chemosterilants, in addition to irradiation, in order to control stored-product pests?

S.V. ANDREEV: Very intensive work in this direction is being done at the Plant Protection Institute with respect to the most harmful orchard and agricultural pests and special attention is being paid to the use of juvenile hormones.

Z.W. SUSKI: What results have been achieved in the Soviet Union with Hyponomeuta malinellus Zell?

S.V. ANDREEV: Work on this pest is being carried out under both laboratory and field conditions. Mass-rearing methods are being developed and studies are being made of the bioeconomic and genetic aspects of applying optimum sterilizing doses. The optimum ratio of sterilized to unsterilized insects is being determined, and significant changes occurring in the reproductory organs of males and females are being investigated.

F.M. WIENDL: Do you have any special programme for the control of stored-product insects, and what is the status of commercial application of irradiation to stored products?

S.V. ANDREEV: A number of scientific centres in the country are actively developing a programme for the control of pests in stored products (grain and its derivatives). Powerful facilities have been built for gamma sterilization of these pests and semi-industrial gamma sterilizers are being used at the Institute of Grain near Moscow. In addition, very promising work is being carried out on the use of 2.5 MeV electron accelerators with beam scanning for sterilizing pests on a large scale.

ECONOMICS OF THE
STERILE-MALE TECHNIQUE
(Session 2, Part 2)

ECONOMICS OF THE STERILITY PRINCIPLE FOR INSECT CONTROL
Some general considerations

F.R. BRADBURY, B.J. LOASBY
University of Stirling,
Stirling, Scotland,
United Kingdom

Abstract

ECONOMICS OF THE STERILITY PRINCIPLE FOR INSECT CONTROL: SOME GENERAL CONSIDERATIONS.

A distinction is made between investment appraisal and cost-benefit analysis, the former being appropriate for private industry investment decisions and the latter for decisions by governments or public agencies. Both evaluation methods are examined and contrasted, and the significance of discounting to allow for the time of money and other resources is argued. The concepts of cost-benefit analysis are explained by means of a simple matrix, and it is concluded that if present estimates of costs and benefits are accepted, sterile-male control or eradication methods (SIRM) will have to be credited with very large positive values for reduction in pollution and toxic hazards to compensate for its much greater costs compared with conventional chemical control methods, if these are to be replaced by SIRM. The question of commercialization of SIRM is discussed, and the view advanced that industrial skills should be harnessed but, because of the nature of the SIRM technology, normal competitive market forces are inoperative and industry involvement must be through government contract. Some areas of uncertainty in the development of SIRM to full operational implementation are identified, and the ways in which industrial skills could contribute to uncertainty reduction are discussed. Finally, it is argued that the future of pest control technology is moving in the direction of greater involvement of government and regional control. This, and the likely spread of insect resistance, may give SIRM better prospects of adoption in the long term than it is believed to have in the short.

INTRODUCTION

The list of topics which was included in the briefing material for this Symposium included (a) procedures for cost-benefit analysis of programmes and (b) commercialization of the sterility method. Both of these are areas in which technology and economics interact strongly and are therefore of interest to the authors as technological economists, and it is to these two aspects of sterile-male control problems that we wish to address our study.

Governments or public agencies, who consider sponsoring sterile-male control or eradication methods (which we shall refer to subsequently by the initials "SIRM"), might be expected to use the methods of cost-benefit analysis to aid their decisions; commercialization implies the decision of private industry to take up the business, and for such decisions investment appraisal is appropriate. We should distinguish between cost-benefit analysis and investment appraisal. There is a practice, too common we believe among non-economists, of using the terms investment appraisal and cost-benefit analysis interchangeably; this practice obscures an important distinction, the understanding of which helps to clarify thinking about economic and social factors governing the diffusion of innovations such as the sterile-male control method. At a conference in Stirling in 1973 on Social and Economic Aspects of Pest Control Methods [1] we discussed the important distinctions

119

between investment appraisal and cost-benefit analysis at some length. In
the following section of this paper we reproduce some of the relevant
comments from this earlier paper.

INVESTMENT APPRAISAL

Investment appraisal is the estimation of the commercial return expected
on an investment, made from the point of view of the investor. A farmer
contemplating the use of a systemic fungicide for mildew control, a firm
contemplating the commercialization of the sterile-male control method no
less than the private citizen contemplating the purchase of company shares,
will make an investment appraisal. To do so the investor will estimate his
costs and his income and relate the balance to the size of the investment.
He may use a simple measure of return on investment:

$$ROI = \frac{S - C}{1} \qquad \begin{array}{l} S = \text{income} \\ C = \text{costs} \\ I = \text{investment} \end{array}$$

or, possibly, the present worth:

$$W = \frac{P_1}{1 + r} + \frac{P_2}{(1 + r)^2} + \ldots \ldots \frac{P_n}{(1 + r)^n}$$

P_1 = cash flow in year 1
P_2 = cash flow in year 2
P_n = cash flow in year n
r = discount rate
W = present worth

which takes account of the time value of money, and is therefore technically
more correct.
Many are familiar with such procedures, but it may be useful to say a
few words about discounting. A full treatment can be found in many texts [2]
The application of discounts to debits and credits to give their present value
reduces the effect of future increments, both negative and positive. A £100
expense to be incurred a year hence has a negative value of:

$$£100/(1 + r)$$

where r is the appropriate interest rate expressed as a fraction. An income
of £100 expected a year hence is likewise equivalent in value to £100/(1 + r)
today, two years hence £100/(1 + r)^2 and so on. The further in the future is
any receipt or payment, the greater, obviously, is the effect of discounting.
The justification for this "bird in the hand is worth two in the bush"
attitude, implicit in discounting, is simply the time value of money. Money
which has been paid back is available for other investment for the creation of

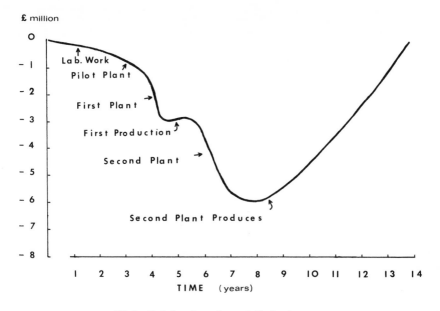

£ million

FIG.1. Cash flow for major pesticide development.

other business (or at worst, for the repayment of bank loans and the consequent saving of interest payments). A quick return can often be an important
criterion of efficient resource conservation. Drawn on a discounted cash flow
(DCF) basis, cash flow diagrams such as Fig. 1 are one of the most informa-
tive ways of exhibiting the financial implications of the investment in a new
project. It should be noted that the costs of research and development, which
come early, are not greatly discounted, and their effect on present value is
therefore heavily weighted, whereas the income from the product of the full-
scale plant, coming late in time, suffers a relatively heavy discount and has
therefore relatively less weight to put into the balance. One can see the great
importance of getting through the development state and gaining the benefits
of application and sales as quickly as possible. One can also see the effect
on the apparent profitability of such a project of the choice of discount rate.
Choosing the right rate is as difficult as it is important, and this is as true
for cost-benefit analysis as for investment appraisal.

There is no need for us to detail the different types of cash flow included
in a formal investment appraisal: capital, working capital, raw materials,
services, manpower, R & D, and so on in the costs or outflow side; or the
inflows coming from sales income, royalties and know-how payments, and,
at the end of the process, credits for sale of working stocks of products,
raw materials and, if it has any residual value, the plant. The point to be
emphasized is that these are costs and income falling on or accruing to the
investor directly. Social costs and social benefits appear in the strict
investment appraisal only in so far as they figure in the cash flow. For
instance, effluent discharge problems will appear in the appraisal only if
they are likely to cost the investor money, either as payments for extra
capital and revenue for effluent treatment by himself, or as payments to a

public authority taking the effluent from him for treatment at an agreed price.
Any effects on the amenities of the area receiving the discharges will not
figure in the investment appraisal; neither will benefits appear in the appraisal
other than those which the consumer or user pays for.

COST-BENEFIT ANALYSIS

The obvious difference between investment appraisal and cost-benefit
analysis is simply that in the latter such social costs and benefits are
explicitly included; and the reason is a change of standpoint of the evaluation.
Profit and loss accounting, modified perhaps by qualitative considerations of
social factors, is the essence of investment appraisal; and this is what is
appropriate for an evaluation from the standpoint of private industry or the
farmer. If we shift the standpoint to that of some larger segment of the
economy, such as the country as a whole, or a region, we may discover
that benefits and costs to individual sectors do not coincide with the benefits
and losses of the whole group of sectors. Cost-benefit analysis attempts the
appraisal of projects from this wider viewpoint.

It was that most distinguished Scottish economist, Adam Smith, who
argued that some invisible hand guided private greed in the direction of
public benefit. If this were always true, the cost-benefit study would be
redundant. In the absence of such a mechanism — which requires a very
special state of affairs which economists call perfect competition as well
as a market in which society could purchase its shared benefits like clean air
and freedom from noise or pollution — there are likely to be mismatches in
the allocation of costs and the enjoyment of benefits, and so the invisible hand
sometimes guides private greed in the wrong direction.

Cost-benefit analysis tries to make good this mismatch by measuring
and allocating all the costs and benefits of an enterprise. Aided by such an
evaluation, it is argued, society (or government acting as the agent of the
public) would make decisions and enact policies to serve the public good more
effectively than would sometimes be the case were these decisions left to the
market place or to Adam Smith's invisible hand. With the great and growing
public interest in the quality of air, water, land and wildlife, we have a
significant number of effects which cannot be bought and sold in the market.
Costs external to investment appraisal, costs not priced in the market, now
demand admittance to the evaluation of pest control methods and their use.
There are also benefits which are not priced in the market, and which therefore do not enter the equations of investment appraisal. It is for such reasons
that cost-benefit analysis is becoming significant today for chemical and other
pest control methods; indeed some of the arguments for replacing insecticides
by SIRM assign social costs to the former, which has the effect of offsetting
to some degree the greater investment and operating costs of the biological
method.

EXTERNALITIES

We refer above to costs "external to investment appraisal". What is
external and what internal in costs and benefits is, of course, relative to
the standpoint of the evaluator. A useful illustration of this point can be seen

in the case of two neighbouring farms. Farmer A in applying herbicide to his crops causes damage to the crops of his neighbour Farmer B. Apart from the loss of good relations and threat of possible lawsuits, this is of no account to Farmer A. The damage to Farmer B's crop is external to A's investment appraisal and will not enter his calculations. If, however, A's son marries B's daughter and the two farms become one, damage to B's crops by A's spraying becomes internalized in the concern AB, and steps will no doubt be taken forthwith to eliminate or reduce the damage (see Mishan [3]).

This parable of Mishan has connotations for a larger problem area. First, it reveals a possible advantage from regionalizing pest control strategies, the virtues of which as a means of internalizing costs and benefits which are external to individual farmers have been noted by Southwood and Norton [4]. Second, it points to the possibility of large shifts in net costs or benefits when the conventional national boundaries of cost-benefit analysis are moved by the impact of political agencies or by new technologies such as SIRM.

A COST-BENEFIT MATRIX

In the paper referred to above [1] we described a matrix representation of cost-benefit analysis. This derives from the theory of Turvey [5] who envisages cost-benefit analysis as a process of identifying and enumerating all costs and benefits likely to arise from the project under review, identifying all parties who may be affected by them, and proceeding to construct a matrix showing the allocation of the costs and benefits among the interested parties. All inputs and outputs (or nearly all) being in commensurable money terms, the algebraic sum of all entries in the matrix is the net appraisal of the proposal.

Figure 2 shows the matrix used in our Stirling paper, which was concerned with chemical pesticides.

Let us attempt to construct matrices for the SIRM. Figure 3 shows the result. In these matrices we have indicated situations where SIRM may be expected to compare unfavourably with chemical methods by "-" and the opposite case by "+"; the blanks in the matrix signify equal outcomes from the alternative methods. The matrices shown here are concept models in which plus and minus entries are used to indicate the more important expected effects of introducing SIRM in place of existing chemical methods. We assume for this argument that the effect of the switch on crop yields is insignificant; we also assume that SIRM is applied regionally and not to individual farms or establishments. It is important to emphasize that entries in our matrices are not quantified; some of the "+" and "-" entries may represent only very small sums in cash terms, others large. Hence there is no question of simply adding algebraically plus and minus signs to decide if an action is favourable or not; the symbols are used simply as an aid in presentation.

Figure 3 represents the expected direction of changes resulting from a switch from insecticide spray to SIRM, sponsored by government or public agency, using its own resources. Costs of the "manufacture" of sterile males and their distribution (rows MFRG/DSTN) will fall on the government giving rise to increased costs in column GOV. If the government or public

No.			Manufacturer	Competitors	Merchant	Farmer	Private user	Food processor	Consumer	Public	Government	Totals
1	Consequences of	Profit from pesticide	░									
2	improved yield	Profit from seed										
3	and / or quality,	Profit from crop										
4	labour saving	Value for money										
5	Toxicity to users											
6	Toxicity to other humans											
7	Toxicity to wildlife											
8	Pollution (various sub-categories)											
9	Other long-term effects											
		TOTALS										░

FIG.2. Cost-benefit matrix for analysis of impact of introduction of a pesticide.

PARTIES \\ C&Bs	M F R R	C N T R	F M R	F D P R	C O N	P U B	G O V	TOTS
M FRG							−	
D S T N							−	
F M C T			+					
F M I N								
Q U A L								
T O X Y						+		
P O L L						+		
C P D M			+					
T O T S			+			+	−	?

FIG.3. Cost-benefit matrix for change from insecticide spray to SIRM.

agency already finances existing regional spraying programmes there is no net gain to the farmer by way of reduced costs. There will be a positive gain to farmers through reduced costs if SIRM applied regionally takes the place of individual farm treatments with insecticidal sprays. There may be expected to be a net gain to the farmers through the specific nature of sterile-male methods and the consequent avoidance of crop or beneficial insect damage by chemicals. We have also credited the public with some advantage accruing from the switch to SIRM in rows TOXY and POLL, due to reduced risks of toxic damage and pollution.

Thus we may expect, given the antecedent and subsequent cases we postulate, that farmers and public are net beneficiaries, government net losers. The grand total in the bottom right-hand corner should decide the public policy.

SOME IMPLICATIONS OF COST-BENEFIT ANALYSIS OF SIRM

The assumption of equal effectiveness of the alternative control methods may be challenged; it was made to simplify the presentation of the case for cost-benefit analysis. If the effectiveness of control differs, the cost-benefit analysis may be even more illuminating, although more difficult to perform. For example, if reduced control of pest population results from a switch to SIRM from insecticides, then the pollution lobby against the chemical method may have to contend with a reaction of farmers who foresee a loss of profit in the change; this makes quantification of the debits and credits in the cost-benefit table important for rational decisions.

It should be noted that farmers do not necessarily suffer if crops are reduced. The critical factor is the responsiveness of price to a change in the quantity available – the price elasticity of demand. The authors of the USDA Economic Research Unit Report [6] on Mediterranean Fruit Fly Control assume an elastic demand. If this is the case, poorer control means reduced profits; but should the demand be inelastic (as with some food crops) the reverse would be true. In such a case both the anti-pollution lobby and the farming lobby might be expected to log gains in the cost-benefit matrix of the effects of a switch to biological from chemical methods! But consumers would then lose, which could be shown if we added a further row entitled "food costs" to the matrix in Fig. 3.

Our reading of the SIRM literature indicates a very great increase in costs, ranging from 25 to 100%, depending on the reference, to be expected as a result of a move from insecticide to SIRM, given the present state of development of the technology. The net benefits to farmers and public will have to make a large contribution to justify the switch. Given good practice in selection and application of chemical insecticides we would not expect the gains in the PUB column to be large enough to offset the increased costs of SIRM. A more likely source of benefits to offset increased costs of SIRM is to be found in the ever present threat of development of resistance of insects to insecticides.

How is the situation changed by commercialization? Apparently very little. If "manufacturer" of sterile-male insects and contractor are to enter the business area, their investment appraisal must show a net positive gain (of sufficient size to earn profit on the investment and compensate for risk). That is to say that the column totals for manufacturer and contractor must be

PARTIES / C&Bs	MFRR	CNTR	FMR	FDPR	CON	PUB	GOV	TOTS
M F R G	+						−	
D S T N		+					−	
F M C T			+					
F M I N								
Q U A L								
T O X Y								
P O L L						+		
C P D M			+			+		
T O T S	+	+	+			+	−	

FIG.4. Investment appraisal by MFRR (manufacturer) and CNTR (contractor) for entry into the SIRM business.

large and positive before commercial production and distribution are ventured upon (see Fig. 4). Although other benefits may accrue − to the farmer and the public − these will not enter the commercial investment appraisal computations any more than will the expenditure by government to whom, we assume, the manufacturer and contractor are contracting.

In order to preserve simplicity in presentation we have so far assumed a switch to SIRM from chemicals to be without effect on manufacturers and contractors. This is unlikely to be true (this Symposium has drawn attention to the large stake that chemical pesticide manufacturers have in cotton boll weevil control). The overall profitability of the pest control markets enjoyed by manufacturers and contractors may be increased or reduced; in any event the switch to a fundamentally different technology such as SIRM is likely to remove profitable business from pesticide manufacturers to government agencies using their own resources (Fig. 3) or to a different sector of industry (in the case where government contracts out the projects to industry (Fig. 4)). Even if such changes have no net overall effect on the cost-benefit analysis, they may well affect the balance of pressures, especially if one sector is predominantly "foreign" and the other "domestic".

It cannot be expected that SIRM will be sold to individual farmers. Commercial appraisal by the farmer of the biological method will be unattractive because the farmer who bears the cost will receive only a fraction of the benefits; control of insects on the neighbouring farms must be seen as an externality by the individual farmer. If the government or agency takes over countrywide or regional treatments the benefits external to the individual farmer become internalized in the large-scale programme, just as they did in the farm merger example quoted earlier.

If the government or other public body is the customer, it and not the commercial supplier must conduct the total cost-benefit analysis. To be sure, the public body may impose conditions and safeguards in its contract, and the costs of meeting these will then enter the commercial investment appraisal. It should perhaps be added that commercial evaluation by a company contemplating entry to this business may actually include a summary cost-benefit analysis, assuming for the purpose a government standpoint, in order to assess the likely government-supported market; just as in normal pesticide business the manufacturer will attempt to assess the benefits farmers may be expected to gain from use of a pesticide. But the commercial decision rests in the end on net benefits emerging at the foot of columns MFRR and CNTR.

It should be pointed out that all costs entered in the column GOV must eventually be passed on to part or all of the public. It is nevertheless proper to include government as an interested party rather than simply as an agent of the public; if only because in cases such as we are discussing here, with a relatively short time horizon, the short-term picture is more relevant than the (very different) long-term equilibrium one.

We conclude this section then by arguing that the nature of the new technology is such as to make normal market forces inoperative, that the public good resulting cannot be mediated by a competitive market of individual suppliers and consumers, and that whether the public agency implements SIRM programmes using its own resources or contracts out the work to commercial organizations, the criteria must be arrived at by cost-benefit study of the interests of all parties and not by the investment appraisal methods common in industry. This does not preclude, of course, the certainty that the manufacturer or contractor accepting the government order will first satisfy themselves by investment appraisal that an adequate profit is likely to be forthcoming.

Furthermore, in the conduct of such cost-benefit analysis, it seems likely that the dimensions of cash equivalents of unmarketable benefits like reduced pollution will have to be very large indeed to justify the implementation of SIRM, until such time as the money costs of the method approach much more closely those of existing chemical approaches. Alternatively there must be real technical advantages (which might emerge if fears of resistance by insects to chemical control methods are realized) to match the increased costs which present estimates impute to SIRM. This we think reveals the potential value of the cost-benefit analysis method. For all its difficulties it guards against the dangers of accepting greatly increased costs against unmeasured, and possibly very small, off-setting virtues of alternative procedures and indicates what changes would be necessary to cause the overall judgement to be revised.

SOME VIRTUES OF EARLY COMMERCIALIZATION OF PARTS OF SIRM PROGRAMMES

The preceding section attempts to stress that the crucial issues facing protagonists of SIRM are costs. How realistic are the existing cost estimates of SIRM? It is a truism that the only really reliable estimate of the cost of any project is to complete it and retain the bills (even that method will not reveal the opportunity costs). But SIRM is a major undertaking involving very large areas and many years of effort to prove its worth. This makes experimentation and costing by large-scale programmes prohibitively expensive.

What can commercialization offer in such a situation? The answer is probably a great deal, essentially by applying development skills and know-how in pushing a research operation to full and effective application <u>at an acceptable time and cost</u>. We believe commercialization (meaning by commercialization the hiring of the industrial resources on contract) is necessary to success in this step because the skills required are complementary to but different from those of government research organizations — precisely because they are profit-motivated. This leads to the use of resources efficiently and in a balanced systematic way, taking into account economic, technological and scientific aspects of the problem, and using all the aids and skills in development which successful industry possesses. Not least in the armamentarium of such an industrial science/technology team is the willingness to take risks which the research scientist may find repugnant. By risk-taking we do not mean, of course, foolhardiness but the ability to grasp opportunities and to drive projects through to completion on adequate but incomplete background information.

QUESTIONS CONCERNING IMPLEMENTATION OF SIRM

Let us look at this aspect of the matter in a little more detail. We have so far been discussing the entries into the cost-benefit matrix and the calculation of the net overall effect. There are important intermediate steps, however, between knowledge of the science which is basic to the method and its successful implementation at full scale.

We might ask these questions:

1.   Have we the basic concepts and scientific knowledge for SIRM?
2.   Have all the problems of production and application been identified? Can they be solved?
3.   Have the diagnosis, timing and monitoring methods been established?
4.   Is the information required for successful scale-up to full-scale application available?
5.   What are the expected costs of full-scale SIRM?
6.   What are the expected benefits?

It could be claimed fairly that questions 1 and 3 may be answered in the affirmative with confidence.[1]   What of the second question; are there unidentified problems or ones which look formidably difficult? One cannot answer this in the negative with confidence. Mr. B. Butt, of the IAEA Seibersdorf Laboratory, has produced a useful checklist of problems in SIRM. These include rearing failure caused by poor diet or disease, low production from equipment failure, mortality during transit or release, and failure to overflood with sterile males. This list in itself shows that many of the problems have been identified. There are others, notably the competitiveness of the sterile males and their fitness to survive outside the laboratory. There is, too, the related problem of possible genetic changes in the wild populations, which might lead to reduced competitiveness

---

[1] This was written before the Symposium began; papers presented on the first day of the meeting raised doubts about both questions.

of the released males. This paper does not attempt to evaluate the biological information but it is fair comment that some of the problems mentioned are big ones and will be costly to solve.

The question of scale-up, the fourth on our list, always presents problems in production processes, and we do not believe that the mass rearing of sterile-male insects will be any exception. The problems of housekeeping and insect health in mass-rearing factories must increase with scale of operation, problems of quality control-sampling and reliable test methods, problems of steady level production and of meeting delivery schedules, raw material supply security and quality, are some obvious aspects of scale-up. There is little doubt that experience will uncover many more.

On questions 5 and 6 we make only one comment; it is to underline the word "expected". The economists' and management scientists' use of the term to indicate the cost times probability of occurrence may be commended. It is all too common in cost-benefit studies to enter costs and benefits as absolute values without reference to the uncertainty which necessarily characterizes such estimates. The USDA Economic Research Unit's field report on medfly in Central America [6] is no exception to this practice, where the estimated cost of an eradication programme is claimed to represent a maximum because the areas requiring treatment may turn out to be smaller than envisaged. Representation of any cost estimate as a maximum is always dangerous, so notorious are over-runs in cost in innovatory projects.

The same caveat holds for values put upon benefits. There must always be a risk of partial or complete failure. We do not wish to labour this point or to be unfairly critical of the estimates. But, since we are dealing with very large and lengthy projects, there will be very expensive errors if the net overall value of the benefits expected to accrue is subject to unexpected variances. Moreover, just as in investment appraisal, future benefits, as well as costs, must be discounted by an appropriate factor, which has the effect of sharply reducing the weights to be put upon benefit accruing several years hence.

The development stage of industrial innovations, pesticides included, is subject to serious over-runs on cost and time. There is a good deal of literature on this subject. We would merely like to refer to the analysis made by Mansfield and co-workers [7] who looked at the statistics of over-runs and errors in estimating development costs for innovations in the pharmaceutical industry. They found a cost factor of 1.78 and a time factor of 1.61, these factors being defined as:

$$\text{Cost factor} = \frac{\text{actual development costs}}{\text{estimated development costs}}$$

$$\text{Time factor} = \frac{\text{actual development time}}{\text{estimated development time}}$$

This analysis, based on approximately 50 innovations, shows that the actual costs and time for development may be expected to be substantially greater than the best estimates made by management at the beginning of the projects. Any value greater than 1 for these factors shows an under-estimate

of cost or time, and it is worth noting that 88% of Mansfield's sample had
ratios for cost factor of > 1 and 68% of cases had ratios for time factor of
> 1. One must not set too much store by average figures of the kind quoted,
of course, and Mansfield himself points out that one of the important
variables affecting the degree of over-run on cost and time is the nature of
the innovation itself; the figures may be much larger in entirely new types of
projects or when larger technical advances are attempted. It is fair to
expect that SIRM, being an entirely new type of project, is likely to experience
large rather than small over-runs. As with future benefits, over-run effects
are exacerbated by discounting. Over-runs on time to full implementation of
projects increase the discount factor applied to the benefits because these are
pushed further into the future; costs, because they occur early in the project's
life, are not so sensitive to reduction by discounting.

The figures extracted by Mansfield for over-runs on time and costs for
industrial innovations reveal some of the pitfalls which await the cost-benefit
analyst when entering figures for cost in his matrix. A factor arguing for the
involvement of industry in SIRM development is the very large resources that
must be applied to such a project in the phase between research and full-scale
use. It is true that much thorough and excellent research has been completed
already on SIRM projects. There is, however, a likelihood that much of the
apparently humdrum, yet critical, aspects of project development, such as
quality control of mass rearing, maintenance of steady production rates,
development of distribution methods, may be in need of much more attention
than they have yet received in the numerous current and previous research
projects of SIRM. But these intermediate steps in innovations are well
understood in industry and known to be vital to the success of any major
innovation project. They are also very costly and need to be conducted
efficiently and economically.

In short, industrialization of SIRM might be expected to lead to an early
attack on some of the major sources of uncertainty, including critical project
appraisal.

It is perhaps germane to refer here to one of the findings of the important
University of Sussex SAPPHO research [8] into innovation success and
failure: that successful innovations are characterized by thorough development, not necessarily rapidly done, compared with failed innovations which
did not, in general, receive the same attention at the development stage.

THE LEARNING CURVE

On the more optimistic side it is reasonable to expect that the operating
costs of SIRM may fall with extensive experience of applying the programmes.
Although one may not expect great economies in capital costs of installations,
it is expected that the operating costs of running a mass-rearing factory and
distributing the sterile males to the target areas will fall dramatically with
numbers released in the same way as most other productive processes. We
refer to the well-known learning curve, publicized by the Boston Consulting
Group [9]. Indeed, a number of papers on SIRM refer to this possibility
including Lindquist [10] and the field report of the Economic Research Unit
of the USDA [6] on the costs of medfly rearing and application.

We may conclude this section of the paper by making three points.
The first (1) that there are important areas between the research and early

field trial findings and the successful full-scale regular treatments that use of SIRM demands. It is our contention that the involvement of industrial establishments with experience of innovation may facilitate and improve the prospects of success in this development stage of SIRM. (2) The estimates of cost of SIRM to be found in the published literature, both capital and operating, are likely to be low, and the almost inescapable under-estimation of costs must be borne in mind when making entries into a cost-benefit analysis matrix for the purpose of taking decisions on whether to go ahead with SIRM projects or not. (3) There is a prospect that the operating costs of SIRM will fall dramatically with accumulated experience through operation of the learning curve.

CONCLUSION

We conclude by drawing attention to the points we see as important from our examination of the SIRM problem.

1. SIRM does not lend itself, because of the nature of the technology involved, to market-mediated competitive commercial development.
2. Investment appraisal methods of the kind exercised by manufacturers are therefore inappropriate for decisions by governments on SIRM programmes.
3. Commercial skills must nonetheless be utilized for the successful development of large-scale SIRM pest control and eradication. They should also be utilized for improving the quality of forecasts for decisions about scale of effort required.
4. Such development would be facilitated by industrial contracting, to government or other public agency, for the mass rearing and distribution of insects.
5. Before placing such contracts the agency should satisfy itself by cost-benefit analysis that there are real prospects of advantages for the community as a whole to be had from a switch from insecticide treatment to SIRM (wholly or in part).
6. The social benefits, such as reduced pollution or toxicity, associated with SIRM have to be very large to justify the decision in favour of SIRM given present development of the technology and its cost relative to existing methods.
7. A development of resistance to insecticides would increase the relative advantages of SIRM.
8. The development of SIRM is and will continue to be handicapped by its essentially large-scale application, which needs a massive government support without the reassurance of mediation through commercial successes and failures which have moderated and guided the development and diffusion among individual users of chemical pesticides.
9. The whole development of SIRM is but one of a number of pest control methods which are moving away from the comparative simplicities of chemical spraying to more sophisticated intrusions into the biological system. These developments seem inexorably to move pest control into regions of plant medicine, plant doctoring and public control of plant health maintenance and disease vector treatment. In so far as pest control moves generally in this direction, SIRM's chances of successful full-scale world-wide use increase.

REFERENCES

[1] BRADBURY, F.R., LOASBY, B.J., in Social and Economic Aspects of Pest Control Methods, S.C.I. Symposium, Stirling, 1973 (to be published by Academic Press, London).

[2] MERRETT, A.J., SYKES, A., Capital Budgeting and Company Finance, Longmans, London (1966).

[3] MISHAN, E.J., Cost Benefit Analysis, George Allen & Unwin Ltd., London (1971) 110.

[4] SOUTHWOOD, T.R.E., NORTON, G.A., Insects Studies in Population Management (GEIR, P.W., et al., Eds), Ecol. Soc. Aust. (Memoirs 1), Canberra (1973) 168.

[5] TURVEY, R., Cost Benefit Analysis (KENDALL, M.G., Ed.), English Universities Press, London (1971) 5.

[6] ECONOMIC RESEARCH SERVICE, Economic Survey of the Mediterranean Fruit Fly in Central America, Economic Research Service, USDA (1972).

[7] MANSFIELD, E., RAPAPORT, J., SCHNELL, J., WAGNER, S.C., HAMBURGER, M., Research and Innovation in the Modern Corporation, Norton & Co., New York (1971).

[8] Success and Failure in Industrial Innovation, Report on Project SAPPHO, C.S.I.I., London.

[9] BOSTON CONSULTING GROUP INC., Perspectives on Experience, Boston Consulting Group Inc., Boston (1968).

[10] LINDQUIST, D.A., Recent advances in insect control by the sterile male technique, Meded. Fakulteit Landb., Gent 38 (1973) 627.

DISCUSSION

D.A. LINDQUIST (Scientific Secretary): Have you done a cost-benefit analysis study on other methods of biological control, such as the use of pheromones or the inundative release of parasites or predators? If so, are the results similar to those for the sterile-male technique?

F.R. BRADBURY: I have not done any cost-benefit analysis studies on biological methods of insect control, but I would expect the same general approach that I made in my paper to apply; the row and column titles might differ and in any real case, including SIRM, there would be many more such rows and columns than I show in my figures.

There is in my opinion an urgent need for a number of detailed cost-benefit studies of biological methods, studies which go much further than the USDA Economic Research Service study in 1972 of the medfly in Central America — perhaps the best study to date — by taking into account and attempting to quantify the social and long-term economic effects associated with the new methods.

L.E. LaCHANCE: I do not wish to detract in the least from your interesting comments; however, I am intrigued as to how you would go about putting a value on such things as a cleaner environment or a disease-free community.

F.R. BRADBURY: All effects which are considered to be important, costs or benefits, must be included in cost-benefit analysis. All such studies are of course comparative, so that whether one considers cleaner environment an advantage of SIRM or contaminated environment a disadvantage of chemical methods is not important when analysing the comparison of the two. The benefits you refer to, cleaner environment and disease-free communities, should not be overwhelmingly difficult to evaluate. Cost-benefit studies in other fields, such as the case for a third London airport, have tackled much more elusive "intangible" values.

D.E. WEIDHAAS: In addition to or in place of a complex cost-benefit analysis, which might be very difficult to quantify, would you consider an alternative approach to be an analysis showing that the investment costs of an alternative method are less than existing costs?

F. R. BRADBURY: Yes. The jargon term for such an approach is "cost-effectiveness" analysis. This is used where a defined target may be approached by alternative routes or programmes. In such a case, where one is not comparing effects but rather different methods of securing a common effect, many of the complexities of cost-benefit analysis are avoided. In the case we are discussing, the use of SIRM, the evaluation would be so simplified if a government were to veto the use of chemicals for pest control. Then we could cease to worry about what social costs to set against use of chemicals and we could look at the rather simpler problem of comparing non-chemical methods of securing pest control – SIRM and others. However, one must avoid falling into the trap of claiming that a chemical-free environment is "a good thing" which need not be priced and then proceeding to launch a very costly SIRM programme on the strength of imputing a very large and unmeasured price to the social benefit, one which might not in fact be great enough to justify or offset the increased costs of the SIRM programme.

R. PAL: A further aspect which might be dealt with in your presentation is the sophistication of SIRM and its feasibility in many countries.

F. R. BRADBURY: Yes, sophistication of treatment methods is a substantial barrier to the diffusion of new control methods. One of my students has encountered this problem in Cyprus in a study of army worm control where the sheer simplicity of applying chemical poisons to the invading insects is an obstacle to persuading farmers to adopt modified procedures.

To what extent such factors should appear in a cost-benefit matrix is debatable. Cost-benefit analysis of any control method presupposes that it is technically and psychologically feasible. The problem of persuading individual farmers will not arise with SIRM, of course, and the necessity for regional use of the method is an advantage in this respect.

SOME ECONOMIC FACTORS AFFECTING CHOICE OF THE STERILE INSECT RELEASE METHOD AS PART OF A PEST CONTROL PROGRAMME

B. G. LEVER
ICI Plant Protection Limited,
Haslemere, Surrey,
United Kingdom

Abstract

SOME ECONOMIC FACTORS AFFECTING CHOICE OF THE STERILE INSECT RELEASE METHOD AS PART OF A PEST CONTROL PROGRAMME.

The choice of sterile insect release as a potential pest control measure depends on an analysis of its technical, economic and organizational merits and disadvantages in relation to alternative pest control techniques and the importance of the insect attack. Generally sterile insect release is more likely to be cost effective when pest population densities are very low, and when the potential economic benefit from control is high. These types of situations could occur in integrated programmes to eradicate disease vector species or in prevention of seasonal migration of a major pest.

INTRODUCTION

Because many species of insect carry disease organisms which may affect humans, animals or plants, and may, through their feeding habits, destroy plant material needed for human and animal food, methods of insect control form a necessary contribution to the welfare of society. The question arises of how best to remove the unwanted pests without causing unacceptable environmental damage or diverting too many resources from other activities, given the skills and organization available within the community. The scope for alternative techniques, chemicals, biological control by predator species or the sterile insect release method (SIRM) will depend on three sets of factors, namely: biological, economic and organizational. Different approaches have different biological efficacy in different situations, need different levels of organizational sophistication for satisfactory application, and give different distributions of costs and benefits within the community.

Some of the economic considerations which bear on a choice between alternatives are here discussed. In no way is an attempt made to make specific statements about what methods are right or wrong in any given circumstance, but some guiding principles are presented upon which some further discussion can be built.

ECONOMICS AND SOCIAL GOALS

We, as human beings, and as members of social groups, try to get as much satisfaction as possible out of those resources available to us. Economics is the study of how to do this most effectively.

Mr. Bradbury has set out[1], in considerable detail, formal methods of calculating whether or not the amount of benefit to be expected from a particular activity is going to be great enough to justify the effort which must be devoted to it. He has made an important distinction between "private" investment appraisal and "social" cost-benefit analysis. "Private" investment appraisal is used to help profit-oriented persons or organizations decide whether they can expect to get enough money back from the project to repay their financial outlay at a suitable rate of return. "Social" cost-benefit analysis attempts to find out whether all the benefits which accrue to the community as a result of the project are great enough to warrant all the sacrifices which have to be made.

For a government organization to undertake a project it must be "socially" profitable. For a commercial organization it must be "privately" profitable.

As Mr. Bradbury argues, SIRM is a technique which can enter a government's range of choice for possible insect control methods, but cannot so easily be considered by private consumers because it is a technique which tends to work best over whole ecologically defined areas rather than for single fields within an agricultural area, or single houses within a pest infestation zone.

In much of developed agriculture, farmers will already be using chemicals because they gain a satisfactory private return from doing so. These chemicals will be available because chemical companies gain a satisfactory private return from producing and selling them. Government pest control only becomes of importance if no satisfactory private pest control systems are operating or if there is some good reason for banning current private practice. The lack of satisfactory private methods of control may be due to lack of technical or economic advance within a low income community, e.g. near subsistence agriculture, or because private persons are in no organizational position to apply private control methods, e.g. locust or tsetse fly control.

CONDITIONS FOR CHOOSING MALE STERILITY AS OPPOSED TO OTHER FORMS OF INSECT CONTROL

If one accepts that the use of SIRM has to be the result of a government decision rather than choice by a private user, and if a need for a government pest control programme exists, then one wants to know the conditions which would need to prevail for administrators to prefer SIRM to other alternatives.

Assuming complete knowledge, the chosen method of solving a particular insect problem would be that which gives the greatest social benefit in relation to its cost, within the practical constraints of the situation.

The "short list" of alternative feasible approaches will be determined by technical considerations, which would include, for example:

1. The availability and properties of available chemical insecticides (their effectiveness and their environmental advantages or disadvantages).

---

[1] BRADBURY, F.R., LOASBY, B.J., "Economics of the sterility principle for insect control; some general considerations", these Proceedings, IAEA-SM-186/51.

2.   The existence of useful predatory insects which could be used as part of an integrated control programme.

3.   The existence of alternate hosts or specific breeding grounds which can be destroyed or treated to break pest life cycles.

4.   The existence of geographical features affecting re-infestation (e. g. isolated islands and valleys forming separate ecosystems).

5.   The density of the pest population, its dynamics and the economic damage which it causes.

6.   Whether, if SIRM is considered, there is a danger of economic damage being caused by the released insects.

This short list will then be subjected to economic and organizational criteria. Which alternative programme gives the greatest social "value for money"? What practical organization, administrative or timing problems would be involved in operating the more attractive techniques?

ECONOMIC CRITERIA

Once a technical analysis of the pest problem has shown that a number of alternative methods of control could be technically feasible, one has to decide which method, or combination of methods, gives the greatest economic benefit per unit of cost.

As I hope the following discussion will show, situations where SIRM is likely to be particularly applicable will tend to differ from those in which pesticides are conventionally used. They will be situations in which, for some reason or another, there is considerable economic merit in controlling insects at very low population densities, when insecticides tend to become less efficient and the searching effect of sterile males becomes particularly valuable.

Vital factors in determining the optimum pest control policy will be knowledge of the "economic threshold" population and the pests population dynamics.

ECONOMIC THRESHOLD

An insect species only causes an "economic problem" worthy of action when the social costs of its crop damage, human or animal health hazard or nuisance value are greater than the cost to society of controlling it over some specified period of time. The extent of damage and cost of control will depend on pest density. This can be illustrated by two simple but contrasting hypothetical examples: one a typical crop pest control situation, the other a disease vector problem.

For each of these examples the hypothetical relationship between the economic return per hectare of land and the density of the particular pest species is shown by Figs 1 and 2. Each of these figures can be reinterpreted

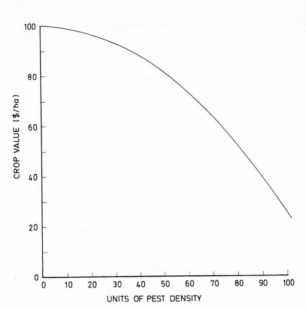

FIG. 1.   The relationship between crop value/ha and pest density.

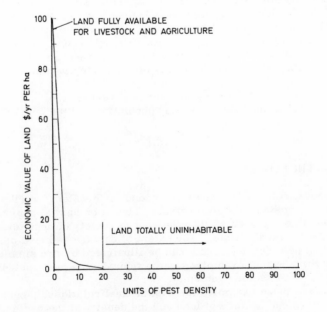

FIG. 2.   The relationship between land value and disease vector density.

to show the loss of potential income resulting from the presence of the
pests (Figs 3 and 4 respectively). These figures are calculated from
income at pest density of zero minus income at actual pest density.

For the crop example, the economic loss from pest incidence diminishes
rapidly with a decrease in pest density. Thus the value of obtaining, say,
a 90% pest kill is very high when the pest density is high, and small when
the pest density is low.

For the disease vector example, the land is practically useless until
the pest density is down below 10 units, and the real development potential
of the land for agriculture and livestock can only be realized when the pest
has been virtually eliminated. The value of a 90% pest kill, therefore, is
consistently high until the pest has been virtually eliminated.

Now suppose the cost of achieving a 90% kill with chemical insecticides
is $20/ha. As shown in Fig. 5 for the crop example, there is no net bene-
fit to be gained by applying a 90% effective treatment once the pest population is below 52 units because, below that level, the treatment cost is
greater than the benefit from extra crop value. (The threshold level will
fall if crop value/ha rises because of price increases, new crop varieties,
etc., or if treatment costs fall. The converse circumstances will raise
the threshold level.)

For the disease vector example, however, a very different pattern
emerges (Fig. 6). At very high levels of infestation, a 90% kill is not good
enough to give real benefits because the area is not sufficiently free of
infestation to enable human or livestock populations to move in. The real
benefits are only achieved when the pest population has been reduced to an

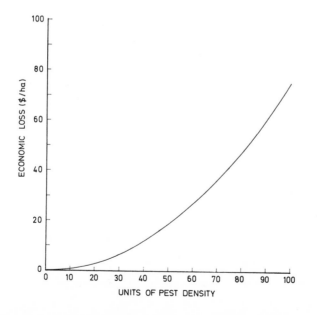

FIG. 3.   The economic loss from pest incidence in the crop example.

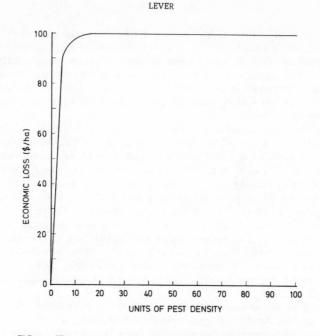

FIG. 4.   The economic loss from pest incidence in the vector example.

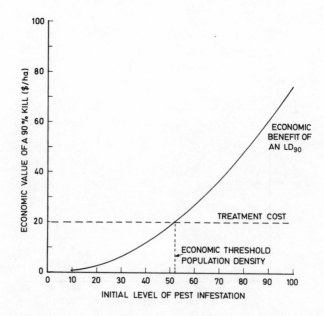

FIG. 5.   The economic value of a 90% pest kill at different levels of initial infestation, and the economic threshold for the crop example.

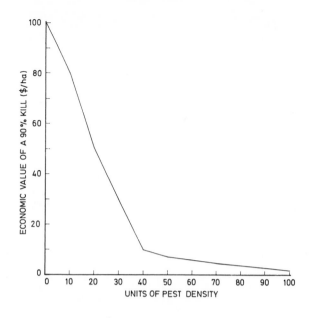

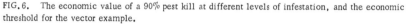

FIG. 6.   The economic value of a 90% pest kill at different levels of infestation, and the economic threshold for the vector example.

extremely low level. For tsetse fly control for example, a final clearance from a few thousand individuals per km^2 down to about 100 km^2 must be achieved before the land can be stocked.

The economic difference between these two situations is of relevance to SIRM. As a fair generalization, biological methods of control, be they natural predators or sterile members of the pest species, tend to be most effective at low pest populations because they depend upon the ratio of numbers of predators or sterile individuals to numbers of pests to be controlled. Pesticides, on the other hand, tend to be more efficient at higher pest densities because a given level of chemical application rate kills a certain proportion of the pest population, e. g. the LD 90 kills 90%, almost irrespective of the pest density. At very low populations, the proportional kill may fall because the individuals manage to avoid contact with the chemical. SIRM could be particularly effective at these low densities because of the natural "searching" behaviour of sterile insects for their native mates.

Suppose the costs per hectare of SIRM to achieve a 100 : 1 sterile/native male population ratio[2] are as shown in Fig. 7 for hypothetical "current" and "improved rearing" situations.[3]   Superimposing this on the curve of "economic value of a 90% pest kill" from Fig. 5 one gets the pattern shown in Fig. 8, where on both the current and improved SIRMs pesticides consistently show a better level of cost effectiveness above the economic

---

[2]   The 100 : 1 sterile/native ratio is purely illustrative. The ratio actually required to give a certain level of control will depend upon the insect species and its current reproduction rate.

[3]   A diminishing cost per insect released with scale of release could be expected due to spreading of overhead costs.

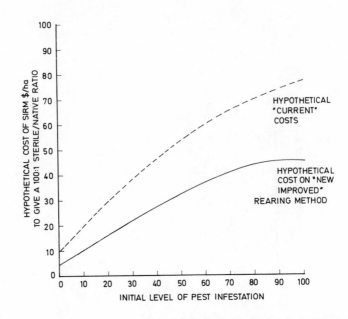

FIG. 7.   The relationship between hypothetical SIRM costs and initial levels of pest infestation.

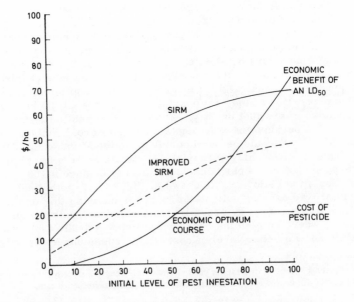

FIG. 8.   Comparison of economic benefits and costs of chemicals and SIRM in the crop example.

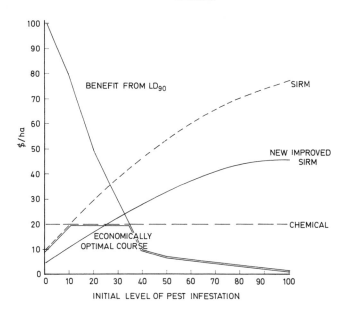

FIG. 9.   Comparison of economic benefits and costs of chemicals and SIRM in the disease vector example.

threshold.   The optimal course of action is to do nothing at pest densities
of less than 52 units/ha and to spray at densities above that.   To make SIRM
cost-affective in the situation, the cost of rearing and release have to be
reduced to an extent which would enable its cost line to fall below the
current "economic optimum course" line (unless there were some major
encironmental reasons for accepting SIRM at a higher cost than pesticides).[4]

A study of the disease vector problem gives a different verdict.   Figure 9
shows the cost of SIRM and chemical control superimposed on the economic
benefit curve for the vector example from Fig. 6.   The optimal course of
action is now more complex.   At a population density of less than 10 units,
SIRM is the most "cost-effective" solution.[5]   On the "new-improved rearing"
technique SIRM has a possibility of becoming the most cost-effective method
at populations of less than 30 units.   Above 35 units, neither SIRM LD 90
nor a chemical LD 90 is justified alone, because neither reduce the population
sufficiently for the benefit to outweigh the costs.   In this situation, the optimal programme is one of "integrated control".   The population is reduced
by chemical means to a level at which SIRM becomes the most cost-effective
way of removing the remaining pests.   Figure 10 shows the comparison of
benefits and costs of an integrated control programme designed to give a
resulting pest population of less than 1 unit per hectare.

---

[4] It should be noted that this example assumes, for simplicity of argument, that chemical treatment
is aimed at one pest species.   In many practical situations the target will be a pest complex, in which case
sterile insect release would not be technically substitutable for a multi-pest chemical treatment.

[5] The hypothetical example has been so constructed that SIRM becomes cheaper than chemicals
at a very low population density.   This does not imply that this will always be the case in practice, but
rather that SIRM is more likely to be relatively cost-effective at low densities than at higher ones.

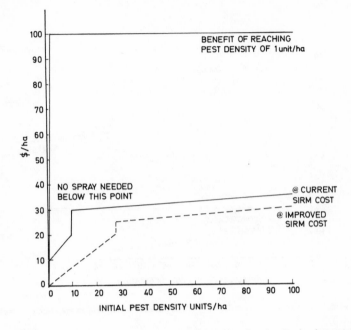

FIG. 10.   Benefit and cost of reaching a population density of less than 1 unit using integrated control.

In practice an integrated control programme may include more than pesticides and SIRM. Effective tsetse fly control, for example, includes bush clearing, game culling and land management.

POPULATION DYNAMICS

The economic threshold examples discussed above showed SIRM at a potential advantage where low populations caused serious problems, whereas chemicals had the potential advantage at high populations. This situation is complicated by the question of population dynamics.

Consider a pest species which emerges in small numbers from its over-wintering state. The adults breed through the spring and summer, building up from a small overwintering population to a large feeding population later in the year. It is the large feeding population which causes crop damage. Figure 11 sets out the population density which would exist each month if no control were to be applied.

If no early control is applied, the economic threshold for pesticide treatment (a population of 52 units/ha) will be reached by mid to late June. At some time, therefore, it is economically advisable to treat the crop for pest control. Using a chemical in this situation, one can treat at any time after the pest becomes evident and be fairly sure that the cost of treatment will be justified by the losses which would have otherwise occurred. There is flexibility of timing, but one would be best advised to treat as early as possible to minimize total damage.

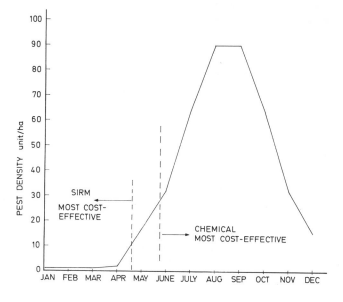

FIG.11. Pest population dynamics in the absence of control measures.

Now there are times, very early in the pest build-up, where populations are low but where the benefits from treatment are high (because the
benefits relate to losses at the high populations which would result from a
lack of control). At these early stages SIRM and insecticides could possibly
become economic rivals. From the previous discussion, the cost of chemical
treatment has been assumed to be $20/ha irrespective of the population
density. From Fig. 7 we have assumed that, on current costs, SIRM becomes
more cost-effective than chemical treatment below a population density of
10 units, and on the hoped for "new improved rearing method" at densities
below 28 units. The choice of SIRM rather than chemicals in this situation
depends on:

1. How critical the timing of SIRM has to be to achieve a large proportion
    (say 90%) of sterile matings at a population density of less than 10 (or
    eventually perhaps 28) units. For control of cherry fruit fly, for
    example, there is a critical period of two weeks in April when treat-
    ment needs to be applied.

2. How sure one can be that, if left untreated, the population would in
    fact exceed the threshold.

    These two questions can only be answered in connection with very
specific situations.
    Similar logic to that described above can be applied to the use of
SIRM as part of a cordon sanitaire policy to prevent new fringe infestations.

THE POSSIBILITY OF SPECIES ERADICATION

Both the discussions of thresholds and population dynamics for the crop example have assumed that benefits and costs occur annually. If there were a reasonable chance that, by reducing the population to minute numbers, the species would die out, then the picture changes. One could be faced with a choice between the high cost of an SIRM eradication programme now and a series of lower annual pesticide costs in the future. Each has to be offset against longer term benefits.

Mr. Bradbury's paper described how discounting can be used to compare costs and benefits occurring over a number of years. Supposing one knew that, left untreated, the pest population would regularly build up to about 90 units annually. Over a period of, say, five years, the present value of the losses incurred would be $232/ha at a 15% DCF rate.

The present value of costs for an insecticide programme would be $75/ha.

If the pest could be eradicated, this would be worth anything up to $76/ha (the cost saved by not having to apply the insecticide repeatedly).

If eradication could be certain for more than five years, because there was nowhere from which new infestations could come, then an eradication programme would be worth more than this for the above example.

THE SOUTH-EASTERN USA SCREW-WORM ERADICATION EXPERIENCE[6]

The classic, economically worthwhile, large-scale practical use of SIRM, the eradication of screw-worm (<u>Callitroga lominivorax</u>) in the south-eastern states of the United States of America exemplifies a number of the theoretical points described above. The screw-worm population which attacked livestock in the states of Mississippi, Alabama, Georgia, South Carolina and Florida overwintered as a small community in southern Florida the bulk of the summer population being unable to survive the winter temperatures prevailing in the rest of the area (Fig. 12). The overwintering site provided an ecologically isolated population for SIRM treatment.

There was no economically satisfactory "private" treatment that livestock owners in the south-eastern states could apply, but cattle losses were important to the livestock economy of the region. By government action in applying an effective treatment in one area, a large "social benefit" could be gained by livestock farmers throughout the region. The cost of control was economically justified per unit of benefit partly because a population was hit when small and restricted in geographical distribution, and partly because, as the pest was eradicated, the benefit of the programme continued to accrue over successive years. Continued government involvement in the form of veterinary inspection and treatment of cattle crossing the Mississippi from the west helps to prevent any reinfestation.

ORGANIZATIONAL FACTORS

If a situation has been identified in which SIRM appears to be an attractive approach, the necessary organization has to be set up. This may

---

[6] KNIPLING, E.F., The eradication of the screw-worm fly, Sci. Am. 203 4 (1960) 54.

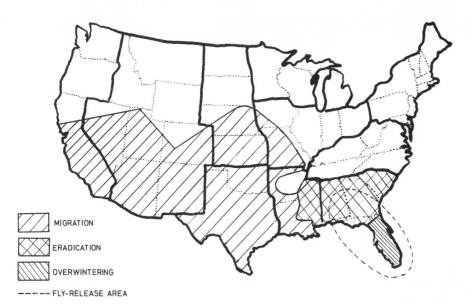

MIGRATION

ERADICATION

OVERWINTERING

━━━━ FLY-RELEASE AREA

FIG.12. The eradication of screw-worm in the United States of America. Range of screw-worm fly originally included the entire southern part of the United States. Release of sterile males in area indicated by broken line eliminated the fly from the southern states east of the Mississippi in 1958. The fly persists west of the Mississippi because it winters in areas adjoining vast areas of infestation in Mexico. Inspection stations along the Mississippi guard the southeastern states against re-infestation by cattle from the southwest. (KNIPLING, E.F., The eradication of the screw-worm fly, Sci. Am. 203 4 (1960) 54.)

include rearing and sterilizing facilities, attractant baits and chemosterilants, pest emergence and density monitoring traps, depending on the programme envisaged in the economic analysis. This programme may be operated with government employees, international agency experts or part sub-contracted to private organizations. As described in Mr. Bradbury's paper, the practicability of an SIRM programme could depend very critically on the efficiency of its operation and on efficiency-oriented research and development.

If it were felt necessary to hire private industrial research, development and operation skills, the government would need to provide a sufficient private financial incentive for the industrial firm to divert resources from other activities. Because sterile insects cannot be "patented" in the same way as chemicals, some longer term R & D contract would need to be negotiated if private capital investment were needed. If some form of "consultancy" was required without capital involvement then there would be considerable scope for short-term co-operation.

CONCLUSION

The rather cursory and hypothetical analysis described above suggests that SIRM could have an economically viable place alongside other insect control methods. Its use, however, will be determined by the needs of specific situations, which could differ in important social, biological and economic ways from those which currently attract large-scale use of chemical insecticides. The most important of these differences is likely to be the economic value of effective control at very low natural population densities. The types of problem where economics of this nature seem most likely to pertain are within an integrated programme for the control of disease vector species, or in maintaining a cordon sanitaire round an area to which a major pest seasonally migrates.

Note: The views expressed here are those of the author and should not necessarily be taken to represent those of ICI.

DISCUSSION

D.W. WALKER: I think that you should include the cost of quarantine measures in the sterile release costs, since you normally only use SIRM when populations are very low and therefore re-introduction problems relatively more important.

B.G. LEVER: I agree that a cost-benefit analysis of a pest control measure must consider the whole system. In the case of the eradication of the screw-worm in the United States of America, this would include the costs of quarantine and inspection of cattle moving eastwards across the Mississippi. Similarly in tsetse fly control, the project analysis would have to include the costs of bush clearing and game culling as well as the direct "pesticidal" measures.

D. ENKERLIN S.: Knowledge of the economic threshold is very important but also very variable under different conditions, even for the same pest. Particularly in the developing countries pests are too often treated with no economic return but many side effects. International organizations and/or governments should put more emphasis on this and make more funds available for research.

As to the economics of the sterile insect technique, it should be remembered that it can be used for suppression or control, for creating barriers or for sanitary control as well as for very important ecological studies. In each case the economics will differ.

B.G. LEVER: In answer to your first remark, there is presumably some trade-off between spending money on research to establish "economic thresholds" and spending money on pest control when it is not necessary. I would think that the correct balance would be to do sufficient work to get an "informed judgement" as to roughly how serious a pest infestation would need to be to justify treatment, and then use control measures when attacks of roughly this magnitude could be expected. I am sure it is not worth trying to establish very exact economic threshold values because such exact knowledge, even if it were gained, would be difficult to use in practice with any degree of accuracy.

I agree that the economic benefit of SIRM would depend upon the aims of the project. For the first types of application you mention — suppression, control or barriers — one is concerned simply with the economic benefits of reduced pest populations and the most cost-effective way of achieving these benefits. For the other case, as an ecological research tool, its use must be justified in the context of the value of the research project as a whole. This will depend upon answers to questions such as: "what is the project designed to investigate?", "what will be done with this knowledge?", "are there more important problems to be solved?"

L. C. MADUBUNYI: In preparing Fig. 9, did you consider the generally accepted concept that SIRM, just like biological control involving the use of parasites, may be initially more expensive than chemical methods but cheaper in the long run? Also, were you envisaging a pest population which does not decrease in spite of the effort to effect such a decrease with SIRM?

B. G. LEVER: I think that you have made the very valid point that Fig. 9 is a two-dimensional representation of a three-dimensional problem. In practice a pest population is dynamic over a season, following the sort of pattern shown by Fig. 11. Chemical control will knock the population down from high numbers back to low numbers, but it may then recover again and perhaps require further treatments. One may hope that an integrated programme would allow SIRM or parasites to reduce markedly the rate of recovery, giving a longer effective period of control.

When one is faced with the existence of a high population, however, I would be surprised if it were not best to use chemicals as the first "quick knockdown" attack on the pest before other methods are introduced.

F. R. BRADBURY: A factor which will certainly be expected to lead to the reduction of costs of SIRM is the operation of the so-called "learning curve" which relates costs of production in many technologies to the volume of products previously made. There is no reason to believe that such an effect would not apply to SIRM operation and production costs once the method has been established at full scale.

B. G. LEVER: I agree that the economic efficiency of most commercial operations improves considerably with experience. Presumably the rate of improvement depends upon the extent to which economic pressures impinge upon the project's management. The greater those pressures, the greater will be the incentive to find better ways of performing the various tasks.

B. NAGY: Your paper contains interesting considerations on the private and governmental management of pests by various techniques, including SIRM. In this connection I should like to draw attention to a significant difference between socialist and non-socialist countries. The system of large state farms and co-operative farms with their huge, homogeneous monocultures may provide a more suitable base for SIRM and other specific control methods than smaller private farms. Have you any comments on this situation?

B. G. LEVER: Certainly sterile insect release requires a large and relatively isolated ecological area for effective application. I would suspect that the management of large state farms would be in a better position than individual smaller farmers to make a choice between SIRM and chemicals on the grounds of cost-effectiveness, because the area which they control more nearly approaches the technical "ecological" requirements. However, when a pest-control programme is proposed over a very large area for an insect which is not effectively controlled by "private" measures, for example the control of screw-worm in the United States of America, then the decision needs to be made at national or regional government level anyway. In these circumstances the farm management organization becomes irrelevant to the cost-effectiveness analysis and policy decision.

ECONOMIC FEASIBILITY OF ESTABLISHING A PILOT PLANT FOR CONTROL OF THE MEDITERRANEAN FRUIT FLY BY DISTRIBUTION OF STERILE MALES *

Y. SPHARIM
Agricultural Research Organization,
Bet Dagan,
Israel

Abstract

ECONOMIC FEASIBILITY OF ESTABLISHING A PILOT PLANT FOR CONTROL OF THE MEDITERRANEAN FRUIT FLY BY DISTRIBUTION OF STERILE MALES.

Control of the Mediterranean fruit fly is today effected by the application of bait from the air. This is a cheap and effective method, which complies with the strict criteria concerning contamination with fruit fly in the countries for which Israeli exports are destined. It is now proposed to examine a new method of controlling the Mediterranean fruit fly by distributing sterile males, within the framework of a pilot plant. Up to now, experiments with the new method were made on a small scale, and results showed that fly populations could be reduced in this way, but control was not complete. Another small-scale experiment is now in preparation; if results are favourable, an experiment on a pilot plant scale would be technically feasible. A discussion is presented, in terms of economic feasibility, of (a) whether a pilot plant should be put up, and (b) the maximum size of such a test. Results of the present investigation show that if export of fruit free from pesticide residues can indeed be envisaged, then the benefit derived from establishing a pilot plant may compensate for any damage involved in a test failure. The maximum scope of a pilot plant, according to the postulates detailed here, could reach a gross of 6-9 thousand hectares.

BACKGROUND

There are 40 000 hectares of citrus plantations in Israel, most of which are planted along a non-continuous strip on the Mediterranean coast. About 60% of the fruit is sold abroad as fresh fruit and the rest is sold for industrial processing and local consumption. The fruit for export is a well-cared-for, waxed and wrapped quality product; much attention is lavished on its outward appearance, although finally it is eaten peeled.

The major potential pest endangering citrus fruit is the Mediterranean fruit fly. This pest is today fully controlled by application of pesticide bait from the air. Israeli fruit exported abroad is entirely free from the fly. In addition to the Mediterranean fruit fly, damage is caused to citrus fruit by mechanical cleaning of the fruit or by spraying with oily emulsions containing suitable chemical additives, according to the extent of damage and the kind of pest. When the fruit is packed, treatments are applied, either directly or to the wrapping, to prevent fungal damage and to increase fruit resistance in transit. Strict criteria exist in many countries concerning infestation of imported citrus fruit by Mediterranean fruit fly. It can be said that generally only fruit that is entirely free from the fly is acceptable.

---

* Contribution from the Agricultural Research Organization, The Volcani Center, Bet Dagan, Israel; 1974 Series, No. 187-E.

For a number of years tests have been made in Israel to control the Mediterranean fruit fly by male sterilization with radioactive radiation. The relatively small-scale tests (up to 1000 hectares) have shown success in exterminating the flies, but not in preventing a new influx of fertilized females from outside. It was then proposed that an experiment on a larger scale should be carried out: a pilot plant project that would make it possible to test the claim that only on a sufficiently large area can new invasions be overcome. Specialists in this field suggested that the area required is about 10 000 hectares gross or 5000 hectares net.

A test on 10 000 hectares involves risking a considerable part of the harvest as well as an investment of more than $ 100 000 to set up a plant for sterilizing male flies. This means that the experiment is an economic operation of considerable proportions, and the project was therefore subjected to the following economic investigations:

1. Comparison of the costs of the control method at present in use — spraying bait — and of the proposed new method — distributing sterilized males. This comparison was made on the assumption that the control by male sterilization would be carried out on a commercial scale and attain a standard of control equalling that achieved with the method at present used, where the fruit is entirely free from flies.

2. Investigation of the possibilities of saving expenditure in the control of other pests, as well as investigation of the prospects of raising the value of the fruit as a result of reducing the use of pesticides by the introduction of control by male sterilization.

3. Examination from an economic point of view of the size of a semi-industrial plant.

COMPARISON OF COSTS OF THE TWO CONTROL METHODS

The method of spraying with bait gives satisfactory results in Israel and Israeli fruit complies with the strict demands of the countries to which the export is destined. The estimate of the costs of controlling the fly by male sterilization rests on the assumption that a rise in contamination by the Mediterranean fruit fly will not occur as a result of adopting the new method. Small-scale experiments with sterilized males have shown that under Israeli conditions the main problem consists of isolating the area where the control operation is carried out. Such isolation could be effected by spraying the vicinity with bait, by distributing sterilized males, by a combination of the two methods and, possibly, partly by using a more novel method, such as creating a male vacuum by using pheromones. In the opinion of experts, the separating belt should be not less than 7 km wide; under Israeli conditions this means a relation of the magnitude of two to one of the gross and the net areas under consideration. Only by an operation of this magnitude can a reasonable chance be assumed to exist that the proposed method will succeed, success in this case meaning fruit suitable for export — zero contamination.

The male sterilization method of control is much more expensive than the method now used in Israel. Tables I and II present the costs of control

TABLE I.   COSTS OF MEDITERRANEAN FRUIT FLY CONTROL BY MALE STERILIZATION

Subject	Estimated cost (Israel pounds) (4.20 I£ = US $1.00)
Building for medfly mass rearing of 100 million weekly. Estimate includes essential equipment (air conditioning, cold rooms and storage area), total area approx. 200 m^2	400 000
Running costs (40 weeks at full production, 12 weeks partial production): Electricity, telephone, water, etc.	10 000
Staff: Professional (2.5) 50 000 Technical (2) 24 000 Daily (6) 58 000 ————— 132 000	132 000
Expendable supplies including fly food, trays, cages, etc. (in addition to existing supplies)	200 000
Transport	6 000
Aerial release (100 km^2)	168 000

TABLE II.   COMPARISON OF COSTS (in US $)

	Bait spray	Sterile males
Cost per hectare: gross area	10.00	13.30[a]
net area	12.00	26.60
Ratio gross/ net area	1.20	2.00

[a] Based on Table I.

by male sterilization and a comparison of costs of the two methods of control. To these calculations must be added the fact that the male sterilization method includes the element of using animals — and these are liable to be tired, ill or simply do not behave as expected — whereas the dependability of the control material in use is both constant and known.[1]

———————
[1] There is also the problem of developing immunity; this is a possibility for both methods of control but it will be many years before it becomes acute.

SAVING OF EXPENDITURE FOR THE CONTROL OF OTHER PESTS AS A RESULT OF THE RESTORATION OF THE BIOLOGICAL EQUILIBRIUM

The Mediterranean fruit fly is controlled by Malathion, which is toxic enough to exterminate the natural enemies of scale insects and mites which are important pests of citrus. It is assumed that the extermination of the Mediterranean fruit fly upsets the biological equilibrium and causes other infestations to increase as populations of their natural enemies decrease. This assumption has not as yet been tested experimentally, but it is supported by incidental results of tests already made with the method of male sterilization and by observation in the field. Control of the Mediterranean fruit fly costs about $10-15 per hectare; control by means of male sterilization will cost about $30 per hectare. However, expenditure for the control of other pests amounts to about 250-300 per hectare (without taking into account the increase in oil prices); it can be argued that partial savings resulting from the restoration of the biological equilibrium will be of a magnitude equalling all additional expenses involved in changing over to the method of control by male sterilization. It seems that this question, too, can be answered only after a semi-commercial operation has been set up.

THE POSSIBILITY OF SELLING FRUIT FREE FROM PESTICIDE RESIDUES

Israeli fruit is selling abroad today because of a combination of characteristics that makes it a quality product. It is beautiful to look at, free from scale insects, and has a long shelf life. Advertising and market competition are directed to emphasize these traits. There are some problems regarding the addition of chemicals to the fruit or the wrapping, but these are not the major problems over which the producers compete. Fashion and technological changes in the production could bring about a change in the consumer's attitude to citrus fruit. If someone were able to produce citrus without toxic materials, such a producer could, in view of the fight against pollution, emphasize this advantage by suitable publicity and compete on this ground.

If spraying of bait against the Mediterranean fruit fly were to cease, this would not yet mean fruit free from pesticide residues. There are scale insects, mites and moths which attack the fruit and against which pesticides are used in one form or another. This could be avoided by spraying with stronger concentrates of oil, which would, however, cause damage to the tree. Abandoning the use of poison bait as a means of combating the Mediterranean fruit fly may cause an increase in the population of natural enemies of the other pests, but this expectation is to a certain extent based on speculation.

It should be investigated from a commercial point of view whether the signs of scale insects and mites on fruit could not be made use of as an indication that the fruit is free from toxic materials, thus killing two birds with one stone, solving the entire problem and increasing consumers' trust in "non-polluted" fruit. Such an attitude, however, entails careful building up of a new citrus fruit image.

Israeli citrus fruit, apart from lemons, are eaten peeled, so that less importance attaches to pesticide residues in respect to this compared with other fruit. Lemons, which as a rule are used with their rind, are sold in Europe with the label "free from toxic materials" and fetch relatively high prices in considerable quantities. Other citrus fruit, sold as free from toxic materials, fetch rather high prices, but only in the small quantities bought by a special and restricted circle of customers. The sale of some of the fruit which is free from toxic material and the emphasis placed on this aspect in the propaganda, may turn out to be harmful to the reputation of the rest of Israeli fruit.

It appears that there are only slim chances of increasing the value of the fruit by adopting the male sterilization method for the control of the Mediterranean fruit fly instead of the method at present in use and that, in any event, this is not a thing of the near future. On the other hand, it can be shown that the relation between the expenditure for the control of the Mediterranean fruit fly and the export value of the fruit is such that a small increase in the value of the fruit, if obtained, could justify a few times over such additional expenses as are involved in the operation of the new method. The relevant magnitudes can be exemplified as follows:

The yield of one hectare of trees is 40 tons.

Of this about 60% or 25 tons go for export.

The wholesale price of the exported fruit abroad amounts to about I£ 650.00 per ton (= US $155.00).

The export value of one hectare according to wholesale prices abroad amounts to about $3900.

The cost of controlling the Mediterranean fruit fly by the present method is $10 per hectare.

With the male sterilization method the costs will amount to $25 per hectare.

The following speculation can be made:

If it is possible to sell 10%[2] of the produce as fruit free of pesticide material at 15% above the price of ordinary fruit, without affecting the price of other fruit, then this could cover 3 to 4 times the additional expense involved in changing over to the proposed control method.[3]

ECONOMIC FEASIBILITY OF A SEMI-COMMERCIAL OPERATION

The preliminary investigation of the chances of marketing fruit free from pesticide residues is not discussed here. Such an investigation must be undertaken by a team of marketing and pest control experts. It is only

---

[2] In this case, fruit could be chosen from areas where infestation with other pests is low, and in these areas no pest control operations with respect to such pests would be carried out.

[3] The sale of 10% of the fruit at a price exceeding that of normal fruit by 15% for a period of 10 years from the day that the new method is fully employed commercially, would bring about an additional income of $3 million at present value.

TABLE III.   THE VARIABLES AND COMBINATIONS INVOLVED IN THE
FEASIBILITY TEST OF THE EXPERIMENT

Symbol	Definition	Assumed value
A_1	Commercially cultivated areas	15 000 hectares
A_2	Pilot plant areas	
P_1	Probability of success based on present information	0.3
P_2	Probability of success after successful operation of pilot plant	0.8
B	Benefit expected from successful full operation	$3-4 million
C_1	Losses expected from failure of full commercial operation	$3 million
C_2	Losses expected from failure of pilot plant	
r	Real interest rate	9%
t	Time until conclusion of pilot plant experiment	3 years
EB	Expected benefit	

$$K = \frac{A_2}{A_1} = \frac{C_2}{C_1}$$

Expected benefit from immediate overall operation would be:

$$EB = P_1 B - (1 - P_1) C_1$$

$$EB = 0.3 \times 3 - 0.7 \times 3 = -1.2$$

Expected benefit from operation in stages [a] would be

$$EB = \frac{P_1 [P_2 B - (1 - P_2) C_1]}{(1 + r)^t} - (1 - P_1) C_2$$

---

[a] Full-scale operation after successful pilot plant.

TABLE IV. EXPECTED BENEFIT FROM OPERATIONS OF VARIOUS
SIZES

EB = 4	EB = 3	K
2.0	1.0	0.1
1.0	0	0.2
0	-1.0	0.3

after such a team has tackled this job, and if it arrives at a positive answer,
that a semi-commercial operation should be started. An economic investigation can answer the following questions:

(a) Assuming that the chances of technical success and of marketing success
     have been investigated and found to be positive, is it economically
     feasible to set up a pilot plant for production of sterile males and a
     semi-commercial operation for testing the method?

(b) What should be the maximum scope of this operation?

Considerations involved in finding answers to these questions can be
exemplified as follows: Let us assume that the probability of success of
the method, as far as we can judge without the pilot plant, amounts to 0.3,
while after successfully operating the pilot plant, the probability of full
commercial success will be 0.8. Technical failure would involve losses
resulting from the destruction of infested fruit. Such losses will be proportionate to the size of the area treated by the new method. Assuming
that in the case of failure up to 10% of the fruit would be attacked and
destroyed, then the damage resulting from full use of the new method
would amount to $3 million. With a pilot plant, the damage would be
proportionate to the size of the experimental area. Success of the method
would mean an additional $3-4 million at present value. Under such
circumstances it can be seen that there would be no sense in employing
the new method at once (i.e. without a pilot plant). In this case the
expectation of benefit would then be negative. A pilot plant would provide
additional information at less risk and therefore the operation in stages
(i.e. first pilot plant) might have a positive expectation of benefit. An
exact analysis detailed in Table III shows that the expectation of benefit
decreases as the scope of the experiment increases, indicating the maximum
scope of a pilot plant which would still have a positive expectation of benefit.
This analysis is independent of the minimum scope determined in the technical programme of the experiment and thus only a comparison of the two
results provides an answer to the feasibility of the pilot plant.

If we substitute estimated values in the equation at the bottom of Table III,
the data presented in Table IV are obtained. Table IV indicates that the
figure for expected benefit is positive, but decreases as the value of K
increases:

$$K = \frac{P_1 \left[P_2 B - (1 - P_2) C_1 \right]}{(1 - P_1)(1 + r)^t C_1}$$

Applying the values assumed, we find the value of K on the two assumptions concerning expected benefit:

B_3 for $3 million present value of benefit

B_4 for $4 million present value of benefit

$$K_{B_3} = \frac{0.3 \times 0.772 \ (0.8 \times 3 - 0.2 \times 3)}{0.7 \times 3} = 0.2$$

$$K_{B_4} = \frac{0.3 \times 0.772 \ (0.8 \times 4 - 0.2 \times 3)}{0.7 \times 3} = 0.3$$

We have thus found that K is located between 0.2 and 0.3, that is to say, that the maximum scope of a pilot plant should be of a magnitude of 3.0-4.5 thousand hectares net or 6.0-9.0 thousand hectares gross, based on the assumptions detailed above with regard to the chances of success or failure, benefit expected and possible damage.

BIBLIOGRAPHY

AGROTECHNICAL DIVISION, Application of the Sterile-Male Technique for Control of Mediterranean Fruit Flies in Israel under Field Conditions, Progress report for the year ending 1973, Biological Control Institute, Agrotechnical Division C. M. B. Israel.

GARDNER, N. K., The appraisal and control of complex development projects, Research Policy 1 2 (April 1972).

US DEPARTMENT OF AGRICULTURE, Economic Survey of the Mediterranean Fruit Fly in Central America, Economic Research Service, US Department of Agriculture (July 1972).

UNITED NATIONS, Fruit Fly Eradication, Report of the evaluation mission, United Nations Development Program (May 1973).

DISCUSSION

F.R. BRADBURY: Will the pilot plant be used to test the market for pesticide-free fruit costing 15% more than ordinary fruit?
Y. SPHARIM: Yes.
F.R. BRADBURY: In your cost analysis I could see no entry for repayment and interest on capital. This is a common omission in SIRM cost calculations but with risk capital at 25% interest the cost of servicing capital can be a very substantial item nowadays.
G.E. CAVIN: Are your cost calculations on the use of the sterile technique based on area-wide control or individual farms?
Y. SPHARIM: They are based on an area of 10 000 ha which is considered to be the optimum scale.
G.E. CAVIN: The cost of $30 per hectare seems high in view of the fact that the cost of rearing medflies in large numbers is now down to about US $10 per million and seems to suggest a high overflooding ratio.
Y. SPHARIM: In Israel the citrus strip is only about 7 km wide and we have to maintain a barrier of about the same width, so the cost per hectare of citrus-growing area is roughly twice the figure calculated on the basis of the total area.

COMPUTER MODELS AND APPLICATIONS OF THE STERILE-MALE TECHNIQUE
(Session 3)

MATHEMATICAL MODELS AND COMPUTER SIMULATION OF INSECT CONTROL*

T.P. BOGYO
Statistical Services,
Washington State University,
Pullman, Washington,
United States of America

Abstract

MATHEMATICAL MODELS AND COMPUTER SIMULATION OF INSECT CONTROL.
Various types of mathematical models are described and illustrated and their use is contrasted with those of computer simulation models used for analysing natural systems, in particular systems of insect control.

INTRODUCTION

The construction of models is central to virtually all investigations in biology or any of the natural sciences. Although the terms "mathematical modeling" and "computer simulation" are often used interchangeably, they are by no means identical terms. It is possible to carry out computer simulation without a mathematical model, although usually some model is involved which does not have to be mathematical. An example of a non-mathematical simulation model is described in the paper of Bogyo and Ting [1]. They used the computer "word" to imitate a chromosome. Various types of simulated chromosomes were created by the manipulation of this computer word through various Monte-Carlo techniques. At the end of the simulation the numbers of different types were counted to obtain an estimate of linkage and selection. In many other simulation problems, however, some mathematical model is used as a background to the simulation process. This is usually necessary for more complex types of biological data where the analysis depends on similarly complex mathematical models, often so complex that their algebraic description and manipulation are only possible for very limited cases. Computer simulation can bridge the gap between the experimentalist dealing with the facts of the real world and the theoretical biologist dealing with extremely complex artificial models.

The confusion (of modeling and simulation) probably arises from the fact that both in modeling and in simulation we can talk about deterministic and stochastic approaches or a mixture of both. The deterministic representation of a real process is based on the assumption that the development of the process is governed by various dynamic laws. Feller [2] shows that in applications abstract mathematical models serve only as tools and that different models can describe the same empirical situations.

DISCUSSION OF DIFFERENT MATHEMATICAL MODELS

A simple example of an exponential deterministic law is the model of cellular development, where $x(t) = x_o e^{bt}$ where s_o (the initial number of cells) for

---

* Washington State University, College of Agriculture Research Center, Pullman, Washington. Project 0161. Supported in part by Research Grant No. GB-35 874 of the National Science Foundation.

a fixed b and t will always produce the same results. In a truly deterministic
model, however complicated the mathematical forms may be, the results will always be the same given the same starting conditions and paths.

Knipling [3], who conceived the original idea of releasing sexually sterile
males into a natural population in order to control or reduce their numbers,
published Table I in his 1966 [4] paper to illustrate the effect of the sterile
male principle. Knipling's model can be expressed as a simple deterministic equation

$$N_{t+1} = N_t rI \tag{1}$$

where N_t is the number of fertile males (and also females) in the population in
generation t, S is the constant number of sterile males released into the population each generation (9 000 000 in Table I), r is the rate of increase in the
population per generation (5 in Knipling's illustration), and I is the inundation
coefficient, $I = N_t/(N_t+S)$. It is evident from the above formula that the population will decrease if the quantity rI is less than one, which is the same thing
as postulating the ratio of overinundation (ratio of steriles to fertiles) to be
greater than (r-1):1. Thus Knipling's 1966 example (Table I) would have worked
to r < 10. In Figure 1, equation (1) was evaluated for r = 5, 7 and 9 for t =
2, ..., 8 assuming N_o = 1 000 000 and S = 9 000 000. His principle is based on

TABLE I.   TREND OF AN INSECT POPULATION SUBJECT TO STERILE
INSECT RELEASES (AFTER Ref. [4])

Generation	Number of insects natural population	Number of sterile insects	Ratio sterile to fertile	Number of progeny
Parent	1 000 000	9 000 000	9 : 1	500 000
F_1	500 000	9 000 000	18 : 1	131 580
F_2	131 580	9 000 000	68 : 1	9 535
F_3	9 535	9 000 000	942 : 1	50
F_4	50	9 000 000	180 000 : 1	0

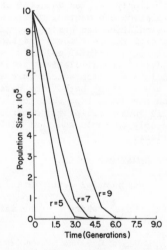

FIG. 1.   Knipling's model with N_0 = 1 000 000, S = 9 000 000 and r = 5, 7 and 9.

becoming smaller and smaller with increasing t. It is also clear from the model that one could simply inundate the population by sterile individuals ($I \to \infty$) and, provided there are no biological differences between steriles and fertiles and that mating between the two types is unlimited, population eradication should be a certainty. However, inundation is economically unfeasible, often technically unachievable, and thus in practice the inundation coefficient will not grow as fast as one would like it. Also r (the rate of increase per generation) is by no means constant in animal species and least of all is it constant in insects. Knipling's model and its variants [5] suffer from treating populations as if generations would not be overlapping. By using deterministic differential equation models, one could overcome this difficulty to some extent, although these models usually have different kinds of difficulties, such as not being equally effective at high and low population levels.

Stochastic processes involve randomly determined sequences of observations each of which is considered a sample of one element with a given probability distribution. Let X_t be a random variable for any t given, and let t be elements of the set T. As t goes over elements of T, we have a family $\{X_t\}$ of random variables. A family $\{X_t\}$, $t\epsilon T$, of random variables indexed by elements of T is called a stochastic process. t is usually referred to as "time" in practical problems. As an example a simple stochastic "Malthusian" birth-death population model would be the following: Let X(t) be the number of fertile males (or females) alive at time t. For every time element t X(t) is a random variable which may be zero or a positive integer. Let K be the carrying capacity of the particular environment in which the insects live. A constant number (S) sterile males are present at any time. Letting the existing population size by N and F the fertility ratio

$$F = N/(N + S)$$

The birth rate b, the death rate d, then

$$P \{ X(t + \Delta t) = N + 1 \mid X(t) = N\} = \Theta_N \Delta t + o(\Delta t) \qquad (2)$$

where

$$\Theta_N = \begin{cases} NbF & \text{if } N < K \\ 0 & \text{if } N \geq K \end{cases}$$

and where

$$\underset{\Delta t \to 0}{\text{limit}} \quad \frac{o(\Delta t)}{\Delta t} = 0$$

Equation (2) simply states that the probability of a unit increase in the population is proportional to the population size, birth rate and the fertility ratio. Similarly,

$$P \{ X (t + \Delta t) = N - 1 \mid X(t) = N \} = \phi_N \Delta t + o(\Delta t) \qquad (3)$$

where

$$\phi_N = dN \qquad \text{for } N = 0, 1, 2, \ldots.$$

or, in words, the probability of a unit decrease is proportional to the death rate and the population size.

In general, a stochastic model is $P(\zeta \leq x) = F(\underline{\Theta}, \underline{X}; x)$ where ζ is a random variable, F is a probability distribution with argument x and vector parameters $\underline{\Theta}$ and $\underline{X}$, with $\underline{\Theta}$ a vector invariant over class K of situations of interest, and $\underline{X}$ a vector which varies over class K [6].

There are many situations which cannot be realistically represented as purely stochastic or purely deterministic processes. "Mixed" or "blended" processes are such that have a basically deterministic structure with a given number of fixed points where random perturbations take place. Biological models often have to include such perturbations to make them more realistic, since such matters as fecundity or the effect of the environment cannot be predicted and fixed parameters would provide unrealistic answers. Instead, the average of a number of runs of a computer simulation model will be a much more reliable estimate, particularly if the distribution of the random perturbations is known or can be reasonably estimated. In many cases comparative investigations show similar results when both stochastic and deterministic models are used.

MODELS IN PEST CONTROL

Several mathematical models of pest control have been proposed recently. Becker [7] developed a model in which the growth of a pest population in a particular habitat is governed by simple birth, death and immigration processes, and the control of the growth is considered over a finite time interval. He developed optimal continuous control of the birth and death rates subject to the cost of control. He also considers the optimal spacing of the control treatment which are applied at discrete time intervals. In Becker's model, the intensity of the controlling action is assumed to be constant throughout the fixed time period. Chatterjee's [8] extension of Becker's model describes the relationship of the frequency, timing and intensity of control on the basic components of birth, death and immigration. Also, in Chatterjee's model the controlling action is assumed to have residual effects. An optimum control program is defined by him as one which minimizes the cost of administering the control program, the loss due to crop damage and also causes minimum harm to the environment. Although his approach is somewhat more practical than that of Becker, both models suffer from oversimplification. This makes the application of these models difficult, particularly when the pest is an insect.

What are, then, the specific difficulties that have to be conquered in a practical insect pest control model? First of all, birth and death processes cannot be easily lumped using some Malthusian parameter. Insect populations have some very specific characteristics, very different from other pests. Most insects have high fecundities subject to extremely low survival probabilities. These survival probabilities depend on a host of other components, such as intra and interspecific competition, predation, parasitism and diseases. Environment affects the pest as well as the host and one has to deal with a complicated feedback mechanism. In many insects the population is measured by the number of adults but the crop is damaged by the immature form of the animal. Different control measures have to be applied for the different stages of the insects' development. The ability of many insects to diapause and thus to overwinter creates a rather unique situation which affects the survival probabilities. One particular problem that has to be considered in modeling of insect population dynamics is spatial heterogeneity. Most mathematical model builders assume that the entire ecosystem with which they are dealing consists of one homologous unit. Wehrhahn [9] discusses this problem in some detail. In motile animals, such as insects, the effect of migration may have to be considered. In a local population of insects one may distinguish between a sedentary nucleus and a group of individuals coming from other demes. This mixture of population maintains crossings and results in an exchange of gametes over a wide territory. Often the birth rate is supplemented through immigration.

It should be noted that the structure of the _biological model_ of an insect population dynamics is usually different from that of a computer simulation model of the same problem. Biological models are mostly built around the life cycle of the species and this is usually based on the history of a single generation. On the other hand, computer models are often book-keeping systems [10] based on time units, usually weeks or days. Thus in a computer model records are kept of the

numbers of the various growth stages of an insect with appropriate mortality
sinks and transition functions. Monro [11] recommends the creation of age groups
with age specific survival and fecundity probabilities. Many model builders [12]
recommend simplifications of the model by lumping a number of transition probabilities into single probabilities on grounds that many of the transitions simply
cancel each other and, therefore, they do not contribute anything to the final
outcome. Thus he considers a "population expansion factor" which is obviously the
end result of a number of stages that one could singly specify. While such an
approach may reduce the cost of simulation, it will not elucidate any of the
problems that may be involved in population expansion. Very often there is no
sound theoretical basis for developing mathematical models to be used for
simulation. In cases like this one would collect data and try to fit a curve to
the empirical data. As an example, in modeling the ecosystem of the codling moth
(<u>Laspeyresia pomonella</u>, L.) the relationship between the proportion of larvae
entering diapause on a particular day and the photoperiod (number of hours of
light) on that day had to be determined. Although experimental data were
available, there was no theoretical basis for the prediction. Using the data, a
curve was fitted (Figure 2) of the form

$$D(t) = C(t)(1 - G(t)/100) \qquad \text{if } 0 \le G(t) \le 100$$

$$= 0 \qquad\qquad\qquad\qquad \text{if } G(t) > 100$$

$$G(t) = \begin{cases} [P(t)/0.5655]^{3.8} & \text{if } P(t) \ge 12.5 \\ 0 & \text{if } P(t) < 12.5 \end{cases}$$

where

    $D(t)$ is the proportion of larvae entering diapause,
    $C(t)$ is the number of cocooning larvae,
    $G(t)$ is the percent of cocooning larvae not entering diapause, and
    $P(t)$ is the length of daylight on day t.

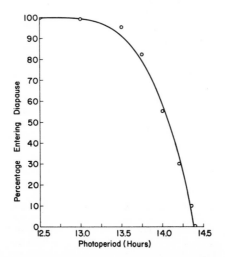

FIG.2.   The relationship between photoperiod in hours and percentage of all larvae entering diapause in
the codling moth. The circles are the values observed in an orchard in the Yakima valley of Washington
and the solid line is the regression curve fitted to the data (see text).

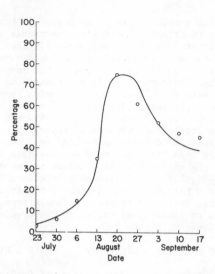

FIG.3. The relationship between days of the year and the percentage of parasitised eggs of codling moth. The circles are the values observed in an unmanaged orchard in the Yakima valley of Washington and the solid line was obtained by fitting a multiple regression curve (see text) using the numbers of eggs laid and the proportion of parasitised eggs as independent variables.

As another example, in Figure 3 the fitted curve to empirical data is shown using a two variate model, where $E_p(t)$ is the percent of parasitised eggs (of codling moth) on the tth day

$$E_p(t) = 1.54 + .325\ E_g(t) + .36\ E_p(t-7)$$

where

$E_g(t)$ is the number of eggs laid on the tth day.

One of the many important problems of model builders is the lack of avail able data on which to build models. The difficulties arise from different sou Many of the available observations come from laboratory studies and in most cases these results cannot readily be applied to field populations. As an ex- ample, the mating probabilities in cages must be very different from those in field. The numbers of eggs laid in captivity are probably very different from numbers obtained in the field. Length of life, the effect of predators, disea the availability of suitable cocooning sites are only some of the observations that cannot be collected in laboratories. In the case of orchard pests even t results of field studies cannot readily be accepted, because they are mostly made in commercial orchards where populations are artificially kept at extreme low levels and where competition for food does not exist, populations are nowh near carrying capacity and thus very little predation occurs and essentially n diseases are present. Certain observations simply cannot be made in commercia orchards; for instance, it is almost impossible to find eggs laid on leaves or fruit when only a microscopic proportion of leaves and fruit have such eggs. These observations can only be obtained from studies in orchards that are un- managed, the life-cycle model has to be based on such unmanaged data and mana ment practices themselves should be part of the model.

Finally some mention should be made of the uses that both the mathematica and the simulation models can be put to.

The great value of mathematical models resides largely in the natural economy and precision of mathematical notation. Construction of a model requires a focusing of attention upon central features of a system and a rejection of anything non-essential or vague. Thus, at least as much benefit often comes from the production of the model as from the finished model itself. If the choice of central features has been made wisely, then the formal model, once it is obtained, may exhibit considerable utility in suggesting additional experiments or measurements and predicting their outcomes. The validity of a model is, of course, conditional on its ability to predict correctly, but even an "unsuccessful" model, i.e., one that leads to erroneous predictions, may be exceedingly fruitful in the sense of leading to interesting experimental approaches or suggesting relationships previously unsuspected. In a certain sense the primary aim of mathematical models is to detect basic relationships between variables of a complex system.

Simulation models on the other hand are mainly written to simulate various natural systems, and in the case of systems analysis they allow us to observe or calculate the outcome of a specific methodology of pest management before it is implemented. In the case of pest management systems Stark [13] and Berryman and Pienaar [14] list four logical steps in the construction and use of simulation models:

1. Specification of simulation objectives. This determines the level of generality or reality the model will achieve, and defines the quality and quantity of data that is needed.

2. Systems definition. The total system is structurally defined by its component parts and their interactions.

3. Model formulation. This step consists of describing system elements and interactions quantitatively, resulting in a series of equations linking together the system components. (After formulation, the simulation model is subjected to validation, which may indicate further data collection, changes in functions, and so on, until it achieves the goals proposed in Step 1.)

4. Systems simulation and manipulation. After valid models have been developed leading to the total system model, this can be used to simulate the "natural" system and to test its behavior and sensitivity to the variety of operational conditions, including management strategies. This analysis of system behavior should eventually enable one to optimize management strategies by insect numbers, socio-economic impact and control costs.

Simulation has its disadvantages. Compared to the functioning of a natural system, it is crude and simplistic. It may also be extremely costly in man-days and computer time. However, it permits analysis of complex ecological processes where other analytical techniques are unavailable or insufficient to the task; its rigorous methodology, if followed, optimizes time and effort and lastly permits direct and complete observation of the dynamic behavior. Both spray-management strategies [15] and sterile-male release strategies could be based on simulation models. An additional advantage could come from understanding the role and importance of each part of the ecosystem which the model describes. Thus one would possibly better appreciate the relationships between the pest and its parasites. A model could, if it existed, help in making a decision in epidemic size outbreak of pest damage, such as in the case of the tussock moth in Washington and Oregon recently. A cost benefit analysis could also result from a simulation model.

REFERENCES

[1] BOGYO, T.P., TING, S.W., Effect of selection and linkage on inbreeding, Aust. J. Biol. Sci. 21 (1968) 45.

[2] FELLER, W., An Introduction to Probability Theory and Its Applications 1, 3rd Edn, John Wiley & Sons, New York (1968).

[3] KNIPLING, E.F., Possibilities of insect control or eradication through the use of sexually sterile males, J. Econ. Entomol. 48 (1955) 459.

[4] KNIPLING, E.F., Some basic principles in insect population suppression, Bull. Entomol. Soc. Am. 12 (1966) 7.

[5] BERRYMAN, A.A., Mathematical description of the sterile male principle, Can. Entomol. 99 (1967) 858.

[6] IOFESCU, M., TAUTU, P., Stochastic Processes and Applications in Biology and Medicine 2, Models, Springer Verlag, Berlin, New York (1973).

[7] BECKER, N.G., Control of a pest population, Biometrics 26 (1970) 365.

[8] CHATTERJEE, S., A mathematical model for pest control, Biometrics 29 (1973) 727.

[9] WEHRHAHN, C.F., "An approach to modelling spatially heterogeneous populations and the simulation of populations subject to sterile insect release programs", Computer Models and Application of the Sterile-Male Technique (Proc. Panel Vienna, 1971), IAEA, Vienna (1973) 45.

[10] MURDIE, G., "Development of a population model of cotton red bollworm as a basis for testing pest control strategies", Computer Models and Application of the Sterile-Male Technique (Proc. Panel Vienna, 1971), IAEA, Vienna (1973) 119.

[11] MONRO, J., "Some applications of computer modelling in population suppression by sterile males", Computer Models and Application of the Sterile-Male Technique (Proc. Panel Vienna, 1971), IAEA, Vienna (1973) 81.

[12] KOJIMA, K.I., "Stochastic models for efficient control of insect populations by sterile-insect release methods", Sterility Principle for Insect Control or Eradication (Proc. Symp. Athens, 1970), IAEA, Vienna (1971) 477.

[13] STARK, R.W., Systems analysis of insect populations, Ann. N.Y. Acad. Sci. 217 (1973) 50.

[14] BERRYMAN, A.A., PIENAAR, L.V., Simulation: A powerful method of investigating the dynamics and management of insect populations, Environ. Entomol. 3 (1974) 199.

[15] GEIER, P.W., HILLMAN, T.J., An analysis of the life system of the codling moth in apple orchards of south-eastern Australia, Proc. Ecol. Soc. Aust. 6 (1971).

DISCUSSION

B. NA' ISA: It seems to me that the computer simulation method of insect control could be used to give a warning of whether there will or will not be an outbreak of insect disease. To be able to do this one would have to feed into the computer known facts, i.e. number of insects at a given time. With some insects (i.e. tsetse flies), however, especially under field conditions, there is no known method of accurately establishing the exact number of the flies at any given time. In such a case, what type of information should one feed into the computer, in order to find out whether there will be an outbreak, for example, of sleeping sickness in an endemic area?

T.P. BOGYO: I think one could develop a method of estimating populations with sufficient accuracy, given enough time and money.

F.R. BRADBURY: I was interested in your reference to the use of mathematical models as an aid to cost-benefit analysis. Would you care to elaborate a little on how the model might help and, in particular, how it could be used to deal with the difficult "intangible" factors of social values, for which input figures are hard to obtain.

T.P. BOGYO: I was referring to the fact that by taking the various parts of the system separately one could more easily compute a cost for the whole system. I have no answer regarding the "intangible" factors, although I do have some ideas.

STUDIES OF POPULATION STRUCTURE AND DYNAMICS OF THE CODLING MOTH FOR THE DEVELOPMENT OF A MODEL OF A STERILE-MALE RELEASE STRATEGY*

R.R. SLUSS, D.N. FERRO, T.P. BOGYO
Statistical Services and Department
  of Entomology,
Washington State University,
Pullman, Washington,
United States of America

Abstract

STUDIES OF POPULATION STRUCTURE AND DYNAMICS OF THE CODLING MOTH FOR THE DEVELOPMENT OF A MODEL OF A STERILE-MALE RELEASE STRATEGY.

The population dynamics of Laspeyresia pomonella L. (the apple codling moth) from an unmanaged, isolated apple orchard and a commercial apple orchard, both located in the Yakima Valley of Washington State, have been compared and considerable differences between the two types of populations were found. In the unmanaged orchard over 90% of individuals are univoltine and only about 10% are multivoltine, whereas in the commercial orchard the situation is the opposite (10% univoltine and 90% multivoltine). Possible causes of this phenomenon are discussed.

INTRODUCTION

The availability of a population simulation model with which various pest control strategies and outcomes can be analysed in advance of field application is needed to enhance the sterile insect release strategy. Accordingly, with the aid of a grant from the National Science Foundation, we undertook to develop such a model. Our studies were conducted on the codling moth because there is an abundance of literature available, there is an active sterile-male release programme in Washington State, and we had the co-operation and assistance of the codling moth research personnel of the Yakima Agricultural Research Laboratory of the United States Department of Agriculture.

Since the strategy of releasing sterilized insects for control involves the removal of insecticide applications, the target insect population will tend towards a situation found in unmanaged ecosystems. Accordingly, we selected an isolated, abandoned and non-irrigated orchard in the Umtanum Canyon between Yakima and Ellensburg, Washington, for our field studies. This orchard consists of about 25 apple trees, most of which are the Jonathan variety with a few trees of the McIntosh variety. The orchard has been unmanaged for approximately 30 years, allowing ample time for stabilization of the unmanaged conditions.

* Washington State University, College of Agriculture Research Center, Pullman, Washington. Project 0161. Supported in part by Research Grant No. GB-35874 of the National Science Foundation.

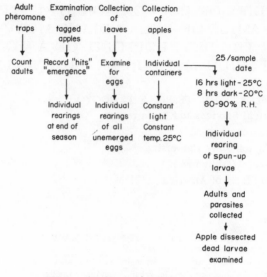

FIG.1.   Sampling and handling procedures.

Unlike many areas in the eastern United States where apple trees mature
readily in the absence of irrigation, the orchards in the Yakima Valley
require irrigation to produce normal fruit.   The abandoned orchard in the
Umtanum Canyon does not receive any irrigation and is completely dependent
on natural precipitation.   During the summer of 1973, little rainfall occurred
hampering normal development with a reduction in the size of the apples.

METHODS

Figure 1 summarizes the routine sampling and procedures used in the
abandoned orchard.   In addition, several large apple and leaf samples were
collected for rearings to assess egg and larval parasitism, and the proportion of the population in diapause.   Also, towards the end of the season, five
trees were banded with cardboard bands to be collected at the next sampling
date, and the larvae placed in a cabinet with a 16-hours-light − 25°C, alternating
with 8 hours of dark − 20°C.   These rearing conditions should permit pupation
if diapause had not been previously induced.
Adults were collected each week from the ten pheromone traps placed in
the study orchard, fresh pheromone was added monthly and the traps were
replaced as deemed necessary.
One hundred small apples were tagged and numbered early in the season
and these were examined during each visit to the orchard.
Leaves and apples were collected on each visit and returned to the
laboratory where they were examined for the presence of eggs.   All non-
hatched eggs were individually reared.   After examination, each apple was
placed in a one-half pint container covered with a fine mesh nylon material,

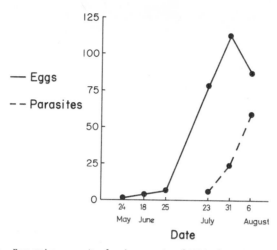

FIG.2.   Eggs and egg parasites found on samples of 100 leaflets plus 100 apples.

and reared in a bioclimatic cabinet programmed for 16 hours light with 25°C, alternating with 8 hours dark with 20°C. All spun up larvae were removed and placed in small petri dishes and returned to the bioclimatic cabinet. After a period of thirty days from the sampling date, the apples were dis- sected and examined for larvae and dead larvae were examined for the cause of death.

RESULTS

Examination of leaves and apples revealed that eggs were very sparse in the field until mid July. From mid July to early August, eggs were found on nearly every leaf and apple. The majority of eggs are deposited on leaf- lets rather than on apples. It is known that females prefer to oviposit on smooth waxy surfaces and the fact that more eggs are laid on leaflets than on apples could be due to random egg laying.

A high degree of egg mortality was attributed to the parasite Trichogramma minutum Riley, and to a lesser degree to some unknown factors. In cases where no evidence of disease could be found, some of this pre-hatch death may have been caused by predation by a piercing-sucking type predator. About 10% of the eggs failed to hatch. This mortality factor seems to be independent of population density. On the other hand, the T. minutum para- sitism indicated a density-dependent response to the egg population level. The percentage parasitism rose with the density of eggs from 7.7% on 23 July to 41% on 6 August (Fig. 2).

Table I summarizes the results of the rearing of larvae in the individual apple containers. These larvae were reared in a bioclimatic cabinet under 16 hours of light per 24-hour period, yet only about 4% of them pupated. It appears that over 90% of the first generation resulting from the overwintering moths entered diapause. The larval mortality due to the two species of

TABLE I.   SEASONAL HISTORY OF LARVAE REARED FROM
25 APPLES SAMPLED AT DIFFERENT TIMES WITHIN THE
APPLE-GROWING SEASON

Date	Emerged			Dead larvae	Mortality (%)	Total No. of hits
	Larva	Adult	Parasites			
18/6	3	0	1^a	0	25	4
2/7	7	0	7^a	7	66	34
16/7	21	0	2^b	2	16	42
23/7	28	1	1^b	13	33	65
6/8	17	2	0	7	29	69
20/8	33	0	1^b	6	18	69
6/9	24	0	1^b	7	25	81
Totals	133	3	13	42	Mean = 30	

[a] Goniozus sp. (Bethylidae).
[b] Ascogaster quadridentata (Branconidae).

TABLE II.   NUMBER OF HITS PER TAGGED APPLE AT
DIFFERENT SAMPLING TIMES

Date	Number of apples			
	1 hit	2 hits	3 hits	4 hits
25/6	4			
2/7	6			
9/7	15			
16/7	38	6	3	
23/7	27	11	10	
6/8	15	5	20	9

parasites and to unidentified causes averaged 30% over the season.   Neither
the parasitic nor other larval mortality factors showed population density
dependence.

Table II summarizes the "hit" data from the tagged apples in the field.
There is a steady increase in the frequency of multiple hits throughout the
season.   Ferro and Harwood [1] have shown that under laboratory conditions
an apple of approximately 4 cm in diameter will carry an average of 2.6 larvae.
They also demonstrated that the more larvae occupying an apple the slower
their development will be.   By 23 July, nearly one half of the apples had
multiple hits, which began to limit the number of larvae successfully completing
development as well as increasing the time of larval development of those
which are able to mature.

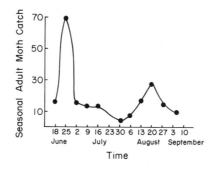

FIG.3.   Seasonal adult moth catch in the unmanaged orchard.

DISCUSSION

Codling moth populations in commercial orchards of Washington are known to consist of two types with respect to diapause behaviour: (1) A multi-voltine type which produces two generations (or more) yearly with most of the second brood larvae entering diapause in mid August in response to photo-period and temperature; and (2) A univoltine type which produces one generation yearly with the first brood larvae entering diapause upon their emergence from the host apples in response to apparently internal mechan-isms. Typical managed orchards in the Yakima area host codling moth populations which consist of approximately 90% multivoltine type individuals and approximately 10% univoltine individuals. The primary mortality action in managed orchards is that of insecticides which maintain codling moth populations of both types at a very low level. Very little secondary mortality results from the action of parasites, predators and disease organisms.

Codling moth population in the unmanaged (abandoned) orchard consisted of approximately 5% multivoltine type individuals and approximately 95% univoltine types (Figs 3 and 4). The primary mortality action in the unmanaged orchard was the result of larval competition for food and space. Significant secondary mortality resulted from the action of parasites, predators and disease organisms.

Normal conditions in the unmanaged orchard exert strong selective pressure favouring the univoltine type of moths. The principal force in this selection is the utilization of available food and space by the first brood which, of course, consists of both types that have successfully overwintered. By mid August when larvae of the multivoltine type are searching for food, there is a scarcity of it. Additional selective pressure favouring the univoltine type moths is exerted by the action of the egg parasite Trichogramma minutum Riley. The population level of this parasite increases directly in proportion to the density of moth eggs. Hence, by early August, when multivoltine type moths are ovipositing, there is a higher level of egg parasitism than that which occur on the eggs deposited earlier by the univoltine type moths.

The actions of larval parasites, predators and diseases remains constant throughout the larval period and accounts for a mortality of about 30%. Since this mortality tends to increase the availability of food by preventing some

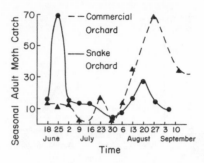

FIG.4.   Comparison of moth activity in unmanaged orchard with that in managed orchard.

apples from being exploited by the moths, it tends to favour the multivoltine type moths. The action probably dampens the selection favouring the univoltine type moths, and hence tends to preserve a limited number of the multivoltine type component of the population.

We have not assessed overwintering mortality, but we realize the importance of doing so at the earliest possible moment. It is likely that a mild winter causing little mortality would favour the univoltine type moths because there would be a large number of first brood larvae utilizing availabl apples. On the other hand, a severe winter causing a great deal of mortality would probably favour the multivoltine type moths, because the resulting small number of first brood larvae would reduce the exploitation of available apples.

The inheritance of photoperiodic reaction has been studied in a number of insects, but not in the codling moth. It seems to be a well established fact that in most species there are local races that can be distinguished from one another by the number of hours of daylight wich induces diapause. In general it seems that in locations further away from the equator, diapause is induced by longer daylength than in races which live nearer to the equator. It seems that in certain species there is more intra-racial variation with respect to thi character than in others. Danilevskii [2] shows that within the same race the exposure of the mite Tetranychus urticae to different lengths of constant temperatures will change the insects' response to photoperiod. Thus, beside photoperiod, temperatures will also influence the onset of diapause. On the other hand, the break of diapause seems to be more dependent on temperature than on photoperiod. Intraspecific crosses between long day and short day strains of Acronycta rumicis (and many other insects) show intermediate response in the hybrids with a slight maternal effect detectable from reciproca crosses. In some other species very little maternal effect is shown (Spilo-soma menthrastri). Distinct heterosis was observed in Acronycta rumicis in the first generation hybrids and this was expressed in increased pupa weight. The intermediate reaction of the hybrids to photoperiod was, more-over, maintained even in the second generation with individual variation being approximately the same as in the first generation and thus filial segregation does not appear to any considerable extent. Consequently, it is very unlikely that photoperiodic reaction would be governed by a small number of genes.

It is the intention of the authors of this paper to study the genetic mechanism of these phenomena in the codling moth.

It seems to be a fact that in commercial orchards around Yakima there is an equilibrium with respect to voltinism with approximately 10% of codling moths being univoltine and about 90% multivoltine. There are a number of genetic conditions which could explain such an equilibrium situation. A heterotic effect of first generation hybrids could greatly increase the fitness of the hybrid genotype, and a strong selection favouring one of the types together with an increased fitness of the hyterozygote could explain an equilibrium situation.

There is very little doubt that in any orchard there are a number of niches where the selection pressures for the different types are dissimilar. Such conditions are also favourable for the creation of a genetic equilibrium.

Selection pressures are not constant from year to year. Differential environmental conditions coupled with varying management techniques in conjunction with varying amounts of food resources from year to year could contribute to such changes, and in turn would maintain enough univoltine types in the population to allow polymorphism.

When the European corn borer (Pyrausta nubilalis) was transported into America about 1920, 99% of the population found around the Great Lakes was univoltine. Within only 30 years the proportion of bivoltine types rose to about 80% [3]. The conditions that favoured this selection in the corn borer are similar to those that favour selection in the same direction in the codling moth — plentiful food supply. Lack of food resources, as experienced in the abandoned orchard, favour the selection of the univoltine type.

We plan to use our simulation programme which we are in the process of developing to study this genetic phenomenon which is so important to the successful prediction of population sizes for sterile-male release programmes. Simultaneously, we also propose to carry out genetic experiments through single pair crossings and isozyme studies with electrophoretic techniques.

REFERENCES

[1] FERRO, D.N., HARWOOD, R.F., Intraspecific larval competition by the codling moth, Laspeyresia pomonella, Environ.Entomol. 2 (1973) 783.
[2] DANILEVSKII, A.S., Photoperiodism and Seasonal Development of Insects, Oliver and Boyd, London (1965).
[3] WRESSELL, H.B., Increase of the multivoltine strain of the European corn borer Pyrausta nubilalis Hb. in south-western Ontario, Annu.Rep.Entomol.Soc., Ont. 83 (1952) 43.

DISCUSSION

E. BOLLER (Chairman): How many years would it take to produce a reasonable predictive model for the codling moth and how many people would be required?

T.P. BOGYO: This depends on how much information you already have available. We were lucky because we had the full co-operation of the USDA Entomological Laboratory in Yakima, Washington. One would need to have the help of scientists in many different fields — entomologists, geneticists, pathologists, etc.

E. BOLLER: How many people participated in your projects?

T. P. BOGYO: There were nine altogether, consisting of an ecologist, three entomologists, a geneticist-statistician, two computer programmers and two laboratory assistants.

K. SYED: You have classed larval competition as the major mortality factor and the action of Trichogramma minutum as only a secondary factor. If you look at this from the practical point of view of a farmer, your primary factor may not have as much significance, because the damage is already done. On this basis would you not consider T. minutum to be the primary factor, especially since, as you have shown in your graph, the action of T. minutum is density-dependent?

T. P. BOGYO: In commercial orchards neither food scarcity nor parasites are important. In the abandoned orchard there is no farmer. Under these conditions competition is the primary factor, whereas diseases and parasites are secondary. In any case Trichogramma parasitizes the egg and not the larva.

L. C. MADUBUNYI: What exactly regulates the codling moth population in the abandoned orchard?

T. P. BOGYO: Availability of food and incidence of disease and parasitism

CONTROL OF FRUIT FLIES
BY THE STERILE-MALE TECHNIQUE
(Session 4)

STERILIZATION AND ITS INFLUENCE ON THE QUALITY OF THE EUROPEAN CHERRY FRUIT FLY, Rhagoletis cerasi L.

E. F. BOLLER, U. REMUND, J. ZEHNDER
Swiss Federal Research Station for
    Arboriculture, Viticulture and
    Horticulture
Wädenswil,
Switzerland

Abstract

STERILIZATION AND ITS INFLUENCE ON THE QUALITY OF THE EUROPEAN CHERRY FRUIT FLY, Rhagoletis cerasi L.

Field releases of sterile cherry fruit flies aiming at the eradication of this important pest in specific target areas were begun in Switzerland in 1972. Experimental data on sterilization and quality control procedures are presented. Unlike other Tephritid species, Rhagoletis cerasi males are more sensitive to gamma irradiation than females when sterilized in the adult stage. Pupal diapause leads to a heterogeneous development of the pharate flies, and makes irradiation at defined developmental stages of the pupae impossible. Pupal irradiation leads therefore to increased somatic damage in the fly population and has stimulated investigations on adult irradiation. The influence of irradiation on important components of the flies' quality, such as longevity, flight characteristics, capability to locate host and sexual partner, mating activity and successful transfer of viable sperm in repeated matings has been investigated. No apparent detrimental effects on these parameters could be detected up to irradiation doses of 10 krad, the dose used in the present SIT programme in northwest Switzerland.

1. INTRODUCTION

Since the principles of eradicating the cherry fruit fly by means of the sterile-insect technique in specific target areas were first described [1, 2], substantial progress has been made in cherry fruit fly research in Europe leading to the first field releases in Switzerland in 1972. Experiments are now in progress in three countries, and the techniques used have been described [3, 4]. This report covers the sterilization and quality control procedures upon which the field programme in Switzerland is based.

The optimal stage for sterilizing Rhagoletis cerasi by gamma irradiation is the early adult stage because the pupal diapause and the resulting prolonged emergence period of adult flies makes an irradiation of a defined pupal developmental stage impossible [5]. Furthermore we have to aim at a certain residual fecundity of the sterile females in order to maintain the typical oviposition behaviour and thus to force the females to arrive on the cherries as rendezvous of the sexes [4].

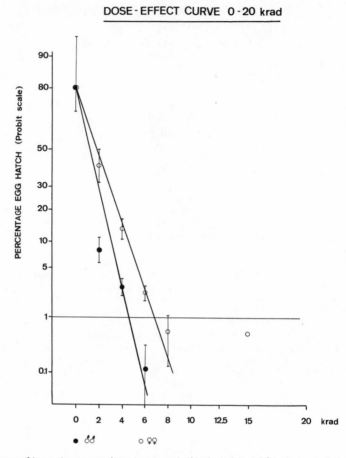

FIG.1.   Influence of increasing gamma dose given to male (black circles) and female (open circles) Rhagoletis cerasi on egg hatch (probit scale).

2.   EFFECT OF GAMMA IRRADIATION ON FERTILITY AND FECUNDITY

Dose-effect curves for fertility and fecundity have been established for adults irradiated 24, 48 and 72 hours after emergence. As no significant differences in the measured parameters could be observed between these three age groups, the presentation of our data is limited to flies irradiated 2 days after emergence — the procedure used in the present programme.

2.1.   Methods

Sterilization is carried out in a SULZER-pool gamma irradiator where eight ^{60}Co bars in a ring configuration are elevated from a 5-m deep protective water pool into the operational position. A dose rate of 300 krad/h

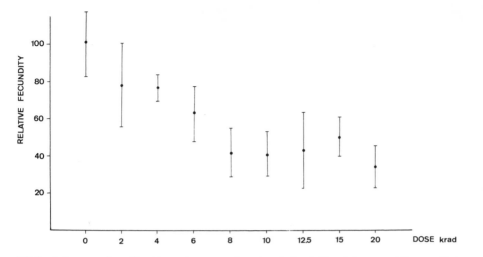

FIG.2.   Influence on fecundity of increasing gamma dose given to female Rhagoletis cerasi at the age of 2 days.

is obtained by observing the appropriate distance between the bars and the containers with the flies. Variation of the doses applied was ± 5%.

All experiments were carried out with Rhagoletis material collected in the field. Diapausing pupae were stored at 2°C for six months and then incubated at 25°C. Freshly emerged flies were sexed daily and held in the rearing room at 25°C, 65 - 70% r.h., 18 h photoperiod, and given a carbohydrate diet (99.9% sucrose, 0.1% dysprosium chloride as marking substance). After irradiation the flies were fed the standard yeast-hydrolisate-sucrose diet (1:4) and provided with oviposition devices [6,7]. Under these conditions oviposition started on the 5th day after emergence.

2.1.1. Irradiation procedures

Males and females of the given age were transferred in their holding cages to a cold room (0°C), chilled and put into 20 ml glass bottles with plastic snap-cover lined with filter paper to absorb condensation water. The bottles were then transferred to the irradiator in an ice-box, exposed at room temperature to the gamma rays and transported back to the laboratory in the cooler-box. Exposure time varied, according to the dose applied, between 12 sec and 4 min.

2.1.2. Experimental design

Dose-effect curves were established for males and females with doses of 0 - 2 - 4 - 6 - 8 - 10 - 12.5 - 15 and 20 krad. The experimental units consisted of 10 males and 10 females whereby the treated sex was combined with the untreated opposite. Five replicates were used for each treatment, and daily records were taken of the number of eggs laid, fertility rate and mortality for 21 days.

2.2. Results

The dose-effect curves with respect to fertility of the eggs are given in Fig. 1; those with regard to female fecundity in Fig. 2.

2.3. Discussion

The data presented indicate that males are more sensitive to gamma irradiation than females as the residual fertility is decreased below the 1% level at 6 krad for males and 8 krad for females. No fertile eggs were found at 10 and 12.5 krad, but the appearance of a low residual fertility at 15 krad (mainly due to the larger number of eggs examined) indicates that complete sterility will probably be reached only at doses beyond our test range. However, we considered a dose of 10 krad as safe enough in our field releases.

The fecundity of the females as affected by irradiation, on the other hand, follows a different pattern than fertility. Fecundity decreases constantly with increasing dose up to 8 krad where it levels off at about 40% of the reproductive potential of untreated females. No significant differences in fecundity were observed when females were irradiated 1 day or 3 days after emergence with 10 krad, although the youngest females showed a tendency to lay fewer eggs. No investigations were carried out on the radiosensitive or resistent stages of oogenesis. However, a residual fecundity of some 40% yielding some 30 - 40 sterile eggs per female under optimal field conditions is considered as acceptable and even desirable from the behavioural and ecological point of view [4, 8].

3. INFLUENCE OF STERILIZATION ON QUALITY COMPONENTS OF THE FLIES

Quality control procedures have been developed in our laboratory to assess the quality of the insects used in field programmes. The tests available so far can be divided into two main categories. The first group of quality control procedures is geared to the assessment and comparison of quality components in various geographic and laboratory strains of the species, and is an important tool for the decision-making process about the strain to be chosen for achieving the highest affinity with the wild population of a given target area [8]. Once the most suitable strain has been found the second category of quality control procedures is applied, namely test procedures that measure such factors as the impact of sterilization, marking and handling on various important charcteristics of the insects to be released. We limit the presentation of data to this second category of quality control that is relevant in the context of this report.

Although developmental research for quality control procedures is still in its very early stages, and no internationally accepted standards or even concepts have been established so far, several methods are applied in our laboratory for monitoring something that is vaguely called quality. We tried to break quality down into several components as follows: The released insect should have normal longevity, match the characteristics of the wild target population as closely as possible, at least with respect to dispersal, host finding, mate finding, mating activity and capability to

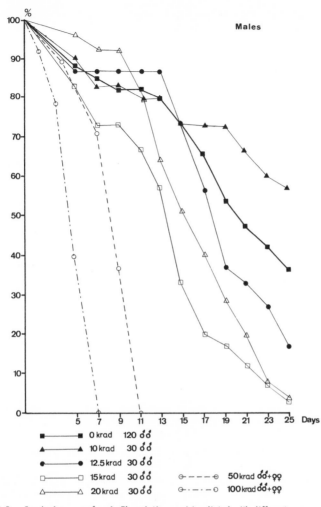

FIG.3.   Survival curves of male <u>Rhagoletis cerasi</u> irradiated with different gamma doses.

transfer viable sperm in that chronological order.   Failure in any given
link of that chain can probably lead to artificially introduced isolating
mechanisms between the two populations, and cause failure of an SIT
programme [8].

<u>3.1.   Influence of irradiation dose on longevity</u>

    The influence of various irradiation doses on the survival of male
<u>Rhagoletis cerasi</u> is demonstrated in Fig.3.   Mortality is increased after 2
weeks in males receiving more than 12.5 krad, whereas 10 krad increases
slightly the longevity in comparison with the untreated check.   From this
point of view there are no objections against the dose of 10 krad used in
our field programme.

BOLLER et al.

TABLE I. INFLUENCE OF IRRADIATION ON ADULTS STERILIZED
2 DAYS AFTER EMERGENCE AND TESTED AT THE AGE OF 6 DAYS

Irradiation dose (krad)	No.	Flight distance in 24 h (m)*	Wing beat frequency (Hz)*
Males			
0	32	1564 ± 262 a	181.6 ± 7.8 a
10	18	1360 ± 320 ab	177.8 ± 7.2 a
20	18	1143 ± 290 ab	175.5 ± 4.5 a
30	18	1124 ± 342 ab	168.2 ± 7.9 ab
40	13	1279 ± 359 ab	157.9 ± 6.3 b
50	21	1008 ± 296 b	159.9 ± 9.1 b
Females			
0	52	2305 ± 291 c	151.9 ± 11.4 bc
10	33	2346 ± 391 c	
20	33	2314 ± 341 c	
30	26	2217 ± 350 c	
40	38	2041 ± 334 c	
50	30	2122 ± 341 c	143.2 ± 6.5 c

* Values followed by the same letter are not significantly different at 5% level.

3.2. Influence of irradiation dose on flight charcteristics

The influence of irradiation on several flight characteristics of sterile
flies is measured with low-friction flight mills [9, and Remund & Boller,
in preparation], and stroboscopes [10] that express flight parameters as
indices of distance flown in a given period of time, flight speed, number
and average length of individual flights as well as wing-beat frequencies.
Table I lists some of these parameters as a function of various irradiation
doses applied. It can be concluded that major somatic damage to the flight
muscles and/or responsible nervous systems occur only at doses far above
the dose of 10 krad used in our SIT programme.

3.3. Capability of finding host and sexual partner

No reliable laboratory test is available so far to measure this important
component of quality. Direct observations in the field in the course of the
releases indicated that the typical behaviour with respect to host finding,
meeting the sexual partner on or near the cherries and to oviposition is
maintained. No abnormal emigration of sterile flies out of the cherry
orchards could be observed.

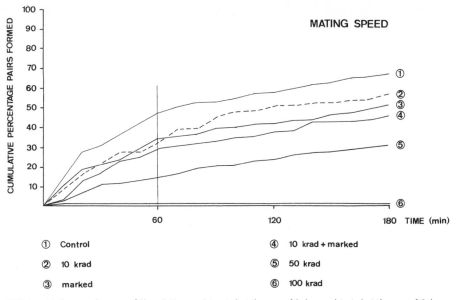

FIG.4.   Mating speed curves of Rhagoletis cerasi treated at the age of 2 days and tested at the age of 6 days (marking with KRYLON paint spray).

TABLE II.   INFLUENCE OF IRRADIATION ON COMPETITIVENESS OF MALES IRRADIATED 2 DAYS AFTER EMERGENCE (10 STERILE MALES: 10 UNTREATED MALES: 10 UNTREATED FEMALES)

Irradiation dose (krad)	0	10	12.5	15	20
No. of eggs examined	2967	641	674	750	872
Observed hatch-rate (%)[a]	72.5 ± 5.5	37.7 ± 1.1	62.1 ± 2.3	61.9 ± 7.5	61.7 ± 8.3
Corrected hatch-rate (%)	100	52.1	85.9	85.6	85.4
Expected hatch-rate (%)	100	50	50	50	50
Competitiveness[b]	1.00	0.95	0.17	0.17	0.17

[a]   Hatch-rates expressed as averages of 5 replicates and S.E. at 5% level.
[b]   Competitiveness calculated by formula of Haisch [5, 11].

3.4.   Influence of irradiation on mating activity

Mating characteristics were measured in the laboratory and expressed as indices for mating speed and mating frequency (Boller, Remund & Katsoyannos, in preparation).   The experimental unit for measuring mating speed consisted of 50 virgin males and 50 virgin females, all 5 days old, that were mixed together in darkness in a plexiglass cage ($20 \times 30 \times 30$ cm).

TABLE III. INFLUENCE OF IRRADIATION (10 krad) OF 2-DAY-OLD
MALES ON THEIR CAPABILITY TO TRANSFER SPERM IN
CONSECUTIVE MATINGS

Chronological number of mating	Number of males in test	Number of eggs laid by same no. of females	Fertility (%)
I	120	2197	19.3
II	93	1444	5.9
III	62	1068	7.7
IV	50	1479	17.8
V	37	1223	14.4
VI	31	1509	2.6
VII	22	1145	9.7
VIII	5	173	0
Check	60	1608	83.4

After 2 hours of complete darkness the experiment was begun by switching
on the light (1500 lx). Formed pairs were removed and counted at 10-min
intervals during 3 hours. The cumulated percentages of formed pairs as
a function of time elapsed are presented as mating-speed curves in Fig. 4.
Every curve is the average of at least three replicates.

 If we consider the flies that copulate within the first 60 min of the test
as fast-mating flies we have to conclude that this portion of the test population is decreased by some 30% in flies receiving 10 krad (curve 2), by
80% in flies receiving 50 krad (curve 5) and by 94% in flies receiving 100 krad
(curve 6) in comparison with the untreated check (curve 1). This loss is
compensated to a certain degree with increasing time lapse, and reaches
after 3 hours 15%, 50% and 96% respectively. As there is a close relationship
between mating speed and mating frequency (as expressed as number of
matings per male during a given period of time), there is a general decrease
of mating activity with decreasing portion of fast-mating flies in a given fly
population (Boller et al., in preparation).

 This observed loss of mating activity of flies sterilized with 10 krad
has apparently no major effect on their competitiveness, as discussed below.
However, the negative influence of the marking procedure (curve 3), and the
apparently additive effect of both sterilization and marking (curve 4), is
under investigation. Whereas no evident problems occur in the field
programme where the overflooding ratio exceeded always 20:1, we concentrate our present attention on the modification of all possible factors that
show a negative influence on mating activity.

3.5. Influence of irradiation on competitiveness of males

 The competitiveness of irradiated males was measured also in standard
competition tests widely used for measuring "vigour" (sterile males:
untreated males: untreated females = 1:1:1).

Some of our data are presented in Table II.

These data indicate no significant loss of competitiveness in males receiving 10 krad, whereas the "vigour" was significantly reduced at higher doses.

3.6. Influence of irradiation on sperm transfer

Sperm transfer can be examined either by dissection and inspection of the spermathecae, by autoradiographic methods or by examination of the fertility rate of the eggs produced by the female after mating. We chose the third approach as we had observed that in two consecutive matings with an untreated and a sterile male the fertility of the second male was always expressed in the hatching rate of the first batches of eggs laid. Therefore untreated virgin females were mated first with an untreated male and the following day with an irradiated male. The sterile males were kept after each copulation individually and mated every day with a fresh female until the sperm supply was exhausted.

Females were transferred immediately after copulation to standard oviposition cages that each received 10 females that had mated with the same type of males, i.e. with males that had performed the xth mating. As methods were not available at that time to handle and egg individual pairs that would have given more reliable information, the calculated hatching rate of the eggs of every individual cage was an average value of 10 females. Investigations in progress in another context, and using mini-cages for individual pairs, have shown that not all observed matings in untreated flies lead necessarily to sperm transfer but this phenomenon applies to some 5 - 10% of the cases (Boller, unpublished). Eggs were collected during a period of 10 days.

Table III gives the average fertility observed in female groups that had mated with sterile males performing their copulation No.I to VIII.

The data indicate that sterile males irradiated at the age of 2 days are capable of transferring viable sperm in up to 8 matings — the highest number of copulations included in our experiments; it will probably never be realized under field conditions. Sperm precedence as indicated in Rhagoletis cerasi is considered as a positive attribute in our SIT programme.

4.   CONCLUSIONS

The experimental data available at present indicate that Rhagoletis cerasi can be irradiated 2 days after emergence with 10 krad to produce a sufficiently high level of sterility and residual fecundity without evident detrimental effects on important components of the flies' quality.

ACKNOWLEDGEMENTS

We gratefully acknowledge the assistance and contributions of V. Katsoyannos, and J. Derron, diploma student of the Swiss Federal Institute of Technology in Zurich.

REFERENCES

[1] BOLLER, E., Neue Gesichtspunkte in der Kirschenfliegenbekämpfung, Die Grüne 97 (1969) 759.
[2] BOLLER, E.F., HAISCH, A., RUSS, K., VALLO, V., Reports of the OILB working group on genetic
      control of the European cherry fruit fly, Rhagoletis cerasi L.: I. Economic importance of the pest,
      the feasibility of genetic control and resulting research problems, Entomophaga 15 (1970) 305.
[3] BOLLER, E.F., HAISCH, A., PROKOPY, J.R., "Sterile-insect release method against Rhagoletis cerasi L.:
      Preparatory ecological and behavioural studies", Sterility Principle for Insect Control or Eradication
      (Proc. Symp. Athens, 1970), IAEA, Vienna (1971) 77.
[4] BOLLER, E.F., REMUND, U., "The application of the sterile-male technique on the European cherry
      fruit fly, Rhagoletis cerasi L., in northwest Switzerland", Panel on the Sterile-Male Technique for Control
      of Fruit Flies, IAEA, Vienna, November 1973 (unpublished).
[5] HAISCH, A., BOLLER, E.F., "Genetic control of the European cherry fruit fly, Rhagoletis cerasi L.:
      Progress report on rearing and sterilization", Sterility Principle for Insect Control or Eradication (Proc.
      Symp. Athens, 1970), IAEA, Vienna (1971) 67.
[6] PROKOPY, J.R., BOLLER, E.F., Artificial egging system for the European cherry fruit fly, J. Econ.
      Entomol. 63 (1970) 1413.
[7] BOLLER, E.F., RAMSER, E., Die Zucht der Kirschenfliege auf künstlichen Substraten, Schweiz. Z.
      Obst- u. Weinb. 107 (1971) 174.
[8] BOLLER, E.F., Behavioral aspects of mass-rearing of insects, Entomophaga 17 (1972) 9.
[9] CHAMBERS, D.L., O'CONNELL, T.B., A flight mill for studies with the Mexican fruit fly, Ann.
      Entomol. Soc. Am. 62 (1969) 917.
[10] SHARP, J.L., CHAMBERS, D.L., HARAMOTO, F.H., Flight mill and stroboscopic studies of Oriental
      fruit flies and melon flies including observations of Mediterranean fruit flies, J. Econ. Entomol. (in press).
[11] HAISCH, A., "Some observations on decreased vitality of irradiated Mediterranean fruit fly", Sterile-
      Male Technique for Control of Fruit Flies (Proc. Panel Vienna, 1969), IAEA, Vienna (1970) 71.

DISCUSSION

G.W. RAHALKAR: Can you suggest any reason for the increased
longevity of males irradiated with a dose of 10 krad?

E.F. BOLLER: I have no explanation for the increased longevity at
10 krad. However, this phenomenon has been reported for quite a number
of other species and two main theories have been put forward. One is that
the metabolic processes are slowed down by irradiation and the other that
regenerative processes are triggered by irradiation.

G.W. RAHALKAR: What is the dose variation in your irradiator?
I am asking this question because Table II shows that with an increase in
the dose of only 2.5 krad the mating competitiveness of males dropped
from 0.95 to 0.17.

E.F. BOLLER: The dose variation was ± 5%.

G. DELRIO: This is the first time that I have ever heard of a
tephritidae species irradiated at sterilizing doses (10 krad) suffering any
loss of sexual competitivity. What is the explanation for this?

E.F. BOLLER: I cannot offer any explanation and I do not think we
necessarily have to look for an explanation. We just measured these
various parameters and found nothing to worry about except a slight
decrease in mating activity due to irradiation and marking.

H. LEVINSON: What is the physiological basis for the reduced mating
speed of Rhagoletis following irradiation with 10 krad? After all 30% is a
significant decrease. Could it be due to reduced or modified chemoreception,
as found by Rahalkar and co-workers for the ♂ Khapra beetle after irradiation,
and/or to reduced pheromone production?

E.F. BOLLER: The proportion of flies copulating within 60 minutes is reduced by 30% but only by 15% after 3 hours in the mating speed test. We have not yet established the reason for this decrease but we are also looking into the possibility that olfactory effects are involved.

K. SYED: Could you tell us more about the isolating mechanisms referred to in your paper?

E.F. BOLLER: So far we have not detected any factor in our released flies that could lead to sexual isolation in the field but, as potential isolating mechanisms can, at least theoretically, develop in any link of the chain of events between release and successful fertilization of the egg, we strongly believe in quality control procedures that cover all aspects or components of quality.

K. SYED: How do you determine whether the pupae you collect are diapausing?

E.F. BOLLER: The European cherry fruit fly is strictly univoltine with pupal diapause. The proportion of pupae that can develop without diapause in the field-collected material is very low, probably less than 1%. This is easily determined as pupae are usually stored at room temperatures for some four months before they are subjected to cold temperatures.

M.S.H. AHMED: Have you observed any recovery in fertility of males after treating them with doses lower than 6 krad as they become older?

E.F. BOLLER: We checked the egg hatch every day for three weeks and did not observe any recovery of fertility within that period. However, at higher doses (12.5 and 15 krad) we observed on several occasions the appearance of small numbers of fertile eggs on day 7 and 8 after irradiation of the males. But this temporary increase in fertility always disappeared after two days.

J.L. MONTY: Is it possible that the good results obtained in the competitiveness tests are due to the wild material used in the tests, as opposed to the laboratory-reared material used elsewhere?

E.F. BOLLER: It could be that the results would have been different if we had tested laboratory-reared flies. However, the object of the tests was not to detect qualitative differences between strains but to monitor possible negative side-effects of various treatments on a given strain. This standard strain happened to be a wild strain.

B. NA'ISA: Could the fact that the males are more sensitive to gamma radiation than the females be the reason why they have a lower flying capacity?

E.F. BOLLER: I do not know whether the effect of radiation on the reproductive organs and flight muscles can be compared. It seems to me that females simply have more reserves due to their higher flight capacities, and that any damage becomes evident at much higher doses than in males.

B. NA'ISA: How does the cost of applying the sterile technique to the cherry fruit fly compare with the damage caused by this pest?

E.F. BOLLER: The calculations are not yet complete but we hope to have figures available soon.

R. GALUN: I would query your use of the term "quality control" as it leads one to expect data on the effect of mass rearing on the quality of the insect rather than the effect of sterilization procedure.

OBSERVATIONS ON THE FLYING BEHAVIOUR OF THE EUROPEAN CHERRY FRUIT FLY (Rhagoletis cerasi L.)*

A. HAISCH
Bavarian State Institute for Soil and
Plant Cultivation,
Munich,
Federal Republic of Germany

Abstract

OBSERVATIONS ON THE FLYING BEHAVIOUR OF THE EUROPEAN CHERRY FRUIT FLY (Rhagoletis cerasi L.).
An experiment has been conducted to gain information on methodological problems in releasing sterile cherry flies in order to control the population. Well-fed flies have been sterilized with gamma rays (9 kR) during the imaginal stage and released. Others, unirradiated, emerged from pupae placed in the soil. The irradiated flies showed a higher distributive movement than the non-irradiated flies. This flying behaviour was explained by the results of laboratory experiments which indicated that the nutrition and the age of the flies influence the flight disposition.

1. INTRODUCTION

Several techniques of genetic control of pest insects are now known. All are based on the principle of the distribution of genetic damage within a target population by releasing insects. The chance of success depends among others essentially on a quick and homogeneous dispersion of the genetic damage. This dispersion, however, is influenced by numerous ecological factors. Bearing in mind the application of the sterile-male technique to control the cherry fly, it seemed advisable that a start should be made on studying the problem involved in the release technique.[1]

It was considered that 1973 would be a favourable year for the experiment, as in the previous year a late frost had completely killed all blossom on the trees at sites of low and intermediate sea levels. Because of the total loss of fruit it could be assumed that the population of the cherry fly had no chance to propagate. Consequently, the population of the following year 1973 was expected to be low. This circumstance was the reason for carrying out the experiment despite the relatively small number of insects available.

---

* This work was supported by the German Ministry of Research and Technology and the Commission of the European Community.

[1] BOLLER, E.F., HAISCH, A., RUSS, K., VALLO, V., Economic importance of Rhagoletis cerasi L., the feasibility of genetic control and resulting research problems, Enotomophaga 15 (1970) 305.

Insecticides are never applied on these rented experimental fields.

The aim of the experiment was to observe the degree of fruit infestation as a measurement of reproduction after the natural decrease of the population during 1973 under following conditions:

(1)   Release of fertile red-labelled flies at site 1,

(2)   Release of sterile and yellow-labelled flies at site 2,

(3)   No release of flies at the control area in site 3.

2.   METHODS

2.1.   Experimental design

Three experimental areas were available for the experiment. Sites 1 and 2, surrounded by woods, were 60 m apart (Fig.1). Site 3 was on the border of the wood at a distance of 450 and 510 m from sites 1 and 2 respectively. Site 1 was the largest with 41 trees, site 2 had 8 trees, and site 3 had 13 trees.

The population of all three sites has been observed for five years. The conditions for living and propagating in these well-protected experimental areas have apparently been very good. In particular the light sandy and never silted soil might have contributed to the high annual infestation. It was also suggested that the flies do not migrate between the sites because of the forests.

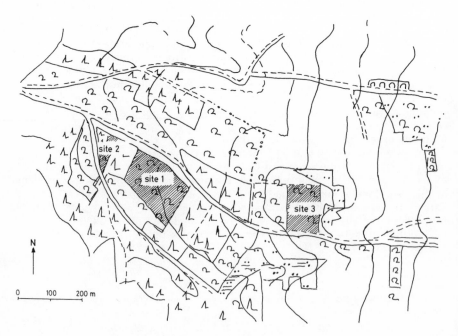

FIG. 1.   Experimental areas.

TABLE I.   TIME TABLE OF THE EXPERIMENT

Time	Site 1	Site 2
April 2	Pupae were placed in the field	Pupae were placed in the field
May 21	Beginning of emergence	Pupae were brought into the laboratory for emergence and the beginning of emergence
June 10		Irradiation and release of 2000 flies; beginning of emergence of the native population
June 18	End of emergence	Irradiation and release of 2900 flies
July 1		End of emergence
July 10	End of catching flies	End of catching flies

2.2.   Material

For the releasing experiment insects were collected from infested fruit coming from sites of higher elevations (500 m). These sites were 10 km to 15 km away from the experimental area. It is not known whether this was a population genetically different from the native population in the experimental area. The collected pupae were stored during the winter in soil in the neighbourhood of the laboratory. On 2 April these pupae were transferred to the experimental field where they were dug in so that the emergence of the wild and collected flies could be synchronized. On 31 May they were removed from the soil and placed in emerging containers (see Table I).

Pupae were placed in silica sand mixed with 5% (wt/wt) dye pigment. The pupae assigned for sterilization were labelled by "saturn yellow" (A-17, Dayglo) and the pupae that would remain fertile were labelled with a "red" pigment (No. 321, Riedel de Haen).

The remaining fertile insects for emergence in the field were divided into 200 pupae to each tree. The other pupae were kept in the laboratory for emerging after 21 May. Here the newly emerged flies received sugar and torula yeast as food. A total of 4900 flies emerged, 2000 of which were irradiated on 10 June and the remainder 8 days later. The sterilization dose was 9 kR of a ^{60}Co apparatus with 2×10^4 R/h. A few hours after irradiation the flies were released at three different points on site 2.

At sites 1 and 2 were 200 pupae at each site in special emergence containers enabling a check to be kept on the course of the emergence.

2.3.   Trapping of flies and evaluation of the results

Released labelled and unlabelled wild flies were caught during the experiment in order to estimate the population size and to discover whether there was emigration.

The flies were caught by visual traps consisting of sticky boards (20 × 20 cm) painted with a fluorescent yellow dye and covered by "Bird tanglefoot". Special care was taken to ensure that the traps were suspended freely on the outer and upper branches of the trees, the southwestern part of the crown being preferred.

The population was estimated by fly catches made on three successive days to prevent mortality and emerging rates from falsifying the results. The recapture period was chosen in such a way that the population might have reached a maximum.

The population size was estimated according to the known formula

$$P = p \frac{n}{m}$$

where P = unknown population
     p = population of the marked released flies
     n = number of unmarked flies caught
     m = number of marked flies caught.

At site 1 where fertile flies have been released the wild population P must have been increased by the released population p of fertile flies.

In order to be able to indicate the standard error of the population estimation the standard error of the ratio between the caught wild and released flies was computed as the standard error of the regression concerned. For this statistical analysis the data (= a) being discrete variates have been transformed into the calculation values ($x = \sqrt{a + 0.5}$). Finally the infestation of the fruit by larvae was established.

3. RESULTS

3.1. Emergence

Emergence of the flies at site 1 began during the last days of May and was finished approximately in the middle of June. The native flies of site 2 emerged during the second half of June, a difference of time of two and one half weeks. The strongly delayed appearance of the flies at site 2 must be explained by its shady conditions caused by the trees of the neighbouring forest. At site 1 this disadvantage does not exist because of the larger area. The time of releasing the sterile flies at site 2 coincided with the emergence of the adults. Subsequent enumeration of emerged pupae in samples showed that at site 1 4750 and at site 2 4920 flies must have emerged.

3.2. Recapture data

The first (red-labelled) flies have been caught at the place of release on 13 June, i.e. at the end of the emergence period. Between the 13 and 15 June 385 red-labelled and 66 unlabelled flies were caught. These figures correspond to a ratio of 1 : 0.17. The confidence limits for this ratio are ± 4% at the 5% level of error probability.

On the basis of these figures the size of the native populations can be computed as follows:

$$P = 4750 \times \frac{66}{385} \pm 4\%$$

$$P \sim 800 \pm 4\%$$

TABLE II. RECAPTURE OF FLIES

Remarks	Site 1	Site 2	Site 3
Flies released	4750	4920	0
Recapture of flies from site 1	594	13	22
Recapture of flies from site 1 per trap	14.5	1.6	2.7
Recapture of flies from site 2	16	12	7
Recapture of flies from site 2 per trap	0.4	1.5	0.5

The native population is much smaller than the released one. Therefore
the total population amounts to $4750 + 800 \sim 5500$.

At site 2, 53 unmarked flies and 12 marked flies were caught (Table II).
The ratio between unmarked and marked flies was therefore $1 : 4.4$ with
a confidence interval of $\pm 2.4\%$. From these figures the following population
can be computed:

$$P = 4920 \quad \frac{53}{12} \pm 2.4\% \sim 22000$$

However, the released flies did not remain at the places of release.
Fertile, red-labelled flies were already caught at a distance of 450 m at
site 3 one day after these flies had been found at the place of release, i.e.
site 1. Site 3 lies in the path of the westerly prevailing wind. At site 2,
in the opposite direction, the flies appeared to move less because 20 days
elapsed before red-labelled flies could be captured at a distance of only
60 m. The movement of sterile flies released at site 2 cannot be explained
fully. These flies were not found before 21 days in the traps of the site
where they were released, although they were discovered after 14 days at
control site 3 510 m distant. It can only be concluded that the flies were
distributed at least within a range of about 500 m.

The fly recapture data confirm this. During the whole flight period
the rate of recapture of the fertile flies at releasing site 1 was 14.5 flies
per trap. At site 3, 2.7 flies per trap were caught, and 1.6 flies per trap
at site 2. In other words, 94% of all flies caught remained at the place of
release and 4% of them moved to site 3 450 m away and 2% to site 2
(Table II). At site 2, where the sterile flies had been released, 12 flies
(34% of all flies caught) were observed at the release site, 16 flies or 46%
of the caught flies moved to site 1 60 m away, and 7 flies or 20% of the
caught flies were found at site 3 510 m away.

A well-marked tendency for the sterile flies to distribute themselves
widely must therefore be noticed. This tendency is not emphasized by the
great difference between the absolute frequencies of the recapture data of
sites 1 and 2 but it is significant as the chi-square test has indicated. The
sterile flies apparently dispersed more extensively than the fertile ones.

3.3. Degree of infestation

At site 2, where the sterile flies had been released, the yield of fruit
from each tree was estimated and multiplied by the infestation rate of two
samples with 100 cherries each from each tree. The weighted mean of
infestation of 18.6% resulted from these figures. The same procedure gave
39% infestation at site 1 and 2.5% at the control site 3.

4. DISCUSSION

A very striking finding is that the infestation rate of site 2, where the sterile flies were released, is far higher than that of control site 3. One of the possible reasons may have been an immigration of mated females in particular from site 1, as 9% of the caught fertile flies moved from site 1 to site 2. Based on 4720 released flies, the total number of flies is 400, or 200 females. The immigration of such a low number of females cannot alone explain the infestation of 18%. Additionally, it must be presumed that a high proportion of the sterile flies left site 2 shortly after having been released. The very low recapture rate of the yellow sterile flies at this site, and certainly also the observation on the dispersion of these flies by means of the recapture data support this assumption. Table III provides additional information on the credibility of the population estimation. It is unlikely that a population as high as 22000 flies is able to infest only 8 kg of fruit whereas 5500 flies (800 wild and 4700 released flies) infested 265 kg of fruit. Thus, the estimate must be incorrect, and is certainly incorrect if the released flies emigrated quickly.

TABLE III. COMPARISON OF INFESTATIONS AT SITE 1 AND 2

Experimental site	Estimated population (flies)	Yield of fruit (kg)	Rate of infestation (%)	Infested fruit (kg)	
				Total	per 1000 ♀♀
1	5500	680	39.0	265	90
2	22000	42	18.6	8	1

It must therefore be concluded that the dispersion of released flies played an important role. There are indications that the dispersion is more intensified in the case of the yellow-labelled irradiated flies than in the case of the fertile flies released. This flight behaviour, if it is correct, is attributed less to the irradiation treatment than to the fact that already emerged and well-fed flies have been released.

In this connection it seems useful to mention two laboratory experiments with the European cherry fruit fly concerning its flying disposition during aging and dependence on satiation or starvation of the flies. A detailed description of the experiment will be published elsewhere, but it may be mentioned that 5 pairs of flies for every experiment were placed in a cage hanging on an electronic balance.[2] In this way it was possible to register every flight of every fly on a recorder as a negative weight peak.

In one experiment lasting three days two populations were compared; each population for twice two hours alternatively was put into the flight registering apparatus. To one of the two populations food was offered for one hour in the evening only, whereas the other was allowed to feed all the time while it was not in the flight registering apparatus.

---

[2] The equipment was supplied by the Deutsche Forschungsgemeinschaft.

The total flights registered were 14060; 7537 of them were made by well-fed flies. These started 15.5% more frequently than the starving flies.

According to the chi-square test, the difference between the two frequencies is significant and the deviations from 15.5% do not approach the 5% level.

Starving flies fly less. It could also be the case that starving flies are more stimulated to fly and to look for food than satiated flies. Despite the fact that nothing reliable is known about the natural food source, it can be assumed that the food supply within a cage is higher than in the field and that the physiological food condition of the caged flies is more favourable than that of the field flies. This could be true, especially immediately after emerging. Thus it can be concluded that newly emerged flies in the field possess only a limited flying radius. This hypothesis, however, agrees well with the result of the releasing experiment in which the flies that emerged in the field displayed a smaller tendency to distribute than those released as well-fed flies.

At the second experiment with the flight registering apparatus the flight frequencies of 5 pairs of flies during 8 hours of the day were counted in the course of 24 days beginning with the day of emergence. The results are shown in Fig.2.

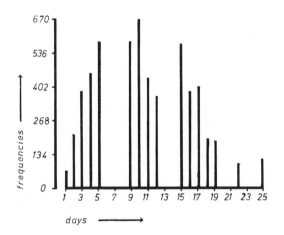

FIG.2.   Flight frequencies of Rhagoletis cerasi L. correlated with age (days).

On the average 355 starts of one fly and day were registered. After the first ten days of the adult stage the flight frequency increased from 67 to 669 per fly and day. The dependence of the disposition to fly upon the age is evident.

If the results of the releasing experiment are considered in respect to this dependence of the flight inclination on age, it must be concluded that the sterilized flies were released when most of the flies had reached a stage of high flight disposition.

The two results obtained from the flight-registering apparatus show that the flies released as adults had a particularly high propensity to fly. This supports the assumption that the sterilized flies released as adults distributed more than the flies put out as pupae.

The experiments revealed that there is a high probability that the European cherry fly has the ability to make relatively long-range flights. Under certain circumstances, such as lack of food or oviposition places, this ability causes the flies to leave their demotop.[3] For this reason it must be concluded that the sizes of the experimental sites were too small for release experiments. Therefore no conclusions about the control effect of the released sterile flies can be drawn.

The consequence which must necessarily be drawn from this experiment, and especially from the observation on the high distributive movement of Rhagoletis cerasi L. is essential for the further development of the sterile-male technique to control this species. Originally the opinion was that there are many small semi-isolated habitats of the cherry fly from which the individuals normally do not emigrate because of its flying behaviour. It was falsely hoped that it would be possible to treat larger cherry orchards as almost isolated demotops. It is necessary to correct this assumption about the size and the borders of a demotop because a reasonable release action can only be performed if the whole demotop is included.

5. SUMMARY

An experiment has been described to show the influence of the release of sterile cherry fruit flies on the build-up of the population after a natural decrease due to a lack of oviposition possibilities one year earlier.

(1) At one experimental site pupae were placed in the field to allow the flies to emerge and to observe the resulting infestation of fruit and the reproduction of the population (consisting of the released and native flies).

(2) At a second site irradiated adults were released to show their influence on the reproduction of the wild flies.

(3) There are strong indications that the released sterile flies left the release site at which not control effect could be observed, and that they had a higher tendency to a distributive movement than the non-irradiated flies put out as pupae.

(4) By means of laboratory observation on the flight disposition of the European cherry fly the reasons for the emigration of the sterile flies were discussed. It seemed that the feeding of the flies before the release as well as the increased age of these flies could have caused the flies to emigrate more quickly and move more extensively than the flies that emerged from the pupae in the soil.

---

[3] Demotop = living space of a population.

DISCUSSION

B. NAGY: You said that sterilized flies dispersed more rapidly than the non-sterilized and I should like to ask what the release conditions were, e.g. the time of day and temperature. We found in the case of the cockchafer (Melolontha melolontha) that sterilized insects released without cooling at midday dispersed more rapidly and more widely than those released in the early morning in cool conditions.

A. HAISCH: The flies were released around noon. Our investigations of the flying behaviour of the cherry fly showed that the flies possess less inclination to fly before noon. A maximum can be observed at 14.00 and a second weaker maximum at about 17.00.

B.S. FLETCHER: It seems possible that the high rate of dispersal of sterilized flies at site 2 was partly due to the small number of trees there. Did you make any estimates of fly density in terms of flies per tree at the different sites?

A. HAISCH: The average population density per tree in years with a normal infestation was about ten times higher than it was during the experiment. The capacity of the smaller experimental site to provide food and oviposition sites was therefore not utilized to the full. Therefore I do not think that the low number of trees was the reason for the increased emigration.

B.S. FLETCHER: Do you have any comparative data on the rates of attraction of wild flies and irradiated flies to yellow traps? It is possible, for instance, that the protein requirements of irradiated females are less because of lower egg production and, if the yellow colour of the traps acts as a feeding cue, there might be differences.

A. HAISCH: We have not done any experiments on this.

POPULATION FLUCTUATION AND DISPERSAL STUDIES OF THE FRUIT FLY, Dacus zonatus Saunders*

Z.A. QURESHI, M. ASHRAF, A.R. BUGHIO,
Q.H. SIDDIQUI
Atomic Energy Agricultural Research Centre,
Tandojam, Sind,
Pakistan

Abstract

POPULATION FLUCTUATION AND DISPERSAL STUDIES OF THE FRUIT FLY, Dacus zonatus Saunders.
Investigations on population fluctuation and the dispersal pattern of the fruit fly, Dacus zonatus Saunders, were carried out in two guava (Psidium guajava L.) orchards. The population fluctuation, as measured by monthly mean catches of male flies in traps baited with a mixture of methyl eugenol + naled, showed that the fly population was at its minimum in January/February and increased gradually to a maximum in March/May. The fly population declined in June/July but started building up again in August and reached a maximum in September. With the onset of winter season the fly population declined. The micro-environment seems to have played an important role in the fluctuation of fly populations. The preliminary data on the dispersal of laboratory-reared sterilized dye-marked flies incidated that most of the marked flies were recaptured from south-westerly and southern directions when the wind was predominantly west-south-west. Marked flies were recovered as far as 25 miles from the release point.

INTRODUCTION

Knowledge concerning population fluctuation and dispersal of insect pests is of considerable importance in control efforts involving the sterile-insect-release technique. Work on population fluctuation and dispersal of the oriental fruit fly, Dacus dorsalis Hendel, the melon fly, D. cucurbitae Coquillett, and the Mediterranean fruit fly, Ceratitis capitata Wiedemann, has been reported by Steiner and associates [1-3]. Similar studies have been reported on the olive fruit fly, D. oleae Gmelin [4-6], the Mexican fruit fly, Anastrepha ludens Loew [7], and the cherry fruit fly, Rhagoletis cerasi L. [8]. However, as no data are available on the population fluctuation and dispersal pattern of the fruit fly, D. zonatus Saunders, in Pakistan the present investigations were conducted.

MATERIALS AND METHODS

Population fluctuation studies

Population fluctuation studies of the fruit fly were made in two guava (Psidium guajava L.) orchards from July 1970 to June 1972. These orchards (8 and 16 acres) were situated approximately 10 miles apart in the vicinity of Tandojam. Trapping was carried out by the Steiner plastic dry trap with

*This research has been partly financed by a grant (FG-Pa-148) made by the US Department of Agriculture under PL-480.

a slight modification using methyl eugenol, a powerful attractant for male flies, in combination with naled. Four mlitre from a mixture of 95% methyl eugenol + 5% naled (by volume) was injected into a cotton wick measuring 5.5 cm long and 1.3 cm in diameter. The wick was pinched between the wire loop inside the trap. Three traps were placed in each orchard and hung on the guava trees 4-6 ft above the ground. Fortnightly observations on the number of flies collected were made during the entire period of study.

Dispersal studies

Mature pupae obtained from stock cultures maintained at the Entomology Laboratory of the Atomic Energy Agricultural Research Centre, Tandojam, were irradiated at 9 kR in a ^{60}Co panoramic source and dye-marked with calco oil blue RA (American Cynamid Co.) at the rate of 0.1 g/1000 pupae. The emergent flies were held in wire gauze cages measuring $60 \times 45 \times 45$ cm and fed on a yeast hydrolysate enzymatic, sugar and water diet. The flies

TABLE I. MONTHLY CATCHES OF MALE Dacus zonatus AND THE CORRESPONDING TEMPERATURE AND RELATIVE HUMIDITY DURING 1970-72

Month	Mean number of males per trap	Mean temperature (°F)		Mean relative humidity (%)
		Max	Min	
1970				
July	78	97.2	80.4	68.7
August	1264	95.4	79.5	70.5
September	4881	94.7	77.7	71.1
October	1619	97.1	70.1	55.3
November	1556	88.7	53.3	39.4
December	668	79.0	47.4	45.9
1971				
January	629	75.4	45.2	44.6
February	60	84.0	44.0	42.0
March	2231	92.8	56.6	38.7
April	1884	103.0	63.9	51.1
May	547	104.9	78.7	69.5
June	646	102.9	81.5	80.6
July	392	98.6	73.0	79.4
August	1743	96.0	78.5	70.0
September	2380	100.0	69.2	50.7
October	1669	92.8	60.9	45.0
November	1153	89.5	55.5	46.5
December	208	80.3	46.4	45.0
1972				
January	45	76.2	43.4	41.0
February	56	75.2	40.0	35.8
March	479	92.9	57.3	38.8
April	1013	99.8	67.7	47.9
May	2212	104.1	76.6	61.4
June	689	103.0	77.9	67.4

were released the following day in a guava orchard near Tandojam. The male flies were recaptured in lure traps placed in various directions and distances up to 25 miles from the release point irrespective of host plants. From the recaptured flies the marked flies were identified by Steiner's method [9]. Observations and counts of flies were made on alternate days for a maximum period of five weeks.

RESULTS AND DISCUSSION

Population fluctuation studies

The results of Table I indicate a fly population maximum in September when an average number of 4881 and 2380 males/trap were captured during 1970 and 1971 respectively. The population gradually declined from October (onset of winter) until February. The lowest fly count of 60 and 45 per trap was recorded in February 1971 and January 1972 respectively. This population reduction may be partially attributed to low temperature (40°-46°F) and non-availability of the preferred guava fruit. Because of the favourable seasonal condition and abundant availability of ripe guavas in March 1971, a rapid increase in fly population was noticed when an average of 2231 males per trap were caught (Fig. 1). In 1972 the increase in fly population was recorded from March with a maximum in May when an average of 2212 males per trap were captured. This could be explained by the seasonal variation and prolonged fruiting season during 1972. The reduction in the fly population recorded during June 1971 and 1972 could probably be caused by the hot and dry season.

Mourikis and Fytizas [6], while studying the population fluctuation of D. oleae in olive orchards during 1966-67, recorded the highest population

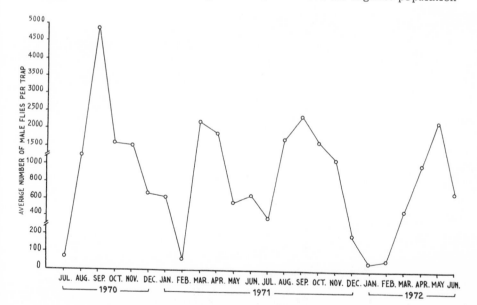

FIG. 1. Population fluctuation of Dacus zonatus during 1970-72.

TABLE II. DISPERSAL OF MARKED STERILIZED Dacus zonatus IN DIFFERENT DIRECTIONS AND DISTANCES FROM RELEASE POINT

Trap direction from release point	Distance from release point (miles)									
	5		10		15		20		25	
	Number of flies captured									
	Total	Marked	Total	Marked	Total	Marked	Total	Marked	Total	Marked
North	613	1	174	4	50	0	19	0	30	0
South	269	5	91	5	126	20	26	0	6	0
East	45	2	27	0	5	0	35	0	15	0
West	20	0	246	0	1133	5	6	0	12	0
Northwest	1094	0	84	0	0	0	0	0	0	0
Northeast	95	0	85	2	11	0	17	0	45	0
Southwest	18	0	196	0	4449	19	17142	23	1847	6
Southeast	244	2	176	8	34	0	6	0	9	0

in October in both the years. They showed that the population of this species as well as the degree of its attack on olive fruits were not identical in different orchards in the same area. They further reported that the greatest attack and population increase occurred chiefly in certain places as a result of climatic conditions of the micro-environment particularly the relative humidity. Steiner and co-workers [3] also reported that for three years the mean monthly catches of the wild oriental fruit fly fluctuated from 14 per trap day in April to 472 in October. In September 1962 the catches reached a monthly average as high as 602 males per trap day. However, the normal trend was upward from early May to a peak in September or October after which the number gradually declined to that in April.

Our results therefore suggest that the fly population increase corresponds most probably with the maturity of the guava fruit, which in turn is governed by the environment. It was also observed that the fly population fluctuated in the same manner in both orchards, with the degree of attack and number of flies captured per trap being variable.

Dispersal studies

An exploratory test was performed to gain knowledge on the general dispersion pattern of sterilized dye-marked flies. For this purpose 31 900 marked flies were released in an 8-acre guava orchard near Tandojam. Trapping was done in north, south, southwest and east directions at various distances ranging from 2 to 15 miles. The marked flies were recaptured from the south and southwest only. In the southwest, a total of 191, 8351 and 47 flies were captured at distances of 2.5, 5 and 15 miles respectively, out of which 29, 375 and 1 flies respectively were found to be dye-marked. In the south, a total of only 5 dye-marked flies were recaptured, one each at 2, 5 and 8 miles and 2 at 15 miles.

Based upon the above findings an experiment was conducted with 40 lure traps placed in eight directions (north, south, east, west, northwest, northeast, southwest and southeast) at 5, 10, 15, 20 and 25 miles from the release point. The data in Table II on the dispersal of marked flies indicated that out of 12 986 marked flies (6493 males) released a total of 102 flies (1.57%) were recaptured in all the traps placed in various directions. Of the total marked flies recaptured, 48, 30 and 10 flies were collected from the southwest, south and southeast respectively compared with the north, east, northeast and west where 5, 2, 2 and 5 flies were captured. No marked fly was recaptured from the northwest direction. The marked fly catch proportionate to the distance appeared somewhat erratically distributed probably because of the placement of traps irrespective of the host areas. Since the wind direction was predominantly west-south-west during the entire test period the movement of the marked flies was more or less upwind.

The results reported on the dispersal and flight range of the fruit flies vary in the literature. Christenson and Foote [10] found that the males of D. dorsalis were readily attracted to methyl eugenol traps against strong trade winds, at least when low-to-moderate wind velocities were involved. They also quoted the results of the tests conducted in Hawaii by Handerson and Keiser and also by Gammon on the effect of the wind direction on D. dorsalis flight, and stated that many marked flies made their way for long distances against the prevailing air flow. Pelekassis and co-workers [5] captured isotope-labelled Dacus oleae flies at 4300 m from the southeastern sector of the treated area when the prevailing wind direction was northeast.

Steiner and co-workers [2] reported that D. cucurbitae, D. dorsalis and C. capitata travelled a distance ranging from 25 to 45 miles from their release point. Moreover, when catches of sterile flies from upwind areas were compared with downwind areas, the former contained 62 - 82% and the latter 23 - 56% of sterile flies. Steiner [1] showed that the populations of D. dorsalis in the leeward direction were four times greater than those in the windward direction during the months when the northeast trade winds were dominant, but became evenly distributed soon after the rainy season began when the wind directions were more variable. He further reported [11] that marked C. capitata, D. dorsalis and D. cucurbitae could move as far as 25, 30 and 67 miles respectively, and that more movements of C. capitata occurred in a downwind direction than in other directions when the wind was predominantly blowing from one direction.

From the results on population fluctuation and dispersal studies it is concluded that the peak population of the fruit fly occurred in September and was lowest in January/February. More flies moved towards the southwest when the wind was predominantly west-south-west. Marked flies were recaptured as far as 25 miles from the release point. To substantiate the dispersal behaviour of the flies detailed studies are planned.

REFERENCES

[1] STEINER, L. F., "Methods of estimating the size of populations of sterile pest tephritidae in release programs", Insect Ecology and the Sterile Male Technique (Proc. Panel Vienna, 1967), IAEA, Vienna (1969) 63.

[2] STEINER, L. F., MITCHELL, W. C., BAUMHOVER, A. H., Progress of fruit fly control by irradiation sterilization in Hawaii and the Marianas-islands, Int. J. Appl. Radiat. Isot. 13 (1962) 427.

[3] STEINER, L. F., MITCHELL, W. C., HARRIS, E. J., KOZUMA, T. T., FUJIMOTO, M. S., Oriental fruit fly eradication by male annihilation, J. Econ. Entomol. 58 (1965) 961.

[4] ORPHANIDIS, P. S., SOULTANOPOULOS, C. D., KARANDEINOS, M. C., «Essai préliminaire avec ^{32}P sur la dispersion des adultes du Dacus oleae Gmel.», Radiation and Radioisotopes applied to Insects of Agricultural Importance (Proc. Symp. Athens, 1963), IAEA, Vienna (1963) 101.

[5] PELEKASSIS, C. E. D., MOURIKIS, P. A., BANTZIOS, D. N., "Preliminary studies of the field movement of the olive fruit fly (Dacus oleae Gmel.) by labelling a natural population with radioactive phosphorus (P^{32})", Radiation and Radioisotopes applied to Insects of Agricultural Importance (Proc. Symp. Athens, 1963), IAEA, Vienna (1963) 105.

[6] MOURIKIS, P. A., FYTIZAS, E., "Review of olive-fly ecology in relation to the sterile-male technique", Sterile-male Technique for Control of Fruit Flies (Proc. Panel Vienna, 1969), IAEA, Vienna (1970) 131.

[7] SHAW, J. G., SANCHEZ-RIVIELLO, M., SPISHAKOFF, L. M., TRUJILLO, G. P., LOPEZ D., F., Dispersal and migration of tepa-sterilized Mexican fruit flies, J. Econ. Entomol. 60 (1967) 992.

[8] BOLLER, E. F., HAISCH, A., PROKOPY, R. J., "Sterile-insect release method against Rhagoletis cerasi L.: Preparatory ecological and behavioural studies", Sterility Principle for Insect Control or Eradication (Proc. Symp. Athens, 1970), IAEA, Vienna (1971) 77.

[9] STEINER, L. F., A rapid method for identifying dye-marked fruit flies, J. Econ. Entomol. 58 (1965) 374.

[10] CHRISTENSON, L. D., FOOTE, R. H., Biology of fruit flies, Annu. Rev. Entomol. 5 (1960) 187.

[11] STEINER, L. F., "Mediterranean fruit fly research in Hawaii for the sterile fly release program", Insect Ecology and the Sterile-Male Technique (Proc. Panel Vienna, 1967), IAEA, Vienna (1969) 73.

DISCUSSION

B.S. FLETCHER: We found a similar high rate of dispersal of immature adults of Dacus tryoni in Australia. Do you have any data on the dispersal rates of flies released when mature?

Z.A. QURESHI: No, the dispersal data reported here relate to the releases of 1 - 2-day-old sexually immature adults.

B.S. FLETCHER: Your trapping data suggest that there is a big drop in population size during the winter. Does breeding continue throughout the winter months or is there a definite overwintering stage?

Z.A. QURESHI: The flies continue to breed throughout the year but their development is appreciably slowed down probably because of seasonal conditions in winter and also the non-availability of the favoured host, i.e. guava.

B.S. FLETCHER: At what stage of sexual maturation do the males of D. zonatus start responding to methyl eugenol, and is there any indication of immature males being attracted, as reported for D. dorsalis in Hawaii?

Z.A. QURESHI: We did not check at what stage of sexual maturation the males of D. zonatus start responding. However, we did catch immature adults in our traps.

A.G. MANOUKAS: You told us that your attractant was a powerful one. Did you compare it with any other attractants?

Z.A. QURESHI: Yes, we tried amyl acetate, molasses and ammonium carbonate but methyl eugenol proved more efficient.

H. LEVINSON: How does the trapping efficiency of methyl eugenol for male peach fruit flies compare with that of the natural sex attractant of the species investigated?

Z.A. QURESHI: We did not use the natural sex attractant.

R. GALUN: How do you analyse your data mathematically in the cases where no marked flies were caught in the traps? Do you assume that there were no marked flies at all in that area?

Z.A. QURESHI: Yes.

B. NA'ISA: Do you find any difference in dispersal range between male and female flies and is the dispersal observed different from that occurring with a normal (non-irradiated) field population?

Z.A. QURESHI: Since the lure used in our traps attracts only male flies, it is impossible to assess the relative dispersal range of males and females. As regards the second part of your question, all we studied was the dispersal pattern of irradiated sterilized flies.

K. SYED: The population fluctuation between summer and winter months has been established simply on the basis of the numbers caught in the traps. Could such a variation not be caused by the traps being differentially attractive in summer and winter? Do you have precise computations of the population densities apart from the trap catches?

Z.A. QURESHI: The traps were equally efficient in attracting the flies in both winter and summer. This was supported by the data on fruit infestation which showed a higher number of larvae per fruit in summer than in winter. We collected about 500 larvae per pound of guava fruit in summer whereas in winter only 1-2 larvae were recovered.

J.L. MONTY: Are the fluctuations in fly population correlated with the seasonal abundance of guava?

Z.A. QURESHI: Yes, the fluctuations depend to a large extent on the availability of ripe guava fruit as well as on the climatic conditions of the micro-environment.

PROSPECTS OF INCREASED EGG PRODUCTION IN THE REARING OF Dacus oleae Gmelin BY THE USE OF CHEMICAL STIMULI*

V. GIROLAMI, G. PELLIZZARI
Istituto di Entomologia agraria dell'Università di Padova,
Padua

E. RAGAZZI, G. VERONESE
Istituto di Chimica Farmaceutica e
    Tossicologica dell'Università di Padova,
Padua,
Italy

Abstract

PROSPECTS OF INCREASED EGG PRODUCTION IN THE REARING OF Dacus oleae Gmelin BY THE USE OF CHEMICAL STIMULI.

An activity of stimulus to oviposition of the Dacus oleae Gmelin has been discovered in oleoeuropeine and demethyloleoeuropeine, phenolic glucosides characteristic of the olive; furthermore, the respective aglucones obtained enzymatically have also proved active, but not the ulterior, already known, products of the cleavage of the molecule. It has been possible to demonstrate that the stimulating activity is linked to products of spontaneous degradation of the glucosides and not to impurities derived from substances present in the olive, not eliminated in the course of the purification of the same. By sprinkling with the glucosides particular oviposition substrata (which are perforated by the ♀♀ without holding back the eggs, which fall and can be collected on the exterior of the network rearing cage), a significant increase in oviposition has been obtained. Research is being carried out to identify the chemical nature of the stimulating principles and also a further improvement in oviposition substrata.

The ♀♀ of Dacus oleae, in the absence of olives in which to deposit eggs, allow them to fall freely in rearing cages or endeavour to deposit them on smooth surfaces, preferably curved. It is thus possible to obtain oviposition inside thin paraffin domes [1] or on the surface of imitation olives [2,3]. Using paraffin domes one obtains a good hatch of eggs deposited inside them after perforation by D. oleae, nevertheless their use cannot be extended to mass rearing of the Diptera, since it is necessary to change the domes themselves frequently, in all the rearing cages, in order to collect a limited number of eggs.

Using cages with a network base [4] on which non-perforable oviposition substrata can be placed [2], it is possible to collect the eggs automatically [5] from beneath the cages themselves.

Although this solves the problem of collection of eggs, such rearing gives somewhat sparse yields because of the limited fertility of the eggs [6], particularly when the adults come from artificial media [7].

Consequently, while problems of another nature concerning the pabulum of larval development and bacterial symbiosis remain open [8], the improvement of oviposition substrata can open up interesting perspectives for the mass production of the olive fly.

---

* Studies on Dacus oleae Gmelin for improving new control methods, No. 9.

It must further be borne in mind that the fecundity of Dacus oleae is influenced drastically by the presence, or otherwise, of suitable oviposition substrata; for example the ♀♀ in the laboratory, if they have undamaged olives at their diposal, lay, on an average, at least five times as many eggs as the number obtained without oviposition substrata, and it is sufficient to foul the surface of the drupes with olive juice for the fecundity to be equally reduced (Girolami, unpublished data) through the presence of a repellent factor [9].

Since this seems to be verified in nature, too, it is reasonable to admit the existence of an important factor of autoregulation of the fecundity of the species which has repercussions also on the laying of eggs in the laboratory.

With this hypothesis research on a chemical stimulus to oviposition present in the olive is extremely interesting, not only in order to improve the yield of the rearing of D. oleae, but also, one hopes, once the repellent principles are known, for the control of the species in nature [10].

Research has therefore been initiated, in collaboration between the Istituto di Entomologia agraria and the Istituto di Chimica Farmaceutica e Tossicologica of the University of Padua, on the presence of chemical principles in the olive which stimulate the oviposition of the D. oleae, particular attention being paid to the action of the specific phenolic glucosides of the olive.

MATERIALS AND METHODS

Adult Dacus oleae were kept in a greenhouse at 23 ± 1°C in network cages [2], and fed by placing in the cages drops, renewed daily, of a solution of 10% of yeast autolysate (Fratelli Piccioni, Brescia, Italy) saturated with saccharose to prevent fermentation.

Small domes obtained by moulding a sheet of paper (Kleenex) lightly impregnated with glycerine to soften it were first used as oviposition substrata.

These paper domes, in addition to the relative simplicity of preparation in regard to paraffin, have the advantage of absorbing rapidly, without disintegrating, the solution of chemical compounds to be examined and, above all, do not hold back the eggs deposited in their interior, which therefore fall and may be collected through the network base of the cages.

The rough surface of the paper can, however, render these oviposition substrata uninviting, for which reason during the course of the experiments domes, made by pouring a mixture of acetone (100 g), celluloid (4 g), talcum (35 g) and green fluorescent pigment (4 g) on to a porcelain mould with numerous concavities, were utilized.

Once this mixture had been spread, the mould was turned upside down, and the excess liquid recovered. By evaporation of the acetone a film was formed which was detached easily by immersing the mould in water. The domes thus formed (corresponding to the concavities of the mould) were placed on the bottom of the cages and were perforated easily by the ovipositor of the ♀♀. In such a case, too, the eggs do not remain adhering to the walls but fall.

The talcum serves to render the substratum friable, and the absorption of the oviposition stimulating solutions is facilitated.

The investigation into the stimulatory action of the various fractions examined has been carried out regularly by placing on a glass slide, situated on the base of various rearing cages, two domes differently treated, and counting the eggs deposited after one day.

For tests of this kind it has proved opportune to place not more than 10 ♀♀ in each cage, to prevent competition for the substratum which takes place among the ♀♀ , thus disturbing oviposition to a greater extent on the most desired domes.

Various fractions of extraction of olives have been tested, and a marked stimulation having been encountered in the solutions containing oleoeuropeine [11], the investigations have been directed principally towards this substance as well as to demethyloleoeuropeine [12, 13] and the products of cleavage of this substance (Fig.1).

Oleoeuropeine is a phenolic glucoside characteristic of the entire olive tree, and is largely responsible for the bitter and astringent flavour of olives, as well as for the hypotensive effect of the extracts of the leaves, bark and root of the tree.

RESULTS AND DISCUSSION

Research directed towards the identification of the stimulating chemical principles to oviposition

The stimulating effect on the oviposition of the Dacus oleae Gmelin of the total extracts of the olive has not proved significant because of the presence, it is supposed, of the repellent principles also, which are formed probably when the juice of the olive comes into contact with the air [9].

Here, therefore, we have concentrated on substances characteristic of the olive, in a state of chromatographic purity, and thus a significant stimulatory activity has been seen in oleoeuropeine [12] (Fig.1; I) preserved for some time in its dry state in a vacuum. An analogous activity is also exercised by demethyloleoeuropeine [12,13] (Fig.1; II).

Since great activity is also known to exist in the product of washing the above-mentioned glucosides with chloroform, as well as a minor activity in the same products recently purified, on the hypothesis that the stimulating principles were due to products of spontaneous degradation of the glucosides, we have tested the principal products of cleavage of the same.

Thus considerable activity has been observed in the aglucones (Fig.1; III; IV) obtained by hydrolysis with beta glucosidase [11] of oleo-europeine and demethyloleoeuropeine.

On the other hand, the successive products of degradation of the molecule of the glucosides [11] proved inactive (Fig.1; V; VI; VII; VIII; IX), as well as tyrosole (Fig.1; X), a phenol characteristic of the vegetation waters of the olive and of virgin olive oil [12, 14].

Consequently an activity has been encountered in molecules, such as the aglucones formed by dihydroxyphenylethanol (Fig.1; IX) and by

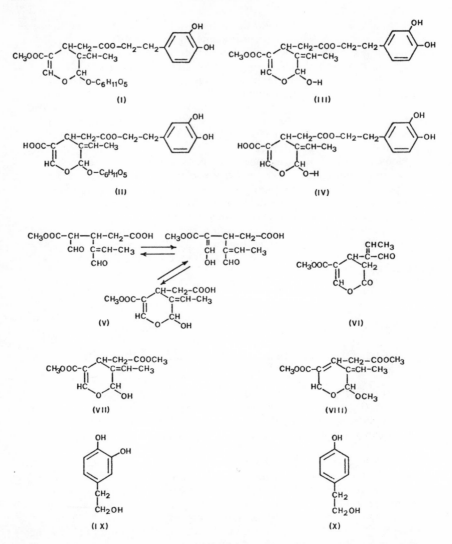

FIG. 1.   I. oleoeuropeine; II. demethyloleoeuropeine; III. aglucone of oleoeuropeine; IV. aglucone of demethyleoleoeuropeine; V. "acid IV" from oleoeuropeine (Ref. [11]); VI. lactone of "acid IV"; VII. methylic diester of "acid IV"; VIII. trimethylic ester-ether of "acid IV"; IX. beta-(3,4-dihydroxy-phenyl-)ethanol; X. beta-(4-hydroxy-phenyl-)ethanol (tyrosole).

"acid IV" (Fig.1; V), whereas the same substances, chemically pure, alone or in a mixture, were shown to be inactive.

On the other hand, even the aglucones do not appear to be by themselves the chemical stimulating principle, in that the product of washing oleoeuropeine (preserved for a long time) with chloroform, and is particularly active, is found to contain, on chromatographic analysis, only traces of aglucone.

TABLE I. OVIPOSITION OF ♀♀ OF <u>Dacus oleae</u> IN TWO DOMES,
ONE TREATED WITH SOLUTIONS CONTAINING AGLUCONE AND
THE OTHER WITH CORRESPONDING QUANTITIES OF OLEOEUROPEINE
WITH WHICH WE STARTED OR RECOVERED (see text): Total layings
of 30 ♀♀ sub-divided into 6 cages

Products tested	Oviposition Daily				Total
Aglucone	29	44	40	31	144
Purified oleoeuropeine	6	5	6	1	18
Aglucone	25	24	19	14	82
Recovered oleoeuropeine	1	0	1	1	3

We have endeavoured also to pick out the stimulating principles of
the product of washing oleoeuropeine, subjecting this last to fractionation
on a thick preparative layer of silica gel, and testing the products of
elution of the 22 zones obtained. The results were not very satisfactory,
possibly because of a similar alteration of the substance in the course of
the chromatography and of the successive recovery of the zones.

This leads to the belief that even in the most active solutions the
quantity of the stimulating principles is in percentage slight. Consequently,
it is reasonable to think that the activity of oleoeuropeine may be caused by
impurities not eliminated in the course of purification. To show that the
stimulating principles were derived from the spontaneous degradation of
the oleoeuropeine and are not caused by impurities extraneous to the
glucoside, while pertaining to the olive, we have tried to verify whether
on artificial aging of the purified oleoeuropeine (in an aqueous solution
at ambient temperature or at boiling point, exposed to light or dark) there
is a corresponding increase of activity, but without obtaining significant
results.

An unequivocal result is, on the other hand, obtained by hydrolysing
oleoeuropeine of recent extraction (coming from olive leaves) with beta
glucosidase.

In particular equal volumes of the three following products were
employed: the above-mentioned oleoeuropeine at a concentration of
100 mg/mlitre, its aglucone obtained by enzymatic hydrolysis, and the
unchanged glucoside; these last two were in their turn dissolved in a
volume of solvent such as to obtain the same ratio of 100 mg/mlitre as
with the total of the oleoeuropeine with which we started.

In Table I are reported the total daily layings in six cages containing
five mature ♀♀; two domes were placed in each cage, one treated with
20 ml of solution containing the aglucone and the other with an identical
volume of solution of the oleoeuropeine with which we started, or
recovered unchanged from hydrolysis.

The laying of eggs has proved significantly greater in the domes
treated with solution containing aglucone; furthermore in none of the
six cages (and thus in 48 tests) has there been a lower deposition in

such substratum. Consequently, either during the hydrolysis, or by
degradation of the products obtained, there has been an increase in
activity and thus a formation of the chemical principles stimulating
oviposition derived, in the end, from oleoeuropeine. Researches are in
progress to ascertain whether in olives, too, the ovistimulating principles
can originate from phenolic glucosides.

Although it is not yet possible to know the exact chemical nature of
the above-mentioned stimulating principles, it is nevertheless established
that, at least in part, they are products of degradation of phenolic glucosides and, with even greater probability, of the respective aglucones.

It is interesting to note that the glucosides characteristic of certain
vegetables seem also to be recognized by insects belonging to different
orders such as <u>Pieris brassicae</u> L. [15] and aphids [16] to enable them
to select the host plants on which they reproduce or feed.

Possibilities of increase in fertility

When the presence had been found in oleoeuropeine of a stimulating
activity which induces the ♀♀ of the <u>D. oleae</u> to lay eggs in substrata
impregnated with such a substance, experiments were initiated to
ascertain whether the presence of glucoside, stimulating oviposition,
also increased the total fertility in the laboratory rearing.

For the experiment three domes were placed on the bottom of the
cages impregnated with about 10 mg of oleoeuropeine, and the oviposition was compared with that of other cages without perforable oviposition
substrata.

In this way (Fig.2) a production of eggs was obtained in each cage
more or less double that in the other cages. Furthermore an intermediate
number of eggs was observed when the domes were placed without
stimulant, and a greater fertility of the eggs deposited in the domes.

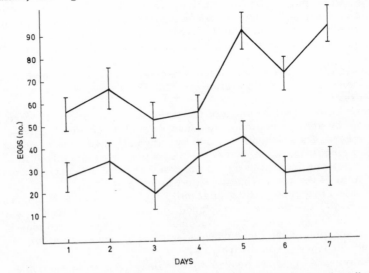

FIG. 2.   Average daily laying of group of 10♀ without egg-laying sites (below), and with small paper
domes impregnated with oleoeuropeine (above). Average values of three replicates ± E.s.

At present we are experimenting with celluloid domes impregnated with aglucone, and the first results are encouraging even if there has not been obtained, at least up to now, an egg production comparable with that in olives.

Furthermore it has been shown that there are numerous factors which influence the oviposition of the D. oleae, and leaving out of consideration the influence exercised by light and by atmospheric pressure (Girolami, unpublished data) that is responsible in part for the great fluctuation verified on different days, even in ambients of constant temperature, there remain nevertheless other complex relations, little or not at all known, between substratum of oviposition and the behaviour of the species.

For example, the stimulating principles present in oleoeuropeine induce the ♀♀ to perforate the substrata impregnated with such a substance, but not always, once perforated, does oviposition follow, probably because further physical or chemical stimuli are needed.

One can observe, in fact, particularly when employing celluloid substrata, that while perforation occurs rapidly, the ♀♀ then take a longer time to deposit the eggs or do not deposit them at all; on an average only a third of the perforations are accompanied by oviposition.

Moistening the substrata with glycerine, sterile deposition diminishes, but in such a case the eggs remain mostly on the walls of the domes, and cannot be collected through the base of the cage.

Nevertheless, although, because of the complexity of the phenomena, it has not yet been possible to achieve an artificial substratum of oviposition that is completely satisfactory, the field of investigation remains extremely reliable, since a wider knowledge of the reproductive behaviour of the species can lead to conclusions that will render the rearing of the adults of the D. oleae comparable economically with that of other Tripetides.

CONCLUSIONS

On the basis of the research carried out, it has been possible to demonstrate that the oviposition of D. oleae is stimulated by particular substances derived from the degradation of the glucosides of the olive oleoeuropeine and demethyloleoeuropeine and probably in greater quantity than from the respective aglucones obtained enzymatically. The successive products have, on the other hand, proved inactive. It has not yet been ascertained whether the stimulating principles can be derived from diverse chemical substances, as it has not yet been possible to separate chemically pure active compounds. It has been possible, on the other hand, to eliminate the doubt that the activity encountered in glucosides, chromatographically pure, is due to contaminants pertaining to the olive, not eliminated during the extraction of the above-mentioned products.

The greater activity encountered in the aglucones with respect to the glucosides may be due to the more marked instability of the molecules of the former; furthermore, one cannot exclude the fact that the active substances are formed even during hydrolysis and, being soluble in chloroform, are collected together with the aglucones.

The use in the rearing of the D. oleae of substances which stimulate egg laying has opened up interesting prospects of improvement; it has, in fact, been possible, at least at laboratory level, to increase fecundity perceptibly.

In addition, the original substrata of oviposition used, in which deposition takes place after an active perforation without the eggs being held back, possibly constitutes an important step towards the realization of an economical collecting method of the eggs without their fertility being impaired.

Prospects of further improvements are also linked to the possibility of having available, in future, pure (or at least dosable) stimulating principles, and to the improvement of the substrata of oviposition, on the basis of a wider knowledge of the factors, physical as well, which induce the species to lay eggs.

D. oleae seems to be attracted by substances present in the olive [17] which also might derive from phenolic glucosides (it is to be noted, by the way, that the olives of the cultivar which contain a minute quantity of demetyloleoeuropeine [18] are also less receptive to the attacks of the olive fly), so that it is not improbable that, on the basis of the researches undertaken, new means of control of the species in nature may be achieved.

REFERENCES

[1] HAGEN, K.S., SANTAS, L., TSECOURAS, A., "A technique of culturing the olive fly, Dacus oleae Gmel., on synthetic media under xenic conditions", Radiation and Radioisotopes applied to Insects of Agricultural Importance (Proc. Symp. Athens, 1963), IAEA, Vienna (1963) 333.

[2] CAVALLORO, R., GIROLAMI, V., Nuove tecniche di allevamento in laboratorio del Dacus oleae Gmel.: I. Adulti, Redia LI (1968) 127.

[3] CAVALLORO, R., DELRIO, G., Incremento della fertilità delle uova di Dacus oleae Gmelin negli allevamenti permanenti, Note ed Appunti di Entomologia Applicata, Perugia XIV (1973) 3.

[4] CAVALLORO, R., Orientamenti sull'allevamento permanente del Dacus oleae Gmelin (Diptera, Trypetidae) in laboratorio, Redia L (1967) 337.

[5] CAVALLORO, R., GIROLAMI, V., Miglioramento nell'allevamento in massa di Ceratitis capitata Wiedman (Diptera, Trypetidae), Redia LI (1969) 315.

[6] TZANAKAKIS, M.E., Rearing methods for the olive fruit fly Dacus oleae (Gmelin), Ann. Sch. Agric. Forestry, Univ. Thessaloniki 14 (1971) 293.

[7] GIROLAMI, V., CAVALLORO, R., Aspetti della simbiosi batterica di Dacus oleae Gmelin in natura e negli allevamenti di laboratorio, Ann. Soc. Entomol. Fr. (N.S.) 8 (1972) 561.

[8] GIROLAMI, V., Reperti morfo-istologici sulle batteriosimbiosi del Dacus oleae Gmelin e di altri Ditteri Tripetidi, in natura e negli allevamenti su substrati artificiali, Redia LIV (1973) 269.

[9] CIRIO, U., Reperti sul meccanismo stimolo-risposta nell'ovideposizione del Dacus oleae Gmelin (Diptera, Trypetidae), Redia LII (1971) 577.

[10] VITA, G., CIRIO, V., FEDELI, E., TIOCINI, G., L'uso di sostanze naturali presenti nell'oliva come prospettiva di lotta contro il Dacus oleae Gmelin, X Congresso naz. di Entomol. Sassari, 20-25 maggio, 1974 (in press).

[11] PANIZZI, L., SCARPATI, M.L., ORIENTE, G., Costituzione della oleoeuropeina, glucoside amaro e ad azione ipotensiva dell'olivo, Gazz. Chim. Ital. 90 (1960) 1449.

[12] RAGAZZI, E., VERONESE, G., Ricerche sui costituenti idrosolubili delle olive: Nota I, Zuccheri e fenoli, Ann. chim. (Rome), 57 (1967) 1386.

[13] RAGAZZI, E., VERONESE, G., GUIOTTO, A., Demetiloleoeuropeina, nuovo glucoside isolato da olive mature, Ann. chim. (Rome) 63 (1973) 13.

[14] RAGAZZI, E., VERONESE, G., Indagini sui componenti fenolici degli oli di oliva, Riv. Ital. Sost. Grasse 50 (1973) 443.

[15] MA WEI-CHUN, SCHOONHOVEN, L. M., Tarsal contact chemosensory hairs of the large white
      butterfly Pieris brassicae and their possible role in oviposition behaviour, Entomol. Exp. Appl. 16
      (1973) 343.
[16] NAULT, L. R., STYER, W. E., Effects of sinigrin on host selection by aphids, Entomol. Exp. Appl. 15
      (1972) 423.
[17] FIESTAS ROS DE URSINOS, J. A., GRACIANI COSTANTE, E., MAESTRO DURAN, R., VASQUEZ
      RONCERO, A., Etude d'un attractif naturel pour Dacus oleae, Ann. Soc. Entomol. Fr. (N.S.) 8
      (1972) 179.
[18] RAGAZZI, E., VERONESE, G., Indagini sulla presenza di demetiloleoeuropeina in olive di cultivar
      diverse, Ann. chim. (Rome) 63 (1973) 21.

DISCUSSION

R. CAVALLORO: Your data from experiments with oleoeuropeine (Fig.2) relate to one week's oviposition. How does this stimulant affect oviposition during the rest of the life-span of adult females?

V. GIROLAMI: It is possible to obtain the same results for 40-50 days; the response of older females to stimulants is not always satisfactory.

R. CAVALLORO: Do you consider control of the olive fruit fly by means of repellents to be a realistic proposition?

V. GIROLAMI: If we can find a selective attractant for the mature females, I think that a repellent substance may prove very effective for concentrating in any desired area, and thereby controlling, the wild population of D. oleae.

B.S. FLETCHER: In your paper you show how glucosides act as an ovipositional stimulant. Do you have any data indicating that they also act as an attractant in the field?

V. GIROLAMI: We have carried out a few field experiments with good results but it is not possible at present to conclude that glucosides also act as an attractant.

PROTEIN HYDROLYSATE-FREE LARVAL DIETS FOR REARING THE OLIVE FRUIT FLY, Dacus oleae, AND THE NUTRITIONAL ROLE OF BREWER'S YEAST

A.G. MANOUKAS
Department of Biology,
"Demokritos" Nuclear Research Centre,
Athens, Greece

Abstract

PROTEIN HYDROLYSATE-FREE LARVAL DIETS FOR REARING THE OLIVE FRUIT FLY, Dacus oleae, AND THE NUTRITIONAL ROLE OF BREWER'S YEAST.

Artificial diets adequate for growth and development of Dacus oleae larvae were elaborated by substituting or eliminating particular constituents in routinely used diets. A larval diet containing only brewer's yeast and olive oil as nutrient sources was formulated which gave pupal yield equivalent to a soy hydrolysate-containing diet. It seems, however, that an essential factor present in soy hydrolysate (most probably free amino-acids) is required for optimum larval development time and maximum pupal weight. Increased levels of brewer's yeast in the protein hydrolysate-free diet resulted in shorter development time, higher pupal yield and heavier pupal weight. Above a minimum level of brewer's yeast larval development time and pupal weight depended upon the content of non-protein amino-acids present in the yeast. It is suggested that protein hydrolysate may be omitted from the diets used at present under the condition that brewer's yeast contains a minimum level of non-protein amino-acids.

INTRODUCTION

Several artificial diets for the olive fruit fly, Dacus oleae (Gmelin), have been reported in the last fifteen years. A review of these diets and their performance is given by Orphanidis and co-workers [1] and Tzanakakis [2]. These diets are unsuitable for certain research purposes and inefficient for mass rearing of the olive fruit fly. All diets, except that of Rey [3], contain among other ingredients a protein hydrolysate source and some of them contain carrots [4-6] or peanuts [6-9]. Rey [3] reported a simplified diet with germinated chick pea seedlings in place of carrots or peanuts and of soy hydrolysate.

All investigators reported variable pupal yield and a mean production of approximately one pupa per gram diet. This may be due either to the fact that these diets did not contain more than two eggs per gram diets [2-7,9] or that the physicochemical properties of the diets did not permit a higher pupal yield. The nutrients of these diets were poorly utilized and nutritional work was difficult and in certain instances impossible, due mainly to the unsuitable physical structure of the diet for larval development [10, 11]. The efficiency of protein utilization of diet N, generally accepted as standard diet by the Athens meeting of the FAO working party on olive pests and diseases (1969), was approximately 1% whereas that of olive fruit was observed to be as high as 20%. In addition diet N contains 100 times the free amino-acid content and 4 times the protein content of the olive fruit based on analyses reported by Manoukas [10, 12, 13]. These results show that diet N contains excessive amounts of protein and of free amino-acids which make the diet inefficient, expensive, more vulnerable to micro-organisms and probably detrimental to the insect [14].

Work aiming at improving diet N resulted in a simplified peanut-free diet developed by Manoukas [10]. This diet was modified with respect to cellulose content and was introduced in this Laboratory for the production of over 5 million flies during 1973 [15]. The peanut-free larval diet was found suitable when offered on cotton towelling [16] instead of cellulose, and promising in testing certain feedstuffs or other inert substances [17]. This larval diet was further improved by using low-cost feed-grade celluloses [10] in the place of the expensive chromatographic ones used today, utilizing propionic acid and/or sodium proprionate [10] instead of nipagin and potassium sorbate, and was found equally efficient when inexpensive local by-products were used (unpublished data of this Laboratory).

Despite recent developments, a diet with less non-protein amino-acid content would be simpler, cheaper, less vulnerable to contamination and possibly less detrimental to the fly [14]. On the other hand, larval diets used in mass rearing of other fruit flies, including Dacus odorsalis and Dacus curcubitae [18], do not contain any protein hydrolysate source. It was therefore the object of this study to determine if and to what extent soy hydrolysate could be removed from the peanut-free diets used at present. This study reports a protein hydrolysate-free larval diet and the effect of graded levels of brewer's yeast upon the performance of D. oleae. In addition it gives comparative results on the nutritional role of three lots of brewer's yeast with or without soy hydrolysate supplementation. Finally the peanut-free, soy hydrolysate-free diets are compared under mass rearing conditions in this Laboratory for performance and dietary efficiency.

MATERIALS AND METHODS

The peanut-free diet [10] was modified by replacing the expensive chromatographic cellulose (Schleicher-Schüll, Federal Republic of Germany) with a three times less expensive feed grade Solka-Floc cellulose (Brown & Co., United States of America). This peanut-free, Solka-Floc diet, is designated as PF in Table I and was used as a control. Furthermore, it was found that sugar may be eliminated from PF diet without any detrimental effect upon larval performance, and that olive oil improves pupal weight and reduces variability when increased to approximately 5% of the diet and Tween-80 appears to be unnecessary (unpublished data). This diet, desig-nated as diet PFS, is also presented in Table I, and constitutes the basic diet for the formulation of the experimental diets reported here. In addition, Table I presents the experimental soy hydrolysate-free diets, designated as PHF, and the diet N.

The eggs used in this study were obtained from adults maintained in this Laboratory according to the procedures described by Tsitsipis [15]. The mixing of the control diet was similar to that for diet N [7] except that boiling is not required. All diets used in this study were made up of three times the quantities listed in Table I for the first four experiments, and of ten times for the last experiment. Diet N was prepared according to the procedure described by Tzanakakis and co-workers [7]. The experimental

TABLE I. CONTROL AND EXPERIMENTAL DIETS USED FOR
REARING OLIVE FRUIT FLY LARVAE

Ingredient	Control diets		Experimental diets	
	N[a]	PF[b]	PFS	PHF
Water (ml)	55.0	55.0	55.0	55.0
Brewer's yeast[c] (g)	7.5	7.5	7.5 to 15.0	5.0 to 20.0
Olive oil (mlitre)	2.0	2.0	5.0	5.0
Soy hydrolysate[d] (g)	3.0	3.0	0.3 to 3.0	-
Sucrose (g)	2.0	2.0	-	-
Peanuts (g)	6.0	-	-	-
Agar[e] (g)	0.5	-	-	-
Tween-80[e] (mlitre)	0.75	0.75	-	-
Potassium sorbate[e] (g)	0.05	0.05	0.05	0.05
Nipagin[e] (g)	0.2	0.2	0.2	0.2
HCl, 2N (mlitre)	3.0	3.0 ± 1.0	3.0 ± 1.0	3.0 ± 1.0
Cellulose Solka-Floc[f] (g)	20.0	27.5 ± 2.5	27.5 ± 2.5	27.5 ± 2.5

[a] Standard diet N reported by Tzanakakis et al. [7].
[b] Peanut-free diet reported by Manoukas [10] and modified by Manoukas (unpublished data).
[c] Fix Co., Athens, Greece.
[d] Nutritional Biochemicals Co., Ohio, United States of America.
[e] Merck A.G., Federal Republic of Germany.
[f] Brown & Co., United States of America, Feed grade BW-40, except for Diet N in which Schleicher-
Schüll, No.123, chromatographic grade was used which is approximately three times as expensive
as Solka-Floc.

N.B. Variable levels of hydrochloric acid were used to adjust pH between 4.1 and 4.4. Cellulose
was also variable to give "desirable appearance" in the diet.

diets were mixed essentially as the control diets. The "desired appearance
and texture" was obtained with the quantity of Solka-Floc cellulose listed
in Table I. Each diet was placed in round plastic containers (of 7 cm lower
diameter, 9 cm upper diameter and 4 cm height) with plastic covers. The
height of feed in each container was 2 to 3 cm, and approximately 60 g of
diet were placed in each container. Exception was made in the final experiments where plastic round trays described by Tzanakakis and co-workers [7]
were used, and approximately 300 g of diet were placed in each tray.
Twenty-four to forty-eight-hour-old eggs were used in all experiments.
The last experiment was performed under the mass-rearing procedures
used in this Laboratory [15]. Six eggs per gram of diet were used to
ensure that the diet was utilized to its maximum capacity, and at the same
time avoiding crowding of the young larvae. The eggs were either placed
directly on the diet of each container and then mixed gently or they were
mixed in the mixer after the cellulose has been added and then the diet
was placed in each container. Hatchability was checked on 20 eggs placed

on tissue paper in each container. An error of approximately 10% is
expected when eggs were counted. Pupae were taken out of the diet on the
12th day following placement of the eggs and each day thereafter. The
day at which 90% of the pupae were obtained was considered as "days to
90% pupation". The pupae were weighed three days after the day on which
90% pupation was obtained. Adult emergence was recorded from a sample
of 100 pupae obtained randomly from each dietary replicate. The diets and
pupae were kept in an incubator set at 25°C, which was sufficiently constant
in temperature but uncontrolled in moisture.

Nitrogen in brewer's yeast and pupae was determined by the micro-
kjeldahl method. Non-protein nitrogen (presumably free amino-acids)
was extracted with 70% alcohol, according to Freeland and Gale [19].
When results were statistically evaluated, the procedures described by
Steel and Torrie [20] were used.

RESULTS AND DISCUSSION

Five experiments were carried out to develop and evaluate the soy
hydrolysate-free diets. Table I presents the formulation of the control
and experimental diets. The diets of the first four experiments were
placed in small plastic containers (five containers per diet), and those of
the last experiment in round plastic trays (three trays per diet). The same
ingredients were used in all experiments, unless otherwise indicated.

The first experiment was designed to compare the performance of the
peanut-free diets PF and PFS with diets in which brewer's yeast replaced
soy hydrolysate. Six diets were formulated for this purpose. Table II
presents the results of this experiment together with brewer's yeast (BY)
and soy hydrolysate (SH) content of each diet. Both diets N and PF were
used as control for the basic diet PFS and for the experimental diets. The
results show that diet N gave a more than twice lower pupal yield than any
other diet tested. Pupal weight and adult emergence for diet N was, however,
the highest. This suggests that diet N does not interfere with normal growth
of the surviving larvae but only with the number of larvae that survive to
pupation. The basal diet PFS gave statistically higher pupal yield and pupal weight
than the same diet without SH. All other diets gave equivalent results.
Both SH-free diets gave pupal weight statistically lower compared with
SH-containing diets. However, pupal yield of diet containing 10.5 g BY/55 ml
water (diet 6) was equivalent to the basic PFS diet which suggests that higher
BY content than that found in basic SH-free diet (diet 4) improves pupal yield.
There was also a delay in days required for 90% pupation for the SH-free
diets. Thus, 17 and 15 days were needed for the SH-free diets containing
7.5 and 10.5 g BY/55 ml water respectively, compared with 13 days for
the other diets.

The second experiment was designed to determine the effect of BY level
upon the performance of D. oleae in SH-free diets. Diet PFS was used as
a control and basic in this experiment. Four additional diets were formulated
containing 5, 10, 15 and 20 g BY/55 ml water. The data of this experiment
(Table II) show that the diet containing 5 g BY (diet 2) depressed significantly
pupal yield, pupal weight and adult emergence compared with other diets.
In addition it took 18 days to 90% pupation for this diet (diet 2) compared
with 16 days for the diet containing 10 g of BY (diet 3) and 14 days for the

TABLE II. PERFORMANCE OF OLIVE FRUIT FLY LARVAE REARED IN PEANUT-FREE DIETS WITH OR WITHOUT SOY HYDROLYSATE AND WITH VARIABLE LEVELS OF BREWER'S YEAST

Diet No.	Basic diet[a]	g/55 ml water		Pupae[b]/ replicate (No.)	Pupal weight (mg/pupa)	Adults (% of pupae)
		Brewer's yeast	Soy hydrolysate			
EXPERIMENT 1						
1	N	7.5	3.0	84A	6.11C	87
2	PF	7.5	3.0	199BC	5.26B	74
3	PFS	7.5	3.0	223C	5.56BC	71
4	PHF	7.5	-	153B	4.22A	62
5	PFS	10.5	3.0	206BC	5.57BC	80
6	PHF	10.5	-	180BC	4.20A	63
EXPERIMENT 2						
1	PFS	7.5	3.0	198B	6.17D	88B
2	PHF	5.0	-	106A	3.28A	53A
3	PHF	10.0	-	173B	4.29B	71B
4	PHF	15.0	-	195B	5.26C	83B
5	PHF	20.0	-	170B	5.43CD	86B
EXPERIMENT 3						
1	PF	7.5	3.0	171AB	5.26B	74
2	PFS	7.5	3.0	202B	5.72B	88
3	PHF	7.5	-	136A	4.04A	73
4	PFS	15.0	3.0	193B	5.61B	96
5	PHF	15.0	-	207B	4.80AB	83

[a] See Table I for formulation of basic diets.

[b] Five replicates (containers) per treatment (diet) were used. Sixty grams of feed and 360 ± 36 eggs were placed in each container. Hatchability was between 68% and 84% of the eggs.

N.B. Numbers in the same column followed by the same capital letter or without a letter do not differ significantly at the 0.05 level of probability.

last two diets. Each increase in yeast content gave a significant increase in body weight for the first three SH-free diets compared with the control and with each other. The last diet (diet 5) gave lower pupal weight than the control diet PFS which failed to be of any significance. Unfortunately diets with higher BY content than this diet (diet 5) reduced pupal yield, probably because of unsuitable texture for larval growth. The data suggest that the SH-free diet containing 15 g BY per 55 ml water was the best because it gave the highest pupal mass (number of pupae × pupal weight).

TABLE III. PERFORMANCE OF OLIVE FRUIT FLY LARVAE REARED
IN SOY HYDROLYSATE-FREE DIETS CONTAINING BREWER'S YEAST
FROM THREE LOTS WITH DIFFERENT EXTRACTABLE NON-PROTEIN
NITROGEN (NPN): EXPERIMENT 4

| Diet No. | Yeast lot | Basic diet[a] | g/55 ml water | | | Pupae[b]/ replicate (No.) | Pupal weight (mg/pupa) | Adults (% of pupae) |
			Brewer's yeast (total)	NPN[c] x6.25	Soy hydrolysate			
1	4	PFS	7.5	0.6	3.0	156C	6.12B	86
2	4	PHF	15.0	1.0	-	150C	5.01A	81
3	5	PFS	7.5	0.4	3.0	60A	5.64AB	88
4	5	PHF	15.0	0.8	-	58A	4.93A	74
5	5	PFS	15.0	0.8	0.3	79A	4.83A	75
6	5	PFS	15.0	0.8	1.0	52A	5.29AB	78
7	6	PFS	7.5	0.3	3.0	142BC	5.91B	83
8	6	PHF	15.0	0.6	-	96AB	4.53A	74
9	6	PFS	15.0	0.6	0.7	121B	5.14A	77
10	6	PFS	15.0	0.6	2.0	165C	5.62AB	85

a See Table I for formulation of basic diets.
b Three replicates per treatment were used. Sixty grams of feed and 360 ± 36 eggs were placed in each
   container. Hatchability was between 72% and 87% of the eggs.
c Non-protein nitrogen (presumably free amino-acids) extracted with 70% alcohol [18] from yeast.

N.B. Numbers in the same column followed by the same capital letter or without a letter do not differ
significantly at the 0.05 level of probability.

        Experiment 3 was designed to test the results of the previous two
experiments and to establish the effect of SH when added to a diet containing
15 g BY/55 ml water. Table II presents the design and the results of this
experiment. The SH-free diet containing 7.5 g BY/55 ml water depressed
significantly pupal yield and pupal weight. This confirms the results of
the first experiment. All diets gave statistically equivalent results, though
pupal weight of the SH-free diet containing 15 g BY was 12% lower than the
same diet supplemented with SH.
        General consideration of the results presented in Table II shows that
10.5 g of BY/55 ml water is sufficient to give a number of pupae and adults
equivalent to the control. Pupal weight, however, is significantly lower
than that of the control. When yeast increases to 15 g, pupal weight
increases but it remains constantly lower than that of the control diets,
though in Experiment 3 this difference was not significant. Higher yeast
level (diet 5, Experiment 2) improved pupal weight but reduced the number
of pupae probably because of unsuitable physical texture. It seems that

SH balances the amino-acid pattern of BY with respect to total amino-
acids or that D. oleae larvae need a larger amount of free and/or peptide-
bound amino-acids for maximum pupal weight than that supplied by BY
at 15 g per 55 ml water. However, it should be mentioned that other factors
may be involved.

The role of non-protein nitrogen (presumably free amino-acids) content
of BY is shown in experiment 4 in which three lots (No. 4, 5, 6) of BY supplied
by the same manufacturer at different intervals were tested (Table III).
Analysis of these samples for total nitrogen and non-protein nitrogen (NPN)
showed that NPN content differed widely contrary to the total nitrogen,
which was fairly constant. The NPN content was 1.10, 0.87 and 0.68%,
and total nitrogen content was 8.53, 8.12 and 8.14% of brewer's yeast for
lot 4, 5 and 6 respectively. Lot 4 was the one used in the previous experiments. Lot 5 gave poor results in our mass-rearing colony and lot 6 gave
poor and variable results when used in SH-free diets. Experiment 4 was
designed to test these results under the same conditions. Ten diets were
formulated in this experiment with and without SH. Soy hydrolysate was
added in BY lot 5 and 6 to make it equal to the NPN found by analysis in
BY lot 4 when 100% of SH nitrogen is considered as NPN, and when 33%
(according to the manufacturer) is free amino-acid nitrogen. Diets 1, 3
and 7 were included as controls. Table III presents the design, diets and
results of this experiment. Lot 4 of BY confirms the results of the previous
experiments. The data of diets containing lot 6 suggest that this yeast is
equivalent to that of lot 4 with respect to other nutrients and properties
but not with respect to NPN. Thus, when 0.7 of SH/55 ml water was added,
pupal yield was improved and when 2.0 g SH was added, yield was equivalent
to the control. On the contrary, lot 5 gave very low results in all diets
which indicates that factors other than NPN were responsible for the low
yield. The results of this experiment suggest that brewer's yeast may
supply the free amino-acids needed for larval development [4] along with
other dietary requirements of the olive fruit fly. They also illustrate the
importance of routine quality control of this ingredient in a mass-rearing
programme.

Experiment 5 was performed to compare the peanut-free and soy
hydrolysate-free diets under mass-rearing practices employed in this
Laboratory [15]. Brewer's yeast lot 4 was used in this experiment.
Table IV presents the results. The number of pupae obtained from diets
PF, PFS and PHF was three times higher that obtained from diet N.
Pupal weight was highest in diet N and lowest in the SH-free diet PHF.
In this experiment a one day delay in pupation time was observed in diet PHF
compared with other diets. Total nitrogen efficiency was 1.1, 4.0, 4.0 and
2.3% for diets N, PF, PFS and PHF respectively. Despite the higher
efficiency of the peanut-free and SH-free diets compared with diet N, it
should be emphasized that protein efficiency of all diets remains very low
which indicates that there is room for considerable improvement. Under
the condition that NPN (mainly free amino-acids) and not intact protein was
exclusively utilized for larval development, efficiency of utilization of NPN
would be 7.8, 20.6, 20.6 and 18.8% for diets N, PF, PFS and PHF respectively
(Table IV, g). Similar calculations for the SH-free diet 4 (Experiment 1)
and for diets 2 and 3 (Experiment 2) show that efficiency of NPN for BY
should fall between 20 and 30%. These figures are rather high if we consider
the low overall dietary utilization (Table IV, b, h) and therefore intact

TABLE IV. COMPARATIVE PERFORMANCE AND DIETARY
EFFICIENCY OF THE OLIVE FRUIT FLY REARED IN DIET N,
PEANUT-FREE DIET PF, PFS AND SOY HYDROLYSATE-FREE DIET
PHF UNDER MASS REARING CONDITIONS: EXPERIMENT 5

Performance and dietary efficiency (mean ± Standard error of the mean)	Diet [a]			
	N	PF	PFS	PHF
Hatchability (% on eggs)	83 ± 13	78 ± 8	85 ± 7	82 ± 9
Pupae (number/g diet)	1.1 ± 0.6	3.2 ± 0.8	3.1 ± 0.6	2.9 ± 0.8
Pupal weight (mg/pupa)	6.1 ± 0.6	5.4 ± 0.5	5.6 ± 0.4	4.8 ± 0.5
Pupal mass (mg/g diet)[b]	6.7	17.3	17.4	13.9
Days to 90% pupation	13	13	13	14
Adults (% on pupae)	84 ± 12	85 ± 14	80 ± 8	76 ± 11
Total nitrogen (mg/g diet)[c]	12.3	9.3	9.3	12.8
Non-protein nitrogen (mg/g diet)[d]	1.8	1.8	1.8	1.6
Pupal nitrogen in mg/g diet[e]	0.14	0.37	0.37	0.30
Total nitrogen efficiency (%)[f]	1.1	4.0	4.0	2.3
Non-protein nitrogen efficiency (%)[g]	7.8	20.6	20.6	18.8
Efficiency of diet (%)[h]	0.9	2.5	2.5	1.8

[a] Diet N reported by Tzanakakis et al. [7]. Diets PF, PFS with 7.5 g brewer's yeast plus 3.0 g soy hydrolysate and PHF with 15.0 g brewer's yeast, per 55 ml water, reported in this study (Table I). Three replicates per diet, 300 g diet per replicate and 6 eggs per g diet were used.

[b] By multiplying pupal weight and number of pupae.

[c] Based on analysis of brewer's yeast, soy hydrolysate and peanuts.

[d] Based on analysis of brewer's yeast and taking 33.3% of soy hydrolysate nitrogen as being amino-acid nitrogen (technical information of manufacturer).

[e] A composite sample of pupae was found to contain 2.15% total nitrogen of fresh pupal weight.

[f] Derived from total nitrogen calculated for diet and pupae. It is assumed that nitrogen utilization is similar whether protein or non-protein.

[g] It is assumed that non-protein nitrogen (mainly free amino-acids) is exclusively utilized by larvae.

[h] Calculated after exclusion of cellulose.

protein that should be utilized by the larvae. Furthermore, nutritional balances made on the soy hydrolysate-free diets for protein and NPN before and after larval pupation indicate that intact protein of BY is utilized by the larvae although it is not clear at present to what extent this is utilized (unpublished results). The dietary efficiency of diets presented in Table IV is 5 to 10 times lower when compared with diets for similar fruit flies [18, and Tanaka, personal communication], which contributes to the high cost of mass-rearing of the olive fruit fly.

ACKNOWLEDGEMENT

I wish to thank J.A. Tsitsipis of this Laboratory for criticism and review of this manuscript.

REFERENCES

[1] ORPHANIDIS, P.S., et al., Ann. Inst. Phytop. 9 (1970) 147.

[2] TZANAKAKIS, M.E., Ann. Sch. Agric. Forestry Univ. Thessaloniki 14 (1971) 293.

[3] REY, J.M., 8th FAO ad hoc Conference on the Control of Olive Pests and Diseases, Athens, Greece (1969).

[4] HAGEN, K.S., SANTAS, L., TSECOURAS, "A technique of culturing the olive fly, Dacus oleae Gmel., on synthetic media under xenic conditions", Radiation and Radioisotopes applied to Insects of Agricultural Importance (Proc. Symp. Athens, 1963), IAEA, Vienna (1963) 333.

[5] SANTAS, L.A., Bull. Prog. Bank, Greece 144 (1965) 1754.

[6] ORPHANIDIS, P.S., et al., 8th FAO ad hoc Conference on the Control of Olive Pests and Diseases, Athens, Greece (1969).

[7] TZANAKAKIS, M.E., ECONOMOPOULOS, A.P., TSITSIPIS, J.A., Rearing and nutrition of olive fruit fly, 1: Improved larvae diet and simple containers, J. Econ. Entomol. 63 (1970) 317.

[8] CAVALLORO, R., Redia L (1967) 337 (in Italian).

[9] PELEKASSIS, C.E., SANTAS, L.A., 8th FAO ad hoc Conference on the Control of Olive Pests and Diseases, Athens, Greece (1969).

[10] MANOUKAS, A.G., Research Report on Nutrition and Chemistry of the Olive Fruit Fly Dacus oleae, GAEC, "Demokritos" Nuclear Research Centre, Athens, Greece, DEMO 72/9 (1972).

[11] SILVA, G.M., Final Report to IAEA: Development of Mass Production Techniques in the Artificial Rearing of the Olive Fruit Fly, Dacus oleae, Gmel., Estacao Agron. National, Oeiras, Portugal, 542/RB (1974).

[12] MANOUKAS, A.G., et al., J. Agric. Food Chem. 21 (1973) 215.

[13] VAKIRTZI-LEMONIAS, C., KARAHALIOS, C.C., MANOUKAS, A.G., "Choline phosphoglycerides in adults and amino acids in larvae of the olive fruit fly", Sterility Principle for Insect Control or Eradication (Proc. Symp. Athens, 1970), IAEA, Vienna (1971) 313.

[14] MANOUKAS, A.G., J. Comp. Bioch. Physiol. 43B (1972) 787.

[15] TSITSIPIS, J.A., FAO/IAEA Panel on the Sterile-Male Technique for Control of Fruit Flies, 1973 (unpublished).

[16] MITTLER, T.E., TSITSIPIS, J.A., Entomol. Experiment. Applic. 16 (1973) 292.

[17] MOURAND, A., FAO/IAEA Panel on the Sterile-Male Technique for Control of Fruit Flies, 1973 (unpublished).

[18] TANAKA, N., et al., J. Econ. Entom. 62 (1969) 967.

[19] FREELAND, J.C., GALE, E.F., Biochem. J. 41 (1947) 135.

[20] STEEL, R.G., TORRIE, J.H., Principles and Procedures of Statistics, McGraw-Hill, London (1960).

DISCUSSION

R. CAVALLORO: How many generations did you rear with this new larval diet and what is the fertility of laboratory colony adults in the successive generations?

A.G. MANOUKAS: We proceeded up to second generation only because pupal weight was low and became even lower in the second generation. We would like to correct pupal weight before we do further studies. Fertility was within normal limits in the first generation, but no fertility data were obtained for the second generation on this new soy hydrolysate-free larval diet. At this stage we are more concerned with pupal weight than fertility.

G. DELRIO: How many generations did you obtain with the larval diets normally used in your laboratory?

A.G. MANOUKAS: I think that over 70 generations have been obtained from peanut-containing diets and over 20 generations from the peanut-free diet.

G. DELRIO: When rearing Dacus oleae do you add wild flies to each generation and, if so, do you not think that you may still be at the first generation?

A.G. MANOUKAS: We keep a small colony of the original strain, from which several generations have been obtained, as reference material as well as our main colony, to which a wild strain is periodically introduced.

H. LEVINSON: Most work on dietetics concerning Dacus and other species has been empirical. Notwithstanding the fact that there could be a better artificial food than the one encountered in nature, one should begin work on developing mass-rearing media with a detailed and very precise chemical and physical analysis of a favoured natural host. An efficient laboratory diet suitable for rearing the above species through numerous generations could then be based on the quantitative and qualitative composition of olives.

Furthermore, recent findings have disclosed that several antibiotics interfere with protein biosynthesis and one should therefore refrain from adding such antimicrobial agents to mass-rearing media as they could postpone pupation.

A.G. MANOUKAS: Your observation is correct. We have already done some work on the quantitative composition of the olive fruit and the results have been published in the Journal of Agricultural and Food Chemistry, Vol. 21, No. 2 (1973) p. 215. Moreover, the chemical composition of the egg, of the larvae and of the adults may also help in formulating a diet for the insect and we have done some work on this too, which has been published in Comparative Biochemistry and Physiology, Vol. 43, No. 4 (1972) p. 787, and in the Journal of Insect Physiology, Vol. 18, No. 4 (1972) p. 683.

With regard to your second comment, I know that antibiotics can interfere with microorganisms which supply vital nutrients to the insect but I have not done any work on this subject.

EFFETS DU FRACTIONNEMENT
DE LA DOSE STERILISANTE
DE RAYONS GAMMA SUR L'EMERGENCE,
LA FERTILITE ET LA COMPETITIVITE
DE LA MOUCHE MEDITERRANEENNE
DES FRUITS, Ceratitis capitata Wied.

I.A. MAYAS
Institut de recherches agronomiques,
Fanar, Liban

Abstract—Résumé

EFFECTS OF FRACTIONATING THE GAMMA-RAY STERILIZATION DOSE ON THE ECLOSION, FERTILITY AND COMPETITIVITY OF THE MEDITERRANEAN FRUIT FLY Ceratitis capitata Wied.

The effects of different ways of fractionating the sterilizing gamma-ray dose (9 krad) on older fruit fly pupae (3, 2 and 1 day before eclosion) have been studied. As a result (a) no significant difference was observed in the adult eclosion rate and male mortality rate between batches subjected to a whole single dose and batches exposed to a fractionated dose; (b) males from pupae subjected to fractionated irradiation show the same sterility rate as those from pupae given a single dose; (c) competitivity between irradiated and unirradiated males tested under laboratory conditions is derived from the percentage of sterile eggs laid by unirradiated females (a ratio of 3 : 1 : 1). Males from pupae that have received the whole single dose are markedly less competitive than males that have been subjected to fractionated irradiation. The reduced competitivity persists 17 days after irradiation; and (d) the different ways of fractionating the sterilizing dose (3, 6 and 9 fractions) did not produce results differing significantly in terms of eclosion and competitivity.

EFFETS DU FRACTIONNEMENT DE LA DOSE STERILISANTE DE RAYONS GAMMA SUR L'EMERGENCE, LA FERTILITE ET LA COMPETITIVITE DE LA MOUCHE MEDITERRANEENNE DES FRUITS, Ceratitis capitata Wied.

Les effets de différentes modalités de fractionnement de la dose stérilisante de rayons gamma (9 krad) sur les pupes âgées de Ceratitis capitata Wied. (3, 2 et 1 jour avant l'émergence) ont été étudiés. a) Il n'a été constaté aucune différence significative quant aux taux d'émergence des adultes et de mortalité des mâles, entre les lots qui ont subi une irradiation continue et ceux qui ont subi une irradiation fractionnée. b) Les mâles provenant de pupes ayant subi un traitement fractionné présentent le même taux de stérilité que ceux qui proviennent de pupes ayant subi un traitement continu. c) La compétitivité entre mâles irradiés et mâles non irradiés, étudiée en laboratoire, est déduite du pourcentage d'œufs stériles déposés par des femelles non irradiées (rapport 3 : 1 : 1). Les mâles provenant de pupes ayant subi un traitement continu montrent une compétitivité nettement inférieure à celle des mâles qui ont subi un traitement fractionné. Cette baisse de compétitivité se maintient encore 17 jours après le traitement. d) Les différentes modalités de fractionnement de la dose stérilisante (3, 6 et 9 fractions) n'ont pas donné de résultats significativement différents en ce qui concerne le taux d'émergence et la compétitivité.

INTRODUCTION

Depuis que la lutte génétique a été conçue par Knipling [1] et appliquée avec succès, pour la première fois contre Cochliomyia hominivorax Coquerel [2], cette nouvelle conception de lutte a suscité le plus vif intérêt. Au fur et à mesure que les recherches progressaient dans ce domaine, il devenait de plus en plus évident que la «technique des mâles stériles», si simple en théorie, nécessite en fait des études de base, dont l'une des plus importantes est celle qui conduit à obtenir des mâles

TABLEAU I.   MODALITES DE FRACTIONNEMENT DE LA DOSE DE RAYONS GAMMA
ADOPTEES POUR LES LOTS TRAITES

Age (jours) \ Lot	-3	-2	-1	Nombre de fractions	Dose totale fournie
A	2 krad 1 fois	3 krad 1 fois	4 krad 1 fois	3	9 krad
B[a]	1 krad 2 fois	$1\frac{1}{2}$ krad 2 fois	2 krad 2 fois	6	9 krad
C[b]	$\frac{2}{3}$ krad 3 fois	1 krad 3 fois	$1\frac{1}{3}$ krad 3 fois	9	9 krad
Dose totale par jour	2 krad	3 krad	4 krad		

[a] Les deux fractions quotidiennes sont fournies avec un intervalle de 8 h.
[b] Les trois fractions quotidiennes sont fournies avec deux intervalles de 4 h chacun.

non seulement stériles, mais également en bon état physiologique, et dont l'agressivité sexuelle n'a été que légèrement affectée par l'agent de stérilisation.

Un certain nombre de travaux, dont ceux de Jefferies sur Sitophilus granarius (Calandra granaria L.) [3], nous ont suggéré l'utilisation du traitement fractionné par les rayons gamma pour maintenir l'agressivité sexuelle des mâles irradiés à un niveau supérieur à celui des mâles qui subissent un traitement continu.

Dans le présent travail nous étudions l'action du traitement stérilisant par les rayons gamma fournis en doses fractionnées et en doses continues aux pupes de la mouche méditerranéenne des fruits, Ceratitis capitata Wied., sur la compétitivité sexuelle des mâles.

MATERIEL ET METHODES

Les pupes utilisées au cours de nos essais proviennent d'un élevage permanent de C. capitata en laboratoire. Les larves sont élevées sur un milieu artificiel à base de germe de blé, mis au point au Laboratoire d'entomologie de l'AIEA à Seibersdorf (Autriche) [4]. L'élevage est effectué dans une pièce climatisée à une température de 25-27°C et à 80% HR.

Les pupes sont irradiées à l'âge de 3, 2 et 1 jour avant l'émergence. Trois lots reçoivent chacun une dose de 9 krad, selon les modalités de fractionnement données dans le tableau I.

Ces trois lots sont comparés à un lot témoin (T_1) non traité, et à un autre lot témoin (T_2) ayant subi 9 krad en dose continue, à l'âge de - 1 jour.

Le choix de l'âge des pupes traitées s'accorde avec les travaux de Katiyar [5], Katiyar et Valerio [6], Féron [7] et Shoukry [8]. Les doses quotidiennes administrées à chacun des trois lots sont justifiées par les travaux de Steiner et Christenson [9] d'une part, et par ceux de Katiyar et coll. [10] d'autre part.

Pour l'accouplement des adultes, nous avons utilisé des cages en matière plastique (13 cm × 13 cm × 17 cm), dont le fond est constitué d'un tissu moustiquaire en fibres synthétiques permettant aux femelles fécondées de déposer leurs œufs à l'extérieur, sur un papier filtre noir fortement imbibé d'eau. Les adultes sont alimentés avec un mélange d'hydrolysat de levure de bière et de sucre (1:3). Les cages sont placées dans une pièce climatisée à une température de 25,0 ± 1°C et 70% HR.

Le développement embryonnaire a lieu dans des boîtes de Pétri, placées dans la même pièce climatisée. Le maintien d'un microclimat fortement humide à l'intérieur des boîtes est assuré par une imbibition permanente des papiers filtres porteurs des œufs.

Les irradiations sont effectuées avec un irradiateur de ^{60}Co de 5000 Ci; le débit est de 6941 rad/min.

Les taux d'émergence des lots irradiés, évalués pour 200 pupes par lot, sont comparés entre eux, et au témoin non traité. A chaque répétition (cinq en tout), les pupes de tous les lots proviennent de larves ayant été élevées sur une même pâte de milieu artificiel. Il en est de même des pupes utilisées dans les autres tests. Après émergence, 50 mâles de chaque lot sont isolés dans les cages en plastique décrites plus haut et, pour chaque lot, le taux de mortalité est établi trois semaines plus tard (cinq répétitions également).

Pour le test de stérilité des mâles, le seul critère retenu est le taux d'éclosion des œufs, compte tenu de l'absence complète, après éclosion, de toute mortalité larvaire consécutive au traitement [7]. A chacune des cinq répétitions nous avons utilisé 25 mâles traités et 25 femelles non traitées par lot. Des échantillons de 300 œufs environ sont prélevés dans chaque lot, trois fois par semaine, et le taux d'éclosion est établi 4 jours plus tard. Le test est poursuivi pendant trois semaines.

Un test de survie d'adultes mâles provenant des différents lots traités a été jugé nécessaire pour interpréter les résultats obtenus au cours du test de compétitivité. Comme pour ce dernier, le test de survie dure trois semaines, au cours desquelles des lots de 100 adultes sont observés et les mortalités notées (cinq répétitions).

La compétitivité sexuelle des mâles irradiés des différents lots est estimée à l'aide de rapports tests [6, 9]. Le rapport 3 : 1 : 1 (36 mâles irradiés : 12 mâles non traités : 12 femelles) choisi pour ce test est calculé d'après la méthode de Fried [11]. Comme pour le test de stérilité, des échantillons d'œufs (en nombre toujours supérieur à 500 par lot) sont prélevés trois fois par semaine, et le taux d'éclosion est établi. Le test de compétitivité est poursuivi pendant trois semaines. Le degré de stérilisation de chaque lot étant établi, la compétitivité des mâles irradiés vis-à-vis des mâles non irradiés est déduite du taux d'éclosion des œufs prélevés. Ensuite, les taux d'éclosion dans les différents lots sont comparés statistiquement entre eux, afin de préciser l'action du fractionnement de la dose fournie.

RESULTATS ET DISCUSSION

Emergence et survie des adultes

L'analyse statistique des taux d'émergence des adultes n'a mis en évidence aucune différence significative entre les valeurs des deux lots témoins (traité et non traité) et celles des trois lots A, B et C. L'absence de toute différence significative entre les deux lots témoins s'accorde bien par ailleurs avec les résultats obtenus par Katiyar [5], Katiyar et Valerio [6] et Féron [7]. Les modalités de fractionnement adoptées au cours de cette étude ne semblent pas affecter les processus qui conduisent à l'émergence de l'adulte.

L'analyse statistique des taux de mortalité des adultes n'a pas non plus mis en évidence une différence significative entre les lots traités et le témoin non irradié. L'analogie des résultats des deux lots témoins s'accorde bien avec les travaux de Katiyar (in Hooper) [12] qui trouve que des doses

TABLEAU II.   MOYENNES DES TAUX D'ECLOSION DES LOTS
TRAITES, CORRIGEES AU TEMOIN NON TRAITE
Entre parenthèses figurent les nombres d'œufs examinés.

Semaine Lot	1	2	3
T_2	1,50% (1533)	1,29% (1474)	0,98% (1521)
A	1,80% (1445)	0,94% (1484)	1,17% (1453)
B	1,49% (1643)	1,68% (1543)	1,46% (1369)
C	1,53% (1503)	1,80% (1604)	1,88% (1382)

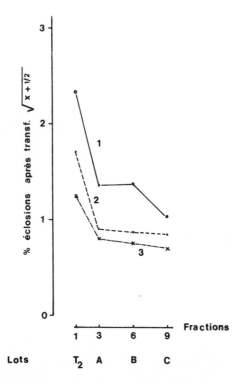

FIG.1.   Variations des taux d'éclosion en fonction du nombre de fractions fournies. 1, 2, 3 représentent
respectivement les première, deuxième et troisième semaines.

allant jusqu'à 20 krad, appliquées à des pupes 3 jours avant l'émergence, n'affectent pas la longévité des adultes des deux sexes. Là encore, les modalités de fractionnement adoptées ne semblent pas affecter la survie des adultes, au moins au cours de la durée de l'essai.

Stérilisation des mâles

Les taux quotidiens d'éclosion dans les lots traités ont varié entre 1,0% et 5,4% au cours de l'essai. En considérant la moyenne des taux d'éclosion au cours des répétitions, ces différences se normalisent pour les différents lots, comme le montre le tableau II.

L'analyse des résultats obtenus ne montre pas de différence significative entre les lots traités, ce qui permet de conclure que l'action stérilisante des fractions administrées selon les modalités adoptées est cumulative dans les conditions de nos essais.

Compétitivité des mâles

La figure 1 représente les variations des taux d'éclosion en fonction du nombre de fractions fournies, après transformation, destinée à normaliser la distribution des variables.

L'analyse de la variance des résultats nous a permis de comparer les lots entre eux par le calcul de la plus petite différence significative

TABLEAU III.   MOYENNES DES TAUX D'ECLOSION DES DIFFERENTS LOTS, APRES TRANSFORMATION DES VARIABLES

Semaines	Lots	% éclosion après transformation $\sqrt{x + 1/2}$	ppds à 0,05
1 (1er - 3e j)	T_2	2,34	
	A	1,36	0,641
	B	1,38	
	C	1,04	
2 (8e - 10e j)	T_2	1,69	
	A	0,90	0,646
	B	0,88	
	C	0,85	
3 (15e - 17e j)	T_2	1,25	
	A	0,81	0,280
	B	0,75	
	C	0,71	

(ppds). Les moyennes de quatre répétitions, corrigées par rapport au témoin non traité, figurent, après transformation des données, dans le tableau III.

Ce tableau fait ressortir une différence significative entre les taux d'éclosion du témoin traité (T_2) et les lots A, B et C. Cette différence s'avère même hautement significative du 1er au 3e jour (à l'exception du lot B), et du 15e au 17e jour après l'émergence des adultes. Par conséquent, il serait permis de conclure que le traitement fractionné favorise significativement la compétitivité sexuelle des mâles par rapport au traitement continu dans les conditions de nos essais. Par contre, les différentes modalités de fractionnement adoptées ne semblent pas favoriser cette compétitivité chez l'un des lots traités par rapport à un autre.

CONCLUSIONS

Dans le présent travail, nous avons essayé d'étudier principalement l'action du fractionnement du traitement stérilisant aux rayons gamma sur la compétitivité sexuelle des mâles de C. capitata Wied.

Il est évident que nos conclusions sont liées au choix de l'âge des pupes et de la modalité du fractionnement adoptée d'une part, et au choix du critère d'évaluation retenu au cours de cette étude, d'autre part.

Nous pensons que les recherches sur l'action du fractionnement doivent s'orienter de manière à déterminer la ou les combinaisons (âges-modalités de fractionnement) pour lesquelles la compétitivité des mâles traités atteint sa valeur maximum. L'application d'autres méthodes d'évaluation de la compétitivité (comparaison de nombres d'accouplements ou de femelles fertilisées, par exemple), apporterait des conclusions qui pourraient compléter celles du présent travail.

REFERENCES

[1] KNIPLING, E.F., J. Econ. Entomol. 48 4 (1955) 459.
[2] BAUMHOVER, A.H., GRAHAM, A.J., BITTER, B.A., HOPKINS, D.E., NEW, W.D., DUDLEY, F.H., BUSHLAND, R.C., J. Econ. Entomol. 48 4 (1955) 462.
[3] JEFFERIES, D.J., in Radioisotopes and Radiation in Entomology (C.R. Coll. Bombay, 1960), AIEA, Vienne (1962) 213.
[4] NADEL, D.J., in Sterile-Male Technique for Control of Fruit Flies (C.R. Groupe d'étude Vienne, 1969), AIEA, Vienne (1970) 13.
[5] KATIYAR, K.P., in Fourth Inter-American Symposium on the Peaceful Applications of Nuclear Energy, Mexico City, 1962 (1962) 211.
[6] KATIYAR, K.P., VALERIO, J., in Fifth Inter-American Symposium on the Peaceful Applications of Nuclear Energy, Valparaiso, 1964 (1964) 197.
[7] FERON, M., Ann. Epiphyt. 17 (1966) 229.
[8] SHOUKRY, A., Thèse Ph.D., Faculté d'agriculture, Université Aïn-Shams, Le Caire, Egypte (1968).
[9] STEINER, L.F., CHRISTENSON, L.D., Proc. Hawaii Acad. Sci. 3 (1956) 17.
[10] KATIYAR, K.P., PERRER, F., VALERIO, J., in The Application of Nuclear Energy to Agriculture, Triennial Report, 1 July 1963 - 30 June 1966, Inter-American Institute of Agricultural Sciences of the OAS, Turrialba, Costa-Rica (1966) 45.
[11] FRIED, F., J. Econ. Entomol. 46 4 (1971) 869.
[12] HOOPER, G.H.S., in Sterile-Male Technique for Control of Fruit Flies (C.R. Groupe d'étude Vienne, 1969), AIEA, Vienne (1970) 3.

DISCUSSION

H. J. HAMANN: In radiation biophysics of cells, dose-splitting effects can very often be explained as a counteraction of damaging and repairing effects. Is it possible that such a mechanism is responsible for the results obtained with your system?

I. A. MAYAS: I am afraid that I cannot say. A detailed cyto-karyologic study would be necessary to answer that question.

PRELIMINARY EXPERIMENTS IN REDUCING THE COST OF MEDFLY LARVAL DIETS

A. MOURAD*
Joint FAO/IAEA Division of Atomic Energy
in Food and Agriculture,
International Atomic Energy Agency,
Vienna

Abstract

PRELIMINARY EXPERIMENTS IN REDUCING THE COST OF MEDFLY LARVAL DIETS.
Brewer's or Torula yeast is used as the primary source of protein and vitamin B in artificial diets for medfly Ceratitis capitata Weid. Many countries conducting large experiments or demonstrations of medfly control by the sterile-male technique must import dried yeast for the larval diet. Shipping charges often exceed the purchase price and both require a substantial outlay in hard currency. For example, brewer's yeast in Austria costs about US $ 0.35/kg; shipping to the Mediterranean region costs about US $ 0.50/kg. The production of 100 million medfly pupae on the standard diet requires 1200 kg brewer's yeast. Because of this cost factor, alternative medfly larval media were investigated using little or no yeast. Also, molasses was evaluated as a substitute for sugar. Several satisfactory diets are reported.

INTRODUCTION

The object of this study was to develop a range of inexpensive larval media for the Mediterranean fruit fly (Ceratitis capitata Wied.). The choice of constituents was such that at least one of the media could readily be made up with materials available in many countries.

Wheat bran and peanuts were used to substitute for brewer's yeast, and molasses for sugar.

MATERIALS AND METHODS

The standard medfly larval diet used at the IAEA Seibersdorf laboratory consists of 52% water, 25% wheat bran, 13% sucrose, 9% brewer's yeast, 0.8% HCl (37%) and 0.2% sodium benzoate (wt: wt) [1, 2].

Diets were prepared by blending successively with water: HCl, aqueous sodium benzoate, yeast, ground peanuts, sucrose or molasses. After thorough blending, wheat bran was mixed in as bulking material. The pH-value of the diet was measured by indicator papers; the range was pH 4 - 4.5. Each diet was placed in petri dishes (15 cm diameter, 1.5 cm deep) at the rate of 150 g/dish. Each diet was replicated three times and so was each experiment.

Approximately 3000 eggs calibrated volumetrically with water were pipetted into each dish, and spread on to the surface.

---

* Present address: Faculty of Sciences, Damascus University, Damascus, Syrian Arab Republic.

TABLE I. MEDFLY LARVAL DIETS TESTED; THREE REPLICATIONS OF THREE PETRI DISHES, EACH CONTAINING 150 g DIET, EACH WITH 3000 FRESH MEDFLY EGGS

Diet	Ingredients (%) [a]				Time to first pupation (days)	Mean pupal wt (mg)	Pupal recovery from eggs (%)	Adult emergence (%)
	Yeast	Sugar	Peanuts[b]	Molasses				
Standard	9.0	13.0	-	-	10.3	8.9	51	90
A	-	13.0	-	-	11.6	7.9	48	80
B	2.25	13.0	-	-	11.6	7.4	49	85
C	4.50	13.0	-	-	10.6	8.1	50	83
D	6.75	13.0	-	-	10.3	8.2	48	93
E [c]	13.5	13.0	-	-	10.6	8.7	34	62
F [d]	18.0	13.0	-	-	11.6	7.7	15	56
G	9.0	3.25	-	-	12.0	5.6	10	75
H	9.0	6.50	-	-	10.6	7.6	45	74
I [c]	9.0	19.50	-	-	10.6	7.9	33	60
J [d]	9.0	26.0	-	-	12.0	6.6	17	50
K	4.50	6.50	-	-	11.0	7.8	45	86
L	-	19.50	-	-	11.0	7.9	54	85
M	9.0	-	-	6.50	11.6	7.7	47	75
N	9.0	-	-	13.0	11.0	8.4	49	73
O [e]	9.0	-	-	19.50	13.0	7.4	40	51
P [d]	9.0	-	-	26.0	-	0	-	0
Q	4.50	-	-	6.50	11.6	7.5	43	80
R	4.50	-	-	13.0	11.0	8.0	46	85
T	-	-	-	13.0	12.3	6.2	45	80
U	-	-	(R) 4.50	13.0	11.3	7.7	47	81
V [e]	-	13.0	(R) 9.0	-	13.3	5.5	7	62
W [d]	-	13.0	(R) 18.0	-	-	-	-	-
X	-	13.0	(B) 4.50	-	12.0	6.3	29	88
Y [f]	-	13.0	(B) 9.0	-	13.0	5.0	7	53
Z [f]	-	13.0	(B) 18.0	-	-	0	-	0

[a] All diets contained 52% water, 25% bran, 0.8% HCl (37%) and 0.2% sodium benzoate, in addition to ingredients tested.
[b] (R) = Raw peanuts, (B) = boiled peanuts.
[c] Diet slightly sticky.
[d] Diet very sticky.
[e] Diet sticky.
[f] Diet greasy.

Dishes were held in the larval rearing room at 25 ± 1°C and 70 - 90% relative humidity. The dishes were covered with lids to maintain the high humidity necessary for hatching. After four days, the covers were slightly opened to permit aeration and heat dispersal. The interior temperature range of the media on the last day of larval development was 28 - 33°C. One day before initial pupation of mature larvae, the covers were removed and the dishes placed in recovery units. These units were paper boxes 25 × 20 × 10 cm. Fine wheat bran was added to the boxes to prevent the larvae from clumping and encourage pupation [3]. Bran and pupae were usually collected once daily during 3 - 4 days. Five days after the last collection the pupae were sieved from the bran, measured and weighed. Mean pupal weight was taken from a random sample of 100 pupae replicates, and the final mean was an average of three replications.

Diets were evaluated according to time elapsed until popping, range of popping, percentage pupal recovery daily, percentage pupae/g of diet, mean pupal weight, pupal period and percentage adult emergence.

RESULTS

Physical characteristics of medfly larval diets influence results [4]. Based on the diets reported here, reasonable variation in the yeast and sugar contents did not seriously affect the physical characteristics of the diet. Molasses tended to make the diets sticky, particularly above the 20% level. Peanuts resulted in "greasy" diets, unless relatively small amounts were used.

The various diets tested and general production figures (Table I) indicate that satisfactory medfly larval diets can be developed using a variety of ingredients and concentrations. The following criteria were arbitrarily established to determine whether a diet was good: (1) pupal weight ≥ 8 mg; (2) no more than 11 days from egg to first pupae; (3) at least 40% yield, egg to pupae; and (4) at least 80% adult emergence. After pupation began, 80% to 95% of the larvae pupated within three days from all diets. Based on these criteria, diets Standard, C, D and R were very good. Molasses can be used in place of sugar. Less yeast than in the standard diet can be used.

It is interesting to note that diet A, which contained no yeast, was quite good. It is obvious that the bran supplies protein and probably vitamin B in the diet.

Raw peanuts can partially substitute for yeast (diets U and X). However, larval development is slow and pupal weight is low. Boiled peanuts were not as satisfactory as raw peanuts.

Diets containing more sugar or yeast than the standard diet were no better than the standard.

The data presented here show that the medfly can be reared on a variety of media and that relatively cheap local ingredients can be used.

REFERENCES

[1] MONRO, J.M., "Insect eradication and pest control group", IAEA Laboratory Activities, Third Report, Technical Reports Series No.55, IAEA, Vienna (1956) 63.
[2] NADEL, D.J., "Mediterranean fruit flies and related species", Advances in Insect Population Control by the Sterile-Male Technique, Technical Reports Series No.44, IAEA, Vienna (1965) 14.

[3] PELEG, B.A., RHODE, R.H., New larval medium and improved pupal recovery method for the
     Mediterranean fruit fly in Costa Rica, J. Econ. Entomol. 63 (1970) 1319.
[4] BAUTISTA, R.C., TANAKA, N., HARRIS, E., A larval rearing diet for the oriental fruit fly utilizing
     Philippine substitutes (unpublished data).

DISCUSSION

J.P. ROS: In our mass breeding laboratory we have performed similar experiments to yours and we found that the egg/larva yield had a decisive effect on all the subsequent data, as the competition between the larvae, when the feed is scarce, has the effect that they do not develop perfectly. Could the same thing not apply in your case also and be the reason why your results are not quite as expected?

A. MOURAD: The larval return depends, among other factors, on the availability of nutrients in appropriate quantity and quality. At high larval densities, nutrient supply can be critical. Since there is a certain amount of nutrient carryover from the mature medfly larva to the adult, competition between larvae can in a limiting situation be expected to have consequences for the adults.

EFFECTS OF STERILIZATION
BY RADIATION AND CHEMICALS
(Session 5)

COMBINED EFFECTS OF RADIATION
AND CHEMICAL AGENTS IN ALTERING
THE FECUNDITY AND FERTILITY
OF A BRACONID WASP

D. S. GROSCH
North Carolina State University,
Raleigh, N.C.,
United States of America

Abstract

COMBINED EFFECTS OF RADIATION AND CHEMICAL AGENTS IN ALTERING THE FECUNDITY AND
FERTILITY OF A BRACONID WASP.
In a search for new methods of control, we should not overlook the advantages in deranging vitellogenesis and thus strike at the basis of the prolificacy characterizing undesirable insects. Species with
merioistic ovarioles are capable of rapid oogenesis because other cells generate molecules and organelles
for the oocyte. Agents which intefere with such activity during vitellogenesis include metal cations, toxic
elements, chelators, enzyme inhibitors, metabolite analogues, base substitution analogues and juvenile hormone
analogues. In combination experiments with ionizing radiation, they prolong the radiation-induced period of
infecundity and decrease the number of eggs subsequently derived from oogonia. Furthermore, hatchability
of the eggs is decreased significantly. The radiation dose chosen was low enough to cause brief infecundity
and thus enable detection of qualitative changes for the different cell types in the ovariole sequence. Contrasted
with the enhancement of radiation effects was the duplication of effort by alkylators, the limited amelioration
from mitotic inhibition, the protection by aminothiols and the subvitalization from low doses of neurotoxic
pesticides.

1. INTRODUCTION

The wasp Bracon hebetor Say, known in the genetic literature as
Habrobracon, has proved to be a useful assay organism with which to identify
agents that attack the specific types of cells of the oogenetic sequence of
the polytrophic type of ovariole. The differential sensitivity of these cell
types enabled our recognition of six kinds of altered oviposition curves
resulting from the differing modes of action by various agents [1]. Data
on unhatched eggs and the effective lethal phase were useful adjunctive
indicators of genetic damage. The purpose of this study is to present results
with an even greater diversity of agents used to alter the reproductive performance of irradiated females. The response with the highest incidence
was a disturbance of vitellogenesis, an aspect of insect vulnerability not
yet exploited for insect control. The nearly unique basis of the prolificacy
that characterizes many undesirable insects will be discussed along with
interpretations from a physiological standpoint.

2. GENERAL METHODS

Wild type strains of Bracon hebetor Say were used. Because the wasp
provides the opportunity to compare parthenogenetically produced offspring
(n) with biparental progeny (2n), induced recessive lethals are detected as

easily as dominant lethal changes. With no block against oviposition by
virgin females or the embryogenesis of unfertilized eggs, production of
haploid offspring is abundant. The ovarian morphology facilitates the identification of vulnerable cell types among the oogenetic sequence. Instead of
the bundles of numerous ovarioles typifying most insects, each ovary has
only two ovarioles of the meroistic type. When eggs are collected in the
sequence of their deposit, discrepancies from the control pattern of oviposition indicate which cell types may be particularly vulnerable to a destructive agent. Verification is afforded by microscopic examination of the
ovariole contents.

For oviposition each wasp was provided with a prestung larva of the
Mediterranean flour moth, Anagasta kuehniella (Zeller). The more flaccid
caterpillar was replaced daily at the time of egg collection. Hatchability
was scored after 48 hours of 30°C incubation in mineral oil. The thin,
transparent chorion enabled visual identification of the effective lethal
phase for the braconid embryo, and stages of death were scored according
to the categories described by Von Borstel and Rekemeyer [2].

In preparation for feeding experiments, wasps were starved for
24-48 hours, that is, until the abdomen appeared slightly shrunken. Wasps
achieving this condition avidly ingest a 10% sugar solution in which a variety
of compounds can be administered. Feeding continued until the expansible
crop was filled. Therefore, a single meal amounted to 0.35 ± 0.05 microlitres

Injections were performed with glass capillary needles attached to a
precision syringe with a micrometric drive. Insertion of the needle between
the second and third abdominal segments was facilitated by using a micro-
manipulator. The amount injected was equivalent to the amount ingested at
a single feeding. The same apparatus was used for topical applications of
0.5 ± 0.1 microlitres to the abdominal surface.

3.   RESULTS AND INTERPRETATIONS

3.1.   Metal toxicity

After studying the effects of a variety of isotopes [3-5], we decided to try
to separate the toxic effect of the element from the damage induced by its
radioactivity. Thirty-six different stable isotopes, predominately metal and
earth elements, were fed to females which were then X-rayed. For palatability, cations were fed either in sulphate or acetate compounds. Non-
metals were fed as simple anions ($NaAsO_2$) or in the acid form (H_2SeO_4).
At the maximum concentration determined by taste, about half the elements
proved toxic to the adult female and deleterious to her reproductive
performance.

Table I presents mean life spans and 20-day egg totals for maximum
palatable doses. Decreases in egg totals were correlated with decreased
life span. Daily deposits were decreased for the subsequent life of the
females and the plotted curves were lower and parallel to that of the controls.
Such a non-selective lowering of egg production has been interpreted as
indicating an altered effectiveness of the somatic tissues concerned with
food assimilation and utilization [1]. Table I shows rare earths to be
ineffective when fed, consistent with the absence of evolved uptake and
transport mechanisms. The most significant differences from control

TABLE I. ELEMENT TOXICITY AS REVEALED BY DECREASED
FECUNDITY AND LIFE SPAN AFTER A SINGLE SUBLETHAL FEEDING

Group	A	Eggs for 20 days		Mean life span (days)	B	Eggs for 20 days		Mean life span (days)
		Mean total	Percent of control			Mean total	Percent of control	
I	Rb	214.3	86.3	18.2	Cu	257.4	82.0	18.2
	Cs	209.9	79.3	19.4	Ag	224.0	84.6	20.2
II	Be	251.7	94.7	23.6	Zn	285.0	90.8	18.5
	Mg	291.5	99.9	24.1	Cd	102.7	65.2	14.0
	Ca	277.3	91.5	19.3	Hg	155.1	79.7	16.4
	Sr	278.6	88.8	18.8				
	Ba	322.3	102.6	20.5				
III	Tl	151.9	55.8	16.6	Y	261.7	102.7	20.2
					La [a]	267.7	107.8	24.3
IV	Sn	216.2	81.6	18.5	Zr	224.6	82.6	17.9
	Pb	250.0	94.4	18.5				
V	As	157.6	68.3	15.6	V	130.8	40.7	11.5
	Sb	196.6	74.3	17.4				
VI	Se	73.1	40.2	14.3	Cr	235.0	94.6	18.2
	Te	87.9	45.4	15.0	Mo	273.1	109.9	23.9
VII	F	256.3	75.0	21.2	Mn	261.9	83.4	16.0
VIII	Fe	289.8	90.8	25.1	[a] Lanthanide earths			
	Co	219.6	69.9	16.4	Ce	192.0	105.5	22.4
	Ni	230.2	73.3	24.1	Pr	209.1	114.9	18.5
					Nd	261.7	143.8	21.9
					Sm	304.2	141.2	18.8
Controls		Many different experiments		19.0 to 24.1	Gd	252.0	116.2	19.3
					Dy	232.1	107.0	21.6

values were obtained with the transition series of the 4th period, with heavy
metals and with known toxicants including As and Se. In equimolar feedings
the order of effectiveness was Ni^{2+}, Co^{2+}, Zn^{2+}, and Cu^{2+}. Depending upon
those used, combinations of metal ions showed either additive or antagonistic effects.

Cell selective decrease in egg production was demonstrated only when
sensory restrictions on concentration as well as the gut barrier were by-
passed by injection. Figure 1 shows that progressive vitellogenesis was

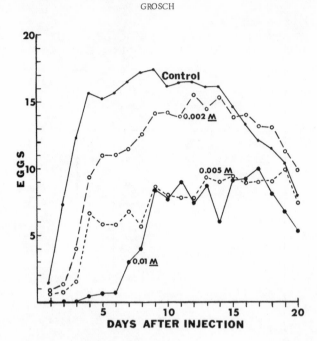

FIG. 1. Average daily egg production per Bracon female. Examples selected from a family of curves obtained by injecting serially diluted NiSO₄ solutions. A higher concentration not shown (0. 015 M) caused death in five days without oviposition.

inhibited in an injected series of $NiSO_4$ solutions which ranged up to eight times the palatability limit. Such a level of treatment is just below a lethal dose.

Table II summarizes hatchability of the eggs laid by females fed a single meal of one of the elements that decreased 20-day egg production. The presentation reflects the initial status of the oogenetic cells destined to give rise to ova during the designated six-day periods. The excellent hatchability during the middle periods for Sn, Pb, Ag and Hg treatment groups demonstrates that decreased life span and fecundity is not necessarily indicative of embryonic lethality. After the 13th day all females were in senile decline, yet of these four metals only Hg depressed hatchability to a degree seen for most of the other experimental groups. Conceivably with palatability limiting the intake level of feeding, the body burden of Sn, Pb, Ag and Hg is relatively low.

Most of the deaths occurred late in embryonic development, but in addition, significantly fewer larvae pupated during the week after treatment. Also there was a slight increase in malformed adult offspring from feeding 0.013 M solutions of the two most effective cations to their mothers:

	Adults		
	Normal	Abnormal	%
Controls	2473	7	0.24
$NiSO_4$	2118	16	0.76
$CoSO_4$	2079	18	0.87

Outcrosses to untreated males did not increase egg production, hatchability or survival to maturity.

3.1.1. Metal toxicity combined with radiation effects

A pattern of damage not achieved in feeding experiments, yet implied by injections of 4th-period cations, became fully evident in combination experiments when a radiation dose adequate to induce a characteristic 7th-day nadir in egg production was delivered to cation-fed females. That is, an extended period of infecundity resulted regardless of an acute or chronic radiation exposure. Figure 2 contrasts results from an X-ray dose given in two fractions separated by a four-hour interfraction period. With Ni^{2+} present in sub-lethal amount, egg production declined promptly and reached zero by the third day. Furthermore the recovery phase (stem cell repopulation) was delayed several days beyond that of females receiving only radiation. Also it never became pronounced.

TABLE II. MEAN HATCHABILITY AS PERCENTAGE HATCHED SUMMARIZED BY THREE PERIODS OF DEPOSIT

Eggs laid	Days after treatment		
	1 - 6	7 - 12	13 - 18
Cell type at treatment	Oocytes	Transitional	Oogonia
Control	97.8 ± 0.8	95.6 ± 0.9	89.6 ± 0.9
Element			
Sn	97.2 ± 1.0	96.1 ± 1.0	90.4 ± 1.1
Pb	95.1 ± 1.2	97.5 ± 2.7	88.2 ± 3.2
As	70.5 ± 2.4	93.2 ± 1.0	80.4 ± 1.9
Sb	92.2 ± 1.1	92.8 ± 0.8	84.0 ± 1.5
Se	92.3 ± 1.1	85.9 ± 6.3	78.9 ± 3.9
Te	86.8 ± 2.4	60.0 ± 3.8	70.4 ± 4.2
V	76.9 ± 4.4	75.4 ± 2.4	61.5 ± 3.1
Fe^{2+}	95.7 ± 0.6	84.3 ± 1.3	67.9 ± 2.0
Co	75.0 ± 2.3	71.8 ± 1.9	75.7 ± 2.0
Ni	65.4 ± 2.7	78.8 ± 1.7	76.9 ± 1.3
Cu	81.2 ± 1.6	90.2 ± 0.9	76.7 ± 1.6
Zn	89.2 ± 1.2	84.3 ± 1.0	73.8 ± 1.2
Ag	95.5 ± 1.4	95.3 ± 0.8	86.5 ± 1.4
Cd	89.7 ± 2.8	85.6 ± 2.1	83.9 ± 3.2
Hg	95.6 ± 1.0	92.8 ± 1.2	63.9 ± 2.3

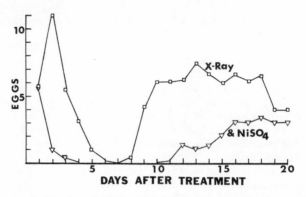

FIG. 2. Braconid egg production after a 2-fraction (2500 R each), 4-hour interfraction exposure to X-rays.
One sample of females ingested 0. 008 M NiSO₄ immediately after the first radiation exposure.

The combination of treatments decreased the egg hatchability significantly. After irradiation only, the mean hatchability of decline phase eggs
was $69.1 \pm 4.1\%$, and $53.2 \pm 1.8\%$ for the recovery phase eggs. For combination experiments hatchability was $49.3 \pm 6.5\%$ and $42.7 \pm 4.4\%$ respectively.

Similar patterns of decreased oviposition and hatchability were obtained
when Co, Cu or Zn feedings were combined with X-ray doses adequate to
induce temporary infecundity.

3.2. Chelating agents

A response similar to an excess of metal ions occurred when chelation
produced a deficiency of metal ions. LaChance [6] demonstrated that ingested
ethylene-diamine-tetra-acetic acid (EDTA) shortens Bracon life span. A
decreasing series of doses provided a sample of females that approached
control values for life span, but showed decreased egg production. In combination with modest radiation doses, EDTA enabled the recovery of more
dominant lethals in the zygotes than found in either irradiated or EDTA fed
groups [6].

Subsequently, feeding experiments were attempted with three other
metal chelating agents in common use: 8-hydroxyquinoline, o-phenanthroline
and diethyl dithio carbamate (DIECA). Solubility of the first two limited
the amounts present in palatable mixture. Only with DIECA were we successful in altering reproductive performance.

Figure 3 shows that a single meal of DIECA caused an obvious deficiency
in eggs laid during the first three days, and a trend to slightly lower egg
production for the remaining three weeks of life. DIECA used together with
the threshold dose for X-ray induced infecundity caused cessation of egg
production four days earlier. Also recovery phase oviposition was consistently lower than that after radiation alone.

As shown on Table III, hatchability was good for the eggs deposited
through youth and middle age of the mothers fed DIECA, but a decline in
hatchability was evident during the senile period. In combination with X-ray,
DIECA did not seem to decrease the hatchability of eggs derived from differentiated oocytes, but few of the ova hatched from the meagre deposits

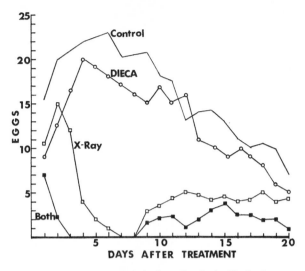

FIG.3. Prolongation of the X-ray-induced period of infecundity obtained by feeding saturated DIECA,
a chelating agent.

TABLE III. HATCHABILITY OF EGGS OF Fig.3:
MEAN PERCENTAGE ± S.E.

Cell type from which derived	Oocytes	Transitional	Oogonia	Oogonia (Senility)
Deposited on days	1 - 5	6 - 10	11 - 15	16 - 20
X-ray	66.8 ± 1.6	67.5 ± 3.2	63.7 ± 1.3	50.2 ± 1.5
DIECA	95.9 ± 0.6	94.9 ± 0.7	77.7 ± 4.4	45.2 ± 3.2
Both	62.9 ± 3.9	15.0 ± 5.6	50.8 ± 1.8	43.3 ± 6.0

of eggs derived from differentiating units. Because of the small sample
of eggs deposited on the 9th and 10th days, enhancement of radiation effects
is not firmly established although the decreased hatchability of the eggs
deposited for the 11th to 15th day period are consistent with this interpretation. On the other hand, the hatchability of the eggs laid during days 16 through
20 seems to reflect the DIECA influence on performance during senility.

3.3. Enzyme inhibitors

Among the "enzyme inhibitors" employed by biochemists to depress
metabolism are compounds that interfere with the production of energy
storage and transport molecules. Sodium azide (NaN_3) inhibits electron
transport over the molecular respiratory chain and decreases ATP synthesis.

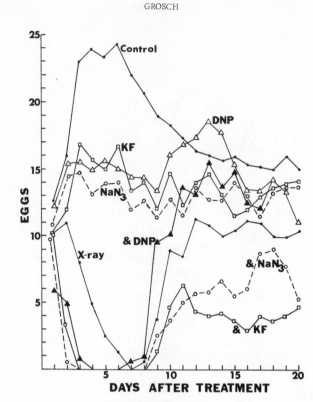

FIG. 4.   The patterns of decrease in daily egg production obtained by feeding 0. 05% solutions of
2, 4-dinitrophenol (DNP), potassium fluoride (KF) and sodium azide (NaN$_3$) alone and in combination
with 300 R of ^{60}Co gamma rays.

When fed to female braconids, it decreased egg production noticeably during
the first two weeks and moderately during the last weeks of life. As shown
in Fig.4, NaN$_3$, plus a modest dose of gamma rays, caused an immediate
sharp drop in eggs per day. The onset of fecundity has been shifted four
days forward from the 7th day nadir expected from the irradiation. Similar
results were obtained also with potassium fluoride (KF) which complexes
with the cytochromes. In addition, a prompt decline was obtained when
2, 4-dinitrophenol, DNP, was used in combination with radiation. The period
of infecundity lasted as long as with the other agents, but subsequent oviposition quickly rose almost to the control level (Fig.4).

The hatchability of the eggs is summarized on Table IV. First-day
mean values for the experimental groups do not show a consistent trend,
and the statistical significance of differences of NaN$_3$ and KF values from
controls may be debated. Hatchability of eggs from the same two groups
are obviously lower than that of controls in the next period of deposit, but
the meagre number of eggs laid provide an inadequate basis for sure interpretation. On the other hand samples were adequate and the data were
sufficiently homogeneous to allow pooling the results for days 8 through 20.

TABLE IV. HATCHABILITY OF EGGS OF Fig.4:
MEAN PERCENTAGE ± S.E.

Days of deposit	Irradiated			Unirradiated		
	1	2 - 7	8 - 20	1	2 - 7	8 - 20
2,4-DNP	53.6 ± 5.7		55.7 ± 2.4	97.3 ± 1.0	95.5 ± 0.5	84.5 ± 1.0
NaN$_3$	67.7 ± 5.9	31.3 ± 6.7	54.5 ± 2.2	97.0 ± 1.7	95.0 ± 0.9	88.0 ± 0.9
KF	76.0 ± 4.3	60.6 ± 8.5	54.2 ± 2.2	92.0 ± 2.9	95.6 ± 0.8	89.4 ± 0.8
Control	64.8 ± 5.1	78.9 ± 9.3	31.0 ± 2.3	98.6 ± 1.4	98.1 ± 0.8	93.3 ± 0.7

The mean values were nearly identical for all three chemical treatments.
Hatchabilities of between 54 to 56% compared with 31.0 ± 2.3% suggest a
radioprotective effect. In view of the slight decrease from control level
in the hatchability of eggs laid during the same period by unirradiated females
fed the same three compounds, the protective effect is all the more impressive.

Combination experiments with another type of agent were attempted.
This was iodoacetate which alkylates SH groups. Females fed iodoacetate
and irradiated died within two to three days, and we have not yet determined
the dose level that will permit at least two weeks of oviposition.

3.4. Metabolite analogues

Our first clear demonstration of infecundity on the third day after treatment was produced by methotrexate, a folic acid antagonist [7]. A prompt
decline in oviposition was observed along with atrophy of nurse cells and
degeneration of oocytes in dissected ovarioles. This characteristic picture
of interference with the process of vitellogenesis has now been obtained
with a variety of other antimetabolites, including sodium fluoroacetate
"1080" [8]. More extreme damage has been reported previously [1]. When
an adequate DON dose was combined with enough radiation to destroy all
oogonia, females were rendered completely infecund.

Under investigation are two compounds related to DON in molecular
structure, azaserine and alanosine. Fed or injected, either one caused less
oocyte destruction than DON, but alanosine provided direct visible evidence
of its ability to depress vitellogenesis. Up to a week after ingesting alanosine,
females laid eggs 33% shorter than normal eggs, a decrease considerably
greater than the 1% caused by 6-mercaptopurine [9].

When alanosine was fed immediately before or after irradiation, the
period of infecundity was prolonged. Figure 5 shows average eggs per day
plotted per female fed alanosine and exposed briefly to ^{60}Co gamma rays.
Plotted curves for the pre-irradiation feeding experiment are nearly
identical. The hatchability of eggs is summarized by successive five-day
periods on Table V. The very low values obtained from combined treatments is a significant enhancement of damage.

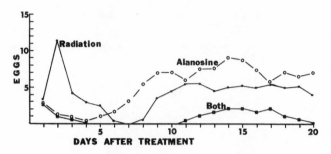

FIG. 5.   Prolongation of the radiation-induced period of infecundity obtained by feeding 0. 0125 M alanosine. 3000 R of ^{60}Co gamma rays were delivered in a one-minute exposure.

TABLE V.   HATCHABILITY DATA FOR ALANOSINE EXPERIMENT: MEAN PERCENTAGE ± S.E.

Days of deposit	1 - 5	6 - 10	11 - 15	16 - 20
Control	94.1 ± 1.1	90.6 ± 0.9	86.1 ± 2.1	73.1 ± 4.3
Radiation	57.3 ± 10.5	18.6 ± 7.4	47.2 ± 2.4	42.0 ± 3.5
Alanosine	34.1 ± 13.9	60.7 ± 14.6	72.3 ± 5.2	53.3 ± 7.2
Both	2.5 ± 1.5		25.8 ± 6.7	12.0 ± 4.9

3.5. Base substitution analogues

Even analogues for the nitrogenous bases of the DNA molecule can interfere with the progress in filling the oocyte with the variety of materials accumulated during vitellogenesis. The most evident consequence of feeding either fluorodeoxiuridine (FUDR) or bromodeoxiuridine (BUDR) was poor egg production for the first four days. Hatchability was decreased although moderately except for the eggs laid on the third day. Furthermore, on this day the eggs laid were shorter than normal by 25%.

When females were irradiated 11 hours after ingesting either FUDR or BUDR, egg production dropped to zero by the third day, and the period of infecundity extended from then until the eighth day. The infecundity induced by the exposure to 4000 R of gamma rays at 100 R/min began on the 6th and ended on the 7th day [10].

The pattern of influence upon embryonic viability revealed in the hatchability data was less consistent. The effects of the two chemical agents differed, and their action in combination with radiation varied with the cell type of the oogenerative series treated. A limit on space prohibits more detailed presentation.

3.6. Other agents for prolonging infecundity

Vitellogenesis and cystocyte formation were impeded with associated decreases in hatchability only when an aromatic ether group was added to

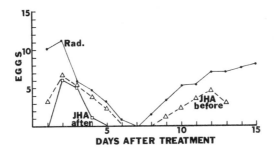

FIG.6. A 2-fraction, 4-hour interfraction experiment with ^{60}Co gamma rays, total dose 5000 R. Also 2.5 micrograms of juvenile hormone analogue RO-20-3600 were applied topically to each female in two samples. One sample received JHA an hour before, the other received JHA a half hour after the initial fraction of the radiation dose.

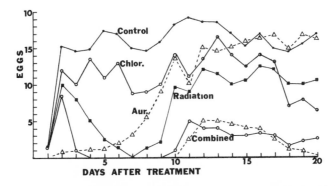

FIG.7. The patterns of decrease in braconid egg production obtained from feeding either 1% chloramphenicol or 1% aureomycin alone and in combination with 3000 R of ^{60}Co gamma rays.

the juvenile hormone molecule. Because this group, piperonyl butoxide, inhibits microsomal oxidations, it may prevent JH from degradation. This may be important for wasps accustomed to feeding upon Lepidopteran larvae rich in JH without obvious repression. Figure 6 shows the prolonged period of infecundity obtained in combination with radiation.

Antibiotics are a heterogeneous group of selectively toxic substances. Among those which prolong the infecund period when used in combination with radiation are aureomycin and chloramphenicol, inhibitors of protein synthesis (Fig.7). Chloramphenicol had the more drastic effect on hatchability.

3.7. Alkylating agents

Of the many types of chemical agents tested on braconid females, only the alkylating agents caused a decrease in egg production similar in pattern to that of radiation damage. The greater effectiveness of the polyfunctional agents was especially evident for the aziridine compounds. Obtaining data with an 8th-day nadir similar to that induced by the hexafunctional apholate

[11] required doses of the monofunctional ethylenimine 50 to 100 times higher than those used for apholate [12]. Even then the induced valley was not so pronounced, and at the high doses somatic toxicity became a problem. Early deaths tended to decrease the size of the sample of treated females. Subsequently we have found trifunctional aziridines intermediate in effectiveness. Sulphonate compounds cause a valley at the same part of the egg production curve but the range of functionality was limited for the available agents. Myleran, only a bifunctional alkylator, was compared with the monofunctional methyl- [13] and ethyl-methanesulphonates [14].

With both classes of alkylating agents, the chromosomal damage responsible for the specific pattern of decline in egg production was borne out by Stage 1 death of non-hatching embryos and by dominant lethal studies using treated males. In a number of other organisms, it is well established that polyfunctional agents are superior to their monofunctional relatives as inducers of chromosome structural change [15].

In combination experiments the situation seems to involve a duplication of effort by two kinds of agents in which that agent given at the most acute dose rate destroys nearly all the sensitive cells present. With most gamma sources and the usual X-ray generators, the demonstrable damage to oogenesis and hatchability in combination experiments is due to the radiation exposure, and little additional damage is caused by the presence of an alkylating agent.

An additional consideration is the relative radiotolerance of the holometabolous insect's adult somatic tissues. In the wasp there is no evidence of somatic damage from ionizing radiations at completely sterilizing doses. Thus with radiation alone, a great range of radiation doses can be employed while in a combination experiment the toxic action of alkylating agents constitutes a limiting factor.

3.8. Protective agents

Despite the almost universal effectiveness of aminothiol radioprotectors on a great range of organisms including viruses, bacteria, plants, protozoa and mammals, positive reports on insects have been the exception [16]. The supposed ineffectiveness was based upon experiments in which adult males were irradiated, but their relatively static mature sperm are unfavourable for demonstrating the phenomenon. Furthermore, the newly laid eggs of many species are unfavourable for demonstrating chemical protection because of the selective impermeability of the shells and membranes.

By using adult females, Grosch [17] established that aminothiols can protect insect tissues, but dividing and differentiating cell types were required for the demonstration. Those available in the distal portion of the ovariole germarium were less vulnerable to radiation if the female had ingested either cysteine or glutathione. Subsequently, injection of the same compounds reduced radiation damage to storage-chamber eggs poised in meiotic metaphase [18]. Dithiothreitor also was an effective protector. On the other hand, when injected immediately after an acute gamma-ray exposure, this agent interfered with the chromosome repair mechanism of eggs in prometaphase but not for oocytes in earlier prophase [18].

3.9. Mitotic inhibition

There are agents that inhibit the macromolecular assembly of the spindle function in quite a different manner from aminothiol protectants.

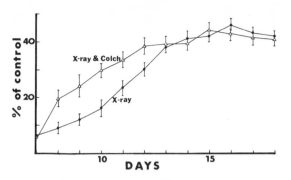

FIG. 8.   Feeding a low concentration of colchicine (5×10^{-4}%) before irradiation improved the recovery
phase of daily egg production after X-ray-induced infecunidity.

Presumably by providing time for post-irradiation repair of potential
chromosome lesions before the tension of kinesis stretches them, a temporary halting of the division cycle can ameliorate the amount of damage.
However with the classic agent, colchicine's notorious animal toxicity has
hampered combination experiments. A sublethal dose high enough to shorten
wasp life span by 25% was required to modify the pattern of radiation-
induced decrease in fecundity. As shown by Fig.8, a modest improvement
in daily oviposition appeared during the second week after combined treatments. Also the hatchability of the eggs from day 7 through day 12 improved;
control hatchability was 92.2 ± 0.9% hatched, eggs from irradiated wasps
64.1 ± 3.5% hatched, eggs from colchicine-fed wasps 89.5 ± 0.7% hatched,
and eggs from wasps receiving both treatments 77.4 ± 1.0% hatched. The
class of cells influenced were those either engaged in or preparing for
division at the time of treatment. These give rise to the eggs laid two
weeks later.

    At higher doses colchicine can influence phenomena other than oogonial
mitosis. A third-day nadir in egg production has revealed attainment of a
dosage high enough to interfere with vitellogenesis [19]. Recently we tested
colcemid (Demecolcine). It proved to be toxic to wasps and offered no
improvement for combination experiments using radiation.

3.10.   Pesticides

    With agents that primarily attack the nervous system, Bracon repro-
duction tests show a modest daily deficit in eggs laid without a significant
decrease in their hatchability.

    When wasps were irradiated with X- or gamma rays either before or
after treatment with organophosphorus or chlorinated hydrocarbon pesticides,
the characteristic pattern of destruction of radiosensitive cells seemed
merely superimposed on the decreased number of eggs laid per unit time.
Also the decrease in hatchability was essentially identical with that obtained
from the radiation dose alone. Only recently, from screening carbamates
and their breakdown products, has a more complex interaction emerged.
Experiments still in progress suggest an antagonistic effect in combination
experiments, somewhat similar to that obtained by mitotic inhibition.

4. DISCUSSION

4.1. Rationale

The ancestral insects from many important orders solved the problem of producing many eggs quickly by evolving meroistic ovarioles capable of strikingly rapid oogenesis. Instead of the amphibian method which employs lampbrush chromosomes and sacrifices time for synthesis, an insect strategy is the use of polyploid nurse cells to generate co-operatively molecules and organelles for the oocyte, while follicle cells accept yolk proteins synthesized by somatic tissues. In our search for new methods of control we should not overlook the nearly unique basis of the prolificacy that characterizes many undesirable insects [20]. The present study demonstrates that a variety of agents can derange the processes operating during the period inclusively termed vitellogenesis. This kind of effect combined with radiation-induced damage to cells of the germarium can cause extended periods of female infecundity and sterility. On the other hand, the alkylating agents which are the most effective types of insect chemosterilants [21, 22] attack the distal part of the germarium. Axiomatic in radiobiology is the inefficient expenditure of energy when cells killed as a consequence of one exposure absorb the damage caused by a second exposure. The concept is easily extended to combination experiments involving radiation and alkylators.

4.2. Modification of the radiation response

The possibility of modifying the radiation response of insect reproductive systems by chemical agents was suggested by notable successes using microbial, plant and mammalian systems, but results in sterile-male experiments were disappointing. With Bracon females we demonstrated the applicability of chemical protection [17] and identified an oxygen potentiation of the response to sparsely ionizing radiation [23]. Intensely desired is a gamete specific sensitizing agent which could enhance the effects of radiation and thus augment the sterilization obtained from a moderate radiation dose. Despite a comprehensive screening, we have not identified such an agent, although we have revealed that two presumably unrelated actions together provide an extended period of female infecundity.

When reactions are similar, combining the two kinds of treatment has not been advantageous. Polyfunctional aziridines plus radiations are not more effective than either used alone. Furthermore, in tests with mixtures of radiations, proportionate doses of two sparsely ionizing radiations were non-additive. The effects from one of the radiations predominated in altering the pattern of oviposition and decreasing hatchability [24]. Additional examples of results from two treatments resembling those from one of the agents used alone are provided by combined chemosterilant and radiation tests on bollworms [25].

4.3. Metal excess and deficiency

Few of the metal elements have been given serious consideration as inhibitors of insect reproduction. Even then organic rather than inorganic compounds have been featured, as for example triphenyl tin [26]. On the other hand, $VOSO_4$ and other inorganic vanadium compounds proved as

effective as organic compounds in reducing oviposition and hatchability
of screw-worm flies [27]. Vanadium, the first of the eight elements classified as essential in animals from set 1 of the transition series [28], is
only one of the effective examples reported above. Five more, including
the Group VIII trio, decreased egg production and hatchability when fed in
excess to wasps.

Our knowledge of insect inorganic nutrition is meagre [29], but conceivably 4th and 5th period interrelationships involve the metallo-enzyme
systems [30]. In addition to the primary effect of a toxic excess of a metal
on cellular enzymes, there is the possibility that metals will alter permeability
barriers and block the uptake of essential nutrients. Thus there are a combination of circumstances which can decrease egg production. In addition,
egg-incorporated burdens help to lower hatchability and survival to maturity.

In Bracon, where over 90% of an ingested 4th-period cation becomes
an abdominal burden, radiotracer experiments suggest modes of action.
Considerable ^{58}Co [5] and ^{63}Ni [31] were retained in the midgut wall, with
lesser amounts in the fat body. Salting out and dialysis of homogenates
revealed strong association with proteins. In view of 4th-period metal
affinity for SH-groups, a prime target would be cysteine and methionine
containing proteins. Furthermore, since thiols readily combine with As
and Sb [32], a previously noted similarity between As- and Ni-induced
decrease in oviposition [1] is understandable if digestive and assimilative
processes are disturbed. Also a basis for enhancing radiation damage to
gonial cells would be the poisoning of natural radioprotective agents like
cysteine. Finally, offspring can receive metal ions via the eggs [33] or via
the toxin used to paralyse the host [31].

Metal binding substances, many of which function by chelation, have
a mode of action obviously related to metal excesses and antagonisms.
Their applicability as antihelminths and fungicides is well proven. Albert
[32] has written at length about the biochemical differences which can
assist selectivity, and the competition by natural metal binding agents in
every living cell. As shown above, the earliest and best known chelating
agents alter reproductive performance in Bracon. DIECA is partial to Cu,
while EDTA ties up iron and several other divalent cations. Nurse cell
pycnosis, gaps from resorption, and posteriorly displaced undeveloped units
are demonstrable from EDTA, as well as prolonged infecundity when used
with radiation [34]. Now a great variety of chelating agents are available
which have not been tested on insects.

4.4. Organic compounds

Further investigations with many different kinds of chemical agents
revealed other types of compounds which added days of infecundity to the
period induced by radiation. The feature common to these agents was an
ability to interfere with vitellogenesis, an involved process in which a
complex of interrelated metabolic pathways and cellular events converge.
Modes of action may be completely different for various agents, as with
NaN$_3$, KF and DNP, nevertheless the end result was similar.

Metabolic analogues can act on other pathways. Early investigations
merely scored insect survival to adulthood [35], but later a reduction in
viable eggs was demonstrated in aminopterin tests [36]. Furthermore,
only the females were effectively sterilized, and this was due entirely to

decreased egg yolk [37]. No differences were demonstrable in males or their sperm. Wasp experiments with aminopterin gave similar results, but methotrexate was even more effective [7]. In adult insects metabolite analogues evidently act primarily on females [21]. Morphological evidence of the antimetabolite influence on vitellogenesis is provided by the under-sized eggs, which have been caused in Bracon by FUDR, alanosine, purine analogues, and modified juvenile hormones. Their highest frequency occurs on the third day after treatment, because vitellogenesis was just starting for them at the onset of treatment.

At sub-lethal doses the insecticides with neurobiological action are known to alter arthropod reproductive potential without direct damage to the gametes [38]. Causes include altered feeding behaviour, decreased appetite or disturbance of some aspect of somatic metabolism. With Bracon these phenomena often are accompanied by a decreased average life span, a recent finding which fits Georghiou's [39] expectancy. On the other hand, sterilizing doses of radiation are well below the level for inducing obvious neurological symptoms. In combination experiments radiations may alter somatic tissues to the extent of either (a) decreasing or increasing the efficiency of cells which can detoxify pesticides, or (b) modify the permeabilit transportation or retention mechanisms for the agent. Species iodiosyncrasie can add special complications such as the Tribolium secretory response to stress encountered when malathion and radiation treatments were combined [40]; but with the possible exception of carbamates, the gonad is not the primary target.

ACKNOWLEDGEMENT

Initially these investigations were supported by USAEC Contract No. AT-(40-1)-1314. In recent years support has come from USPHS Grant ES-00044, National Institute of Environmental Health Sciences.

REFERENCES

[1] GROSCH, D.S., "The response of the female arthropod's reproductive system to radiation and chemical agents", Sterility Principle for Insect Control or Eradication (Proc. Symp. Athens, 1970), IAEA, Vienna (1971) 217.
[2] VON BORSTEL, R.C., REKEMEYER, M.L., Genetics 44 (1959) 1053.
[3] GROSCH, D.S., SULLIVAN, R.L., LaCHANCE, L.E., Radiat. Res. 5 (1956) 281.
[4] GROSCH, D.S., Nature (London) 195 (1962) 356.
[5] GROSCH, D.S., Nature (London) 208 (1965) 906.
[6] LaCHANCE, L.E., Radiat. Res. 11 (1959) 218.
[7] GROSCH, D.S., Science 141 (1963) 732.
[8] SMITH, G.J., MS Thesis, North Carolina State University, USA, 1973.
[9] CASSIDY, J.D., GROSCH, D.S., J. Econ. Entomol. 66 (1973) 319.
[10] SMITH, R.H., MS Thesis, North Carolina State University, USA, 1962.
[11] VALCOVIC, L.R., GROSCH, D.S., J. Econ. Entomol. 61 (1968) 1514.
[12] VALCOVIC, L.R., Mutat. Res. 15 (1972) 67.
[13] KRATSAS, R.G., MS Thesis, North Carolina State University, USA, 1972.
[14] HOFFMAN, A.C., GROSCH, D.S., Pest. Biochem. Physiol. 1 (1971) 319.
[15] LOVELESS, A., Genetic and Allied Effects of Alkylating Agents, Pennsylvania State University Press (1966) 190.
[16] BACQ, Z.M., ALEXANDER, P., Fundamentals of Radiobiology, 2nd Edn, Pergamon Press, New York (1961) 465.

[17] GROSCH, D.S., Radiat. Res. 12 (1960) 146.
[18] HOFFMAN, A.C., Mutat. Res. 16 (1972) 175.
[19] GROSCH, D.S., Ann. Entomol. Soc. Am. 52 (1959) 294.
[20] DAVIDSON, E.H., Gene Activity in Early Development, Academic Press, New York (1968) 181.
[21] SMITH, C.N., LaBRECQUE, G.C., BORKOVEC, A.B., Annu. Rev. Entomol. 9 (1964) 269.
[22] LaBRECQUE, G.C., SMITH, C.N. (Eds), Principles of Insect Chemosterilization, Appleton-Century-Crofts, New York (1968).
[23] GROSCH, D.S., CLARK, A.M., Nature (London) 190 (1961) 546.
[24] GROSCH, D.S., Atompraxis 5 (1959) 290.
[25] KLASSEN, W., NORLAND, J.F., BRIGGS, R.W., J. Econ. Entomol. 62 (1969) 1204.
[26] KENAGA, E.E., J. Econ. Entomol. 58 (1965) 4.
[27] CRYSTAL, M.M., J. Econ. Entomol. 63 (1970) 321.
[28] MATRONE, G., "Chemical parameters in trace element antagonisms", Trace Element Metabolism in Animals, Am. Chem. Soc. Symp., in press, 1974.
[29] DADD, R.H., Annu. Rev. Entomol. 18 (1973) 381.
[30] MEDICI, J.C., TAYLOR, M.W., J. Nutrition 93 (1967) 307.
[31] GROSCH, D.S., LIN, J., SMITH, R.H., Radiat. Res. 25 (1965) 194.
[32] ALBERT, A., Selective Toxicity, 5th Edn, Chapman and Hall, London (1973).
[33] GROSCH, D.S., Nature (London) 195 (1962) 356.
[34] LaCHANCE, L.E., PhD Thesis, North Carolina State University, USA, 1958.
[35] GOLDSMITH, E.D., TOBIAS, E.B., HARNLEY, M.H., Anat. Rec. 101 (1948) 93.
[36] GOLDSMITH, E.D., FRANK, I., Am. J. Physiol. 171 (1952) 726.
[37] MITLIN, N., BUTT, B.A., SHORTINO, T.J., Physiol. Zool. 30 (1957) 133.
[38] MORIARTY, F., Biol. Rev. Camb. Philos. Soc. 44 (1969) 321.
[39] GEORGHIOU, G.P., J. Econ. Entomol. 58 (1965) 58.
[40] KINKADE, M.L., ERDMAN, H.E., J. Econ. Entomol. 66 (1973) 967.

DISCUSSION

K. SYED: I think you have done a very good thing using a hymenopterous wasp as a test insect. This should be an eye-opener for those conventional entomologists who restrict themselves to the same old species of insects.

D.S. GROSCH: Thank you, I also work on salt-water arthropods.

D.W. WALKER: Are the accessory cells connected with the oocytes?

D.S. GROSCH: Yes. Reliable studies using electron microscopy have been performed on this subject by J.D. Cassidy.

D.W. WALKER: Was a combination treatment tried with the boll weevil?

D.S. GROSCH: No. My laboratory does not study boll weevils, and, as far as I know, combinations of radiation with ant-vitellogenic substances have not been investigated in weevil laboratories. Combination experiments using alkylating agents with radiations are cited and discussed in my manuscript.

L.E. LaCHANCE: I believe that the combination of radiation and metals may be a useful approach in sterilizing boll weevils. However in the boll weevil the problem also involves two different types of cells, but vastly different ones from those discussed by Mr. Grosch. In the weevil we are dealing with sperm and spermatogonial cells and the mid-gut epithelial cells, and we are trying to induce damage in the one cell type without damaging the other.

LABORATORY STUDIES ON STERILIZATION
OF THE MALE RED PALM WEEVIL,
Rhynchophorus ferrugineus Oliv.

G.W. RAHALKAR, M.R. HARWALKAR,
H.O. RANANAVARE
Biology and Agriculture Division,
Bhabha Atomic Research Centre,
Trombay, Bombay,
India

Abstract

LABORATORY STUDIES ON STERILIZATION OF THE MALE RED PALM WEEVIL, Rhynchophorus ferrugineus Oliv.

When 1 - 2-day-old males of the red palm weevil(Rhynchophorus ferrugineus) were X-ray radiated with a dose of 1.5 krad, 90% sterility was induced without affecting their survival. Higher doses, however, significantly reduced male survival, although near complete sterility was obtained. Some of the sperms present at the time of irradiation were more radioresistant and contributed towards egg viability during the initial period of oviposition. Treatment with metepa and hempa failed to induce the desired level of sterility without affecting their survival. Metepa proved more toxic than hempa. A combination of any radiation dose and hempa concentration was inadequate to induce near complete sterility without affecting male survival; an overall sterility of 87% only was obtained following exposure of irradiated males (1 krad) for 1 hour to hempa residue of 158 $\mu g/cm^2$. These males, however, exhibited to some extent improved mating competitiveness compared with those sterilized to the same level by radiation alone.

Red palm weevil, Rhynchophorus ferrugineus Oliv., is a serious pest of coconut (Cocos nucifera) and other cultivated palms in India [1]. It is also a serious pest of coconut in Sri Lanka [2], Indonesia [3] and the Philippines [4]. Eggs are laid in the soft portions of the crown or in damaged parts of the petioles and trunk. The hatching grubs worm their way through the soft weed into the trunk and feed voraciously causing destruction of the palm. Timely action for the control of this insect is made difficult by the absence of any visual symptoms of infestation. By the time the insect attack becomes discernible irreparable damage has already been done.

We are currently investigating the feasibility of the sterile-male release approach towards controlling this insect, and this report deals with laboratory studies on sterilization of male weevils and their mating competitiveness. X-ray radiation, chemosterilization and a combination of both treatments have been studied.

MATERIAL AND METHODS

Insects were reared in the laboratory at 30° ± 1°C using sugar-cane stem tissue as larval and adult food. In all the experiments, adults, 1-2 days after their emergence from cocoons, were used. The radiation source was a Siemen's "Stabilipan" X-ray unit operated at 250 kV, 15 mA tube current and 2 mm Al filtration. This produced a dose rate of 300 rad/min at a

distance of 35 cm between tube and the subject.   Insects were localized in the
radiation field by confining them in a perforated plastic box (5.0×3.5×1.5 cm).
The chemosterilants used were metepa and hempa of technical grade.   For
treatment with chemosterilants, males were exposed for one hour to varying
concentrations of sterilant residues deposited on a glass surface.   Metepa
residues ranged from 54 to 217 $\mu g/cm^2$ and those for hempa from 79 to
316 $\mu g/cm^2$.

Treated insects were caged individually in plastic jars (7.5 cm high
and 6.5 cm diam.) along with normal virgin females of the same age.   Sugar-
cane stem pieces were provided as food and site for oviposition.   Eggs were
harvested on alternate days, placed on moist towelling in petri dishes and
held at 30° ± 1°C.   At this temperature they hatch within 5 days.   Therefore,
eggs that failed to hatch even after 8 days were considered non-viable.

EXPERIMENTAL RESULTS AND DISCUSSION

Radiation sterilization

To determine the sterilizing dose, males were irradiated with doses of
1, 1.5, 1.75, 2, 2.5 and 3 krad and crossed with virgin females.   Data
obtained from 10 replicates showed that egg viability decreased with increase
in the dose (Fig. 1).   The reduction in egg hatch was 54% with a dose of
1 krad and almost complete with a dose of 3 krad.   However, survival of
males radiated with doses of 1.75 krad and above was considerably reduced
(Fig. 2).

In all the dose treatments, any egg viability observed was during the
initial period of oviposition, which decreased with the increase in the dose.
In the case of the 1.5 krad dose treatment, viable eggs were laid over a
period of 26 days.   During this period the female laid on an average 154 eggs
of which 9.6% proved viable.   In the subsequent period of oviposition, an
average of 42 eggs per female were laid and none of them was viable.   The
weevil, after eclosion, remains in the cocoon for about a week before
emergence.   Mating takes place immediately after the emergence.   The

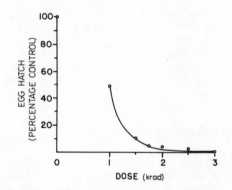

FIG.1.   The relationship of the X-ray dose to induced sterility in the red palm weevil (radiated males crossed
individually with virgin females).

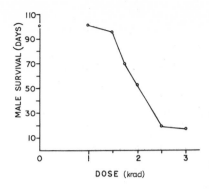

FIG.2.   Effect of X-ray radiation on the survival of red palm weevil males.

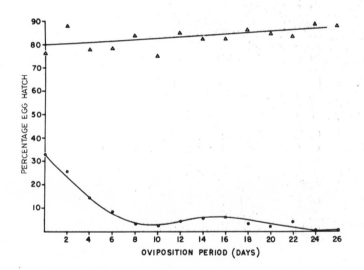

FIG.3.   Viability of eggs when unirradiated virgin females were crossed with X-ray irradiated males (O) or unirradiated males (Δ); males irradiated with a dose of 1.5 krad.

initial egg viability, therefore, could be attributed to higher radio-resistance of mature sperm and those that were in advanced stages of spermiogenesis at the time of irradiation.   The data on egg viability during the initial oviposition period supports this assumption; the proportion of viable eggs decreased as oviposition progressed (Fig. 3).   It is thus evident that 1.5 krad is an optimum sterilizing dose for this insect.

Chemosterilization

In treatments with both metepa and hempa, the egg hatch decreased progressively with the increase in residue concentration (Table I).   At the highest level of metepa residue, nearly 24.2% of the eggs laid were viable.

TABLE I.   STERILITY INDUCED IN MALE RED PALM WEEVILS
AFTER EXPOSURE FOR 1 h TO METEPA AND HEMPA RESIDUES

Chemosterilant	Residue (μg/cm^2)	No. of pairs	Total eggs scored	Egg hatch (%)	Total mortality (%) of treated males at the end of	
					30 days	60 days
Metepa	Control	18	3013	85.9	0	0
	54	15	1731	64.6	0	0
	108	14	1663	54.3	28.5	28.5
	195	20	2670	27.5	45.0	45.0
	217	10	1192	24.2	50.0	50.0
Hempa	Control	18	2804	84.1	0	0
	79	9	2540	48.0	0	0
	158	20	4400	25.6	5	5
	237	20	2842	13.6	65	75
	316	20	2822	4.9	90	95

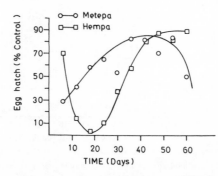

FIG.4.   Viability of eggs when untreated females were crossed with males treated with metepa (195 μg/cm^2)
and hempa (158 μg/cm^2); males exposed for 1 hour to sterilant residue.

However, 50% of the treated males died within 30 days and there was no
further mortality. In the two lower dose treatments this pattern of survival
was also noticed.

In the case of hempa, although the egg hatch was 4.9% with a residue
dose of 316 μg/cm^2, 90% of the treated males died within 30 days and at the
end of 60 days mortality was 95%. At the lowest concentration of 79 μg/cm^2,
there was no mortality but the egg hatch was 48%. When the hempa residue
was 158 μg/cm^2, the egg hatch was 25.6% and only 5% of the treated males
died within 30 days without any further increase in mortality.

In the case of metepa treatment with residue of 195 μg/cm^2, the residual
fertility was observed during the entire observation period of 60 days (Fig.4).

However, there was a specific pattern in the levels of fertility.   Beginning
with about 27% egg hatch on the 6th day, it increased gradually up to 80% on
44th day and then tended to decline.   In the case of hempa (158 $\mu g/cm^2$) the
egg hatch of about 70% noticed initially dropped to as low as 3% by the 18th day,
and progressively increased thereafter, reaching a level of about 88%.

Both sexes of the azuki bean weevil, Callosobruchus chinensis L,
were sterilized by dipping in a solution of metepa [5] or by topical application [6].   Hempa has also been shown to be effective against the azuki bean
weevil [7,8] and the cotton boll weevil, Anthonomus grandis [9].   Our
investigations, however, demonstrate that metepa and hempa were inadequate
for inducing a reasonable level of sterility in the male of the red palm weevil
without any adverse effects on its survival.

Combination of radiation and sterilant treatments

In the case of the boll weevil [10] and the tobacco budworm [11], an
additive effect has been recorded following combined radiation and chemosterilant treatments.   The present studies have shown that hempa was less
toxic than metepa to the red palm weevil.   The effectiveness of a combination
of X-ray radiation and hempa treatment in inducing sterility in this insect
was therefore examined.   In this study X-ray radiation preceded exposure
to hempa residue, and the interval between the two treatments was 1 hour.

It is evident from the data (Table II) that an additive effect on sterility
was induced when males were treated with sublethal treatments of radiation
and chemosterilant.   In the case of a combined treatment involving a 1.5 krad
radiation dose and a hempa residue of 158 $\mu g/cm^2$, the egg hatch was reduced
to 1.9% compared with 8.7 and 22.2% respectively when the same treatments
were administered separately.   However, survival of males receiving a
combined treatment was considerably reduced.   Nearly 60% of the treated
males survived for an average period of 20 days only compared with 80 days
of normal insects.   Reduction in the radiation dose or chemosterilant residue

TABLE II.   STERILITY INDUCED IN RED PALM WEEVIL MALES
AFTER X- RAY RADIATION AND/OR EXPOSURE FOR
1 h TO HEMPA RESIDUES

Treatment		No. of replicates (single pair)	Total No. of eggs scored	Egg hatch (%)
Radiation dose (krad)	Hempa residue ($\mu g/cm^2$)			
1	-	20	4342	30.8
1.5	-	25	4000	8.7
-	79	9	2540	48.0
-	158	30	5457	22.2
1	79	20	4633	19.3
1	158	20	4334	10.2
1.5	158	30	4482	1.9
Control		30	5055	85.3

TABLE III.   MATING COMPETITIVENESS OF RED PALM WEEVIL
MALES STERILIZED BY A COMBINATION OF RADIATION AND
HEMPA TREATMENT.   MALES X-RAY RADIATED WITH A DOSE OF
1 krad AND THEN EXPOSED FOR ONE HOUR TO HEMPA RESIDUE OF
158 $\mu g/cm^2$

Ratio			No. of	Total No. of	Egg hatch
TM	NM	NF[a]	replicates	eggs scored	(%)
0	1	1	30	5055	85.30
1	0	1	20	4334	10.18
4	1	1	6	1016	36.0
8	1	1	13	2105	12.35
1[b]	0	1	10	1409	12.26
4[b]	1	1	10	1152	49.07
8[b]	1	1	10	1300	16.67

[a] TM = Treated male; NM = Normal male; NF = Normal Female.
[b] Males were sterilized by X-ray irradiation with a dose of 1.5 krad alone.

produced no adverse effects on male survival, but there was no improvement
in the level of induced sterility over that obtainable when the radiation dose
of 1.5 krad was administered alone.   In the case of a combined treatment
involving X-ray radiation with a dose of 1 krad followed by exposure for 1 hour
to a hempa residue of 158 $\mu g/cm^2$, the egg hatch was 10.2% and is comparable
with the hatch observed in the case of the 1.5 krad dose treatment.

Mating competitiveness of sterile males

    For studies on mating competitiveness males were X-ray radiated with
a dose of 1 krad and then exposed for 1 hour to a hempa residue of 158 $\mu g/cm^2$
For comparison, males were also irradiated with a dose of 1.5 krad alone.
Newly emerged virgin females were caged individually along with treated and
normal males in the ratios of 4:1 and 8:1.   The number of eggs laid and
hatched were scored as mentioned above.
    When 8 males that received a combined treatment were competing with
one normal male, only 12.35% of the eggs laid proved viable (Table III).
This figure compares well with the egg hatch obtained when a similarly
treated male alone was caged with a female.   In the case of males sterilized
by radiation alone, competition between 8 sterile and one normal male
reduced the egg hatch to 16.67% compared with 12.26% egg viability when an
irradiated male alone was caged with a female.   It is thus evident that males
receiving a combination of radiation and chemosterilant treatments exhibited
some improvement in their mating competitiveness compared with those
that were only irradiated with a dose of 1.5 krad.

REFERENCES

[1]   GHOSH, C.C., Mem.Dept.Agric.India Ent.Ser. 2 (1911) 193.

[2]   BRAND, E., Trop.Agric.Mag.Ceylon Agric.Soc. 49 (1917) 22.

[3]   LEEFMANS, S., Inst.Plantenziekten Meded. 43 (1920) 1.

[4]   VIADO, G.B.S., BIGORNIA, A.E., Philipp.Agric. 33 (1949) 1.

[5]   SHINOHARA, H., NAGASAWA, S., Entomol.Exp.Applic. 6 (1963) 263.

[6]   NAKAYAMA, I., NAGASAWA, S., Jap.J.Appl.Entomol.Zool. 10 (1966) 192.

[7]   NAGASAWA, S., SHINOHARA, H., SHIBA, M., Botyu-Kagaku 31 (1966) 108.

[8]   BORKOVEC, A.B., NAGASAWA, S., SHINOHARA, H., J.Econ.Entomol. 61 (1968) 695.

[9]   HAYNES, J.W., HEDIN, P.A., DAVICH, T.B., J.Econ.Entomol. 59 (1966) 1014.

[10] KLASSEN, W., NORLAND, J.F., BRIGGS, R.W., J.Econ.Entomol. 62 (1969) 1204.

[11] GUERRA, A.A., WOLFENBARGER, D.A., LUKEFAHR, M.J., J.Econ.Entomol. 64 (1971) 804.

DISCUSSION

R. GALUN: Why is the red palm weevil so sensitive to radiation and how does it compare with other Coleoptera?

G.W. RAHALKAR: This weevil is no doubt more sensitive to radiation than other coleopteran insects. There could be many reasons for this but I do not have any positive information available at the moment.

K. SYED: Is sugar-cane an alternate host plant of this pest in nature? If not, does rearing on sugar-cane have any effect on the fecundity, longevity, etc., of this pest?

G.W. RAHALKAR: Sugar-cane is not an alternate host plant of the red palm weevil. This insect has not been reported to attack sugar-cane, although the insect has been found breeding in bagasse dumps. Rearing of the weevil on sugar-cane does not affect its development, fecundity or longevity.

B. NA'ISA: What is the reason for the sharp drop and then the progressive increase in egg hatch with hempa treatment, as shown in Fig. 4? Is it due to the toxicity of the chemosterilant?

G.W. RAHALKAR: There are two possibilities: (a) there is a critical stage in the process of spermiogenesis which is most sensitive to hempa or (b) hempa might have been metabolized so that it was not available throughout the males' life-span. I should mention that in this case spermiogenesis is a continuous process and takes place during the major portion of adult male life.

J.L. MONTY: Was there any damage to the mid-gut epithelium, as occurs for example in the boll weevil treated with gamma radiation?

G.W. RAHALKAR: There was no apparent damage to the mid-gut epithelium. Irradiated weevils (1.5 krad) survived for periods comparable with those of the controls.

EFFECT OF SUBSTERILE IRRADIATION DOSES ON THE PROGENY OF TREATED BEAN WEEVIL ADULTS
(Acanthoscelides obtectus Say, Col., Bruchidae)

Á. SZENTESI
Research Institute for Plant Protection,
Budapest, Hungary

Abstract

EFFECT OF SUBSTERILE IRRADIATION DOSES ON THE PROGENY OF TREATED BEAN WEEVIL ADULTS (Acanthoscelides obtectus Say, Col., Bruchidae).

The effect of substerile doses (2, 4, 6, 8 and 10 krad) of gamma radiation applied to the parental generation of Acanthoscelides obtectus Say on the descendants' fertility and fecundity was investigated throughout four generations. The decrease in sterility of the successive generations was rather rapid, and showed differences between "closed-bred" and outbred lines. No distortion was observed in the sex ratio. Treated female parents did not transmit any sterility to their progeny. The release of substerile adults in the genetic control of the bean weevil is not recommended.

INTRODUCTION

The effect of administering substerilizing doses was first observed on the codling moth by Proverbs [1]. Since then a series of investigations on the type of mutagens have shown the characteristics and advantages, as well as the practical use, of delayed sterility. In a review of the genetic control methods of insect populations, Smith and von Borstel [2] stressed the importance of using low (substerile) radiation doses to induce a few dominant lethal mutations, but also many damaged chromosomes which therefore produced an imbalanced chromosome set.

It was also shown that the existence of this type of sterility, transmitted throughout generations, is connected with the holokinetic chromosomes of Lepidoptera [3] and Heteroptera [4]. Coleopterous and dipterous insects, not possessing such a type of chromosomes, cannot inherit an increased level of sterility during successive generations from their parents treated with substerile doses. It is worth remarking that the higher sterility of the progeny of treated leipdopterous parents was not found in all cases [5]. The possible use of inherited sterility for control purposes was demonstrated theoretically by Knipling [6] and also in practice in some cases of the taxa mentioned above [7].

Very few investigations have been carried out on coleopterous pests using substerile doses to study the alteration in induced sterility during successive generations. Laviolette and Nardon [8], examining the effect of radiation on Sitophilus sasakii Takahashi adults and their progeny, observed that the effect of 5000 rad of gamma radiation was transmitted through generations, and the fertility of the descendants became the same as that of the control only in F_4. Administration of 8000 rad caused a similar but even more delayed effect.

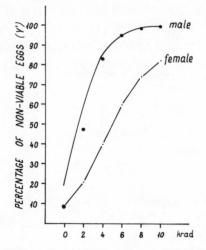

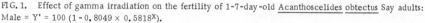

FIG. 1.   Effect of gamma irradiation on the fertility of 1-7-day-old <u>Acanthoscelides</u> <u>obtectus</u> Say adults:
Male = Y' = 100 (1 - 0. 8049 × 0. 5818^X).

Cornwell and co-workers [9] used substerilizing doses of gamma radiation for disinfestation of <u>Sitophilus</u> <u>granarius</u> L. in grain. Doses of 10-14 krad suppressed the weevil population to a very low level and prevented their increase for 4 to 8 months.

Jacklin and co-workers [10] found that a radiation dose (8 krad) giving full sterility to <u>Conotrachelus</u> <u>nenuphar</u> Herbst adults reduced longevity. In order to increase competitiveness a lower dose (6 krad) was applied. In a population suppression experiment using substerile males, a reduction of 90% in the number of the F_1 generation was achieved.

Substerilizing doses were also used to study the possibility of irradiation resistance in <u>Callosobruchus</u> <u>maculatus</u> F. [11].

To find out whether substerilizing doses could be used in the sterile release technique against the bean weevil, experiments have been carried out with substerilizing doses to determine the character of sterility transmitted during successive generations. The cytogenetic background was not investigated.

MATERIAL AND METHODS

The laboratory strain of the bean weevil was maintained and virgin adults were obtained as described in a previous paper [12].

One- to seven-day-old bean weevil adults were irradiated in air with substerile doses (2, 4, 6, 8 and 10 krad) of gamma radiation (radiation source: LMB-gamma-IM; ^{137}Cs source; dose rate 235 krad/h).

Ten males and 10 females were placed in glass jars (12 × 9.5 cm) containing 100 g of white beans and covered with a linen cloth. After the adults had died and the small larvae had entered the beans, all the empty egg-shells and the non-viable eggs were separated from the beans and were counted. The larvae continued to develop and from the time of emergence of the first descendant adults (after about 25 days), the latter were sifted out daily. Ten males and 10 females (1-3 days old) were picked out at random and placed in a new glass jar to begin a new generation. The breeding of the successive generations has been maintained up to the F_4 generation at 28°C and about 50% r.h.

Experiment 1

A preliminary experiment was carried out to obtain information on the main trend in the alteration in sterility. The following variants were used:
1. Treated males and untreated females reared in the closed-bred[1] line during generations;
2. Untreated males and treated females reared in closed-bred line;
3. Untreated males and females reared in closed-bred line (control).

The doses applied were 2, 4, 6, 8 and 10 krad of gamma radiation. The number of replicates were two.

Experiment 2

A comparison was made between the sterility of offspring when treated males were to mate with females originating from the same parental generation and when males were outcrossed to virgin colony females. The following variants were used:
1. Treated males and untreated females in closed-bred line;
2. Treated males and untreated females in the outbred line;
3. Untreated males and untreated females reared in closed-bred and outbred lines (control).

The doses applied were 4, 6 and 10 krad. The number of replicates were five.

The total number of eggs and the number of non-viable eggs were counted. The number of emerged adults per female was also registered. The percentage of non-viable eggs showing visible embryonal development was determined at the F_3 generation.

RESULTS

The males and the females of <u>Acanthoscelides obtectus</u> Say can be sterilized with comparatively low doses (Fig.1). Males require 10 krad of gamma radiation to become sterile, but induction of high frequencies of dominant lethals in females requires about double doses (about 17 krad). Doses necessary for sterilizing bean weevil adults are higher than required for dipterous and lower than needed for lepidopterous insects. (Further details of this question have been discussed in a previous paper [13].)

---

[1] The term "closed-bred" indicates a laboratory population, beginning with 10 males and 10 females as the parental generation. The term "outbred" is applied to crossings using in each generation females of the laboratory mass rearing.

TABLE I. EFFECT OF SUBSTERILE DOSES OF GAMMA RADIATION
ON THE FERTILITY AND FECUNDITY OF PROGENY OF TREATED
MALE BEAN WEEVIL PARENTS (CLOSED-BRED LINE)
(2 REPLICATES)
Experiment 1

Generations	Doses applied to parental males (krad)	No. of eggs laid/ female (mean)	Non-viable eggs (%)	No. of adults emerged per female	
				♂♂	♀♀
P	0	43.1	8.0 ± 0.4	21.1	19.9
	2	43.9	47.4 ± 0.7	11.2	11.3
	4	44.2	83.0 ± 4.5	4.0	2.9
	6	49.0	94.8 ± 1.5	1.1	1.1
	8	47.8	98.5 ± 1.3	0.3	0.3
	10	58.8	99.5 ± 0.1	0.1	0.1
F₁	0	76.7	4.8 ± 1.0	21.4	20.5
	2	73.3	16.1 ± 15.0	20.8	20.7
	4	64.1	49.1 ± 8.2	11.9	11.6
	6	68.7	64.3 ± 30.2	3.5	2.6
	8	90.3	69.3 a	8	7
	10	73	100 a	-	-
F₂	0	50.4	10.7 ± 5.4	13.4	13.4
	2	46.1	4.9 ± 1.6	15.9	14.0
	4	54.5	32.5 ± 16.5	14.2	13.7
	6	42.0	39.4 ± 17.1	5.3	3.7
	8	44	59.5 a	4.5	4.6
	10	-	-	-	-
F₃	0	67.2	6.5 ± 4.3	24.5	25.7
	2	74.8	3.5 ± 1.5	30.1	26.0
	4	68.7	19.4 ± 11.1	21.8	21.3
	6	60.6	28.0 ± 16.3	11.1	12.4
	8	52	50.9 a	8.1	8.4
	10	-	-	-	-
F₄	0	67.7	4.9 ± 1.7	24.5	21.1
	2	75.9	6.5 ± 3.4	23.5	23.5
	4	65.0	6.4 ± 1.6	21.9	22.5
	6	62.3	30.0 ± 31.8	11.9	11.3
	8	60.5	54.5 a	6.2	6.3
	10	-	-	-	-

a Only one date.

TABLE II. EFFECT OF SUBSTERILE DOSES OF GAMMA RADIATION
ON THE FERTILITY AND FECUNDITY OF PROGENY OF TREATED
FEMALE BEAN WEEVIL PARENTS (CLOSED-BRED LINE)
(2 REPLICATES)
Experiment 1

Generations	Doses applied to parental females (krad)	No. of eggs laid per female (mean)	Non-viable eggs (%)	No. of adults emerged per female ♂♂	♀♀
P	0	43.1	8.0 ± 0.4	21.1	19.9
	2	48.0	21.1 ± 3.4	15.8	14.9
	4	35.7	40.0 ± 2.2	14.9	11.3
	6	26.0	59.9 ± 8.2	6.5	6.8
	8	31.3	74.2 ± 2.0	3.4	4.9
	10	33.1	82.3 ± 4.6	2.9	2.8
F_1	0	76.7	4.8 ± 1.0	21.4	20.5
	2	75.4	6.5 ± 3.6	25.8	26.7
	4	71.6	8.9 ± 6.6	27.0	29.6
	6	63.2	6.9 ± 3.1	24.1	24.5
	8	75.3	5.2 ± 3.9	30.7	30.6
	10	78.8	12.0 ± 8.6	27.7	28.0
F_2	0	50.4	10.7 ± 5.4	13.4	13.4
	2	55.7	4.7 ± 3.8	21.5	22.5
	4	55.3	6.6 ± 3.0	19.2	21.2
	6	45.1	8.2 ± 3.8	18.2	16.9
	8	49.4	4.8 ± 0.06	14.8	14.0
	10	48.2	3.9 ± 0.09	20.6	17.5
F_3	0	67.2	6.5 ± 4.3	24.5	25.7
	2	70.0	5.7 ± 3.9	28.4	26.2
	4	60.0	5.5 ± 3.7	22.6	22.5
	6	61.3	10.5 ± 7.4	23.7	20.8
	8	75.9	5.0 ± 4.8	25.6	23.9
	10	64.0	4.5 ± 2.6	25.6	25.0
F_4	0	67.7	4.9 ± 1.4	24.5	21.2
	2	69.6	7.8 ± 3.4	26.2	23.0
	4	63.7	6.6 ± 6.4	18.7	21.4
	6	64.4	14.0 ± 3.5	24.1	22.2
	8	64.2	11.6 ± 1.7	21.3	19.6
	10	64.9	10.8 ± 7.3	24.0	22.8

TABLE III. DECREASE OF STERILITY IN SUCCESSIVE GENERATIONS ORIGINATING FROM BEAN WEEVILS TREATED WITH SUBSTERILE IRRADIATION DOSES (5 REPLICATES)
Experiment 2

Generations	Doses applied to parental populations (krad)	Non-viable eggs (%)	
		T ♂ × U ♀ Outbred line	T ♂ × U ♀ Closed-bred line
P	0	5.1 ± 1.8	4.3 ± 2.6
	4	94.8 ± 0.8	93.0 ± 2.7
	6	95.5 ± 1.7	96.0 ± 1.7
	10	99.6 ± 0.2	99.5 ± 0.3
F_1	0	4.0 ± 0.5	8.3 ± 7.2
	4	59.7 ± 17.7	47.8 ± 11.0
	6	52.1 ± 19.0	81.6 ± 6.8
	10	94.5 [a]	-
F_2	0	5.0 ± 2.9	4.9 ± 3.2
	4	23.9 ± 6.9	17.8 ± 11.5
	6	27.4 ± 18.6	40.5 ± 8.3
	10	-	-
F_3	0	4.1 ± 1.7	5.4 ± 2.4
	4	6.5 ± 3.4	19.3 ± 17.7
	6	13.6 ± 13.0	30.9 ± 7.8
	10	-	-
F_4	0	7.1 ± 4.9	7.0 ± 4.7
	4	6.6 ± 1.0	21.1 ± 10.3
	6	12.0 ± 9.8	33.5 ± 2.7
	10	-	-

T = treated.
U = untreated.
[a] only one date.

Experiment 1

The sterility of P-F_4 generations of T♂ × U♀ (T = treated, U = untreated) closed-bred line is summarized in Table I. In the P generation even 2 krad of gamma radiation causes almost a 50% frequency in dominant lethals, but this sterility is completely eliminated by F_2. Only those lines inherit a relatively high level of sterility, even in F_4, in which very high frequencies of dominant lethal mutations were induced (6, 8 and 10 krad). In the case of treatment with 10 krad, the sterility of F_1 was higher than that of the parents, only one male and female emerged, mated and laid completely non-viable eggs. There was no distortion observed in the sex ratio, and the number of eggs laid during successive generations.

Data summarized in Table II show the alterations in sterility of four generations induced in female parents. The level of induced dominant

TABLE IV. EFFECT OF SUBSTERILE DOSES OF GAMMA RADIATION
ON THE FECUNDITY OF PROGENY OF TREATED MALE BEAN WEEVIL
PARENTS REARED IN CLOSED-BRED AND OUTBRED LINES
(5 REPLICATES)
Experiment 2

Generations	Doses applied to male parents (krad)	No. of eggs per female		No. of adults emerged			
		T ♂ × U ♀ outbred (mean)	T ♂ × U ♀ closed-bred (mean)	T ♂ × U ♀ outbred		T ♂ × U ♀ closed-bred	
				♂♂	♀♀	♂♂	♀♀
P	0	48.3	42.2	20.8	19.3	18.7	16.7
	4	48.2	48.4	0.6	0.5	0.4	0.6
	6	45.1	48.7	0.4	0.4	0.4	0.3
	10	50.7	50.4	0.02	-	0.1	-
F$_1$	0	51.5	38.5	24.1	21.1	15.1	15.2
	4	48.2	40.1	8.9	8.0	5.9	6.7
	6	45.5	37.5	9.4	8.1	0.2	0.2
	10	29.2	-	0.1	0.3	-	-
F$_2$	0	53.8	56.6	24.3	23.4	26.4	23.9
	4	54.2	61.3	17.8	15.4	16.6	16.7
	6	51.3	54.4	14.9	13.8	14.6	13.3
	10	-	-	-	-	-	-
F$_3$	0	53.1	68.1	24.5	22.8	30.3	26.7
	4	51.4	70.6	23.2	20.8	19.6	20.6
	6	56.7	47.7	21.8	21.6	11.7	10.4
	10	-	-	-	-	-	-
F$_4$	0	48.3	71.2	22.2	21.8	27.8	27.9
	4	49.7	67.1	21.5	21.1	21.6	21.4
	6	49.8	67.6	19.5	18.9	17.9	16.9
	10	-	-	-	-	-	-

T = treated.
U = untreated.

lethals is low at the highest dose applied. Sterility was quickly eliminated.
The F$_1$ sterility was as low as that of the control. There was no effect
on sex ratio or the number of adults emerging in each generation. The
parental fecundity was influenced slightly by the irradiation.

From the preliminary experiment it could be stated that the sterility,
transmitted from generation to generation, is connected with the paternal
line.

Experiment 2

Table III summarizes the alteration in sterility of four generations
of treated males and untreated females reared in closed-bred and outbred
lines. The very rapid decrease in sterility can be observed in the outbred
line, almost twice as fast as in the closed-bred line. There are significant
differences between the sterility in F$_1$ and F$_4$ generations of the two lines

respectively. The sterility is about four times as high as in the control, even in F_4. There is no change in the sex ratio registered (Table IV).

Among the non-viable eggs found in the F_3 generation of the control, none showed visible signs of embryonal development. In the F_3 generation of the closed-bred line of males exposed to 6 krad, about 27% of the eggs were dead with signs of differentiation (pigmentation of head), and 2.6% contained fully developed, first instar larvae dead inside the egg-shells.

DISCUSSION

There are significant differences in the frequencies of induced sterility in Acanthoscelides obtectus males and females at the same irradiation doses. The progeny populations of the treated females did not inherit any significant sterility, the sterility of F_1 generation being equal to that of the control population. The control populations did not accumulate any sterility during successive generations either in the closed-bred or outbred lines.

The sterility transmitted through generations is connected with the paternal line. Low irradiation doses (2 and 4 krad) induced a low level of damage in the sperms, and the sterility of the progeny decreased to that of the control level by F_2, both in the closed-bred or in the outbred lines. The higher doses (8 and 10 krad), still not sufficient to cause total sterility, resulted in a considerable level of sterility even at F_4. The sterility of the progeny decreased more rapidly in the outbred lines. These lines made it unlikely that specimens heterozygous to dominant lethals would meet.

It can be supposed that one of the reasons for the relatively rapid decrease of sterility in successive generations may be the repair processes of the spermatogenesis in parents in the case of low irradiation treatment.

The application of substerile doses did not affect the sex ratio of the descendant populations.

There was about 30% of non-viable eggs containing dead embryos, a few of them with fully developed first instar larvae.

It can be postulated that the transmission of sterility is not the lepidopterous-type inherited sterility, and also has no similar cytogenetic background with Acanthoscelides obtectus Say.

The relatively rapid decrease of sterility in the progeny indicates that the release of substerile adults would decrease the effectiveness of the sterile release technique against the bean weevil.

ACKNOWLEDGEMENTS

I wish to thank T. Jermy, director of the Research Institute for Plant Protection, Budapest, and G. Bencze, of the Institute of Genetics, Biological Research Centre, Hungarian Academy of Sciences, Szeged, for their critical review of the manuscript.

REFERENCES

[1] PROVERBS, M.D., Progress on the use of induced sexual sterility for control of the codling moth, Carpocapsa pomonella (L.), Lepidoptera: Olethreutidae, Proc. Entomol. Soc. Ont. 92 (1962) 5.

[2] SMITH, R.H., von BORSTEL, R.C., Genetic control of insect populations, Science 178 (1972) 1164.

[3] NORTH, D.T., HOLT, G.G., "Genetic and cytogenetic basis of radiation-induced sterility in the adult male cabbage looper Trichoplusia ni", Isotopes and Radiation in Entomology (Proc. Symp. Vienna, 1967), IAEA, Vienna (1968) 391.

[4] LaCHANCE, L.E., DEGRUGILLIER, M., LEVERICH, A.P., Cytogenetics of inherited partial sterility in three generations of the large milkweed bug as related to holokinetic chromosomes, Chromosoma 29 (1970) 20.

[5] STIMMANN, M.W., GOUGH, D.G., Inherited sterility among progeny of cabbage loopers treated with tretamine, J. Econ. Entomol. 65 (1972) 994.

[6] KNIPLING, E.F., Suppression of pest Lepidoptera by releasing partially sterile males, Bioscience 20 (1970) 465.

[7] TOBA, H.H., KISHABA, A.N., NORTH, D.T., Reduction of populations of caged cabbage loopers by release of irradiated males, J. Econ. Entomol. 65 (1972) 408.

[8] LAVIOLETTE, P., NARDON, P., "Influence de l'irradiation sur les adultes des Sitophilus sasakii Takahashi (Curculionidae) et leurs descendants", Radiation and Radioisotopes Applied to Insects of Agricultural Importance (Proc. Symp. Athens, 1963), IAEA, Vienna (1963) 431.

[9] CORNWELL, P.B., BULL, J.O., PENDLEBURY, J.B., "Control of weevil populations (Sitophilus granarius (L.)) with sterilizing and substerilizing doses of gamma radiation", Entomology of Radiation Disinfestation of Grain (CORNWELL, P.B., Ed.), Pergamon Press, Oxford (1966).

[10] JACKLIN, S.W., RICHARDSON, E.G., YONCE, C.E., Substerilizing doses of gamma irradiation to produce population suppression in plum curculio, J. Econ. Entomol. 63 (1970) 1053.

[11] BROWER, J.H., Inability of populations of Callosobruchus maculatus to develop tolerance to exposures of acute gamma irradiation, Ann. Entomol. Soc. Am. 67 (1974) 287.

[12] SZENTESI, Á., JERMY, T., DOBROVOLSZKY, A., Mathematical method for the determination of sterile insect population competitiveness, Acta Phytopathol. Acad. Sci. Hung. 8 (1973) 185.

[13] SZENTESI, Á., Effects of gamma- and X-ray irradiation on different developmental stages of the bean weevil (Acanthoscelides obtectus Say) (Col., Bruchidae), Folia Entomol. Hung. (in press).

DISCUSSION

D.W. WALKER: Can you explain the mechanism of transmission of this genetic damage to subsequent generations?

Á. SZENTESI: I do not know anything about these mechanisms, as my work has not included any cytogenetic investigations.

L.E. LaCHANCE: Have you any information on the time of death of the embryos formed from the union of an irradiated sperm and a normal egg?

Á. SZENTESI: The development of normal eggs requires about 4-5 days at 28°C. We observe typical larval organs in various stages of development in non-viable eggs, and thus it would appear that the embryos die at different times, mostly after one or two days.

L.E. LaCHANCE: Are the males of this species the heterogametic sex?

Á. SZENTESI: I cannot say.

H. LEVINSON: Have you any idea what mechanism is responsible for the surprisingly rapid recovery of spermatogenesis?

Á. SZENTESI: I have no explanation for the rapid recovery that occurs in irradiated males at the second mating. This question is currently being investigated.

G. DELRIO: Have you carried out experiments on separate pairs of insects to investigate the degree of sterility in generations succeeding the irradiated parents? I think this sterility may be due to chromosomal translocations.

Á. SZENTESI: I have not carried out any such experiments.

INFLUENCE DE MODIFICATIONS DE COURTE DUREE DE LA TEMPERATURE D'ELEVAGE SUR LA CAPACITE REPRODUCTRICE DE LA BRUCHE DU HARICOT
Acanthoscelides obtectus Say (Coléoptère, Bruchidae)

J.C. BIEMONT
Laboratoire d'écologie expérimentale,
Université François-Rabelais,
Tours, France

Abstract—Résumé

INFLUENCE OF SHORT-TERM MODIFICATIONS OF THE REARING TEMPERATURE ON THE REPRODUCING CAPACITY OF THE BEAN WEEVIL (Acanthoscelides obtectus Say, Coleoptera, Bruchidae).

Acanthoscelides obtectus males which are normally reared at 27°C were kept at 36°C during the nymphal or imaginal stage for variable periods. After this thermal treatment, they were mated with females reared at 27°C. The ovarian production of the females that were mated with males kept at 36°C for 6 or 7 days during the nymphal stage, or for 10 days during the imaginal stage, was lower than that of the control females. After this treatment, the fertility was very reduced. A long thermal treatment seems to perturb the paragonia secreting activity, which produces spermatophore. The spermatophores are smaller, their structure is often altered and they have a reduced stimulating effect upon oogenesis. Moreover, a long thermal treatment perturbs the spermatozoa gathering into spermatodesms. The influence of temperature stress on females has also been studied. Half-an-hour exposure of virgin females to high temperature (45°C) at different ages causes a significant increase in egg mortality on the first day of oviposition after the treatment. Chorion-covered oocytes stored in lateral oviducts were more sensitive than young growing oocytes.

INFLUENCE DE MODIFICATIONS DE COURTE DUREE DE LA TEMPERATURE D'ELEVAGE SUR LA CAPACITE REPRODUCTRICE DE LA BRUCHE DU HARICOT Acanthoscelides obtectus Say (Coléoptère, Bruchidae).

Des mâles de la bruche du haricot, Acanthoscelides obtectus, élevés à 27°C sont placés à 36°C pendant des durées variables au cours du stade nymphal ou du stade imaginal. Après ce traitement thermique, ils sont accouplés avec des femelles vierges élevées continuellement à 27°C. La production ovarienne des femelles accouplées avec les mâles maintenus à 36°C pendant 6 ou 7 jours au cours du stade nymphal, ou pendant 10 jours durant la vie imaginale, est inférieure à celle des témoins. Leur fertilité est également réduite. Le traitement thermique prolongé perturbe l'activité sécrétrice des glandes annexes mâles. Les spermatophores produits sont plus petits et de structure anormale; leur effet stimulant sur l'ovogenèse est réduit. De plus, la formation des spermatodesmes est altérée. L'influence de «stress thermiques» est également étudiée chez les femelles. Lorsque des femelles sont exposées à 45°C pendant une demi-heure à des âges différents, le pourcentage de mortalité des œufs émis le premier jour après le traitement augmente. Les ovocytes chorionés maintenus en rétention dans les oviductes latéraux sont plus sensibles que les ovocytes en cours de vitellogenèse.

INTRODUCTION

Les variations de la température d'élevage peuvent influencer l'activité reproductrice des insectes. Les mâles semblent particulièrement sensibles à ces variations et une faible modification de la température peut provoquer une stérilité importante chez <u>Drosophila melanogaster</u> [1], chez <u>Heliothis Zea</u> et <u>H. virescens</u> [2], chez <u>Carpocapsa pomonella</u> [3] ou chez <u>Hypera postica</u> [4].

Nous avons examiné au cours de ce travail les conséquences d'une augmentation de la température sur la capacité reproductrice des mâles d'<u>Acanthoscelides obtectus</u> sans nous limiter à l'étude des effets stérilisants. Chez cet insecte, en effet, la copulation n'apporte pas seulement des spermatozoïdes dans les voies génitales femelles, mais peut stimuler l'ovogenèse, l'amplitude de l'effet dépendant de l'état physiologique du mâle au moment de la copulation [5,6]. Dans ces conditions, l'augmentation de la température risque d'influencer non seulement leur pouvoir fertilisant (déterminé par la fertilité des femelles avec lesquelles ils ont copulé) mais aussi leur pouvoir fécondant (déterminé d'après la production ovarienne de ces mêmes femelles).

Chez les femelles, les variations de température interviennent sur le taux d'éclosion des œufs et leur production [7,8].

Nous nous sommes limités ici à l'étude des conséquences de «stress thermiques» sur le pourcentage de stérilité journalière des pontes chez <u>A. obtectus</u>, dans des conditions de plus ou moins longue rétention des ovocytes.

1. MATERIEL ET METHODES

L'action de la température est testée, soit au cours de la nymphose, soit au cours de la vie imaginale, chez des bruches de la lignée II sélectionnées par Labeyrie [9]. Dans ces deux cas les insectes mâles, qui sont élevés à 27°C et 70% d'humidité relative, ont été placés pendant une période de durée variable à 36°C et 60% d'humidité relative. D'autre part, les femelles vierges sont soumises, à l'état adulte, à des chocs thermiques (passage de courte durée à température élevée: une demi-heure à 45°C) à des âges divers.

1.1. Action de la température sur les mâles au cours de la nymphose

Les mâles sont soumis, avant l'émergence, à un traitement thermique à 36°C pendant 3, 5, 6 ou 7 jours (fig.1). Ils sont maintenus pendant 4 jours à 37°C avant d'être mis en présence de femelles.

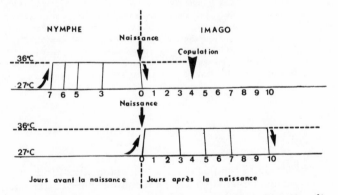

FIG. 1. Schéma résumant les différents traitements thermiques subis par les mâles.

1.2. Action de la température sur les mâles au cours de la vie imaginale

Les mâles sont placés, dès la naissance, à 36°C pendant 3, 5, 7 ou 10 jours. Après ces différents traitements, ils sont de nouveau placés à 27°C et mis en présence de femelles (fig.1).

Dans ces deux séries d'expériences, les femelles vierges, élevées à 27°C, sont accouplées à l'âge de 4 jours. Après une copulation, elles sont isolées pendant 20 jours en présence de grains de haricot; leur fécondité (nombre d'œufs pondus) et leur fertilité (nombre d'œufs éclos) sont déterminées quotidiennement. Les femelles sont ensuite disséquées de manière à dénombrer les ovocytes maintenus en rétention (ce qui permet de calculer la production ovarienne[1]) et à prélever le spermatophore.

Les spermatophores émis par les mâles soumis aux différents traitements thermiques sont fixés au Bouin puis étudiés sur coupe histologique de 5 μm après coloration au trichrome de Masson.

1.3. Action de la température sur les femelles adultes

Deux types d'expérimentations ont été effectués:
— une température peu élevée (36°C) est appliquée pendant 3 jours sur des femelles vierges âgées de 9 jours; ces femelles sont ensuite accouplées au 12e jour avec des mâles de 4 jours non soumis au traitement thermique;
— des «stress thermiques» (passage de courte durée: une demi-heure à 45°C) sont appliqués à des femelles vierges, isolées de la plante-hôte pendant des durées croissantes; dans ces conditions les ovocytes formés ne sont pas émis et sont maintenus en rétention dans les oviductes latéraux pendant des durées plus ou moins élevées; la fertilité de ces femelles est étudiée pendant 10 jours.

2. PRODUCTIONS OVARIENNES DES FEMELLES ISOLEES APRES UNE COPULATION AVEC DES MALES SOUMIS AUX DIFFERENTS TRAITEMENTS THERMIQUES

2.1. Mâles placés à 36°C au cours de la nymphose

L'ensemble des résultats du tableau I a été soumis à une analyse de variance à un facteur de classification. Les moyennes observées sont ensuite classées au moyen de la méthode de Newman et Keuls [10].

Les productions ovariennes de femelles accouplées aux mâles «3 j à 36°C» sont les plus élevées. Elles sont supérieures à celles des témoins. Cependant il n'y a pas de différence significative entre les productions ovariennes des femelles témoins et celles des femelles accouplées aux mâles «5 j à 36°C».

Dans les deux autres expériences (femelles $\times$ mâles «6 j à 36°C» et femelles $\times$ mâles «7 j à 36°C»), les productions ovariennes sont inférieures à celles des témoins, cependant elles sont supérieures à celles des femelles vierges.

---

[1] Production ovarienne = nombre d'œufs pondus + nombre d'ovocytes maintenus en rétention à la fin de la période de ponte.

TABLEAU I. ACTIVITE REPRODUCTRICE DES FEMELLES AYANT
COPULE AVEC DES MALES TRAITES PENDANT 3, 5, 6 OU 7 JOURS
A 36°C AU COURS DE LA NYMPHOSE

	Nombre de femelles	Fécondité	Production ovarienne
Témoins vierges	43	48, 7 ± 8, 1	64 ± 5, 6
Témoins (1 copulation)	44	102, 5 ± 6, 3	107, 2 ± 5, 9
♀ × ♂ 3 j à 36°C	42	127, 3 ± 8, 2	131, 4 ± 3, 1
♀ × ♂ 5 j à 36°C	43	103, 1 ± 9, 4	112, 9 ± 8, 5
♀ × ♂ 6 j à 36°C	38	88, 5 ± 7, 5	95, 5 ± 6, 8
♀ × ♂ 7 j à 36°C	45	70, 4 ± 6, 3	74, 2 ± 6, 4

TABLEAU II. ACTIVITE REPRODUCTRICE DES FEMELLES AYANT
COPULE AVEC DES MALES TRAITES PENDANT 3, 5, 7 OU 10 JOURS A
36°C LORS DE LA VIE IMAGINALE

	Nombre de femelles	Fécondité	Production ovarienne
Témoins vierges	43	48, 7 ± 8, 1	64 ± 5, 6
Témoins (1 copulation)	44	102, 5 ± 6, 3	107, 2 ± 5, 9
♀ × ♂ 3 j à 36°C	46	112, 8 ± 7, 8	119, 3 ± 7, 5
♀ × ♂ 5 j à 36°C	50	101, 0 ± 7, 5	106, 2 ± 7, 3
♀ × ♂ 7 j à 36°C	49	93, 2 ± 7, 9	96, 1 ± 7, 9
♀ × ♂ 10 j à 36°C	48	63, 8 ± 6, 1	67, 6 ± 5, 6

Une copulation avec un mâle maintenu pendant 6 ou 7 jours à 36°C
provoque donc une certaine stimulation de l'ovogenèse, mais son influence
est nettement plus faible que celle d'une copulation avec un mâle continuellement élevé à 27°C. Par contre, l'effet stimulant est plus important lorsque
les femelles s'accouplent avec des mâles maintenus pendant 3 jours à 36°C.

2.2. Mâles placés à 36°C au cours de la vie imaginale

Les données du tableau II ont été soumises à la même analyse statistique
que précédemment.

Les productions ovariennes les plus élevées sont observées chez les
femelles accouplées aux mâles traités pendant 3 jours à 36°C à l'état
imaginal. Un tel traitement stimule donc le pouvoir fécondant des mâles.
Cependant, les mâles maintenus pendant 5 ou 7 jours à 36°C conservent
le même pouvoir fécondant que les témoins. Les productions ovariennes des
femelles accouplées aux mâles placés 10 jours à 36°C ne diffèrent pas
significativement de celles des femelles vierges. Dans ce cas, la copulation
ne provoque aucune stimulation de l'ovogenèse.

2.3. Spermatophores émis par les mâles soumis aux différents traitements thermiques

Lorsque les mâles ont été placés à 36°C pendant 3 ou 5 jours soit au stade nymphal, soit au stade imaginal, ils émettent des spermatophores apparemment identiques à ceux des témoins.

Lorsque les mâles ont été placés pendant 5, 7 ou 10 jours à 36°C lors de la nymphose, ils émettent, dans la plupart des cas, des spermatophores de petite taille qui distendent partiellement la bourse copulatrice.

On observe trois types de spermatophores:
— des spermatophores de structure apparemment normale,
— des spermatophores totalement dépourvus de membrane externe,
— des spermatophores dans lesquels la sécrétion des glandes médianes, qui forme la membrane [6], se trouve rassemblée dans la région centrale et mêlée aux autres sécrétions.

Un séjour prolongé à 36°C semble donc perturber l'activité des glandes annexes mâles qui produisent le spermatophore [11].

3. FERTILITE DES FEMELLES ISOLEES APRES UNE COPULATION AVEC DES MALES TRAITES THERMIQUEMENT

3.1. Mâles placés à 36°C pendant la nymphose

La distribution des femelles en fonction du pourcentage de fertilité de leurs pontes est représentée figure 2.

Chez les témoins, la fertilité est élevée et comprise entre 80 et 100%.

Dans l'expérience où les femelles copulent avec des mâles traités 3 jours à 36°C, on observe une certaine modification dans la répartition: le nombre d'individus ayant une fertilité supérieure à 90% devient très réduit. On remarque d'autre part l'apparition de femelles à faible fertilité.

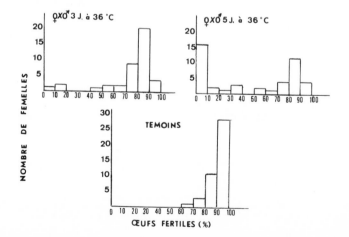

FIG. 2. Distribution, en fonction du pourcentage d'œufs fertiles, des femelles témoins et des femelles accouplées à des mâles soumis pendant 3 ou 5 jours à 36°C au cours de la nymphose.

Si le séjour des mâles à 36°C est de 5 jours, la distribution des femelles devient très étalée et peut être décomposée en deux groupes:
— des femelles dont le pourcentage d'œufs fertiles est supérieur à 50%, chez lesquelles on trouve en fin d'expérience de nombreux spermatozoïdes actifs dans la spermathèque; ces femelles se sont donc accouplées à des mâles peu sensibles à l'action de la température et qui produisent dans la plupart des cas des spermatophores analogues à ceux des témoins;
— des femelles dont le pourcentage d'œufs fertiles est inférieur à 50% et qui ne possèdent pratiquement plus de spermatozoïdes dans la spermathèque en fin d'expérience; ces femelles ont donc copulé avec des mâles très sensibles à l'action de la température et qui produisent des spermatophores de petite taille contenant de nombreux spermatozoïdes actifs.

Lorsque la durée du séjour à 37°C dépasse 5 jours, les spermatozoïdes ne sont plus groupés en spermatodesmes et sont inactifs; ils sont déposés dans le spermatophore mais ne migrent pas dans la spermathèque. Ces mâles n'ont plus aucun pouvoir fertilisant.

3.2. Mâles placés à 36°C pendant la vie imaginale

La figure 3 montre que:
— un traitement thermique de 3 ou 5 jours modifie peu le pouvoir fertilisant des mâles; cependant, il apparaît quelques insectes ayant une fertilité réduite;
— la sensibilité des mâles à un traitement thermique de 7 ou 10 jours varie de façon notable selon les individus.

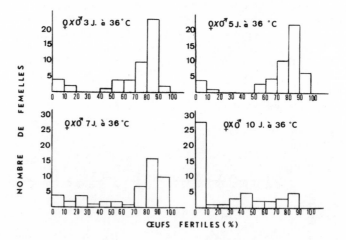

FIG. 3. Distribution, en fonction du pourcentage d'œufs fertiles, des femelles accouplées à des mâles traités pendant 3, 5, 7 ou 10 jours à 36°C au cours de la vie imaginale.

4. EFFETS DE LA TEMPERATURE SUR LES FEMELLES

Le maintien pendant 3 jours à 36°C de femelles vierges âgées de 9 jours provoque une importante diminution de la fertilité de leurs pontes (tableau III).

Cependant, il nous a paru plus intéressant d'appliquer sur les femelles
des chocs thermiques de courte durée (une demi-heure) à température
élevée (45°C).

Ainsi, lorsque les femelles sont soumises à un choc thermique dès la
naissance, la fertilité de leur ponte n'est pas modifiée (le pourcentage
d'œufs non éclos du premier jour de ponte est de 9,17 ± 2,09). Cependant,
lorsqu'un traitement analogue est appliqué sur des femelles vierges âgées
de 8, 12 ou 20 jours (et isolées en l'absence de haricot pendant cette période),
le pourcentage d'œufs non éclos du premier jour de ponte devient d'autant
plus élevé que la durée de rétention des œufs a été plus longue (fig.4). On
remarque donc selon le moment d'application du traitement une sensibilité
différente en ce qui concerne la ponte du premier jour après l'accouplement et l'introduction des grains de haricot. Ce phénomène se retrouve,
quoique moins accentué, en l'absence de tout traitement thermique.

TABLEAU III. EVOLUTION JOURNALIERE DU POURCENTAGE D'ŒUFS
NON ECLOS (FEMELLES DE 9 JOURS SOUMISES A 36°C PENDANT
3 JOURS)

	J 1	J 2	J 3	J 4	J 5	J 6 à 10
Nombre d'œufs émis	693	193	338	194	193	736
Pourcentage d'œufs non éclos	84,8 ± 2,6	63,2 ± 6,8	43,8 ± 5,3	23,7 ± 6,0	26,4 ± 6,2	21,5 ± 3,0

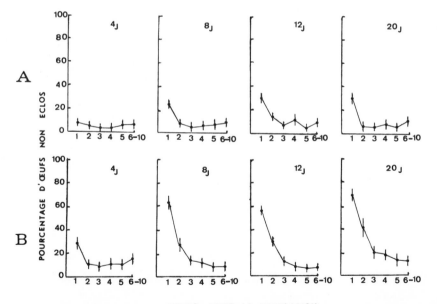

FIG.4. Evolution du pourcentage d'œufs non éclos en fonction de la durée de la rétention (4, 8, 12 ou
18-20 jours) chez les femelles témoins (A) et chez les femelles soumises à un choc thermique (B).

D'autre part, nous avons pu remarquer que les ovocytes qui sont maintenus en rétention chez les femelles vierges de 4 jours sont émis dans les
24 heures qui suivent la copulation et l'introduction des grains de haricot [12].
De plus, des femelles vierges âgées de 12 jours pondent en moyenne au
cours des 24 heures qui suivent la copulation 32,4 œufs. Cette valeur n'est
pas significativement différente du nombre moyen d'ovocytes maintenus en
rétention par les femelles vierges âgées de 12 jours ($\epsilon = 1,59$ au seuil 0,05).

On peut donc conclure de l'ensemble de ces résultats que les ovocytes
qui sont maintenus en rétention dans les oviductes sont plus sensibles au
traitement thermique que les ovocytes qui sont en cours d'évolution. Des
observations publiées ultérieurement et concernant les effets des irradiations
sur les femelles de la bruche du haricot avaient permis d'aboutir aux mêmes
conclusions [12].

5.   DISCUSSION

5.1. Sensibilité thermique des mâles

L'augmentation momentanée de la température d'élevage au cours de
la nymphose ou de la vie imaginale modifie les pouvoirs fécondant et
fertilisant des mâles.

La diminution du pouvoir fertilisant des mâles observée après un
séjour à 36°C semble due à l'altération de la structure des spermatodesmes
et peut-être à une modification de l'organisation ou de la physiologie des
spermatozoïdes à l'intérieur des testicules [11]. Cependant, dans tous les
cas, les spermatozoïdes sont émis dans le spermatophore au moment de
la copulation.

Chez les nymphes, le traitement thermique agit au moment du groupe-
ment des spermatozoïdes en spermatodesmes et provoque des altérations
importantes [11].

L'augmentation de la durée du traitement thermique a un effet défavorable
et le pouvoir fécondant décroît. Il devient même nul lorsque les insectes
sont maintenus pendant 10 jours à 36°C. Deux facteurs peuvent être
responsables de cette décroissance du pouvoir fécondant des mâles:
— la composition chimique des sécrétions émises par le mâle a pu être
   modifiée, les synthèses enzymatiques pouvant être altérées par l'action
   prolongée de la température; les sécrétions émises dans le spermatophore
   auraient alors une efficacité limitée et ne provoqueraient qu'une faible
   augmentation de la production ovarienne;
— l'activité sécrétrice des glandes annexes mâles est plus réduite après
   un traitement thermique prolongé puisque les spermatophores sont de
   petite taille et ont une structure modifiée par rapport à celle des mâles
   témoins.

Deux glandes seraient particulièrement sensibles à l'élévation de
température:
— les glandes tubuleuses produisant la substance active responsable de la
   stimulation de l'ovogenèse qui suit la copulation [6,13];
— les glandes médianes qui élaborent normalement la membrane du
   spermatophore; cette membrane peut être absente lorsque les mâles ont
   séjourné 6 à 7 jours à 36°C au cours de la nymphose.

D'autre part, l'ordre d'émission des différentes sécrétions au cours de la sécrétion du spermatophore pourrait être modifié par le traitement thermique.

5.2. Sensibilité différentielle des ovocytes maintenus en rétention

L'action des chocs thermiques appliqués sur des femelles d'âge croissant montre que les ovocytes mûrs chorionés maintenus en rétention dans les oviductes latéraux au moment de l'application du traitement sont particulière-ment sensibles, tandis que les potentialités de développement des ovocytes en cours de vitellogenèse ne sont pas altérées. Un tel phénomène, qui se retrouve chez les femelles témoins, pourrait être attribué à des altérations métaboliques des gamètes femelles en particulier au niveau de leurs réserves au cours de la rétention [14]. Ces ovocytes, qui sont les premiers formés, ont vieilli et probablement évolué au cours de leur séjour dans les oviductes.

Ces ovocytes mûrs retenus dans les oviductes sont également sensibles aux rayons X [12]. L'ensemble de ces constatations laisse supposer que la rétention ne correspond pas à un simple stockage des ovocytes et qu'elle peut avoir une importance non négligeable sur le pouvoir reproducteur d'un insecte. Ces résultats soulignent la nécessité de réaliser une analyse chronologique de la stérilité induite par un stress et de ne pas se limiter à l'examen global d'une moyenne calculée sur la ponte de plusieurs jours.

REFERENCES

[1] DAVID, J., ARENS, M.F., COHET, Y., C.R. Acad. Sci. Paris 274 (1971) 3102.
[2] GUERRA, A.A., J. Econ. Entomol. 65 (1972) 368.
[3] PROVERBS, M.D., NEWTON, J.R., Can. Entomol. 94 (1962) 225.
[4] LE CATO, G.L., PIENKOWSKI, R.L., J. Econ. Entomol. 65 (1972) 146.
[5] HUIGNARD, J., C.R. Acad. Sci. Paris 268 (1969) 2938.
[6] HUIGNARD, J., Thèse Sciences naturelles, Université de Tours (1973) 206 p.
[7] LAUGE, G., C.R. Acad. Sci. Paris 277 (1973) 1545.
[8] DATERMAN, C., Can. Entomol. 104 (1972) 1387.
[9] LABEYRIE, V., C.R. Soc. Biol. Poitiers 162 (1968) 2203.
[10] SOKAL, R.R., ROHLF, F.J., Biometry, W.H. Freeman, San Fransisco (1969).
[11] HUIGNARD, J., BIEMONT, J.C., Ann. Zool. Ecol. Anim. (1974) (sous presse).
[12] BIEMONT, J.C., Ann. Zool. Ecol. Anim. 5 (1973) 581.
[13] HUIGNARD, J., Ann. Sci. Nat. (1974) (sous presse).
[14] MERLE, J., DAVID, J., C.R. Acad. Sci. Paris 265 (1967) 2070.

DISCUSSION

B.S. FLETCHER: Have you established whether a chemical is involved in promoting fecundity of the female or is it the physical presence of the spermatophore in the female reproductive tract?

J.C. BIEMONT: The presence of the spermatophore causes distension of the bursa copulatrix which inhibits sexual receptivity for a fairly long time. Observations by Huignard (1973) show that a substance emitted by the male in the spermatophore appears to be responsible for stimulating oogenesis. Laboratory investigations are now being carried out to confirm these observations.

J.L. MONTY: Does oogenesis occur even in the absence of mating?

J.C. BIEMONT: In the virgin females of the strain used here the oocytes are produced and stored in the lateral oviducts, if the insects cannot find their preferred laying base, the bean.

A. FELDMANN: Can the differential radiosensitivity observed for different periods of oocyte retention be correlated with different stages of meiosis?

J.C. BIEMONT: This is something which we have not investigated.

B. LEIGH: I should just like to mention that there are two separate effects of oocyte storage in Drosophila; firstly, there is a short-term effect related to the meiotic process, and secondly there is a cytoplasmic effect which can be detected after about ten days' storage.

D.S. GROSCH: The same sort of thing can be observed in the wasp Bracon. There are two different reasons for oocyte sensitivity, one nuclear and the other cytoplasmic.

A METHOD OF DETERMINING
THE GAMMA RADIATION DOSES
FOR THE STERILIZATION
OF STORED PRODUCT INSECTS*

F.M. WIENDL
Department of Entomology and Centre
of Nuclear Energy in Agriculture

J.M. PACHECO, J.M.M. WALDER, R.B. SGRILLO,
Rachel E. DOMARCO
Centre of Nuclear Energy in Agriculture,
Piracicaba, Brazil

Abstract

A METHOD OF DETERMINING THE GAMMA RADIATION DOSES FOR THE STERILIZATION OF STORED PRODUCT INSECTS.
    The method consists in the irradiation of infested stored grains or other products with increasing doses of gamma irradiation. The samples were weighed weekly to determine loss caused by insect attack. This loss in weight, which shows that sterilization has not occurred, can be easily detected after six to seven cycles. Differences were found in the sensitivity of insects to different environments, i.e. temperature, humidity and substrate. On maize (Zea mays L.) the Sitophilus zeamais Mots. was sterilized with 8 krad at 30°C (62 to 82% r.h.). The same insect, at 25°C (65 to 95% r.h.) was sterilized with 7 krad, and at a temperature varying between 22° and 35°C (20 to 90% r.h.), the required dose was 7 krad also. In the case of Zabrotes subfasciatus (Boh.) on beans (Phaseolus vulgaris L.), the sterilizing dose was found to be 10 krad, at a temperature of 27°C (62 to 78% r.h.). Also, on beans (Vigna sinensis Endl.) the species Callosobruchus maculatus (Fab.) was sterilized with 8 krad at 30°C (70 to 75% r.h.). The species Sitophilus oryzae (L.), reared at 27°C (62 to 78% r.h.) in various substrates had varying sensitivity to radiation. On maize (Zea mays L.) and rice (Oryza sativa L.) the sterilizing dose for this insect was 7 krad, and on macaroni 5 krad.

1. INTRODUCTION

    Pest infestation causes great losses to stored grains throughout the world. This is especially true in the developing countries where the technology is less advanced, and climatic conditions are extremely favourable for the development of pests.
    A review of the literature indicates that many authors have already carried out work on the control of pests by gamma, X-ray and/or accelerated electron irradiation (see Bibliography). However, very little work has been done on pest control under actual store conditions, taking into consideration the ecology of the insects and the type of grain stored.
    In view of this, the present method was developed. It consists of irradiation of the insects in their habitat, i. e. in the stored grain, with increasing gamma-radiation doses. The samples are then kept under the desired environmental conditions, preferably in places where the temperature and humidity are ideal for the development of the species.

Text continued on p. 303

* The work was carried out with aid from the Brazilian Nuclear Energy Commission.

TABLE I.   LOSS IN WEIGHT (IN PERCENTAGES) CAUSED BY Sitophilus oryzae (L.) IN RICE (Oryza sativa L.) FOR 46 WEEKS; IRRADIATION DOSES: 0, 6, 7 AND 20 krad; TEMPERATURE 27°C

Week	Dose (krad)			
	0	6	7	20
0	0.00	0.00	0.00	0.00
1	0.78	0.30	0.74	0.76
2	0.28	0.21	0.02	0.17
3	0.68	0.01	0.21	0.36
4	1.18	0.20	0.40	0.53
5	1.37	0.07	0.27	0.38
6	1.92	0.20	0.39	0.50
7	3.68	0.62	0.62	1.07
8	5.47	1.04	1.19	1.31
9	8.98	1.33	1.43	1.54
10	12.49	1.64	1.78	1.79
11	16.90	1.67	1.63	1.73
12	22.87	1.71	1.59	1.69
13	30.94	1.93	1.62	1.75
14	36.44	2.68	1.70	1.82
15	41.91	2.76	1.70	1.85
16	49.31	3.45	1.74	1.93
17	52.93	4.31	1.69	1.78
18	60.32	5.89	1.78	1.80
19	63.61	7.05	1.80	1.83
20	65.85	8.03	1.50	1.50
21	68.49	9.60	1.91	1.92
22	70.11	10.03	2.07	2.11
23	71.25	12.17	2.19	2.21
24	72.04	13.42	2.39	2.40
25	72.27	14.22	2.47	2.48
26	72.31	15.21	2.46	2.49
27	72.39	15.98	2.41	2.50
28	72.47	16.41	2.40	2.50
29	72.59	17.83	2.52	2.61
30	72.64	18.13	2.65	2.69
31	72.64	18.87	2.68	2.79
32	72.67	19.71	2.72	2.79

TABLE I.(cont.)

Week	Dose (krad)			
	0	6	7	20
33	72.71	20.45	2.83	2.86
34	72.71	21.17	2.87	2.89
35	72.74	21.88	2.91	2.88
36	72.80	22.63	3.07	3.11
37	72.84	23.84	3.16	3.18
38	72.84	23.85	3.10	3.11
39	72.86	24.46	3.13	3.11
40	72.74	24.86	2.93	2.92
41	72.69	24.18	2.74	2.73
42	72.70	28.63	2.63	2.60
43	72.69	26.16	2.55	2.51
44	72.68	26.97	2.43	2.31
45	72.69	26.97	2.39	2.27
46	72.64	26.44	2.22	2.12

TABLE II. LOSS IN WEIGHT (IN PERCENTAGES) CAUSED BY Sitophilus
oryzae (L.) IN MAIZE (Zea mays L.) FOR 46 WEEKS; IRRADIATION
DOSES: 0, 6, 7 AND 20 krad; TEMPERATURE 27°C

Week	Dose (krad)			
	0	6	7	20
0	0.00	0.00	0.00	0.00
1	0.63	0.54	0.61	0.58
2	0.01	0.02	0.02	0.05
3	0.17	0.14	0.14	0.06
4	0.41	0.29	0.28	0.22
5	0.32	0.14	0.13	0.06
6	0.54	0.24	0.22	0.16
7	1.25	0.62	0.60	0.50
8	1.94	1.00	0.96	0.85
9	2.25	1.15	1.16	1.01
10	3.44	1.29	1.26	1.14
11	3.96	1.09	1.22	1.08
12	4.49	0.96	1.08	0.97
13	5.26	1.02	1.12	1.01
14	5.99	1.12	1.18	1.10
15	6.79	1.17	1.17	1.08
16	7.80	1.25	1.16	1.05
17	8.92	1.29	1.17	1.05
18	10.52	1.57	1.20	1.06
19	11.46	1.70	1.22	1.07
20	12.11	1.66	0.98	0.79
21	13.32	2.29	1.36	1.17
22	14.26	2.80	1.53	1.38
23	15.06	3.42	1.70	1.47
24	25.89	4.15	1.92	1.70
25	16.40	4.63	2.03	1.81
26	16.74	4.98	2.03	1.82
27	17.14	5.32	2.05	1.83
28	17.44	5.72	2.06	1.83
29	17.59	6.02	2.12	1.92
30	17.99	6.55	2.23	1.97
31	18.14	6.79	2.31	2.06
32	18.26	7.01	2.31	2.04

TABLE II.(cont.)

Week	Dose (krad)			
	0	6	7	20
33	18.42	2.26	2.42	2.12
34	18.55	7.46	2.46	2.12
35	18.66	7.61	2.45	2.11
36	18.86	7.86	2.59	2.37
37	19.08	8.14	2.76	2.38
38	19.03	8.14	2.65	2.35
39	19.17	8.30	2.67	2.33
40	19.06	8.21	2.49	2.16
41	18.96	8.11	2.29	1.97
42	18.95	8.07	2.13	1.79
43	18.93	8.09	1.99	1.68
44	18.99	8.15	1.89	1.59
45	19.09	8.26	1.89	1.50
46	19.07	8.36	1.75	1.36

TABLE III.   LOSS IN WEIGHT (IN PERCENTAGES) CAUSED BY Sitophilus oryzae (L.) IN MACARONI FOR 46 WEEKS; IRRADIATION DOSES: 0, 3, 4 AND 20 krad; TEMPERATURE 27°C

Week	Dose (krad)			
	0	3	4	20
0	0.00	0.00	0.00	0.00
1	-1.09	-1.22	-1.96	-1.34
2	-0.33	-0.45	-1.20	-0.53
3	-0.04	-0.09	-0.91	-0.34
4	-0.14	-0.53	-1.27	-0.73
5	0.44	-0.10	-0.90	-0.34
6	1.03	0.65	0.48	0.65
7	2.34	1.25	0.45	0.84
8	2.48	1.17	0.37	0.72
9	2.65	1.11	0.34	0.61
10	2.83	1.16	0.32	0.67
11	2.84	1.02	0.18	0.54
12	3.24	1.18	0.30	0.72
13	3.45	1.29	0.39	0.80
14	3.59	1.25	0.29	0.73
15	3.75	1.24	0.27	0.58
16	3.95	1.33	0.31	0.57
17	4.45	1.53	0.41	0.58
18	4.66	1.78	0.53	0.71
19	4.34	1.30	0.05	0.06
20	5.55	2.46	1.05	1.03
21	5.78	2.55	1.01	1.03
22	6.26	2.88	1.29	1.22
23	6.76	3.30	1.64	1.48
24	6.96	3.33	1.65	1.41
25	6.99	3.41	1.68	1.42
26	7.21	3.48	1.72	1.42
27	7.41	3.56	1.74	1.43
28	7.81	3.23	1.84	1.55
29	7.96	2.97	2.00	1.61
30	7.97	4.09	2.05	1.67
31	8.05	4.17	2.07	1.64
32	8.18	4.32	2.18	1.73

TABLE III.(cont.)

Week	Dose (krad)			
	0	3	4	20
33	8.20	4.35	2.17	1.72
34	8.21	4.36	2.17	1.68
35	8.53	4.74	2.48	2.07
36	8.49	4.69	2.46	1.94
37	8.39	4.59	2.33	1.83
38	8.44	4.64	2.35	1.80
39	8.03	4.20	1.91	1.43
40	7.92	4.06	1.77	1.28
41	7.89	4.04	1.73	1.22
42	7.80	3.93	1.61	1.13
43	7.80	3.89	1.58	1.02
44	7.80	3.87	1.55	0.90
45	7.63	3.66	1.36	0.72
46	7.71	3.65	1.35	0.73

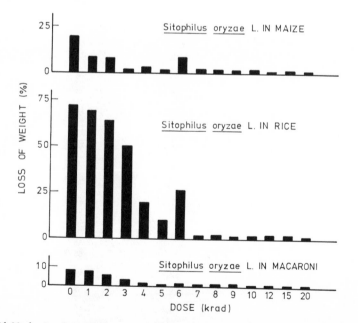

FIG.1. 46th Week. Loss in weight caused by Sitophilus oryzae L. in maize (Zea mays L.), in rice (Oryza sativa L.) and in macaroni, all at 27°C, in percentage, on all doses.

TABLE IV. LOSS IN WEIGHT (IN PERCENTAGES) CAUSED BY Sitophilus zeamais Mots. IN MAIZE (Zea mays L.) FOR 53 WEEKS; IRRADIATION DOSES: 0, 7, 8 AND 20 krad; TEMPERATURE 30°C

Week	Dose (krad)			
	0	7	8	20
0	0.00	0.00	0.00	0.00
1	3.37	0.67	0.61	0.71
2	1.08	1.15	1.06	1.08
3	1.64	1.54	1.28	1.31
4	2.44	2.04	1.62	2.01
5	3.62	2.36	2.10	2.65
6	4.68	2.83	2.64	2.99
7	5.47	3.00	2.83	3.07
8	6.18	2.92	2.64	2.95
9	7.06	2.64	2.47	2.61
10	8.71	2.91	2.81	2.79
11	10.74	3.03	2.68	2.85
12	12.97	2.95	2.68	2.84
13	15.48	2.89	2.51	2.85
14	18.26	2.95	2.52	2.92
15	22.19	2.82	2.37	2.60
16	22.54	2.64	2.13	2.39
17	30.13	2.75	2.05	2.29
18	34.20	2.95	1.95	2.19
19	38.19	3.49	2.03	2.32
20	42.19	4.02	2.15	2.30
21	46.02	4.60	2.12	2.18
22	49.64	5.25	2.12	2.05
23	54.51	6.1	2.17	2.18
24	58.45	7.28	2.39	2.07
25	62.24	7.61	1.91	1.94
26	65.40	9.61	1.75	1.75
27	68.14	10.45	1.62	1.59
28	71.22	11.48	1.65	1.45
29	73.42	12.59	1.68	1.52
30	75.26	13.13	1.53	1.40
31	76.52	14.23	1.53	1.38
32	78.96	15.47	1.43	1.23

TABLE IV.(cont.)

Week	Dose (krad)			
	0	7	8	20
33	80.29	16.65	1.09	1.02
34	81.46	18.47	0.91	0.91
35	82.46	20.81	0.62	0.75
36	83.16	22.54	0.64	0.69
37	83.84	24.07	0.55	0.55
38	84.43	25.52	0.48	0.48
39	85.07	26.59	0.46	0.39
40	85.74	27.93	0.71	0.69
41	86.08	29.05	0.62	0.65
42	86.54	30.47	0.89	0.66
43	87.04	31.65	1.16	0.90
44	87.51	32.95	1.56	1.16
45	87.99	34.52	1.85	1.37
46	88.38	35.72	1.94	1.84
47	88.53	36.60	1.97	1.98
48	88.31	37.47	2.68	1.99
49	89.12	39.00	2.56	2.74
50	88.83	39.69	2.62	2.83
51	89.02	40.62	2.70	2.88
52	88.82	40.97	2.75	2.73
53	89.02	40.81	2.69	2.65

TABLE V.   LOSS IN WEIGHT (IN PERCENTAGE) CAUSED BY Sitophilus
zeamais Mots. IN MAIZE (Zea mays L.) FOR 53 WEEKS; IRRADIATION
DOSES: 0, 6, 7 AND 20 krad; TEMPERATURE 25°C

| Week | Dose (krad) | | | |
	0	6	7	20
0	0.00	0.00	0.00	0.00
1	0.36	0.35	0.34	0.37
2	0.64	0.58	0.58	0.62
3	1.07	0.91	0.93	0.95
4	1.77	1.41	1.44	1.44
5	2.54	1.86	1.85	1.85
6	3.22	2.15	2.17	2.15
7	3.76	2.20	2.17	2.12
8	4.47	2.28	2.14	2.11
9	5.35	2.33	2.11	2.08
10	6.84	2.66	2.39	2.37
11	8.47	2.87	2.54	2.50
12	10.51	3.08	2.71	2.63
13	12.60	3.22	2.53	2.45
14	15.56	3.70	2.50	2.45
15	20.02	4.06	2.28	2.29
16	22.97	3.97	1.86	1.86
17	26.65	4.24	1.71	1.69
18	30.21	5.09	1.90	1.86
19	33.35	6.11	1.97	1.95
20	36.27	7.32	1.99	1.95
21	39.44	8.62	2.13	2.06
22	42.22	9.65	2.07	2.00
23	44.66	10.51	2.14	2.05
24	46.73	11.08	2.10	2.02
25	48.61	11.49	1.92	1.85
26	50.53	11.82	1.67	1.59
27	52.32	12.20	1.53	1.45
28	54.19	12.53	1.34	1.27
29	55.97	12.92	1.30	1.22
30	57.57	13.17	1.08	1.01
31	59.36	13.67	1.13	1.07
32	60.68	13.83	0.90	0.84

TABLE V.(cont.)

Week	Dose (krad)			
	0	6	7	20
33	61.94	13.89	0.58	0.53
34	63.44	14.31	0.49	0.45
35	64.97	14.41	0.27	0.22
36	66.13	14.70	0.25	0.21
37	67.36	15.00	0.19	0.15
38	68.53	15.38	0.26	0.22
39	70.89	16.45	0.93	0.90
40	73.56	17.60	1.68	1.63
41	76.29	18.46	2.06	2.08
42	79.09	19.37	2.41	2.36
43	80.79	19.93	2.63	2.56
44	82.76	20.63	2.90	2.83
45	84.70	21.35	3.19	3.12
46	86.04	21.89	3.38	3.31
47	86.89	22.21	3.40	3.34
48	87.66	22.40	3.25	3.18
49	88.37	22.74	3.30	3.23
50	88.75	22.84	3.18	3.11
51	89.15	22.76	3.13	3.06
52	89.51	23.06	3.13	3.08
53	89.62	23.11	3.16	3.11

TABLE VI.   LOSS IN WEIGHT (IN PERCENTAGES) CAUSED BY Sitophilus
zeamais Mots. IN MAIZE (Zea mays L.) FOR 53 WEEKS; IRRADIATION
DOSES: 0, 6, 7 AND 20 krad; TEMPERATURE FROM 22°C TO 35°C

Week	Dose (krad)			
	0	6	7	20
0	0.00	0.00	0.00	0.00
1	-0.17	-0.12	-0.14	-0.10
2	-0.08	-0.06	-0.07	-0.10
3	0.03	0.02	0.02	0.06
4	0.18	0.14	0.16	0.18
5	0.40	0.32	0.36	0.35
6	0.73	0.57	0.65	0.63
7	0.76	0.55	0.62	0.60
8	0.94	0.63	0.68	0.66
9	1.33	0.87	0.93	0.88
10	1.86	1.24	1.23	1.22
11	2.35	1.50	1.52	1.44
12	2.99	1.81	1.84	1.74
13	3.30	1.74	1.74	1.64
14	4.00	2.08	2.01	1.97
15	4.77	2.24	2.20	2.08
16	5.24	2.14	2.05	1.91
17	6.15	2.38	2.15	1.97
18	6.90	2.77	2.38	2.10
19	9.04	3.45	3.01	2.51
20	10.86	3.84	3.33	2.69
21	13.11	4.23	3.84	3.03
22	16.06	4.77	4.09	3.08
23	20.07	5.60	4.56	3.39
24	23.44	6.07	4.64	3.28
25	27.26	6.57	4.84	3.21
26	31.30	7.14	5.12	3.13
27	34.47	4.17	5.33	3.08
28	37.34	8.18	5.63	2.99
29	40.08	9.00	5.94	3.04
30	43.25	9.67	6.21	3.03
31	46.99	10.56	6.72	3.23
32	50.18	11.22	6.94	3.18

TABLE VI.(cont.)

Week	Dose (krad)			
	0	6	7	20
33	52.87	11.58	6.93	2.95
34	56.09	12.18	7.12	2.94
35	58.94	12.60	7.04	2.71
36	61.48	13.09	7.14	2.65
37	66.88	13.57	7.17	2.75
38	66.91	14.14	7.40	2.77
39	69.52	14.56	7.59	3.00
40	72.36	15.29	7.81	2.95
41	74.81	15.60	7.80	2.95
42	77.13	16.06	7.86	3.01
43	78.60	16.36	7.92	3.12
44	80.15	16.77	8.08	3.26
45	81.60	17.18	8.27	3.36
46	82.50	17.44	8.34	3.18
47	82.96	17.51	8.22	2.97
48	83.28	17.44	8.02	3.00
49	83.72	17.57	8.06	2.80
50	83.90	17.47	7.84	2.75
51	84.13	17.51	7.82	2.74
52	84.40	17.55	7.85	2.76
53	84.50	17.67	9.92	2.81

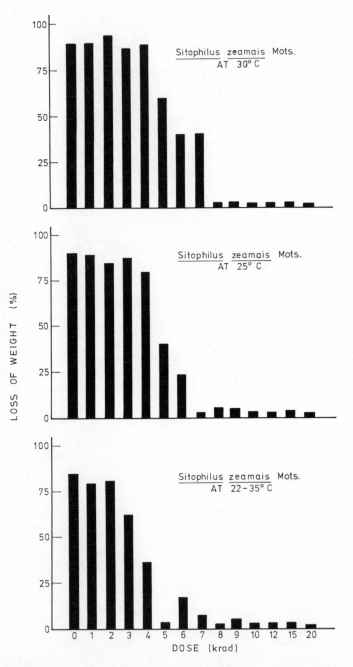

FIG.2. 53rd Week. Loss in weight caused by <u>Sitophilus zeamais</u> Mots. in maize (<u>Zea mays</u> L.), at 30°C, 25°C and at 22°-35°C, in percentage, on all doses.

This is done to obtain a greater reaction from the insects, and also makes possible the determination of the maximum dose necessary to sterilize the insects without changing the natural environment.

2. COMMON METHODOLOGY FOR ALL EXPERIMENTS

The insects used were obtained from a colony reared and kept under optimum environmental conditions for at least 70 generations, which resulted in a high index of homozygosis.

The substrates used were maize (Zea mays L., var. HMD-7974), rice (Oriza sativa L., var. Jaguari), bean (Phaseolus vulgaris L., var. Rosinha) and bean (Vigna sinensis, var. Seridó). Commercial macaroni was also used. All substrates were kept for at least 30 days at a temperature of -20°C to eliminate all latent infestation before the beginning of the experiments.

Radiation doses utilized were 0 (control), 1, 2, 3, 4, 5, 6, 7, 8, 9, 10, 12, 15 and 20 krad. Each dose was replicated four times, i.e. in flasks of approximately 250 ml vol. or about 150 g of grains. The macaroni, however, because of its size, weighed about 60 g.

The replications, consisting of the substrate and initially with 20 insects, were irradiated at a rate between 57.65 and 66.52 krad/h. The temperature in the irradiation chamber was maintained at about 28°C and relative humidity at about 65% r. h.

After irradiation, the replicated treatments were kept under optimum environmental conditions for the development of the insects. The samples were weighed every 7 days, over a period sufficient for the development of at least eight generations, in order to check any changes in weight. The development of populations until their extermination through lack of food was also determined.

3. METHODOLOGY FOR Sitophilus oryzae (L.)

The main objective was to observe the influence of the substrate in relation to the sterilizing doses. The substrate used was maize, Zea mays L., var. HMD-7974, husked rice, Oryza sativa, var. Jaguari, and commercial macaroni. For each dose and each substrate. four replications were used and weighed every 7 days for 46 weeks (322 days).

After irradiation the experiment was maintained in an environment with a temperature at 27°C ± 1°C, and relative humidity between 62 and 78% r. h.

4. METHODOLOGY FOR Sitophilus zeamais Mots.

This experiment aimed mainly at checking the influence of temperature in relation to sterilization doses.

Maize was used, Zea mays L., var. HMD-7974, as substrate. However, 12 replications were made for each dose, and after irradiation four were kept at a temperature of 30°C ± 1°C, and relative humidity between 62 and 82% r. h., four other replications were kept at 25°C ± 1°C, and relative humidity between 65 and 95% r. h., and the remaining four replications were

TABLE VII.   LOSS IN WEIGHT (IN PERCENTAGES) CAUSED BY
Zabrotes subfasciatus (Boh.) IN BEANS (Phaseolus vulgaris L.) FOR
36 WEEKS; IRRADIATION DOSES: 0, 9, 10 AND 20 krad;
TEMPERATURE 27°C

Week	Dose (krad)			
	0	9	10	20
0	0.00	0.00	0.00	0.00
1	2.36	2.15	2.09	2.25
2	4.08	3.74	3.59	3.81
3	6.13	5.43	5.17	5.46
4	7.11	6.31	5.94	6.28
5	7.99	7.19	6.68	7.06
6	9.37	7.97	7.41	7.80
7	13.94	8.68	8.16	8.43
8	15.55	9.20	8.48	8.83
9	16.74	9.53	8.81	9.16
10	22.34	9.75	8.97	9.34
11	29.48	10.03	9.19	9.58
12	32.86	10.26	9.34	9.74
13	34.41	10.25	9.37	9.68
14	36.98	10.65	10.45	10.02
15	39.20	11.16	9.86	10.22
16	42.03	11.52	9.91	10.26
17	42.37	11.59	9.86	10.21
18	42.83	11.48	9.68	10.01
19	43.31	12.32	9.56	9.88
20	43.42	12.88	8.97	9.22
21	43.49	13.10	8.68	8.94
22	43.65	13.25	8.62	8.90
23	43.66	13.52	8.46	8.77
24	43.65	13.71	8.32	8.62
25	43.56	13.77	8.05	8.46
26	43.48	13.79	7.81	8.26
27	43.57	13.94	7.81	8.24
28	43.95	14.38	8.11	8.53
29	43.86	14.38	8.06	8.45
30	43.74	14.38	8.01	8.35

TABLE VII.(cont.)

Week	Dose (krad)			
	0	9	10	20
31	43.48	14.14	7.66	8.00
32	43.43	14.07	7.54	7.87
33	43.27	13.89	7.33	7.63
34	43.25	13.75	7.19	7.50
35	43.56	13.93	7.32	7.62
36	43.56	14.17	7.45	7.76

maintained at a temperature varying between 22°C and 35°C. As in the previous experiment, a weekly weighing was made, but for a period of 53 weeks (371 days).

5. METHODOLOGY FOR Callosobruchus maculatus (Fabr.)

For this species the only objective was to determine the sterilization doses in an environment ideal for its development.

Five replications were made for each dose, and only 10 insects were used in each replication at the beginning of the experiment. After irradiation the samples were kept for 28 weeks (196 days) in an environment with a temperature of 30°C and relative humidity between 70 and 75% r.h., and were also weighed weekly.

The substrate used was bean, Vigna sinensis Endl., var. Seridó.

6. METHODOLOGY FOR Zabrotes subfasciatus (Boh.)

As in the previous case, the aim for this species was also to determine the sterilization dose in an environment ideal for its development.

Five replications were used for each dose, with 40 insects in each at the beginning of the experiment.

After irradiation, the samples were kept for 36 weeks (252 days) under optimum conditions for the development of the species, i.e. at a temperature of 27°C ± 1°C, and relative humidity between 62 and 78% r.h.

The substrate used was the bean Phaseolus vulgaris L., var. Rosinha, which was susceptible to infestation, like the substrates in the other experiments.

7. RESULTS

The results obtained from about 10 000 weighings are summarized in the tables as a percentage of loss in weight compared with the first weighings.

TABLE VIII.   LOSS IN WEIGHT (IN PERCENTAGES) CAUSED BY
Callosobruchus maculatus (F.) IN BEANS (Vigna sinensis Endl.) FOR
28 WEEKS; IRRADIATION DOSES: 0, 7, 8 AND 20 krad;
TEMPERATURE 30°C

Week	Dose (krad)			
	0	7	8	20
0	0.00	0.00	0.00	0.00
1	0.50	0.47	0.37	0.36
2	3.35	0.80	0.71	0.70
3	6.08	1.03	0.92	0.89
4	9.21	1.26	1.15	1.10
5	19.45	1.28	1.07	1.03
6	28.20	1.46	0.93	0.87
7	33.77	1.68	1.02	0.98
8	39.79	2.41	0.99	0.96
9	45.20	7.72	1.08	1.01
10	50.07	11.13	1.19	1.13
11	52.42	12.68	1.20	1.15
12	55.88	16.65	1.66	1.63
13	57.85	19.86	1.63	1.59
14	58.77	21.79	1.63	1.60
15	59.29	23.72	1.57	1.55
16	59.57	26.44	1.49	1.47
17	59.86	28.96	1.64	1.62
18	59.98	31.13	1.73	1.72
19	59.93	33.08	1.58	1.55
20	59.89	35.07	1.48	1.49
21	59.82	36.33	1.32	1.37
22	60.01	37.87	1.40	1.40
23	60.01	39.97	1.43	1.45
24	60.12	40.61	1.49	1.50
25	60.07	41.03	1.37	1.39
26	60.03	41.34	1.24	1.25
27	60.00	41.59	1.12	1.15
28	60.06	41.90	1.01	1.01

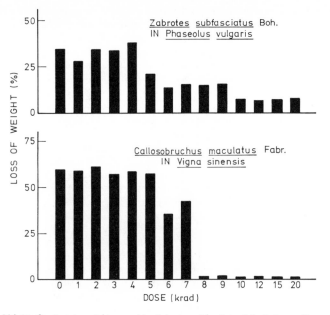

FIG.3. Top: 36th Week. Loss in weight caused by Zabrotes subfasciatus Boh. in beans (Phaseolus vulgaris L.) at 27°C.
Bottom: 28th Week. Loss in weight caused by Callosobruchus maculatus F. in beans (Vigna sinensis Endl.) at 30°C. Loss in weight in percentage, on all doses.

As it was impossible to present all the results in this paper, only treatments which clearly show the difference between sterilizing and non-sterilizing doses are presented. Also for comparison purposes the results from the 0 krad dose were included as they represent the untreated substrate (control) whereas the results from the 20 krad dose show changes in weight due to humidity changes alone.

Tables I-III indicate the percentage loss in weight occurring after irradiation of rice, maize and macaroni samples infested with Sitophilus oryzae L. and kept at the same temperature, 27°C. It can be noticed that during the first 20 to 25 weeks, in the samples where sterilization did not occur, the loss in weight was characteristic. This loss was due to an increase in insect population and subsequent attack on the substrate. The damage caused by S. oryzae was greatest in rice and much lower in maize and macaroni. Figure 1 shows the percentage loss in weight at the 46th week on all three substrates and at the different doses used.

Tables IV-VI indicate the percentage loss in weight on a weekly basis after irradiation of maize samples infested with Sitophilus zeamais Mots. at the three different temperatures used. It can be seen that a higher dose is required for samples stored at the higher temperature range. This is further clarified in Fig. 2 which shows the loss in weight at different doses and at three different temperatures observed at the 53rd week.

Finally Tables VII and VIII show the percentage loss in weight after irradiation of bean samples infested with Zabrotes subfasciatus (Boh.) and

Callosobruchus maculatus (F.) at 27°C and 30°C respectively. Figure 3
indicates percentage loss in weight in substrate infested with Zabrotes
subfasciatus (Boh.) and subjected to different doses at the 36th week. It
also shows the loss in weight of the substrate infested with Callosobruchus
maculatus (F.) and subjected to different dose rates at the 28th week. The
difference in susceptibility of insects to radiation treatment can be shown
clearly in this figure.

8.   CONCLUSION AND DISCUSSION

In view of these results we concluded that the method described above
to determine the dose necessary for the sterilization of insects that attack
stored grains was satisfactory. It takes into account environment,
temperature, humidity and substrate, in relation to the permanently
sterilizing dose.

The method is also useful because it is a very simple one. Samples
could be weighed after a period of 20 to 25 weeks, i.e. after a time
sufficient for the development of 6 or 7 generations.

Finally, it may be pointed out that visual observation alone of the
samples tested might be sufficient to determine the best sterilizing dose for
these kind of insects. When sterilization occurs, the flasks were clean
and free of infestation, whereas when the desired sterilization did not
occur, infestation by both insects and mould was easily observed.

BIBLIOGRAPHY

ANDREEV, S.V., MARTENS, B.K., MOLCHANOVA, V.A., SALDAN, L.N., Radiation sterilization in
controlling insect pests, Vest. sel'khoz. Nauki, Mosk. 1 (1967) (in Russian).

ANDREEV, S.V., MARTENS, B.K., MOLCHANOVA, V.A., SALDAN, L.N., Radiosterilization in the fight
against insect pests, Vest. sel'khoz. Nauk., Mosk. 1 (1967) 48 (in Russian).

ANDREEV, S.V., MARTENS, B.K., Calculation of dose distribution in a grain gamma irradiator and
determination of its parameters for the "Disinsection" flowline method, Radiobiologiya 5 (1965) 605
(in Russian).

ANDREEV, S.V., MARTENS, B.K., Au Laboratoire de biophysique, Zashch. Rast. Vredit. Bolez. 11
(1967) 25.

ANDREEV, S.V., MARTENS, B.K., MOLCHANOVA, V.A., Development of research in sterilization techniques
for agricultural insect pests in the USSR, Sterile Male Technique for Eradication and Control of Harmful Insects
(Proc. Panel Vienna, 1968), IAEA, Vienna (1969) 57.

BADITSING, C., "A Study on the life history and the effects of radiation on rice weevil (Sitophilus oryzae L.),
Insect Eradication by Irradiation, Bangkok, Thailand, 28-29 June, 1966, 1341.

BAKER, V.H., TABOADA, O., WIANT, D., Lethal effects of electrons on insects which infest wheat and
flour: I., Agric. Eng. 34 (1953) 755.

BAKER, V.H., TABOADA, O., WIANT, D.E., Some effects of accelerated electrons or cathode rays on
certain insects and on the wheat and flour they infest: I, Mich. St. Coll. Agric., Exp. Sta. Quart. Bull. 36
(1953) 94.

BAKER, V.H., TABOADA, O., WIANT, D.E., Electron gun aimed at insects, Food Eng. (Philadelphia)
26 4 (1954) 64.

BRUEL, W.E., BOLLAERTS, D. Van Den, Resistance of Sitophilus granarius and Sitophilus oryzae at different stages of their development to gamma irradiation from cobalt 60, Bull. Inst. Agron. Stns Redr. Gembloux, Extra Vol. 2 (1960) 883.

BULL, J.O., CORNWELL, P.B., "A comparison of the susceptibility of the grain weevil Sitophilus granarius (L.) to accelerated electrons and ^{60}Co gamma radiation", Entomology of Radiation Disinfestation of Grain (CORNWELL, P.B., Ed.), Pergamon Press, Oxford (1966) 157.

BULL J.O., WOND, T., CORNWELL, P.B., A Comparison of the Susceptibility of the Grain Weevil (Sitophilus granarius L.) to Accelerated Electrons and ^{60}Co Gamma Radiation, Rep. AERE-R-3895, U.K. Atomic Energy Authority, Research Group, Isotope Research Division, Wantage, Berks, England (1961).

CAVALLORO, R., BONFANTI, G., Possibilita dell'impiego di radiazioni ionizanti contro Acanthocelides obsoletus Say (Col. Bruchidae), A defesa dei legumi conservati, Boll. Zool. Agrar. Bachiolt. II (1967) 8.

COGBURN, R.R., TILTON, E.W., BROWER, J.H., Bulk grain gamma irradiation for control of insects infesting wheat, J. Econ. Entomol. 65 (1972) 818.

CORNWELL, P.B., The disinfestation of foods, particularly grain, Int. J. Appl. Radiat. Isot. 6 (1959) 188.

CORNWELL, P.B.,"Insect control by ionizing radiations – An appraisal of the potentialities and problems involved", The Technological Use of Radiation (Proc. Conf. Technological Use of Radiation, Sydney, Australia, 23-25 May 1960), Australian Atomic Energy Commission (1961) 104.

CORNWELL, P.B., BULL, J.O., Insect control by gamma-irradiation, an appraisal of the potentialities and problems involved, J.Sci. Food Agric. 11 (1960) 754.

CORNWELL, P.B., BURSON, D.M., Grain weevils Calandra granaria L. and C. oryzae L., reared on irradiated wheat, Nature (London) 181 (1958) 1747.

CORNWELL, P.B., BURSON, D.M., PENDLEBURY, J.B., MARTIN, V.J., BULL, J.O., Control of Weevil Populations Sitophilus granarius L. with Sterilizing and Substerilizing Doses of Gamma Radiation, Rep. AERE-R-3892, U.K. Atomic Energy Authority, Research Group, Isotope Research Division Wantage, Berks, England (1962).

DENNIS, N.M., The effect of gamma ray irradiation on certain species of stored products insects, J. Econ. Entomol. 54 (1961) 211.

DENNIS, N.M., "Effect of cathode ray irradiation on the rice weevil in wheat", US Agricultural Marketing Research Rep. 531 (1962) 1.

DENNIS, N.M., SODERHOLM, L.H., WALKDEN, H.H., "The effects of cathode ray irradiation on the rice weevil in wheat", US Dep. Agric. Marketing Research Rep. 531 (1962) 14.

DePROOST, M., L'Effort Belge en matière d'irradiation des aliments, Meded. bouwhogesch. Opzoekingsstn. Staat Gent 30 (1965) 667.

DePROOST, M., "Belgium contributions to the data on irradiation of substances", Conf. 65081-1, Centre d'Etude de l'Energie Nucléaire Mol (Belgium), and 2nd Int. Symp. Appl. Nuclear Sci. in Agron. and Veter. Medicine, Gent, Belgium (1969).

DUCOFF, H.S., BOSMA, G.C., Response of Tribolium confusum to radiation and other stresses, XVI International Congress on Zoology, Washington 2, 20-27 Aug. 1963 (MOORE, J.A., Ed.), Washington, D.C., 83.

DUCOFF, H.S., BOSMA, G.C., The influence of pupal age on sensitivity to radiation, Biol. Bull. 130 (1966) 151.

DUCOFF, H.S., BOSMA, G.C., Acute lethality after X-irradiation of Tribolium confusum adults, Entomologia Exp. Appl. 10 (1967) 153.

DUCOFF, H.S., WALBURG, H.W., Response of Tribolium larvae to X-irradiation (Abstr. 140), Anat. Rec. 137 (1960) 351.

ERDMAN, H.E., "Effects of irradiating single and mixed species of beetles", Hanford Biology Research Annual Report for 1960, Rep. HW-76000, General Electric Co., Hanford Atomic Products, Richland, Washington (1961).

ERDMAN, H.E., Arrested development in X-rayed larvae of Ephestia kuehniella Zeller (Lep., Phycitinae), Rep. HW-SA-2281, General Electric Co., Hanford Atomic Products, Richland, Washington (1961).

ERDMAN, H.E., "Developmental arrest of irradiated Ephestia larvae", Hanford Biology Research, Annual Report for 1961, Rep. HW 72500, General Electric Co., Hanford Atomic Products, Richland, Washington (1962) 102.

ERDMAN, H.E., "X-ray tolerance of two related species of beetles", Hanford Biology Research, Annual Report for 1961, Rep. HW 72500, General Electric Co., Hanford Atomic Products, Richland, Washington (1962) 156.

ERDMAN, H.E., Beginning of Reproduction by Age of the Female Flour Beetle, Tribolium confusum (Col. Tenebrionidae), Rep. HW-SA-2576, General Electric Co., Hanford Atomic Products, Richland, Washington (1962).

ERDMAN, H.E., "X-ray effect on different life stages of two flour beetle species (Tribolium confusum, T. castaneum), Summer Meeting of American Soc. Zoologists, Oregon State Univ., Corvallis, Oregon, Aug. 1962, Am. Zool. 2 (1962) 407.

ERDMAN, H.E., Beginning of reproduction determined by age of the female flour beetle, Tribolium confusum (Col. Tenebrionidae), Naturwissenschaften 49 (1962) 248.

ERDMAN, H.E., Comparative X-ray sensibility of Tribolium confusum and T. castaneum (Col. Tenebrionidae) at different developmental stages during their lifecycle, Nature (London) 195 (1962) 1281.

ERDMAN, H.E., Arrested development in X-rayed larvae of Ephestia kuehniella Zeller (Lep., Phicitinae), Am. Midl. Nat. 69 (1963) 34.

ERDMAN, H.E., "Effects of radiation on ecological systems" (Abstr. E1A774), Research and Development in Progress: Biology and Medicine, No. 2, Rep. TID-4201, Div. Tech. Inform., AEC, Oak Ridge (Nov. 1963).

ERDMAN, H.E., The differential sensitivity of flour beetles, Tribolium confusum and T. castaneum to X-ray alteration of reproductive abilities, induced dominant lethals, biomass, and survival, J. Exp. Zool. 153 (1963) 141.

ERDMAN, H.E., Age, temperature, coexistence, and X-radiation effects on flour beetles productivity", Hanford Biology Research, Annual Report for 1963, Rep. HW-80550, General Electric Co., Hanford Atomic Products, Richland, Washington (1964) 144.

ERDMAN, H.E., "Fast neutron effect of flour beetles", Hanford Biology Research, Annual Report for 1963, Rep. HW 80500, General Electric Co., Hanford Atomic Products, Richland, Washington (1964) 147.

ERDMAN, H.E., Dominant lethal proportions modified by X-radiation; temperatures and cohabitation in single and mixed species populations of flour beetles Tribolium confusum and T. castaneum, Radiat. Res. 22 (1964) 187 (Abstr. 59).

ERDMAN, H.E., "Effects of radiation on competitive insects" (Abstract), Research and Development in Progress: Biology and Medicine, No. 3, Rep. TID-4203, Div. Tech. Inform., AEC, Oak Ridge (1964) 277.

ERDMAN, H.E., Fast Neutron Effects on Productivity of Young and Old Flour Beetles, Tribolium castanenum and Alterations due to Temperature and Sex Exposed, Rep. HW-SA-3537, General Electric Co., Hanford Atomic Products, Richland, Washington (1964).

ERDMAN, H.E., Differential Response of Germ Cells in Flour Beetles Tribolium castaneum Herbst, due to X-Ray Dose, Hypothermia, Sex Exposed and Age, Rep. HW-SA-3747, General Electric Co., Hanford Atomic Products, Richland, Washington (1964).

ERDMAN, H.E., Reproductive Performance of X-Rayed Single Species and Mixed Species Cultures of Tribolium confusum and T. castaneum reared at Different Temperatures, Rep. HW-SA-3748, General Electric Co., Hanford Atomic Products, Richland, Washington (1964).

ERDMAN, H.E., "X-ray and fast neutron effects on reproductivity of flour beetles", Hanford Biology Research, Annual Report for 1964, Rep. BNWL-122, Battelle Northwest, Richland, Wash., Pacific Northwest Lab. (1965) 137.

ERDMAN, H.E., Dose ratio of X-rays to fast neutrons in producing dominant lethals in flour beetles, Tribolium castaneum, Nature (London) 205 (1965) 99.

ERDMAN, H.E., Fast neutron effects on productivity of young and old flour beetles, Tribolium castaneum Herbst. and alterations at different temperatures and after exposures of either or both sexes, Int. J. Radiat. Biol. 9 (1965) 305.

ERDMAN, H.E., Modifications of productivity in flour beetles, Tribolium castaneum Herbst, due to X-ray dose, hypothermia and sex exposed, Radiat. Res. 25 (1965) 341.

ERDMAN, H.E., X-radiation and temperature modification of reproductive performance on single species and mixed species cultures of Tribolium confusum and T. castaneum, Physiol. Zool. 39 (1966) 160.

ERDMAN, H.E., Effects of X-rays on metamorphosis and adult life span of flour beetles (T. confusum), Nature (London) 211 (1966) 1427.

ERDMAN, H.E., Modifications of fitness in species and strains of flour beetles due to X-ray and DDT, Ecology 47 (1966) 1066.

HASSET, C.C., Lethal radiation for stored product insects, Pest Control 25 11 (1957) 13.

HASSET, C.C., JENKINS, D.W., Use of fission products for insect control, Nucleonics 10 12 (1952) 42.

HOEDAYA, M.S., HUTABARAT, D., SASTRADIHARDJA, S.I., SOETRISNO, S., "Radiation effects on four species of stored rice and the use of radiation disinfestation in their control", Radiation Preservation of Food (Proc. Symp. Bombay, 1972), IAEA, Vienna (1973) 281.

HOOVER, D.L., FLOYD, E.H., RICHARDSON, H.D., Effects of 300 kV X-ray radiation on Sitophilus oryzae, J. Econ. Entomol. 56 5 (1963) 584.

HUNTER, W.D., Results of experiments to determine the effect of roentgen rays upon insects, J. Econ. Entomol. 5 (1912) 188.

HUQUE, H., "Preliminary studies on irradiation of some common stored-grain insects in Pakistan", Radiation and Radioisotopes applied to Insects of Agricultural Importance (Proc. Symp. Athens, 1963), IAEA, Vienna (1963) 455.

JEFFERIES, D.J., "The effects of continuous and fractionated doses of gamma-radiation on the survival and fertility of Sitophilus granarius (Calandra granaria L.)", Radioisotopes and Radiation in Entomology (Proc. Symp. Bombay, 1960), IAEA, Vienna (1962) 213.

JEFFERIES, D J., The Susceptibility of the Saw-Toothed Grain Beetle, Oryzaephilus surinamensis L. to Gamma Radiation, Rep. AERE-R-3891, U.K. Atomic Energy Authority Research Group, Isotope Research Div., Wantage, Berks, England (1962).

JEFFERIES, D.J., "Effects of continuous and fractionated doses of gamma radiation on the survival and fertility of Sitophilus granarius (L.), Entomology of Radiation Disinfestation of Grain (CORNWELL, P.B., Ed.), Pergamon Press, Oxford (1966) 41.

JEFFERIES, D.J., BANHAM, E.J., The Effects of Continuous and Fractionated Doses of Gamma Radiation on the Survival and Fertility of Sitophilus granarius (Calandra granaria), Rep. AERE-R-3503, U.K. Atomic Energy Authority Research Group, Isotope Research Div., Harwell, Berks, England (1961).

JEFFERIES, D.J., BANHAM, E.J., "The effect of dose rate on the response of Tribolium confusum, Oryzaephilus surinamensis, and Sitophilus granarius to ^{60}Co gamma radiation", The Entomology of Radiation Disinfestation of Grain (CORNWELL, P.B., Ed.), Pergamon Press, Oxford (1966) 177.

KAHN, Z.A., MATIN, A.S.M.A., "The susceptibility of different stages of rice weevil, Sitophilus oryzae L., to gamma radiation", Rep. AECD-AG-22, Atomic Energy Centre, Dacca, Pakistan (1967) 15.

KUMAGAI, M., Influence of post treatment humidity on the irradiated rice weevil adult, Sitophilus zeamais Mots. (Col. Curculionidae), Appl. Entomol. Zool. 2 1 (1967) 51.

KUMAGAI, M., Effects of irradiation on the eggs of the Adzuki bean weevil, Callosobruchus chinensis L. (Col. Bruchidae), Appl. Entomol. Zool. 4 1 (1969) 9.

LAUDANI, H., TILTON, E.W., BROWER, J.H., USDA research program and facilities for the use of gamma irradiation in the control of stored product insects, Food Irrad. 6 (1965) A6.

LAVIOLETTE, P., NARDON, P., "Influence de l'irradiation sur les adultes de Sitophilus sasakii takahashi (Curculionidae) et leurs descendants", Radiation and Radioisotopes applied to Insects of Agricultural Importance (Proc. Symp. Athens, 1963), IAEA, Vienna (1963).

LAVIOLETTE, P., NARDON, P., Actions des rayons gamma du cobalt 60 sur la mortalité et la fertilité des adultes d'un charançon du riz, Bull. Biol. 97 (1963) 305.

312   WIENDL et al.

MAHROUS, M.A., ROSTON, Z.M.F., Effect of γ-rays on the duration of life of the rice weevil, Sitophilus oryzae (L.), Int. J. Radiat. Biol. 5 (1962) 191.

MATIN, A.S.M.A., "Susceptibility of adult rice weevil, Sitophilus oryzae (L.), to gamma radiation", Proceedings of the Agricultural Symposium, May 1966, Dacca, Pakistan, Atomic Energy Centre, 133.

MATIN, A.S.M.A., Susceptibility of Adult Rice Weevil, Sitophilus oryzae (L.), to Gamma Radiation, Rep. AECD-AG-15, Atomic Energy Centre, Dacca (1966).

MARTIN, V.J., BURSON, D.M., BULL, J.O., CORNWELL, P.B., The Effect of Culture Environment on the Susceptibility of the Grain Weevil Sitophilus granarius L. to Gamma Radiation, Rep. AERE-R-3893, U.K. Atomic Energy Authority Research Group, Isotope Research Division, Wantage, Berks, England (1962).

MATSUYAMA, A., "Recent advances in food irradiation research in Japan", Food Irradiation (Proc. Symp. Karlsruhe, 1966), IAEA, Vienna (1966) 767.

MATSUYAMA, A., "Present status of food irradiation research in Japan with special reference to microbiological and entomological aspects", Radiation Preservation of Food (Proc. Symp. Bombay, 1972), IAEA, Vienna (1973) 261.

MONTE, G. dal, Utilizatione dei Raggi Ionizanti per la Conservazione dei Cereali, Molini d'Italia 10 (1959) 29.

NARDON, P., Répercussion de l'influence des rayons gamma dans la descendance de Sitophilus sasakii Takahashi (Col. Curculionidae), C.R. Acad. Sci., Paris 254 (1962) 2454.

NARDON, P., Les possibilités d'emploi des radiations dans la lutte contre les insectes, Phytoma 144 (1963) 7.

NEHARIN, A., CALDERON, M., YACOBI, O., Susceptibility of Callosobruchus maculatus to High Dose Gamma Irradiation, Rep. IA-1010, Israel Atomic Energy Commission, Soreq Research Establishment, Rehovot, Israel (1965).

NEIDINGER, J.W., BARATZ, R.A., MOOS, W.S., Oxygen uptake of three life stages of Tenebrio molitor prior and subsequent to massive doses of low energy X-rays, Atompraxis 11 (1965) 564.

NICHOLAS, R.C., WIANT, D.E., Radiation of important grain infesting pests: Order of death curves, and survival values for the various metamorphic forms, Food Technol. 13 (1959) 58.

PAPADOPOULOU, C.P., "Disinfestation of dried figs by gamma radiation", Radiation and Radioisotopes applied to Insects of Agricultural Importance (Proc. Symp., Athens, 1963), IAEA, Vienna (1963) 485.

PENDLEBURY, J.B., "The influence of temperature upon the radiation susceptibility of Sitophilus granarius (L.)", Entomology of Radiation Disinfestation of Grain (CORNWELL, P.B., Ed.), Pergamon Press, Oxford (1966) 27.

PENDLEBURY, J.B., JEFFERIES, D.J., BANHAM, E.J., BULL, J.O., Some Effects of Gamma Radiation on the Lesser Grain Borer (Rhizopertha dominica F.), Tropical Warehouse Moth (Cadra (Ephestia) cautella Wlk.), Indian Meal Moth (Plodya interpunctella Hbn.), and the Cigarette Beetle (Lasioderma serricorne F.), Rep. AERE-R-4003, U.K. Atomic Energy Authority Research Group, Isotope Research Div., Wantage, Berks, England (1962).

PENDLEBURY, J.B., JEFFERIES, D.J., BANHAM, E.J., BULL, J.O., "Some effects on Rhizopertha dominica (F.), Cadra cautella (Wlk.), Plodia interpunctella (Hbn), and Lasioderma serricorne (F.)", Entomology of Radiation Disinfestation of Grain (CORNWELL, P.B., Ed.), Pergamon Press, Oxford (1966) 134.

PESSON, P., "Travaux de recherches utilisant les isotopes et les rayonnements nucléaires en entomologie appliquée en France et dans les pays associés", Radioisotopes and Radiation in Entomology (Proc. Symp. Bombay, 1960), IAEA, Vienna (1962) 297.

PESSON, P., Utilisation des radiations ionisantes (⁶⁰Co) pour la protection des denrées contre les insectes nuisibles: Recherches relatives à la détermination des doses utiles pour assurer la stérilité des insectes, Industr. Agric. 80 (1963) 211.

PESSON, P., Some experimental data on cobalt-60 radiation doses capable of arresting insect infestation of cereals and flour, Food Irrad. 3 4 (1963) A18.

PESSON, P., GIRISH, G.K., Sensibilité des divers stades de développement de Sitophilus zeamais Mots. (S. oryzae L.) aux radiations ionisantes: Etude des stades endogènes par radiographie et enregistrement actographique, Ann. Epiphyt. 19 (1968) 513.

PESSON, P., OZER, M., Utilisation d'un actographe à détecteur électro-acoustique pour l'étude des insectes des grains: Actographie du développement larvaire de Sitophilus granarius; Détection des effets immédiats des radiations ionisantes, Ann. Epiphyt. 19 (1968) 501.

PESSON, P., VERNIER, J.M., La protection des denrées contre les insectes ravageurs par l'emploi des radiations ionisantes en vue d'obtenir la stérilité des insectes adultes; Etude particulière de la réaction des gonades de Sitophilus granarius, Ann. Nutr. Aliment. 17 6 (1963) B-487.

PO-CHEDLEY, D.S., Effects of X-Rays on Meal Worm Embryos, Progress Report TID-19463, D'Youville College, Buffalo (1963).

PROCTOR, B.E., GOLDBLITH, S.A., Food processing with ionizing radiations, Food Technol. 5 (1951) 376.

QURAISHI, M.S., METIN, M., "Radiosensitivity of various stages of Callosobruchus chinensis L.", Radiation and Radioisotopes applied to Insects of Agricultural Importance (Proc. Symp. Athens, 1963), IAEA, Vienna (1963) 479.

QURESHI, Z.A., WILBUR, D.A., "Effect of sublethal gamma radiation on eggs, early, intermediate and last instar larvae of the Angoumois grain moth, Sitotroga cerealella Oliv.", Proc. Agric. Symposium, May 1966, Dacca, Pakistan, Atomic Energy Centre (1966) 122.

RASULOV, F.K., ANASTASIEV, S.A., The control of warehouse pests by gamma radiation, Vestn. S-kh. Nauki (Moscow) 9 (1963) 34 (in Russian).

REVETTI, L.M., "Preservacion de maiz (Zea mays L.) y caraotas (Phaseolus vulgaris L.) por irradiación gamma", Radiation Preservation of Food (Proc. Symp. Bombay, 1972), IAEA, Vienna (1973) 339.

RUNNER, G.A., Effect of Roentgen Rays on the Tobacco or Cigarette Beetle and the results a new form of Roentgen Tube, J. Agric. Res. 6 (1916) 383.

SHIPP, E., "Susceptibility of Australian strains of Sitophilus and Tribolium species to gamma radiation disinfestation of grain", Radiation and Radioisotopes applied to Insects of Agricultural Importance (CORNWELL, P.B., Ed.), Pergamon Press, Oxford (1966).

SLATER, J.V., AMER, N.M., TOBIAS, C.A., "Modification of radiation response during embryonic development by the use of elevated temperatures" (Abstr.), 2nd International Congress on Radiation Research, Harrogate, Yorks, England, 1962, Silver End Documentary Publications Ltd., London (1962) 241.

TABOADA, O., Some Effects of Radiant Energy on the Beetles, Tribolium confusum, Duv., Sitophilus granarius (L.) and Acanthocelides obtectus (Say), MS Thesis, Dept. of Entomology, Michigan State College, East Lansing (1953).

TILTON, E.W., BROWER, J.H., Irradiation Studies with Insects Infesting Bulk-Grain and Packaged Commodities, Rep. TID-22414, Stored Product Research and Development Lab., Savannah (1965).

TILTON, E.W., BROWER, J.H., Sexual competition of gamma sterilized male cowpea weevils, J. Econ. Entomol. 64 (1971) 1337.

TILTON, E.W., BROWER, J.H., "Status of US Department of Agriculture research on irradiation disinfestation of grain and grain products", Radiation Preservation of Food (Proc. Symp. Bombay, 1972), IAEA, Vienna (1973) 295.

TILTON, E.W., BURKHOLDER, W.E., COGBURN, R.R., Sterilizing effects of gamma radiation on eight insect and one mite species that infest stored products, Bull. Ent. Soc. Am. 10 (1964) 163.

TILTON, E.W., BURKHOLDER, W.E., COGBURN, R.R., Notes on effect of preconditioning confused flour beetles with temperature variations or carbon dioxide prior to gamma radiation, J. Econ. Entomol. 58 (1965) 179.

TILTON, E.W., BURKHOLDER, W.E., COGBURN, R.R., Effects of gamma radiation on Rhizopertha dominica, Sitophilus oryzae, Tribolium confusum and Lasioderma serricorne, J. Econ. Entomol. 59 (1966) 1363.

VEREECKE, A., PELERENTS, C., De invloed van gammastralen op de fecunditeit en fertiliteit van Tribolium confusum Duval, Meded. Landbouwhogesch. Opzoekingsstn. Gent 30 2 (1965) 1017.

VEREECKE, A., PELERENTS, C., De invloed van gammastralen op de levensuur van Tribolium confusum Duval. Meded. Landbouwhogesch. Opzoekingsstn. Staat Gent 30 (1965) 1824.

VIADO, G.B., MANOTO, E.C., "Effects of gamma radiation on three species of Philippine insect pests", Radiation and Radioisotopes applied to Insects of Agricultural Importance (Proc. Symp. Athens, 1963), IAEA, Vienna (1963) 443.

WALDER, J.M.M., WIENDL, F.M., Influencia da radiação gama na longevidade e oviposição de Callosobruchus maculatus (F.) (Col. Bruchidae), Anais Soc. Entomol. Bras. (1974) (in press).

WALDER, J.M.M., WIENDL, F.M., Efeitos de altas doses de radiação gama na longevidade de adultos de Callosobruchus maculatus (F.), Bol. Cient. Centro de Energia Nuclear na Agricultura, Piracicaba, No.13 (1974).

WATTERS, F.L., MacQUEEN, K.F., Effectiveness of gamma irradiation for control of five species of stored product insects, J. Stored Prod. Res. 3 (1967) 223.

WIENDL, F.M., Alguns Usos e Efeitos das Radiações gama em Zabrotes subfasciatus (Boh.) (Col. Bruchidae), Doctoral Thesis, Piracicaba (1969).

WIENDL, F.M., Some gamma-irradiation effects on survival, longevity and reproduction of Zabrotes subfasciatus (Boh.), Sterility Principle for Insect Control or Eradication (Proc. Symp. Athens, 1970), IAEA, Vienna (1971) 525.

WIENDL, F.M., Efeitos da Radiação gama em Sitophilus zeamais Mots (Col. Curculionidae), "Livre Docente" Thesis, Piracicaba (1972).

ZAKLODNOI, G.A., Effect of the dose rate of gamma irradiation on the life span of the grain weevil Sitophilus granarius, Radiobiologiya 6 (1966) 478.

DISCUSSION

E.F. BOLLER: Since I'm rather ignorant as to how stored-product pests can be controlled by the sterile-male technique, would you please tell us why you do not simply kill the pests directly with lethal doses?

F.M. WIENDL: For a population any sterilization dose is lethal. However, if you wish to cause immediate death, you have to apply much higher doses, about 500 krad, and this could produce cancerigenous or mutagenic substances. Another important factor is the cost, and the aim in grain irradiation is to achieve the greatest effect, i.e. total sterilization and total disinfestation, for the lowest cost (i.e. minimum dose).

H. LEVINSON: Have you any comparative data for the doses required for complete sterilization in the absence and presence of a food substrate, and also for different food substrates?

F.M. WIENDL: I do not have any data on the irradiation of insects lacking food. However, absence of food is of course another stress on the insect, so I think that it would result in a lower dose being required to achieve sterilization. As far as different substrates are concerned, a "good" food, for S. oryzae such as rice, induces a higher resistance to irradiation and sterilization, than a "bad" food for this insect, such as macaroni.

R. PAL: I should like to ask Mr. Rahalkar if he knows whether any experiments have been carried out on the possible health hazard to humans from consuming irradiated grain?

G.W. RAHALKAR (Chaiman): The radiation dose recommended for grain disinfestation (20 krad) is not very large. Irradiation of wheat with this dose has been found to have no adverse effects on the chemical composition and wholesomeness of wheat and its food preparations. Extensive feeding trials over a number of generations using rats and mice showed no deleterious effects on them. Even under conditions of malnutrition, feeding on irradiated wheat had no adverse effects on laboratory animals.

F. M. WIENDL: We have carried out experiments with insects reared on irradiated food and found that at 20 krad fertility increased by 25% and longevity also increased.

A. G. MANOUKAS: I am intrigued by the great difference in effects between 7 and 8 krads. Have you applied any statistical evaluation to your data? How many replicates did you have and what was your standard deviation? I wonder if any other factor is involved in your results.

F. M. WIENDL: Yes, we had four replicates, i.e. four mason jars per dose, and in each we had 20 insects at the beginning of the experiment. Our aim was to establish the lowest dose that would cause the death of the population, and we only considered that total sterilization had occurred when all four jars were clean and without insects. The standard deviation was about 10%.

I. D. DE MURTAS: What type of radiation source did you use?

F. M. WIENDL: We used a Gammabeam 150 radial irradiator from Canada with about 650 Ci at the time of utilization.

B. NA'ISA: Your paper gives the impression that a lethal dose rather than a sterilizing dose was applied. We think of the sterile-male technique in terms of achieving a decrease in fertility. You spoke of extinction of the flies which suggests that a lethal dose is being applied. Perhaps sterility here is being confused with sterility in medical terms, where it means killing all germs.

F. M. WIENDL: Sterility in entomology means no more reproduction of the insects in the same way as in medicine it means no more reproduction of germs. But the sterilization dose for insects is about 1000 times less than that for germs. Furthermore a dose which causes immediate death to individual insects is about 50 times higher than the population killing dose, i.e. the sterilizing dose.

A.-F. M. WAKID: You have shown that distinct differences occur between 6 and 7 and 7 and 8 krads in some insects. How many generations have developed by this stage?

F. M. WIENDL: The final results were for about the 6th to 8th generations, but one can see this effect even in the 4th or 5th generation.

B. AMOAKO-ATTA: Your level of pest infestations is very high. Is the infestation mainly of Coleoptera origin, and if not, did you take into account the response of any Lepidoptera species to the sterilizing doses.

F. M. WIENDL: Yes, in Brazil our infestation level is always high, and involves mainly Coleoptera. Sitophilus, Callosobruchus and Zabrotes account for about 20% of our losses. Fortunately, these species can be sterilized with 8 to 10 krad. However, as you observed, Lepidoptera also exist and they can be sterilized only with 30 to 50 krad. Lepidoptera species like Sitotroga, Ephestia, etc., were the subject of other studies involving field releases of sterile insects, since the higher doses needed for disinfestation could induce undesired changes in the grain. Fortunately Lepidoptera are not a serious silo pest in Brazil and we have established that pests in stored grain can be more or less completely controlled with 10 krad.

MATING COMPETITIVENESS OF THE RADIOSTERILIZED MALE ARMYWORM, Spodoptera exigua Hb., AND THE MALE MOSQUITO, Aedes aegypti L., IN FIELD CAGES

S. LOAHARANU, S. CHIRAVATANAPONG,
M. SUTANTAWONG, P. KAOCHNONG
Entomology Section,
Office of Atomic Energy for Peace,
Bangkhen, Bangkok,
Thailand

Abstract

MATING COMPETITIVENESS OF THE RADIOSTERILIZED MALE ARMYWORM, Spodoptera exigua Hb., AND THE MALE MOSQUITO, Aedes aegypti L., IN FIELD CAGES.

Artificially reared male armyworms, Spodoptera exigua, were exposed to gamma rays at the late pupal stage, and subsequently caged with untreated females at 1:1 ratio. It appeared that in the colony which contained males subjected to 30, 40 and 50 krad of gamma rays, the viability of eggs deposited was reduced to 46.31, 9.16 and 3.54% respectively, compared with 91.95% of egg-hatch in the untreated colony. Further studies were undertaken by introducing males irradiated at the doses mentioned into field cages containing untreated males and untreated females at 10:1:1 ratio. From evaluating the viability of eggs deposited, it could be reported that males that had been exposed to 50 krad of gamma rays were least competitive. Male mosquitoes, Aedes aegypti, were irradiated in the early adult stage and then mated with untreated females. It was found that males exposed to very low radiation doses could greatly reduce the viability of eggs deposited. Those exposed to 12 krad led to complete infertility in eggs. Mating and cytological studies showed that males sterilized with 12 krad of gamma rays could mate effectively.

1. RADIOSTERILIZED MALE ARMYWORMS

The armyworm, Spodoptera exigua, has been reported as a serious pest of several agricultural products, including citrus fruit, cotton and tobacco [1, 2]. Studies on the effects of X-rays and gamma rays have been carried out [2-4].

In Thailand, this insect is a serious pest of shallot, Allium ascalonicum Linn., and onion, Allium cepa Linn. For control purposes, male sterilization of this armyworm was carried out at our research centre. The following parameters were investigated:

(1) To investigate sterilizing dosage for the male armyworm.
(2) To investigate the effectiveness of sterile males by mating.

1.1. Materials and methods

The adult moths were fed on beer. The eggs were surface disinfested by the method of Mangum and co-workers [5]. Larvae were reared on an

artificial diet modified from Shorey and Hale [6]. The compositions of the
diet are as follows:

(a)	Soaked mungbean	106	g
(b)	Brewer's yeast	16	g
(c)	Ascorbic acid	1.6	g
(d)	Methyl-P-hydroxy benzoate	1.0	g
(e)	Formaldehyde (40%)	1.0	mlitre
(f)	Agar	6.4	g
(g)	Water	320	g

In sterilization, the male pupae (24-36 hours before emerging into
adults) were irradiated at 30, 40 and 50 krad of gamma rays from a
30 000 Ci ^{60}Co source. Upon emergence, they were caged with non-
irradiated females (40 pairs in each replicate), along with the untreated
group. The sterility and mortality of the treated moths were investigated.
    Males irradiated at the above mentioned doses were combined with
untreated males and untreated females at a ratio of 10 :1 :1 (actual 300 :30 :30)
in 6 × 2 × 2-ft double-layered screened cages in the field (temperatures
ranging from 27°C - 32°C). Eggs deposited on onion grown in pots inside
the cages were sampled to study the percentages of hatch as related to the
competitiveness in mating of the treated males.

1.2. Results

    The development of the armyworm in the laboratory is shown in Table I.
The percentage of eggs developed to pupae, and from pupae to adults, was
66. 39 and 88. 10 respectively.
    Table II shows that at a 1 :1 ratio the viability of eggs deposited by
females caged with treated males was progressively reduced as the irradiation dose increased.
    Table III reveals that only 5.61% of pupae irradiated at 50 krad failed to
develop into adults. The longevity of the treated male moths was not significantly shorter than those untreated.

TABLE I. DEVELOPMENT OF SIX SUCCESSIVE GENERATIONS OF THE
ARMYWORM, S. exigua, IN THE MUNGBEAN MEDIUM

	Percentage development	
Generation	Eggs to pupae	Pupae to adults
1	68.00	86.63
2	72.89	87.36
3	63.33	87.63
4	68.44	88.93
5	62.00	90.67
6	66.89	87.41
Average	66.9 ± 3.9	88.1 ± 1.48

TABLE II. DECREASE IN VIABILITY OF EGGS DUE TO TREATED
MALES MATED WITH UNTREATED FEMALES AT 1:1 RATIO

Dose applied to males (krad)	Number of eggs examined (average)	Percentage egg hatch
0	609	91.95
30	596	46.31
40	593	9.16
50	594	3.54

TABLE III. EFFECT OF RADIATION ON THE EMERGENCE OF
TREATED PUPAE, S. exigua

Doses applied to pupae (krad)	Percentage of pupae that failed to emerge into adults (corrected)	Longevity of male moths (days)
0	0	7.2
30	1.12	7.3
40	3.37	7.1
50	5.61	7.2

TABLE IV. HATCHABILITY OF EGGS DEPOSITED AS RELATED TO
THE COMPETITIVENESS OF STERILIZED MALE S. exigua WHEN
CAGED WITH UNTREATED MALES AND UNTREATED FEMALES AT
THE RATIO OF 10:1:1

Males sterilized with doses (krad)	Percentage of egg hatch[a]
0 (untreated)	92.17
30	59.83
40	58.67
50	87.17

[a] Calculated from about 600 eggs sampled from each replicate, on each of the first three days of egg laying.

Results of treated males caged with untreated males and untreated
females at a 10:1:1 ratio are shown in Table IV. In the colony that contained
males treated at 30, 40 and 50 krad of gamma rays, the percentage of egg-
hatch was 59.83, 58.67 and 87.17 respectively. In the untreated colony the
percentage of egg hatch was 92.17.

1.3. Discussion

At the ratio of 10:1:1, males irradiated at 50 krad appeared to be not
as competitive, based on the hatchability of eggs deposited. Results

suggested that sterilizing doses of 30 to 40 krad would be more appropriate
for reducing the egg hatch of the moth.

Ratios other than 10:1:1, sterilization at an early adult stage, and F_1
sterility would be taken into consideration when drawing up a suppression
programme.

2.   RADIOSTERILIZED MALE MOSQUITOES, Aedes aegypti L.

The mosquito A. aegypti L. has been known as the haemorrhagic fever
(dengue virus) carrier in Thailand and in other Asian countries [7]. It is a
domestic species. The adult female gives a painful bite both day and night.
Sterilization in males, A. aegypti, irradiated in the immature stage has been
carried out by many scientists in foreign countries, for instance by Fay and
co-workers and George [8, 9]. Results of cytological studies on sperm of
sterilized male A. aegypti were described by LaChance and co-workers [10]

In exploring the possibility of radiosterilization for control of this pest,
we irradiated the males at an early adult stage with the following objectives:

(1) To determine the suitable sterilizing dose.
(2) To determine the competitiveness of the treated males by mating
     and sperm transfer observation.

2.1. Materials and methods

In rearing, larvae of A. aegypti were provided with a mixture of guinea-
pig chow and brewer's yeast. The adults were provided with sucrose solution
and banana slices. For egg production, females were fed on the blood of
guinea pigs. The method of rearing was modified from that of Fay and co-
workers [8].

The 4 - 40-hours-old adult males were irradiated at 1, 3, 6 and 12 krad
of gamma rays. The irradiated males (400 males in each replicate for each
dosage) were subsequently caged with non-irradiated females at the ratio
1:1 in 1 × 1 × 1-m double-layered screened cages, along with a non-irradiated
colony. The percentages of egg hatch were calculated.

In competitiveness studies, males irradiated with 12 krad of gamma rays
were caged with non-irradiated females in 3 × 1 × 1-m cages in the field at
ratios of 5:1:1, 10:1:1 and 20:1:1 (actual 1000:200:200, 2000:200:200,
4000:200:200). Eggs deposited from each combination on 11th and 12th days
of mating (about 600 eggs on each day) were sampled to evaluate the reduction
of egg hatch.

In cytological studies, the testes of males irradiated at 12 krad and the
genital tracts of mated females (along with those of untreated groups) were
removed and put in a drop of saline solution. Following the squashed
technique, the motility of sperms and sperm transfer were observed through
a phase-contrast microscope. An average number of 30 individuals of each
sex from each of the total of nine replicates of each group were examined.

2.2. Results

From Table V it can be seen that very low radiation dosages could
greatly reduce the viability of the eggs deposited. A dose of 12 krad caused
complete infertility in eggs.

TABLE V. DECREASE IN VIABILITY OF EGGS DEPOSITED BY
FEMALES, A. aegypti, MATED WITH TREATED MALES AT 1:1 RATIO

Dose applied to males (krad)	Percentage of egg hatch
0	85.50
1	63.17
3	8.33
6	0.95
12	0.0

TABLE VI. MORTALITY OF MALE A. aegypti IRRADIATED AT
4 - 40 HOURS OLD

Irradiation dosage (krad)	Percentage mortality after irradiation	
	8 days	12 days
0	9.2	12.3
1	9.2	12.8
3	11.8	13.2
6	13.8	14.3
12	14.5	17.2

TABLE VII. VIABILITY OF EGGS AS RELATED TO THE COMPETITIVE-
NESS OF MALE A. aegypti STERILIZED AT 12 krad

Ratio of T. males[*]: N[**] males: N[**] females[a]	Percentage of egg hatch
0 : 1 : 1	85.83
5 : 1 : 1	12.83
10 : 1 : 1	6.50
20 : 1 : 1	4.83
1 : 0 : 1	0.0

[a] * T = treated.
** N = non-irradiated.

Table VI shows the effect of gamma rays on the mortality of the male
A. aegypti. It can be seen that a dose of 12 krad did not greatly affect the
longevity of the males.

Table VII shows that when males irradiated at 12 krad were caged with
untreated males and untreated females at 5:1:1, 10:1:1 and 20:1:1 ratios,
the viability of eggs deposited was greatly reduced. At a 20:1:1 ratio, the
percentage of egg hatch was reduced to 4.83 whereas that of the normal
colony was 85.83.

TABLE VIII.   RESULTS OF SPERM ACTIVITY IN MALE A. aegypti AND
INSEMINATION STUDIES

Sex observed	Percentage of individuals in which active sperm was found
Untreated males	83.33
Females (that mated with untreated males)	84.44
Treated males	85.56
Females (that mated with treated males)	83.33

Results of sperm motility and sperm transfer are shown in Table VIII.
More than 80% of treated males (exposed to 12 krad) and mated females were
found to contain active sperms. Results suggested that transfer of sperms
took place.

2.3. Discussion

Males of A. aegypti could be sterilized with very low radiation dose at
an early adult stage.   Those sterilized with 12 krad were found to be
competitive. However, more detailed studies as well as work on the ecology
and biology of this mosquito in field conditions would be required.

3.   CONCLUSIONS

When sterilized male armyworms, S. exigua, were caged with untreated
females at a 1:1 ratio, the viability of eggs deposited was progressively
reduced as sterilizing dosage increased. However, when these treated males
were introduced into a colony which contained untreated males and untreated
females at a 10:1:1 ratio, it was found that males sterilized with 50 krad
were least competitive. A sterilizing dose of 30 to 40 krad could be more
suitable for control of this pest.

Very low radiation dosages could sterilize the males of A. aegypti,
treated at the early adult stage. When males irradiated at 12 krad were
caged with untreated males and untreated females at a 20:1:1 ratio, the
viability of eggs was greatly reduced. Taken together with cytological
studies, it could be concluded that the sterilized males were competitive.

ACKNOWLEDGEMENTS

Appreciation is expressed to N. Eikarat and M.F. Sullivan of the SEATO
Medical Research Laboratory for providing the original A. aegypti L. for
our research use.

REFERENCES

[1] ATKINS, E.L., The beet armyworm, Spodoptera exigua, an economic pest of citrus in California, J. Econ. Entomol. 53 4 (1960) 616.

[2] ANWAR, M., Quelques effets de l'irradiation aux rayons gamma sur Spodoptera (Laphygma) exigua Hb. (Lepidoptera: Noctuidae), Isotopes and Radiation in Entomology) (Proc. Symp. Vienna, 1967), IAEA, Vienna (1968) 109.

[3] LaBRECQUE, G.C., KELLER, J.C. (Eds), Advances in Insect Population Control by the Sterile-Male Technique, Technical Reports Series No.44, IAEA, Vienna (1965) 43.

[4] RAJASEKHARA, K., Effect of gamma radiation on the beet armyworm, Spodoptera exigua Hb. (Lepidoptera: Noctuidae), Univ. Calif. Microfilms 71 (1970).

[5] MANGUM, C.L., RIDGWAY, W.O., OUYE, M.T., Production of fourth instar pink bollworm larvae from sodium hypochlorite treated and untreated eggs, J. Econ. Entomol. 62 2 (1969) 515.

[6] SHOREY, H.H., HALE, R.L., Mass rearing of the larvae of nine noctuid species on a simple artificial medium, J. Econ. Entomol. 53 3 (1965) 522.

[7] SCOTT, H.G., Dengue, Plates of Vector-borne Disease in Vietnam, US Dept. of Health, Education and Welfare, National Communicable Disease Center, Atlanta (1967) 14.

[8] FAY, R.W., McCRAY, E.M., Jr., KILPATRICK, J.W., Mass production of sterilized male Aedes aegypti, Mosqu. News 23 3 (1963) 210.

[9] GEORGE, J.A., Effect of mating sequence on egg-hatch from female Aedes aegypti L. mated with irradiated and normal males, Mosqu. News 27 1 (1967) 82.

[10] LaCHANCE, L.E., SCHMIDT, C.H., BUSHLAND, R.C., "Radiation-induced sterilization", Pest Control, Academic Press, New York (1967) 147.

DISCUSSION

F. OGAH: How do you measure the competitiveness of irradiated males in the presence of unirradiated male Aedes aegypti?

S. LOAHARANU: I investigated the competitiveness of irradiated males by combining them with non-irradiated males and non-irradiated females at different ratios and evaluating the viability of the eggs.

R. PAL: I would just like to suggest that, if you added another column to Table III showing the expected egg hatch, this would provide a direct measurement of the competitiveness ratio.

S. LOAHARANU: Thank you for the suggestion.

D. ENKERLIN S.: How successful have you been in mass-rearing?

S. LOAHARANU: Up to now we have carried out only small-scale experiments but, if the sanitation of the rearing room is improved, I am quite sure that satisfactory armworm production will be achieved.

A.-F.M. WAKID: In Aedes aegypti, have you followed the hatched larvae from the competing population to the pupal or adult stage or have you just calculated the percentage mortality of the eggs?

S. LOAHARANU: I calculated the sterility of the eggs only.

SPERMATOGENESIS AND OOGENESIS OF
Ceratitis capitata Wiedemann,
Dacus dorsalis Hendel AND
Dacus cucurbitae Coquillett WHEN
SEXUALLY STERILIZED WITH TEPA
AFTER ADULT EMERGENCE

H. A. NAVVAB GOJRATI
Department of Plant Protection,
College of Agriculture,
Pahlavi University,
Shiraz, Iran

I. KEISER
Entomology Research Division,
Agricultural Research Service,
US Department of Agriculture,
Honolulu, Hawaii,
United States of America

Abstract

SPERMATOGENESIS AND OOGENESIS OF Ceratitis capitata Wiedemann, Dacus dorsalis Hendel AND Dacus cucurbitae Coquillett WHEN SEXUALLY STERILIZED WITH TEPA AFTER ADULT EMERGENCE.

Three species of fruit flies were exposed to varying doses of tepa in drinking water immediately following emergence, and also in the middle of their life span, for 48 hours. Histological preparations were made of the ovaries and testes at daily intervals thereafter. It was noticeable that sperm production ceased in the anterior end of the testes, with a general necrosis of the germinal epithelium in this area. Ovaries were considerably reduced in size, very small Feulgen-positive clumps of chromatin being noted, indicating a complete breakdown of the nurse cells, oocytes and follicle cells in the ovaries. Egg laying was completely suppressed when flies were egged about 20 days after exposure to the chemosterilant.

EXPERIMENTAL PROCEDURE

The adult Mediterranean fruitfly, Ceratitis capitata[1], the Oriental fruitfly, Dacus dorsalis[1], and the melon fly Dacus cucurbitae[1] were exposed to tepa in drinking water in two different series of experiments:
I. Three species of fruit flies were exposed to two different concentration of chemosterilant immediately after emergence for 48 hours. For each species six cages with about 1200 flies in each cage were set up as follows:
   (a) 2 cages for control (no chemosterilant).
   (b) 2 cages for low concentration of chemosterilant (0.2 g for Ceratitis capitata and 0.25 g for the other two species).
   (c) 2 cages for higher concentration of chemosterilant (0.4 g for C. capitata and 0.05 g for the other two species).

---

[1] Diptera: Tephritidae.

II. In the second series of experiments, the three species of flies were exposed to chemosterilant for 48 hours in the middle of their life span, as follows:

   (a) Ceratitis capitata: 20 days after emergence
   (b) Dacus dorsalis: 40 days after emergence
   (c) Dacus cucurbitae: 60 days after emergence

For both experiments oviposition sites were placed in the cages for 24 hours, after which they were removed and the eggs were counted. From the control 100 eggs only were placed on a piece of wet paper to check the hatchability. From the treated group all the eggs (if there were any) were used for the hatchability test.

Ovaries and testes of five males and five females from treated and normal adults were dissected at daily intervals thereafter.

Histological preparations were made of the ovaries and testes using Feulgen stain. Permanent mounts were made of each preparation. Microscopic examination of these slides revealed the physiological effects of tepa on the reproductive tissues. The extent, pattern and general picture of damage sustained by flies treated immediately after emergence and by those treated in the middle of their life cycle were the same.

RESULTS

Ovaries

   (a) Reduction in size; noticeable change occurred in about 3 - 4 days.
   (b) After 3 - 4 days, the treated ovaries ceased increasing gradually.
   (c) Ovarioles remained small with apical cells more prominent.
   (d) The middle section of each ovary was densely tracheated, but the top and the bottom sections were relatively clear.
   (e) Ovarioles showed very small Feulgen-positive clumps of chromatin, indicating complete breakdown of nurse cells, oocytes and follicle cells in the ovaries.

Spermatheca

No difference in the size of the spermatheca was observed between the treated and untreated flies. The absence of sperm in the spermatheca of the treated females for a few days more than the untreated ones is an indication of a delay in mating.

Testes

   (a) Cessation in the sperm production.
   (b) General necrosis of the germinal epithelium.
   (c) Progress of necrosis with age.

Egg laying

Egg laying was completely suppressed when flies were egged at about 20 days after exposure to the chemosterilant. Among three species of flies, untreated C. capitata and Dacus cucurbitae laid 14 000 and 18 000 respectively, and only treated C. capitata gave less than 10% hatch (control 85%). When the flies were egged 10 days later, very few eggs were laid for some species, and none of the eggs hatched. The results of egging and the percentage of hatch were as follows:

Flies exposed to chemosterilant immediately after emergence

I. 10 days after exposure

		No. of eggs	Hatch (%)
1.	C. capitata		
	(a) Control	16 000	86
	(b) Treated	15	6.6
2.	D. dorsalis		
	(a) Control	13 500	88
	(b) Treated	0	0
3.	D. cucurbitae		
	(a) Control	19 500	85
	(b) Treated	12	0

II. 20 days after exposure

1.	C. capitata		
	(a) Control	14 000	85
	(b) Treated	11	9.0
2.	D. dorsalis		
	(a) Control	13 500	86
	(b) Treated	0	0
3.	D. cucurbitae		
	(a) Control	18 000	85
	(b) Treated	15	0

III. 30 days after exposure

1.	C. capitata		
	(a) Control	10 000	85
	(b) Treated	8	0
2.	D. dorsalis		
	(a) Control	12 000	88
	(b) Treated	0	0

3. <u>D. cucurbitae</u>

(a) Control	16 000		86
(b) Treated	9		0

BIBLIOGRAPHY

FOOD AND AGRICULTURE ORGANIZATION and INTERNATIONAL ATOMIC ENERGY AGENCY, Nuclear Techniques for Increased Food Production, Basic Studies Series No.22, FAO (1969).

FYE, R.L., Screening of chemosterilant against the house fly, J. Econ. Entomol. 60 (1967) 605.

KILGORE, W.W., "Chemosterilants", Pest Control (KILGORE, W.W., DOUTT, R.L., Eds), Academic Press, New York and London (1967).

KNIPLING, E.F., Possibilities of insect control or eradication through the use of sexually sterile males, J. Econ. Entomol. 48 (1955) 459.

LaBRECQUE, G.C., SMITH, C.N., Principles of Insect Chemosterilization, North Holland, Amsterdam (1968).

LINDQUIST, A.W., New ways to control insects, Pest Control 29 6 (1961) 9, 36.

WILSON, J.A., HAYS, S.B., Histological changes in gonad and reproductive behavior of house fly following treatment with chemosterilants, J. Econ. Entomol. 62 (1969) 960.

DISCUSSION

Z.A. QURESHI: How serious are <u>Ceratitis</u> <u>capitata</u> and <u>Dacus</u> <u>dorsalis</u> in Iran and what are their major host plants?

H.A. NAVVAB GOJRATI: <u>Ceratitis</u> <u>capitata</u> is becoming serious after having been one of our quarantine pests for many years but <u>Dacus</u> <u>dorsalis</u> is not a serious pest today. The major host plants for <u>D.</u> <u>dorsalis</u> are several members of the Cucurbitaceae and Cruciferae families.

EFFECT OF INSECT CHEMOSTERILANTS
ON THE DEVELOPMENT OF THE
RED COTTON BUG, Dysdercus koenigii F.

M.R. HARWALKAR, G.W. RAHALKAR
Biology and Agriculture Division,
Bhabha Atomic Research Centre,
Trombay, Bombay,
India

Abstract

EFFECT OF INSECT CHEMOSTERILANTS ON THE DEVELOPMENT OF THE RED COTTON BUG, Dysdercus koenigii F.
   Studies were carried out to examine the influence of metepa, apholate and hempa on moulting and metamorphosis in the red cotton bug Dysdercus koenigii F. Following exposure for 1 hour to a metepa residue of 43 $\mu g/cm^2$, moulting and metamorphosis of the 4th and 5th instar nymphs treated within 8 - 10 hours of moulting was completely inhibited. However, when the same treatment was administered during the terminal period, the 5th instar nymphs metamorphosed into adults and 4th instar nymphs moulted into 5th instar but failed to metamorphose. The inhibitory effect could be reversed to a certain extent by withholding normal food. It was observed that inhibition of moulting or metamorphosis was not caused by the failure of moulting hormone production. Topical application of apholate to early 5th instar nymphs produced effects similar to metepa but the same was not seen with hempa.

A variety of chemicals have been shown to be effective in inducing sexual sterility in a diverse group of insects [1]. During the course of our investigations on the sterilization of red cotton bug, Dysdercus koenigii F., a pest on cotton and Hibiscus esculentus, it was observed that an hour's exposure of late 5th instar nymphs to a metepa residue of 43 μg per cm^2 induced near complete sterility in the developing adult males. However, when the same treatment was administered to freshly moulted 5th instar nymphs, there was complete inhibition of their metamorphosis into adults (unpublished data). We therefore investigated the effects of metepa, as well as two other known chemosterilants, apholate and hempa, on moulting and metamorphosis of D. koenigii and our findings are here reported.

MATERIALS AND METHODS

The insects were reared in the laboratory on germinated cotton seeds at 29° ± 1°C. The various instars were separated on the basis of their body size and wing pads [2], and both sexes were used in these experiments. For treatment with metepa insects were exposed to the sterilant residues deposited on the inner surface of 12-dram glass vials. In each vial 10 insects were exposed at a time. In the case of treatments with apholate and hempa, acetone solutions of desired concentrations were prepared and 1 μlitre of the solution was topically applied to each insect on the dorsal side of the

abdomen using an "agla" micrometer syringe. Treated insects were placed
in glass jars containing a layer of moist sand at the bottom and germinated
cotton seeds were provided as food. Observations on mortality, moulting
and metamorphosis were recorded daily. All the experiments were
performed at a temperature of 29° ± 1°C.

EXPERIMENTAL RESULTS

To study the effects of metepa on moulting and metamorphosis, the
early as well as the late stage of the 3rd, 4th and 5th instar nymphs were
treated by exposing them for 1 hour to a residue of 43 $\mu g/cm^2$. The early
stage is 8-10 hours after moulting and late stage 24 hours before the next
moult. It was observed (Table I) that when nymphs were treated during
the early period of an instar, they failed to moult into the next stage.
However, when the same treatment was given during the terminal period
of an instar, the nymphs moulted into the next instar simultaneously along
with the control insects, but their subsequent moulting and/or metamorphosis
was adversely affected. In the case of 3rd instar nymphs treated during
the terminal period, 96.7% moulted into 4th instar and of these none into
the 5th instar. Similarly, all the treated late 4th instar nymphs moulted
into 5th instar but only 17% of these metamorphosed into adults. At the
experimental temperature, the duration of the 4th and 5th nymphal instar
ranged from 3-4 and 5-7 days respectively. The 4th and 5th instar nymphs
treated during the early period survived, on an average, for 8.3 and 11.2 days
respectively, without undergoing a moult. Further detailed studies were
however restricted to 5th instar nymphs.
In order to determine whether inhibition of metamorphosis was dose
dependent, early 5th instar nymphs were exposed for 1 hour to graded
concentrations of metepa residue. It was observed that with an increase
in concentration there was a progressive decrease in the number of nymphs
metamorphosing into adults (Table II). However, these adults exhibited
varying degrees of morphological abnormalities and were unable to mate.
The same pattern was discernible when the nymphs were exposed for varying
periods to a residue of 43 $\mu g/cm^2$ (Table III).

TABLE I. EFFECT OF METEPA ON MOULTING AND META-
MORPHOSIS IN D. koenigii (nymphs exposed for 1 hour to a residue of
43 μg per cm^2)

Stage treated	No. of insects	Percentage that moulted to 4th instar	Percentage that moulted to 5th instar	Percentage that developed into adults
Late Third	60	96.7	0.0	
Early Fourth	60		0.0	
Late Fourth	60		100.0	17.0
Early Fifth	60			0.0
Late Fifth	60			95.0

TABLE II. EFFECT OF METEPA ON METAMORPHOSIS OF 5th
INSTAR NYMPHS OF D. koenigii (early stage nymphs exposed for one
hour to different residues)

Residue ($\mu g/cm^2$)	Number of insects	Percentage developed into adults
0	50	100^a
10	50	74^b
21	50	14^c
32	50	4^c
43	50	0

[a] Normal adults.
[b] 20% adults showed abnormalities.
[c] All adults showed severe abnormalities.

TABLE III. EFFECT OF METEPA ON METAMORPHOSIS OF 5th INSTAR
NYMPHS OF D. koenigii (early stage nymphs exposed for varying periods
to a residue of 43 μg per cm^2)

Period of exposure (min)	Number of insects	Percentage metamorphosed to adults
0	50	98.0^a
10	50	34.0^b
20	50	14.0^c
30	50	2.0^d
40	50	2.0^d
60	50	0.0

[a] Normal adults.
[b] Nearly 50% adults showed severe abnormalities.
[c] Nearly 86% adults showed severe abnormalities.
[d] All adults showed severe abnormalities.

The developmental period up to which nymphs were susceptible to the
treatment was then determined. For this, freshly moulted 5th instar nymphs
were allowed to feed on germinated cotton seeds, and were treated at varying
periods by exposing them for 1 hour to a metepa residue of 43 $\mu g/cm^2$.
Treated insects were allowed to feed on germinated cotton seeds. It was
observed (Table IV) that nymphs that had access to food for 24 hours only
before treatment failed to metamorphose. In comparison, in the nymphs
that had an opportunity to feed for 48 hours or more before treatment,
metamorphosis was similar to that in the controls.

TABLE IV. INFLUENCE OF NYMPHAL AGE ON METEPA-INDUCED
INHIBITION OF METAMORPHOSIS IN D. koenigii (fifth instar nymphs
exposed for 1 hour to a residue of 43 μg per cm^2)

Age of nymph (days)	Number of insect	Percentage metamorphosed into adults
1	30	0.0
2	30	93.3
3	30	93.0
5	30	95.0

TABLE V. EFFECT OF POST-TREATMENT FEEDING ON
METAMORPHOSIS OF METEPA-TREATED NYMPHS OF D. koenigii
(early fifth instar nymphs exposed for one hour to a residue of 43 μg per
cm^2)

Duration in days of post-treatment feeding on sucrose solution	Number of insects	Percentage metamorphosed into adults	
		Treatment	Control
0	50	0	96.67
7	60	11.6	87.5
14	80	11.25	91.25
21	60	36.67	88.0
32	60	46.67	81.6

In this insect feeding on germinated cotton seeds was found essential
for initiation of the processes leading to moulting or metamorphosis. On
the other hand, when the nymphs were fed sucrose solution, they survived
for a considerable period without undergoing moulting or metamorphosis
and eventually died. However, they could moult or metamorphose if normal
food was restored (unpublished data). An experiment was therefore performed to ascertain whether the inhibitory effect of metepa could be modified
by delaying the initiation of the moulting process. Early 5th instar nymphs
were treated by exposing them for 1 hour to a metepa residue of 43 μg/cm^2
and they were the allowed to feed on a 10% sucrose solution. After varying
periods of feeding on sucrose, they were provided with germinated cotton
seeds. It was observed that with the increase in the period of feeding on
sucrose solution, progressively greater number of nymphs metamorphosed
into adults (Table V). Metamorphosis of treated insects was 11.6, 11.2,
36.6 and 46.6% respectively when feeding on sucrose solution lasted for
7, 14, 21 and 32 days. Feeding on sucrose even up to 32 days had no
appreciable effect on the metamorphosis of untreated insects after normal
food was restored.

FIG.1.   Wing deformities in the adult red cotton bug, Dysdercus koenigii, when nymphs were treated with metepa.

TABLE VI. EFFECT OF APHOLATE AND HEMPA ON METAMORPHOSIS OF D. koenigii NYMPHS (one μlitre of acetone solution topically applied to early 5th instar nymphs)

Dose (μg/nymph)	Number of insects	Percentage metamorphosed into adults	
		Apholate	Hempa
0	40	95.0	97.5
10	40	15.0	97.5
20	40	0.0	97.5
30	40	0.0	97.5
40	40	0.0	90.0
50	40	0.0	97.5
100	40	-	73.7

Whenever the treated insects metamorphosed into adults, the latter exhibited varying degrees of abnormalities (Fig. 1). These were in the form of wing deformities, failure to extricate themselves from the nymphal skin, inability to retract aedegus, etc.

As in the case of metepa, topical application of apholate to early 5th instar nymphs adversely affected their metamorphosis (Table VI).

With a dose of 10 μg/nymph only 15% developed into adults and none when
the dose was doubled. Topical application of hempa, however, had no
inhibitory effect on the metamorphosis of 5th instar nymphs. Even with
a dose of 100 μg/nymph, 73.7% of the treated insects developed into normal
adults.

DISCUSSION

The general similarities between ionizing radiations and chemosterilants
in inducing sexual sterility in insects have been well documented [3, 4].
The present investigations have demonstrated that various developmental
and morphological abnormalities are produced in Dysdercus by treatment
with metepa and apholate. These closely resemble those observed by
Harwalkar and Nair [5] in X-irradiated D. koenigii nymphs and by
Economopoulos [6] in tretamine-treated Oncopeltus fasciatus.

In hemipterous insects, nymphal epidermal cells exhibit intense
mitotic activity at a particular period after the previous moult. In Rhodnius
distension of the abdomen after ingestion of a blood meal has been shown
to release a stimulus necessary for the initiation of the moulting process [7].
In Dysdercus, too, feeding on the natural food is essential for the initiation
of this process which, once initiated, generally proceeds to completion
without any need for further stimuli (unpublished data). It is a well-
established fact that initiation of moulting is triggered by the moulting
hormone. The fact that the nymphs treated during the terminal period of
an instar successfully moulted into the next stage, whereas those treated
during the early period failed to do so, suggests that the chemosterilant
in some way interferes with the production of moulting hormone or some
step in the process of moulting or metamorphosis. Kuzin and co-workers [8]
have shown that inhibition of pupation in gamma-irradiated larvae of Ephestia
kuhniella was not because the DNA of the hypodermis was damaged but
because of the absence of moulting hormone. In the present case, if it
was merely the question of inhibition of moulting hormone production one
could expect to reverse the inhibitory effect by providing extraneous hormone.
However, when we injected beta-ecdysone into treated insects, the inhibitory
effect could not be revoked. When early 5th instar nymphs were treated,
formation of adult cuticle was seen in many instances but nymphs failed
to ecdyse (Fig. 2). Thus, the chemosterilant treatment appears to affect
the process of ecdysis and not the production of moulting hormone.

When late 3rd and 4th instar nymphs were treated with metepa they
moulted into the next instar but failed to undergo subsequent moulting or
metamorphosis. In this case nymphs were treated within 24 hours of their
expected next moult. It is then likely that sufficient quantities of the sterilant
were carried over to the instar into which the treated nymphs moulted, and
these thus received treatment during the early period. These nymphs were
therefore unable to undergo subsequent moulting or metamorphosis.

A greater number of 5th instar nymphs treated during the early period
metamorphosed into adults when the initiation of the moulting process was
progressively delayed by witholding the normal diet. It is likely that during
the period of feeding on sucrose solution metepa was gradually metabolized,
and was therefore not available in adequate quantities at a time when restoration of normal food initiated the moulting process.

FIG.2.   Adult cuticle visible in 5th instar Dysdercus koenigii nymphs about to metamorphose;   nymphs treated with metepa during early period of the instar.

Saxena and Williams [9] have shown that when 4 µg of the "paper factor" now identified as juvenile hormone analogue was topically applied at the outset of the 5th instar in D. koenigii, most of the nymphs failed to extricate themselves from the old cuticle during moulting and died in the process. A small number moulted into supernumerary 6th instar nymphs that showed absence of any imaginal characters.   They further observed that with reduced doses, progressively greater number of nymphs metamorphosed into adults but the wings failed to attain adult size and form.   Failure to ecdyse and the wing deformities observed in the present investigation might suggest a similarity in the action of metepa, apholate and the "paper factor". However, this does not appear to be the case.   In none of our experiments did supernumerary 6th instar nymphs appear and also the adults, whenever they developed, failed to show any nymphal characters.

Hempa, a non-alkylating agent, has been shown to be an insect chemosterilant [10].   However, it had no adverse effects on the metamorphosis of Dysdercus as seen with metepa and apholate.   Chang and co-workers [11] have shown that in the male housefly hempa is rapidly demethylated to pentamethyl phosphoric triamide, and only 2.3% of the applied dose is retained at the end of 24 hours.   It is possible that in Dysdercus, also, hempa is similarly metabolized.

REFERENCES

[1]   BORKOVEC, A.B., "Advance in pest control research", Insect Chemosterilants 7 (METCALF, R.L., Ed.), John Wiley, New York (1966).

[2]   SRIVASTAVA, U.S., BAHADUR, J., Ind. J. Entomol. 20 (1958) 228.

[3]   PROVERBS, M.D., Annu. Rev. Entomol. 14 (1969) 82.

[4] LaCHANCE, L.E., NORTH, D.T., KLASSEN, W., "Cytogenetic and cellular basis of chemically induced sterility in insects", Ch. 4, Principles of Insect Chemosterilization (LaBRECQUE, G.C., SMITH, C.N., Eds), Appleton-Century-Crofts, New York (1968).

[5] HARWALKAR, M.R., NAIR, K.K., Ann. Ent. Soc. Amer. 61 (1968) 1107.

[6] ECONOMOPOULOS, A.P., "Effects of tretamine on 4th- and 5th instar Oncopeltus fasciatur nymphs: development of eggs from tretamine-treated females", Sterility Principle for Insect Control or Eradication (Proc. Symp. Athens, 1970), IAEA, Vienna (1971) 259.

[7] WIGGLESWORTH, V.B., The Physiology of Insect Metamorphosis, Cambridge University Press, London (1954).

[8] KUZIN, A.M., KOLOMIJTSEVA, I.K., YUSIFO, N.I., Nature (London) 217 (1968) 743.

[9] SAXENA, K.N., WILLIAMS, C.M., Nature (London) 210 (1966) 441.

[10] LaBRECQUE, G.C., "Laboratory procedures", Ch. 3, Principles of Insect Chemosterilization (LaBRECQUE, G.C., SMITH, C.N., Eds), Appleton-Century-Crofts, New York (1968) 41.

[11] CHANG, S.C., TERRY, P.H., WOODS, C.W., BORKOVEC, A.B., J. Econ. Entomol. 60 (1967) 1623.

DISCUSSION

K. SYED: In the case of the agents which caused disruptions to the moulting pattern, was there any effect on the maturing adults? If there was not, then all that has been done is to prolong the life-span of the insects and thus increase the feeding time on the host plant.

G.W. RAHALKAR: As you can see from Table I, only the late fifth instar nymphs, among those treated with metepa, metamorphosed into adults. These showed no morphological abnormalities. Adults that were formed when late fourth instar numphs were treated exhibited varying degrees of morphological abnormalities. Whenever the nymphs were treated in their early state they failed to moult into the next instar. Furthermore, the longevity of these treated nymphs was not significantly prolonged compared with that of the control insects.

G.R. RENS: Why have you not used topical applications for the chemosterilants in your tests? Wouldn't that have given you more accurate estimates of critical dosages?

G.W. RAHALKAR: We applied apholate and hempa topically, as you can see from the paper, but did not do so in the case of metepa. Variability in the results was avoided by ensuring uniformity of treatment, i.e. keeping the temperature constant and keeping the insects moving on the residue by periodically disturbing them.

I.A. KANSU: As I understand it, you tested the effects of some factors such as dosage, exposure time and nymphal instar stage separately. May I ask why you did not carry out a polyfactorial experiment, as this would have yielded statistically more valuable results.

G.W. RAHALKAR: Our first experiment established that when metepa treatment was given in the early stage of an instar, the treated nymphs failed to undergo an immediate moult. On the other hand, when the same treatment was given during the late stage, the nymphs moulted into the next instar but their subsequent moulting and/or metamorphosis was inhibited Our primary interest then was to ascertain the factors that cause inhibition of moulting and metamorphosis. For this, we used fifth instar nymphs for experimental convenience. I do not feel that the picture would be different with other nymphal instars.

APPLICATIONS OF A JUVENILE HORMONE ANALOGUE AS A STERILANT ON Dysdercus cardinalis Gerst., Dysdercus fasciatus Sign., Dysdercus nigrofasciatus Stal AND Dysdercus superstitiosus F.

G.R. RENS
National Agricultural Laboratories,
Nairobi, Kenya

Abstract

APPLICATIONS OF A JUVENILE HORMONE ANALOGUE AS A STERILANT ON Dysdercus cardinalis Gerst., Dysdercus fasciatus Sign., Dysdercus nigrofasciatus Stal AND Dysdercus superstitiosus F.
Laboratory and field cage tests were carried out in Kenya to evaluate the potential of ethyl farnesoate dihydrochloride, a juvenile hormone analogue with specific action on Pyrrhocoridae, as a sterilant of four species of Dysdercus, which are important pests of cotton and other Malvaceous crops. The juvenile hormone analogue was applied by topical application on young Dysdercus males. A dosage of 100 µg technical material was required for 100% male sterility. No significant differences in susceptibility were found between Dysdercus cardinalis Gerst., D. nigrofasciatus Stal and D. superstitiosus F., but D. fasciatus Sign. was slightly more susceptible than the three other species. In further tests only D. cardinalis was used. The sterility effects were permanent at a dosage rate of 125 µg or more. By copulation with treated Dysdercus males, sterility could be passed on to untreated mating partners, but the induced infertility was not permanently 100% effective, unless high dosages were used with negative effects on the longevity and mating vigour of the treated insects. Release of sterilized males in field cage colonies of Dysdercus cardinalis caused a considerable reduction in the multiplication rate and a subsequent suppression of the normal population growth. A 60% decline in pest population was recorded over one generation cycle in a field cage colony, which had started with a 1:1 ratio between sterilized and normal male insects.

INTRODUCTION

A juvenile hormone mimetic substance, "paper factor", was found [1, 2] to have inhibitory effects on the embryonic development of eggs of Pyrrhocoris apterus L. The methyl and ethyl esters of trans-dihydro-dichloro-farnesenic acid are substances with juvenile activity [3], and it was shown by Masner et al. [4] that these substances had strong sterilizing effects on Pyrrohocoris apterus and could be transmitted from treated males to mated females, causing sterility in the females [5]. Experiments by Bransby-Williams [6] and Critchley and Campion [7] showed that ethyl and methyl farnesoate dihydrochloride were also effective on Dysdercus cardinalis and D. fasciatus.

Studies are here described, carried out in the laboratory at Nairobi and in the field at Makueni (central Kenya), which were conducted with ethyl farnesoate dihydrochloride (DEF) (1) to test the action of this material as a sterilant of Dysdercus spp., (2) to assess the effectiveness and persistence of the sterility effects, (3) to investigate the possibilities for transmitting the hormone material from insect to insect in successive copulations, and (4) to study the effects of the release of sterilized male Dysdercus on the development of field cage populations of Dysdercus.

MATERIALS AND METHODS

The juvenile hormone analogue used was ethyl farnesoate dihydrochloride 62% trans, prepared and supplied by the Zeocon Corporation, California, United States of America. The test insects were one to two days-old adults of Dysdercus cardinalis, D. fasciatus, D. nigrofasciatus and D. superstitiosus the four common "cotton stainers" in Kenya. The hormone material was administered in one ml droplets of various concentrations of DEF in acetone by topical application to the caudal abdominal sternites. The apparatus used was an Arnold automatic micro-droplet applicator [8], fitted with a one ml Agla syringe. The insects were kept in containers, made up of cylindrical lantern glasses of kerosene lamps (diameter 11 cm and height 12 cm), covered with a weighted piece of muslin cloth and resting on a glass plate, lined with moist Whatman No. 1 filter paper. Cottonseeds were provided as food and the moisture of the filter papers was the source of water for the insects. The temperature in the laboratory ranged from 20°C at night to 27°C during the day.

Dosage tests

The topical application treatments were administered to one-day-old male insects. Two days after treatment the insects were transferred to individual containers and partnered with virgin females. Each pair of insects was kept together in their glass cage for the duration of the experiment. Eggs were laid in batches at intervals of three to four days. They were removed the day after they had been deposited and transferred to moist Whatman No. 1 filter paper for incubation. The total number of eggs was recorded for each of the five different dosage treatments. The results were expressed as percentages and plotted on logarithmic probability paper to calculate the critical dose for sterilizing 100% of the treated insects. The differences in susceptibility between the four different Dysdercus species were subjected to statistical analysis [9].

Persistence tests

One-day-old Dysdercus cardinalis were treated with the hormone analogu by topical application and then caged with virgin females in a glass container. A number of consecutive mating partners were provided by replacing the females with other untreated females each time an egg clutch had been deposited. The inter-oviposition period was 3 to 4 days.

Transfer tests

Untreated female Dysdercus cardinalis were allowed to mate for 24 hours with a male treated with various doses of DEF. The pair of Dysdercus was kept caged separately in a glass container. After 24 hours the treated male was removed from the cage and replaced with an untreated male. Subsequently males were removed and replaced by other untreated males each time oviposition had taken place. The cages were kept under observation to make sure that copulation had taken place with all the males provided.

Sterile-male technique applied to field cage populations

   Known numbers of one-day-old adult Dysdercus cardinalis were enclosed
together with batches of sterilized male Dysdercus in large field cages,
measuring 2 × 2 × 1 m.   The field cages were made of timber frames, provided with wire gauze walls and roofs; the bottoms were left open.   Entry
could be gained into the cage through a door in the roof.   The cages were
placed in a cotton field near Makueni in such a way that they covered two
rows of cottonplants over a distance of 2 m.   The enclosed six to eight stands
of cottonplants provided a natural habitat for the caged insects.   The insects
fed on the bolls of the cottonplants, except for a period during the early
stages of the experiment, when the cottonbolls had not yet been produced and
cottonseeds had to be supplied as food.   The weather was unusually dry,
and wet cottonwool wicks were provided when needed as a source of water for
the insects.   Three cages were used at the same time, each one starting with
a colony of 20 untreated virgin female and 20 untreated male Dysdercus
cardinalis.   In addition 20 males, treated with 125 μg DEF, were released in
one cage, and 10 males, to which the same treatment had been applied, were
added to the colony of another cage.   The third cage contained the control
population into which no treated insects had been released.   The numbers of
the various instars of Dysdercus were recorded twice weekly for the duration
of the experiment.   Despite the precautions taken, some nymphs managed to
escape through the gauze wire, but losses caused in this way were estimated
to be no higher than 20%.   The experiments were conducted during January/
February 1974 for a period of 8 weeks, covering one full generation cycle of
Dysdercus, and were repeated later during March/April.

RESULTS

Dosage tests

   Dosage tests were carried out to determine the minimum dose of DEF
needed to obtain complete sterility in a pair of Dysdercus of which the male
partner had been treated.
   Using the totalled results of Table I, it was calculated that a dose of
100 μg material would cause sterility in 100% of the male insects treated.
Statistical analysis of the data showed that D. fasciatus was significantly (at
the 5% level of probability) more susceptible than the three other species.
There were no significant differences in susceptibility between Dysdercus
cardinalis, D. nigrofasciatus and D. superstitiosus. It was therefore thought
that conclusions based on experiments using one of these three species of
Dysdercus as the only test insect might be considered likely to apply also to
the other species of Dysdercus.

Persistence of sterility effects

   Table II shows the effects of various doses of DEF on the viability of
eggs, fertilized by treated male Dysdercus cardinalis. The fertility is given
of the first egg batches laid by young females after copulation with one treated
male.   The percentages shown in the table are the means of ten tests per
dosage treatment.   The number of eggs produced was not significantly affected
by the treatments.

TABLE I.   STERILITY EFFECT OF ETHYL FARNESOATE
DIHYDROCHLORIDE ON MALE Dysdercus cardinalis, D. fasciatus,
D. nigrofasciatus AND D. superstitiosus, MATED WITH
UNTREATED FEMALES

Dose (µg)	No. males tested	Mean number of hatched eggs from one pair of insects			
		D. card.	D. fasc.	D. nigro.	D. super.
125	4 × 20	0	0	0	0
25	4 × 20	92	85	97	111
5	4 × 20	193	148	202	184
1	4 × 20	256	227	269	243
0	4 × 20	271	216	285	257

TABLE II.   PERCENTAGE OF FERTILE EGGS LAID BY FIVE
CONSECUTIVE MATING PARTNERS OF MALE Dysdercus cardinalis
TREATED WITH ETHYL FARNESOATE DIHYDROCHLORIDE

n = 10 Dosage	0 µg (%)	25 µg (%)	125 µg (%)	625 µg (%)
Mate 1 (1-4 days after treatment)	74	28	0	0
Mate 2 (5-8 days after treatment)	82	22	0	0
Mate 3 (9-12 days after treatment)	73	24	0	0
Mate 4 (13-16 days after treatment)	76	35	0	0
Mate 5 (17-20 days after treatment)	72	40	0	0

TABLE III.   FERTILITY OF UNTREATED FEMALE Dysdercus cardinalis,
CAGED WITH NEW, UNTREATED MATING PARTNERS, AFTER
COPULATION WITH A MALE TREATED WITH ETHYL FARNESOATE
DIHYDROCHLORIDE

n = 10 Dosage	125 µg (%)	500 µg (%)	1000 µg (%)
Egg batch No.1 (1st untreated male)	0	0	0
Egg batch No.2 (2nd untreated male)	10	0	0
Egg batch No.3 (3rd untreated male)	29	21	0
Egg batch No.4 (4th untreated male)	67	41	0

The data show that, compared with the average 75% viability of the eggs of the control (untreated) insects, low dosages of 25 µg DEF caused a considerable reduction initially in reproductive success, but that after some time the insects were able to recover their fertility partially. The sterility of treated male Dysdercus was permanent when dosages of 125 µg or more were used.

Transfer of sterility effects

Table III presents the variations in the fertility of eggs laid by Dysdercus cardinalis females which, after copulation with a male treated with DEF, had been provided with new untreated mating partners. The four egg clutches were produced after copulation with four different untreated male insects. The percentages shown in Table III are the means of ten tests per treatment.

The indications from Table III are that some measure of sterility had been transmitted from the treated male Dysdercus to the untreated female by the close contact during copulation. However, the effects of applications of 125 and 500 µg material were not permanent. It was noticed that the speed with which the females recovered their full reproductive potential varied greatly between individual insects. Permanent sterility was obtained when the test insects mated with males treated with 1000 µg DEF. This very high dosage had, however, a toxic effect on the treated males, reducing the longevity after application of the DEF treatment to an average period of two to three days and affecting the mating instinct of many males.

Release of DEF-sterilized males in cage populations

The release of males sterilized with DEF had a marked effect on the growth rate of Dysdercus cardinalis colonies enclosed in large field cages. Figure 1 shows a gradual natural decline in the populations of adult insects during the period of six weeks after they had been caged on cottonplants.

The release of sterilized males in two of the three field cages had only a negligible effect on the death rate of the first cage generation of insects. The number of nymphs produced during the same period show that the three colonies had different rates of reproduction, which might be attributed to differences in the fertility of eggs laid by the 20 females in each of the three Dysdercus colonies.

The effects of the reduced birth rate on the growth of Dysdercus cardinalis colonies are shown by a graphical representation of each of the three populations under observation for a period of 60 days (Fig. 2). In the untreated colony the population grew by more than 250% within one generation cycle. There was hardly any population growth in the colony where sterile males had been added to an untreated population in a 1:2 ratio between sterile and normal males. A substantial decrease in numbers was found in the colony which had started with equal numbers of sterile and normal male insects.

DISCUSSION

Dysdercus spp., commonly known as "cotton stainers", are pests of cotton and other Malvaceous crops in many parts of Africa [10]. In Kenya cotton stainers are widespread and rank as one of the most damaging pests

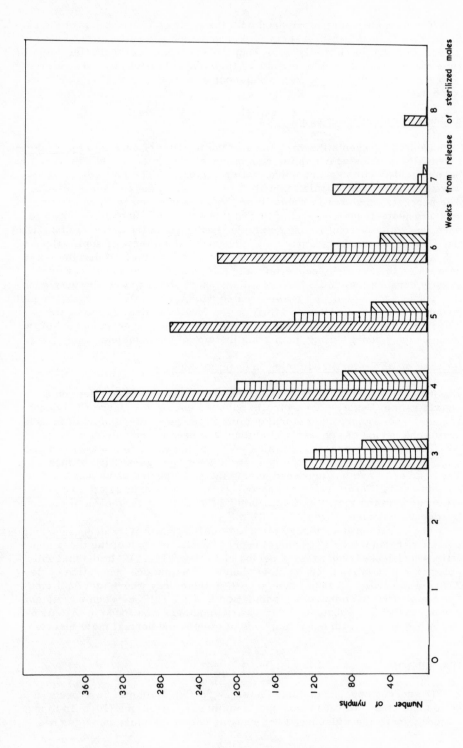

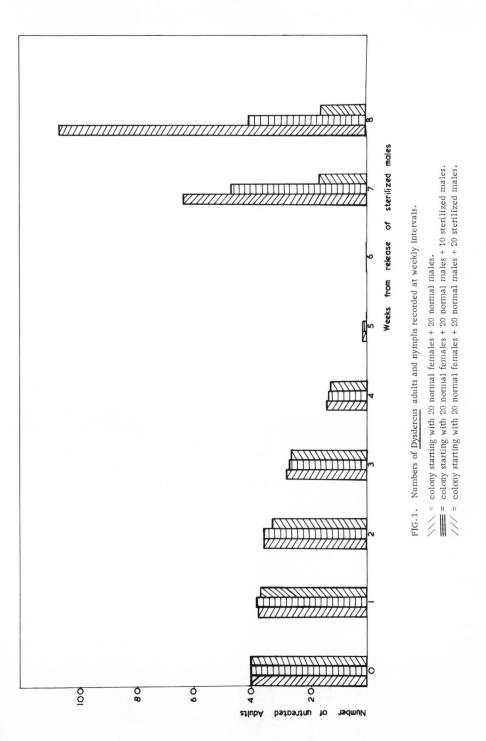

FIG.1.   Numbers of Dysdercus adults and nymphs recorded at weekly intervals.

\\\ = colony starting with 20 normal females + 20 normal males.
≡ = colony starting with 20 normal females + 20 normal males + 10 sterilized males.
/// = colony starting with 20 normal females + 20 normal males + 20 sterilized males.

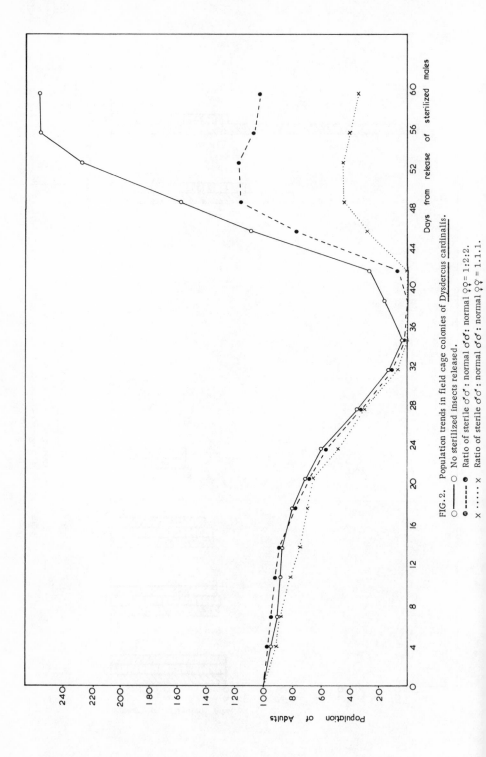

FIG. 2.   Population trends in field cage colonies of <u>Dysdercus cardinalis.</u>

○ ─── ○   No sterilized insects released.

◑ ---- ◑   Ratio of sterile ♂♂ : normal ♂♂ : normal ♀♀ = 1:2:2.

× · · · · · ×   Ratio of sterile ♂♂ : normal ♂♂ : normal ♀♀ = 1.1.1.

of cotton, causing bollrot and discolouration of cottonlint. Population densities of over 250 000 adults per hectare have been counted in unsprayed cottonfields [11] causing yield losses of up to 100% of the cottoncrop. Effective chemical control is possible, but insecticide applications have to be repeated very frequently to control the continuous influx of cotton stainers migrating from their wild hostplants into bolling stands of cotton [12, 13]. More efficient control of cotton stainers might be possible, if the control measures could be applied before the insects begin to move into the cottonfields. Application of chemical control measures on the numerous wild hostplants of Dysdercus spp. is impossible, but the use of a sterilant such as ethyl farnesoate dihydrochloride with selective action on Pyrrhocoridae [14] might be effective, especially if the sterility-causing substance could be transmitted from insect to insect by copulation, as was found possible in Pyrrhocoris apterus [15], an insect closely related to Dysdercus spp.

The work reported here shows that the four species of Dysdercus that are important pests of cotton in Kenya can be sterilized using small quantities of ethyl farnesoate dihydrochloride. Attempts to transfer the sterility effects by copulation failed to produce permanent results. A programme, aimed at the eradication or suppression of Dysdercus spp. by the mass-release of DEF-sterilized males, can therefore not rely on a diffusion of sterility-causing substances and will be successful only when the number of sterilized insects constitute a large proportion of the entire Dysdercus population.

Ecological investigations [16] have revealed that there are very favourable conditions to attempt a temporary suppression of Dysdercus populations in selected cotton areas during the last part of the dry season, when the cotton is not yet sown. The Dysdercus populations are then at very low density levels and confined to wild Malvales, concentrated in parts of the habitat that are least dry. It was pointed out by Borkovec [17] that under conditions of very low insect density the sterile insect release method might be particularly effective. In the adverse conditions prevailing late in the dry season, the Dysdercus populations consist of adult, non-mating insects only, and widespread starvation has considerably reduced the mating competitiveness of the resident male insects [18]. The release of insectary-reared, sterilized males may under these circumstances have extra-proportional effects on the natural populations. It is hoped that suppression of the wild hostplant populations of Dysdercus will cause so much delay in the subsequent infestation of the cottonfields that serious crop losses can be prevented by planting early.

Testing of the sterile-male release method for the control of cotton stainers is being continued with a field experiment on a small island off the Kenya coast in which several releases will be made of DEF-sterilized males on known numbers of Dysdercus.

ACKNOWLEDGEMENTS

The author is grateful to the Director of Agriculture, Ministry of Agriculture, Kenya, for permission to publish this article, and to Professor de Wilde, Wageningen University of Agriculture, for critical reading of the manuscript.

REFERENCES

[1] SLAMA, K., WILLIAMS, C.M., "Paper factor" as a inhibitor of the embryonic development of the European bug, Pyrrhocoris apterus L., Nature (London) 210 (1966) 329.

[2] CARLISLE, D.B., ELLIS, P.E., Abnormalities of growth and metamorphosis in some Pyrrhocorid bugs: the paper factor, Bull.Ent.Res. 57 (1967) 405.

[3] ROMANUK, M., SLAMA, K., SORM, F., Constitution of a compound with a pronounced juvenile hormone activity, Proc.Natl.Acad.Sci. U.S. 57 (1967) 349.

[4] MASNER, P., SLAMA, K., LANDA, V., Natural and synthetic materials with insect hormone activity: IV. Specific female sterility effects produced by a juvenile hormone analogue, J.Embryol.Exp. Morphol. 20 (1968a) 25.

[5] MASNER, P., SLAMA, K., LANDA, V., Sexually spread insect sterility induced by the analogues of juvenile hormone, Nature (London) 219 (1968b) 395.

[6] BRANSBY-WILLIAMS, W.R., Juvenile hormone activity of ethyl farnesoate dihydrochloride with the cotton stainer Dysdercus cardinalis Gerst., Bull.Ent.Res. 61 (1971) 41

[7] CRITCHLEY, B.R., CAMPION, D.G., Effects of a juvenile hormone analogue on growth and reproduction in the cotton stainer Dysdercus fasciatus Say., Bull.Ent.Res. 61 (1971) 49.

[8] ARNOLD, A.J., A high speed automatic micrometer syringe, J.Scient.Instrum. 42 (1965) 350.

[9] SNEDECOR, G.W., COCHRAN, W.G., Statistical Methods, 6th Edn, Iowa State University Press, Ames, Iowa (1967).

[10] PEARSON, E.O., The Insect Pests of Cotton in Tropical Africa, Empire Cotton Growers Corp.& Commonwealth Inst.Ent., London (1958).

[11] DEPARTMENT OF AGRICULTURE, KENYA, Annual Report 1967, 2, Record of Investigations (1967) 29.

[12] RENS, G.R., Cotton Spraying, Agricultural Information Centre, Nairobi (1970).

[13] BROWN, K.J., RENS, G.R., TVEITNES, S., AAKEBAKKEN, O.N., Cotton Growing Recommendations for Kenya, Agricultural Information Centre, Nairobi (1972) 6/6.

[14] BAGLEY, R.W., BAUERNFEIND, J.C., in Insect Juvenile Hormones (MENN, J.J., BEROZA, M., Eds), Academic Press, New York (1972) 113.

[15] MASNER, P., SLAMA, K., ZDAREK, J., LANDA, V., Natural and synthetic materials with insect hormone activity: X. A method of sexually spread insect sterility, J.Econ.Entomol. 63 (1970) 706.

[16] RENS, G.R., Unpublished data.

[17] BORKOVEC, A.B., Insect Chemosterilants, Advances in Pest Control Research 7, Interscience, New York (1966).

[18] ARORA, G.S., A Study of the Oocyte Development in the Adult Pyrrhocorid Bugs of the Genus Dysdercus, MSc thesis, University of Nairobi, 1971.

DISCUSSION

J.E. SIMON F.: You said that you are going to release insects in cotton fields. One of the limitations of SIRM is that a pest insect cannot be released in the state in which it causes damage. How do you propose to tackle this problem?

G.R. RENS: It is our intention to try and control Dysdercus on their wild hostplants, long before the cotton plants start fruiting and become targets of infestation by cotton stainers. We do not think it likely that we could apply the sterile-male release method on small, scattered holdings. However, areas with a heavy concentration of cotton cultivation could be protected. We shall carry out our first field application of the SIRM method on a 1000-hectare cotton irrigation area, situated in dry savannah land. It is thought that Dysdercus living on wild hostplants, cotton seeds and volunteer cotton plants can be eradicated and remain eradicated by repeated releases of sterile males on this area and on a surrounding strip of bush about three miles wide. On the basis of flight range measurements, we consider three miles sufficient to prevent rapid re-infestation of the area.

L. C. MADUBUNYI: Have you assessed the possible effects of laboratory colonization on the field performance of the Dysdercus with which you plan to artificially infest the island in your proposed pilot programme?

G. R. RENS: After the artificial infestation of the island with Dysdercus, we intend to wait about eight weeks before releasing sterilized insects. This period will give the introduced insects time to settle down and breed. The insects that will be targets for control by the SIRM method will therefore be the offspring of the introduced stainers. The target insects will therefore not be laboratory animals but insects that have developed on the wild host-plants and cotton fields of the island.

Generally speaking, we expect insectary-reared insects to be more vigorous than those in the "wild" populations, especially in the dry season when the wild Dysdercus are close to starvation.

B. BENNETT-ŘEŽÁBOVÁ: You have stated that juvenile hormone analogues can be transferred by copulation. Do you have any evidence that chemosterilants cannot be thus transferred.

G. R. RENS: No, I have not worked with chemosterilants. However, even if chemosterilants can be transferred, juvenile hormone analogues may still be preferable since they do not pose health hazards, as do chemosterilants such as tepa, metepa and apholate.

R. PAL: What is the rationale of using a juvenile hormone analogue as a sterilant rather than conventional methods?

G. R. RENS: Juvenile hormone analogues can be transferred by copulation from one insect to another. In the case of Dysdercus we found a limited potential for the transfer of the sterilizing effects. This is a very useful asset since our aim is to sterilize the whole population. In addition, the conventional chemosterilants are known to have some undesirable effects on mammals.

H. LEVINSON: How do you intend to administer the ethyl farnesoate dihydrochloride to the Dysdercus spp. before release?

G. R. RENS: We are now testing methods for the mass application of the material. One promising technique is to administer it via the drinking water provided for the insects.

H. LEVINSON: It is well known that feeding juvenoids is definitely less effective than topically applying juvenoids. Perhaps an aggregation pheromone could be used for this purpose.

G. R. RENS: We provide drinking water through filter paper which at the same time forms the bottom of the breeding cages. The insects therefore come into regular contact with the juvenile hormone material. I was not aware of the existence of an aggregation pheromone. It might certainly be very useful for ensuring effectiveness of application. Can you tell me whether it works on the nymphal or the adult stages?

H. LEVINSON: It works on both.

J. L. MONTY: It would seem that you have an excellent opportunity here to sterilize the natural population without releasing any insects. Do you have an attractant which could be used to apply the juvenile hormone analogue to the natural population?

G. R. RENS: No, I do not know of any food- or sex-attractant for Dysdercus. It would be difficult to apply the sterilizing materials to the natural population during the dry season, when the conditions for control are most favourable as the insects take refuge from the heat in inaccessible places such as under fallen leaves, behind the bark of trees etc. to protect themselves from dehydration.

B. NAGY: Although your experiments were apparently performed mostly in field cages, I wonder if you could tell us anything about the parasites of Dysdercus in your district and whether JHA has any effect on these parasites?

G. R. RENS: The most important parasites of Dysdercus are Tachinidae; they are much more common than hymenopterous parasites. We have found more than fifteen species of tachinids as parasites of Dysdercus, which under certain conditions cause a parasitization rate of 30 to 40%. However, at the end of the dry season the rate of parasitization is much lower. We have therefore not thought it necessary to test the material on tachinids. Bagley and Bauernfeind (Ref. [14] in the paper) have listed the results of tests using ethyl farnesoate dihydrochloride on a large number of insects, including Diptera. I do not recall if this list mentions tachinids.

A NEW DEVICE AND CONCEPT
TO ATTRACT AND STERILIZE PEST
INSECTS IN THE FIELD, PARTICULARLY
APHIDS (Homoptera: Aphidina)

A.W. STEFFAN
Institut für Zoologie,
Biologische Bundesanstalt für Land- und Forstwirtschaft,
Berlin (Dahlem)

Abstract

A NEW DEVICE AND CONCEPT TO ATTRACT AND STERILIZE PEST INSECTS IN THE FIELD, PARTICULARLY APHIDS (Homoptera: Aphidina).

On the basis of earlier suggestions for the application of the sterility principle to the control of aphids, an apparatus has been constructed and a new concept is presented here which should enable the eradication or at least the continuous suppression of harmful species of this pest group. The new device will use yellow colour tables to attract winged aphids of the sexupara and sexual generations in the field, and to suck them into a lockable cabin where they may be treated with chemical or radiation sterilants. As the oogenesis in the sexual females as well as the spermatogenesis in the males takes place already during their embryogenesis within the bodies of their mothers, the sterilants have to be applied in sub-sterilizing dosages to the gynoparae, androparae or sexuparae. By this means, the reproduction and migration potency of the winged pre-bisexual generation members that will be captured, sterilant-treated and afterwards released, will not be reduced. The sexual females or the males should, however, be sterilized by interrupting their oogenesis or spermatogenesis while they develop in the bodies of the pre-bisexual generation to be treated. The advantage of the proposed technique compared with the usual sterile-male method is, firstly, the avoidance of expensive and time-consuming rearing and release programmes and, secondly, the fact that each sterilant-treated individual will give rise to about half a dozen sterile sexual females or males.

1. INTRODUCTION

The basic principles of the eidonomy, ethology and ecology of aphids involved in an application of the sterility principle to the elimination or continuous suppression of serious plant pest species of this insect group have been discussed in earlier papers [1-5]. It also has been taken into consideration that any method used to induce sterility would involve an expensive mass-rearing process under controlled laboratory conditions to produce seasonally large numbers of sexuparae or sexuals. To reduce both the labour and costs, experiments have been made to rear parthenogenetic generations of heterogenetic aphid species on artificial diets, and by special techniques to sterilize the sexuals which are then released in the field [6].

With the aim of producing simpler and more economical procedures, proposals are made here to combine the application of the sterility principle with optical and olfactorial attraction techniques. Further, it will be pointed out that, because of the early completion of spermatogenesis as well as of oogenesis, the sterility of sexuals should be achieved by administering sub-lethal doses of gamma radiation or chemosterilants to the sexuparae (androparae and gynoparae), in the bodies of which the embryogenesis of males and females takes place.

2. CONSTRUCTION AND MODE OF ACTION OF A FIELD APPARATUS TO ATTRACT AND STERILIZE PEST APHIDS

The most important aim in the economical application of the sterility principle in aphids is to avoid expensive rearing of the specimens for sterilization.   Therefore an apparatus has been constructed for use in the field to attract winged aphids, and either kill them instantly or release them again after they have gone through a sterilization procedure. This apparatus should be used preferably during the autumnal migration period of the sexuparal or bisexual generations of the aphid species that are to be controlled. It has to be set up in the field close to the plantations of the primary host plants of the species concerned [7].

The construction and method of working of this apparatus is shown in Fig.1. The lower part of the structure is fitted with yellow tables, to which winged males, or winged sexuparae and androparae, migrating from the summer to the winter hosts, will be attracted, and afterwards sucked inside the apparatus (B) by a suction fan (L) electrically operated (M). Such a fan, unfortunately, restricts the usefulness of the apparatus, as it can only be set up where electric power is available.

For use in the field where power is not available, the following sucking mechanism may be used: The roof of the apparatus consists of transparent plastic plates (C), with metal plates (D) beneath. Sunbeams penetrating the plastic plates will encounter the iron plates and heat them up, thus producing a thermal air current within the interior of the apparatus (B). Insects that have landed on the yellow plastic plates will try to continue their migration when they realize that they have not reached their host plants, but they will be hindered from escaping by the protruding roof projections (E), and will be sucked into the interior by the thermal air flow, having first passed through the special size meshes of the inlets (F).

Afterwards the insects will be carried further by the thermal air current, through the interior of the apparatus, and finally into the sterilization chamber at the top of the structure. This passive transport, in the case of males can possibly be supported by the use of pheromones for inducing the trapped aphids also to find their way by active movement.

The sterilization chamber (G) is a lockable and removable cylindrical compartment, with an inlet (H) at the bottom and an outlet (I) at the top. Its interior is equipped with a coil (K), the spirals of which must be separated from each other to allow a steady and strong circulation of the air current, on the one hand, and on the other, to force the insects into contact with the walls of the spiral.

The spiral walls must be impregnated with a chemosterilant. The distance that the insects have to cover on this surface must be determined by separate experiments for each species to be sterilized, so that contamination by a given amount of the chemical compound within a given time will not cause death but induce sterilization.

Another sterilization technique that might be used is radioactive substances. These could be placed in the lockable sterilization chamber instead of chemosterilants, in which case the interior of this chamber has to follow another design, and its mantle should consist of lead. The radiation dose must be large enough also to produce a break in the spermato-genesis or oogenesis of the aphids, but not to diminish their competitiveness.

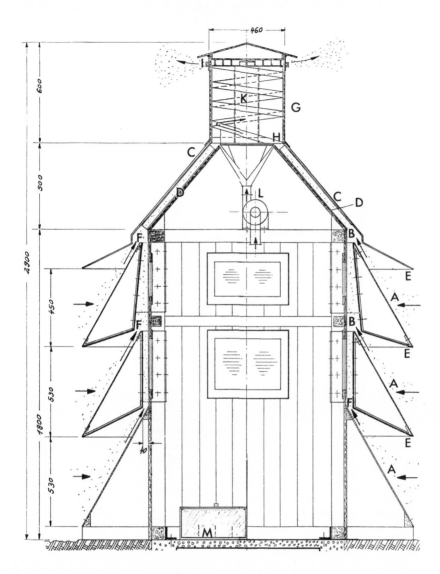

FIG. 1. Workshop drawing (vertical section and elevation combined) of a new device for capturing and
sterilizing winged aphids in the field: (A) Yellow plastic tables for optical attraction; (B) Interior air
channels leading from the inlets (F) to the sterilization chamber (G); (C) Transparent plastic plates;
(D) Metal sunbeam-absorbing heating plates; (E) Roof protrusions preventing the aphids that have
landed from flying off again; (F) Inlets equipped with fine meshed screens to allow the selective capture
of aphids; (G) Chamber for sterilization with an inlet (H) at the base and an outlet (I) at the top; (K) Coil,
the spiral walls of which are impregnated with chemosterilants; (L) Suction fan; (M) Electric motor or
battery (optional).

After sterilization by chemical or radiation agents, the captured aphids should find their way out of the apparatus through the upper outlet of the sterilization chamber. They could then continue their regular migration to the winter host plants and there compete with the fertile members of the pest population.

3. A NEW CONCEPT FOR APHID STERILIZATION WITH REGARD TO OOGENESIS AND SPERMATOGENESIS IN THE SEXUAL GENERATION

To make use of this apparatus for the capture and sterilization of aphids, attention must be paid to an essential phenomenon of their development. Earlier studies on members of the aphid family Adelgidae have proved that spermatogenesis takes place already in the early larval stages [8,9]. Examination of males of several species of the aphid family Aphididae have shown that adult winged males hold only spermatophores ready for insemination. Even in the early larval stages, the spermatogenesis is almost finished. Spermatogonia are to be found only in the embryos that develop within the bodies of the viviparous androparae. This means that chemo- or radiation sterilants do not have any effect on the sterilization of males when applied to their winged adult stage.

Because of these peculiarities, aphid species of the type that has winged migrating males (Fig.2) cannot be treated in the way proposed earlier [1,4,5]. However, there is another and possibly even greater possibility for eliminating or at least reducing the fertility of the sexual generation. One or two weeks before the winged males migrate to their winter hosts, the winged gynoparae migrate in parallel. When they arrive on winter host trees, they viviparously produce wingless females. Aphid species of this migratory type could be treated in the following manner. The winged gynoparae are captured by the apparatus described above and exposed to a non-lethal dose of chemo- or radiation sterilants. The oogenesis in the embryos of the sexual female generation, developing in their bodies, should be interrupted by this application. One or two weeks later, the then migrating males could also be captured and not treated with sterilants, however, but killed instantly. By this means, one part of the males of the pest population would be completely eliminated. The other part, that is, those males that have not been captured by the apparatus, would at least partly be excluded from reproduction by their mating with sterilized females. Only those females would be able to produce offspring which have been borne by non-captured gynoparae, and which furthermore could mate with the few non-captured fertile males.

In a second type of heterogenetic aphid species, the sexual females are wingless, and the males are either winged or also wingless. The females are borne by winged gynoparae and the males by winged androparae. Here again, the males and females are not treated directly with sterilizing agents, but by means of their viviparous mothers. The winged gynoparae as well as the winged androparae should be captured by the aid of the apparatus on their route from summer to winter host plants. As both types do not migrate at the same time, it should be easy to catch them separately. Then, either the gynoparae or the androparae would be treated with sterilants, depending on whether sterilization of the sexual females or of the males proves to be more effective. The other of the two parthenogenetic forms,

FIG. 2. Generation cycle of the mealy apple aphid, Dysaphis plantaginea (Passerini 1860). (1) Fundatrix;
(2) Civis-virgo-aptera; (3) Civis-virgo-allata = Allata-migrans; (4) Exsulis-virgo-aptera; (5a) Exsulis-
virgo-aptera; (5b) Gynopara-allata, which must be treated with sterilants; (6a) Sexualis-♂-allata, which
must be killed; (6b) Sexualis-♀-aptera, which must be sterilized.

producing viviparously one of the two sexuals, should, however, be killed.
By this means aphid species of this second migration type can be treated
in the same manner and presumedly with an efficiency equal to that of the
first type.

There is a third migratory type of aphids (Fig.3), to which the application
of the sterility principle hitherto did not seem to be feasible [1-3]. In this
type, sexuparae exist instead of gynoparae and androparae, each individual
being capable of producing sexual females and males also. The winged
sexuparae migrate from summer to specific winter host plants, and there
they give birth to wingless sexual females and wingless males. Since these

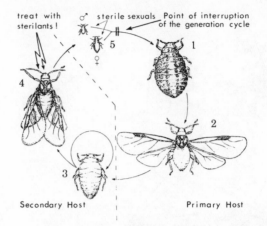

FIG. 3. Generation cycle of the lettuce root aphid, Pemphigus bursarius (Linnaeus 1758). (1) Fundatrix; (2) Civis-virgo-allata = Allata-migrans; (3) Exsulis-virgo-aptera; (4) Sexupara-allata, which must be treated with sterilants; (5) Sexualis-♂-aptera and sexualis-♀-aptera, one of which should be sterilized and the other killed.

cannot be captured with the aid of the proposed apparatus, and since reared, sterilized and then released wingless males would not be able to cover great enough distances to find and paralyse fertile females, the members of this type have been neglected in earlier studies. On the basis of the new finding, however, that aphid sexuals should not be exposed to sterilants directly but by the way of their viviparous mothers, the application of the sterility principle can also be considered in this case. The winged sexuparae should be captured on their migration route from summer to winter hosts with the aid of the proposed apparatus, and therein exposed to sterilants. By this means, the spermatogenesis in the males as well as the oogenesis in the sexual females, both developing in their bodies, could be interrupted. After being released, the sexuparae would continue their migration, and give rise to sterile sexual males and females on the winter host trees. These sterile insects, however, would not only compete with the fertile ones, borne by the uncaptured and untreated sexuparae, but it would be possible also for one sterile sexual to mate with another. Thus the efficiency of their sterilization might be reduced. Further investigations are necessary to discover whether the development of sexual females and males within the body of a sexupara takes place at the same time, and whether the oogenesis and spermatogenesis of such sisters and brothers reaches a phase in which sterilization is possible, exactly in parallel or with a time interval. When further knowledge has been obtained on these embryological and cytological problems, an efficient application of the sterility principle might be advisable also in this third aphid type.

4.   CONCLUSIONS AND RECOMMENDATIONS

The use of the proposed apparatus in the field would enable the sterility principle to be applied for the eradication or at least continuous suppression of many pest species in aphids. In every instance it must first be ascertained

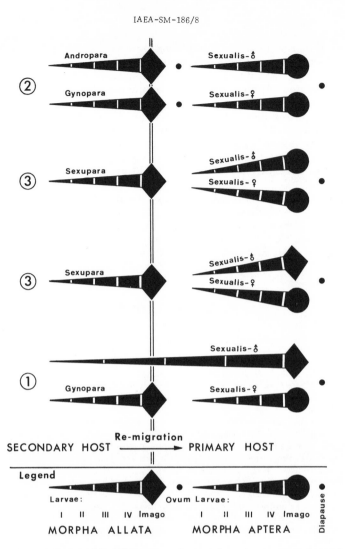

FIG. 4. Schematic representation of the different types of autumnal re-migration generations and their offsprings. (1) Re-migration performed by winged sexualis-♂♂, and winged gynoparae producing wingless sexualis-♀♀. (2) Re-migration performed by winged androparae producing wingless sexualis-♂♂, and winged gynoparae producing wingless sexualis-♀♀ . (3) Re-migration performed by winged sexuparae producing winged or wingless sexualis-♂♂, and wingless sexualis-♀♀ also.

to which migration type of aphids (Fig.4) the species to be controlled belongs. On this depends which one of the generations and morphs has to be exposed to substerilizing agents to cause sterility in either its sexual female or male offsprings.

The result of sterilization could be enhanced by capturing and killing the sexual counterpart also with the aid of the proposed apparatus. Furthermore, this device might be used during the spring and summer to capture and kill the winged individuals of other generations, and thus achieve a further reduction in the pest population.

The number of devices which have to be used, and the correct distance between them, will vary for each species to be controlled. The handling of the apparatus, and its charging with chemical or radiation sterilants, should be carried out only with the co-operation of the official plant protection services.

The use of this apparatus and technique for controlling harmful aphid species would entail a lower expenditure in cost and labour than conventional insect release methods, which require expensive rearing procedures. The proposed device and technique would also permit a rather selective control. When applied correctly the technique would not produce any damage to the environment.

Before the successful application of the proposed method can be ensured, however, much analytical and experimental research will be necessary on the ecological, physiological and cytological peculiarities of the aphid species concerned. The results of some earlier studies [8-15] might be used as a basis and starting point for this work.

REFERENCES

[1] STEFFAN, A.W., Möglichkeiten genetischer Bekämpfung von Blattläusen (Homoptera: Aphidina), Z. Angew. Entomol. 70 (1972) 267.

[2] STEFFAN, A.W., Sind mitteleuropäische Blattlausarten genetischen Bekämpfungsverfahren zugänglich?, Nachrbl. dtsch. Pflsch. d. (Braunschweig) 24 3 (1972) 33.

[3] STEFFAN, A.W., Proposals for the application of genetical control techniques in the eradication of harmful aphid species (Homoptera: Aphidina), Abstr. 14, Int. Congr. Entomol., Canberra 1972, 8 (1972) 234.

[4] STEFFAN, A.W., Zur Anwendung des Sterilpartner-Verfahrens bei der Bekämpfung von schädlichen Blattlausarten im Forst, Z. Waldhygiene 9 5/8 (1973) 247.

[5] STEFFAN, A.W., KLOFT, W.J., "The possibilities of using the sterile male technique for aphid control - a theoretical discussion", Computer Models and Application of the Sterile-male Technique (Proc. Panel Vienna, 1971), IAEA, Vienna (1973) 129.

[6] ARNOLD, M.-R., Haltung und Aufzucht zweier weiterer Aphiden-Arten auf künstlichen Diäten unter besonderer Berücksichtigung der Erzeugung ihrer Sexuales, Manuscript, diploma-thesis for the Mathematisch-naturwissenschaftliche Fakultät der Rheinischen Friedrich-Wilhelms-Universität, Bonn (1974).

[7] STEFFAN, A.W., Ein neues Gerät zur Anlockung und Freiland-Sterilisierung der geflügelten Geschlechts-tiere schädlicher Blattlausarten (Homoptera: Aphidina), Mitt. Biol. Bund.Anst. Ld- u. Forstw. 151 (1973) 315.

[8] STEFFAN, A.W., Evolution und Systematik der Adelgidae (Homoptera: Aphidina): Eine Verwandtschafts-analyse auf vorwiegend ethologischer, zytologischer und karyologischer Grundlage, Zoologica, Stuttg. 115 (1968) 1.

[9] STEFFAN, A.W., Zur Zytologie und Zytotaxonomie der Adelgidae (Homoptera: Aphidina), Verh. 13. Int. Congr. Entomol. Moskau 1968, 1 (1971) 349.

[10] STEFFAN, A.W., Problems of evolution and speciation in Adelgidae (Homoptera: Aphidoidea), Can. Entomol. 96 (1964) 155.

[11] STEFFAN, A.W., Zum Generations- und Chromosomenzyklus der Adelgidae (Homoptera: Aphidina), Verh. Dtsch. Zool. Ges., Heidelberg 1967, 31 (1968) 762.

[12] STEFFAN, A.W., Zur Karyologie und Chromosomen-Evolution der Blattläuse (Homoptera: Aphidina), Verh. Dtsch. Zool. Ges., Innsbruck 1968, 32 (1969) 558.

[13] STEFFAN, A.W., Generative Parallelreihen in den Entwicklungszyklen der Fichtengallenläuse (Homoptera: Aphidina: Adelgidae), Umsch. Wiss. Tech. 69 (1969) 843.

[14] STEFFAN, A.W., Die eidonomischen und zytologischen Grundlagen bie der Entstehung anholozyklisch-parthenogenetischer Adelgidae-Species (Homoptera: Aphidina), Z. Angew. Entomol. 65 (1970) 444.

[15] STEFFAN, A.W., "Aphidina-Blattläuse", Die Forstschädlinge Europas: Ein Handbuch in fünf Bändern (SCHWENKE, W., Ed.), Parey, Hamburg, Berlin (1972).

DISCUSSION

E.F. BOLLER: Have you already constructed a prototype of this device and, if so, from what distance are aphids attracted to it?

A.W. STEFAN: A prototype of the apparatus has been built, but so far it has been used only to test the possibility of using solar heat to generate a thermal air current. We are currently carrying out investigations to see from what distance mealy apple aphids and some grain aphid species can be induced to land on sticky yellow tables. We are also trying to establish what percentage of migrating aphids, landing in a given area, can be caught with this device.

A. HAISCH: A stimulus for a landing response for aphids must come from below. Aphids are, I believe, unable to direct their flight very accurately, yet the size of the yellow plates in your apparatus seen from above seems to be very small. I cannot understand why the plates are arranged one above the other, as the lower plates do not increase the effective, attractive yellow area of the apparatus.

A.W. STEFFAN: The yellow tables would, of course, have a greater attraction effect if they were arranged horizontally but there would not then be any possibility of sucking the aphids, which have landed on the tables, into the apparatus. Therefore, the tables are arranged like roof tiles, each one partly covering the other. All of them thus add to the extension of the total attractive surface. Between each two, however, there are inlets through which the aphids can be sucked in. Furthermore, each table mounted one above the other acts as a guard, preventing the aphids which take off again from escaping.

G. NOGGE: Can you control the duration of stay of the aphids in the sterilization chamber of your apparatus? This will obviously be very important for the sterilization effect.

A.W. STEFFAN: The time for the trapped aphids to pass through the spiral of the sterilization chamber is dependent on the speed of the thermal air current. This speed, again, is a function of the air temperature within the apparatus and the diameters of the whole channelling system. It will be possible to regulate the speed of the air current by making the diameters, especially of the inlets and outlets of the spiral of the sterilization chamber, larger or smaller. Thus the time that the aphids are inside the spirals of the sterilization chamber can be precisely established. Optimum sterilization can be achieved within this given time by (a) varying the concentration of the sterilizing agent or (b) extending or reducing the length of the impregnated part of the spiral.

B. NAGY: Assuming the optimum case, what percentage of the flying aphid population do you expect to be attracted and sterilized by your system at a given plantation?

A.W. STEFFAN: Aphids do not find their specific host plants by searching for them directly but rather by the trial and error method. Thus, it does not seem necessary to capture those members of the migrating generations which do not find their winter hosts, as they are doomed anyway. However, those which reach the plantation containing their winter hosts, must be captured in a high ratio. Therefore, traps of the proposed kind need to be set up in the vicinity of these plantations. The number of traps and their distance apart will depend on the distance from which the aphids can be attracted. This might be different for each aphid species concerned and thus one of our main tasks will be to establish these distances.

EFFECTS OF AN ANALOGUE OF THE JUVENILE HORMONE ON OVARIAN DEVELOPMENT AND FECUNDITY IN THE HOUSEFLY, Musca domestica L.

B. BENNETT-ŘEŽÁBOVÁ, V. LANDA
Institute of Entomology,
Czechoslovak Academy of Sciences,
Prague, CSSR

Abstract

EFFECTS OF AN ANALOGUE OF THE JUVENILE HORMONE ON OVARIAN DEVELOPMENT AND FECUNDITY IN THE HOUSEFLY, Musca domestica L.

The effects of a juvenile hormone analogue (methyl ester 2, 7, 11-trimethyl 7, 11 dihydro-dichloro-2 dodecenoic acid) on ovarian development in Musca domestica L. and its fecundity were investigated. The compound was applied in doses of 10 - 200 µg to larvae of various instars, pupae of different ages and teneral adults. Apart from examination of mortality and development of individual stages, attention was focused on ovarian development. It has been found that the effects of this analogue are similar to those of chemosterilants, i.e. certain changes occur in the follicular epithelium of egg chambers, the changes depending on the time of treatment. Tumour-like proliferation of the follicular epithelium was found in all egg chambers only when the analogue of the juvenile hormone had been administered to the last larval instar or a newly formed pupa. The ovaries then degenerated.

Within a broader study of the effects of analogues of the juvenile hormone on the ontogeny of insects these compounds have been tested in our laboratory for effects on the growth and development of the ovaries of Musca domestica L. Preliminary experiments made in recent years showed certain effects of juvenoids on the fecundity of Musca domestica. The compounds were administered as a substance to newly emerged adults. It was found that some compounds have definite qualitative effects on ovaries without considerably affecting fecundity. More pronounced effects were discovered by detailed examination of different developmental stages.

Results are presented of the study of effects of one analogue of the juvenile hormone, methyl ester 2, 7, 11-trimethyl 7, 11 dihydro-dichloro-2 dodecenoic acid, administered to different developmental stages, on the ovarian development and fecundity of the housefly.

A WHO standard strain of Musca domestica was used. The juvenile hormone analogue was synthesized in the Institute of Organic Chemistry and Biochemistry of the Czechoslovak Academy of Sciences. The compound, either in acetone solution or as a substance, was applied as follows:

1. Treatment of teneral adults

Ten to 200 µg of the analogue either as a substance or in acetone solution were applied to the dorsal side of the thorax of females. The females were then dissected at three-day intervals and the degree of development of

individual egg chambers was examined as well as the condition of the follicular epithelium and nutritive cells. Hatchability of the laid eggs, metamorphosis of larvae into pupae and the number of emerging adults were also
recorded.

2.   Treatment of larvae of different ages

    One hundred and twenty-five µg of the analogue as a substance or in
acetone solution were applied to larvae 24 - 186-hours-old at 12-hour
intervals. Metamorphosis of the larvae into pupae, the number of emerging
adults, the number of laid eggs and percentage of their hatchability were
examined as well as the sex ratio of adults. Developmental morphology of
the ovaries was minutely examined as in the previous series of experiments.

3.   Treatment of puparia of different ages

    One hundred — two hundred µg of the analogue either as a substance or
in acetone solution were applied to the anterior part of puparia either newly
formed or 12-hours-old, and 24 - 48-hours-old. The number of emerging
adults, sex ratio and hatchability of their eggs were recorded. As in the
previous series of experiments, morphological examination concentrated on
the condition of the follicular epithelium of individual egg chambers. For
histological examination the dissected ovaries were fixed in Carnoy's solution and stained with modified Heidenhain's haematoxyline and Masson's
trichrome.

    The dose of 10 µg per female of the analogue in acetone solution (first
series of experiments) did not affect vitellogenesis of the first egg chamber.
Sixty per cent of these eggs hatched. The second and third egg chambers
developed normally (77% and 86% hatchability). The dose of 100 µg of the
analogue sometimes disturbed vitellogenesis in the first egg chamber. The
plasma of nutritive cells turned opaque and eggs with opalescent to yellow
content developed. Three per cent of these eggs hatched. The second egg
chamber was much less subjected to such changes (25% hatchability). The
third egg chamber also developed.
    Administration of 200 µg of the analogue as a substance affected
vitellogenesis of the first egg chamber so much that some eggs were laid
half empty. No changes in follicular epithelium were noticed with any of the
concentrations given above.
    Application of 125 µg of the analogue (second series of experiments) in
acetone solution to larvae 24-hours-old did not affect their survival and
pupation; adults emerged normally and hatchability of their eggs was reduced
by 40%. When the analogue had been administered to older larvae
(30 - 60-hours-old) their mortality reached 100%. Pupation of larvae treated when
78 - 132-hours-old was not affected, adult emergence was also normal.
Most of their eggs hatched normally. A different situation developed when
the analogue was applied to larvae 150 - 186-hours-old. Pupation was
reduced by 30 - 40%, adult emergence was not affected, but hatchability of
eggs was reduced by 70 - 80%, particularly after treatment of larvae
138 - 168-hours-old. The ovaries of females in this experiment displayed
characteristic changes. Examination in anoptral contrast showed some of

the first and second egg chambers to be opaque and about one-third to one-half of the size of a mature egg. Some ovarioles contained normal eggs, in others vitellogenesis had been affected so much that although the first egg chamber was as large as a mature egg only a tiny amount of yolk was present and irregularly deposited. Histological examination of the ovaries revealed proliferation of follicular epithelium. On the 5th day after emergence double nucleoli and later nuclei were found in the cells of the follicular epithelium, which further divided (some without subsequent cell division) producing an irregular multi-layered follicular epithelium. The nuclei later underwent pycnosis and some of them disintegrated. Similar changes were found in the second egg chamber. If the third egg chamber descended it also went through proliferation changes. Nutritive cells did not divide; only chromatin in their nuclei clustered. The content of the egg chambers was completely resorbed by the time development finished (about the 18th day).

Application of 100 - 200 μg of the analogue to newly formed or 12-hour-old pupae (third series of experiments) reduced eclosion of adults by 40% and hatchability of their eggs by 40 - 50%. Treatment of older pupae was ineffective.

Morphological examination revealed that some egg chambers in females that had emerged from pupae treated when 0 - 12-hours-old underwent changes resembling those described after treatment of larvae 138-168-hours-old. The proliferation of the follicular epithelium spread to the second and third egg chambers, and later the whole ovariole degenerated. The changes described occurred approximately in the same number of cases as after treatment of 138 - 168-hour-old larvae.

These results indicate a certain sterilizing activity of some analogues of the juvenile hormone on the housefly. Comparison with the effects of conventional chemosterilants has shown that as much as 50 times more of the analogue is necessary to produce the same degree of sterility. The time of treatment for induction of certain morphological changes also differs. The conventional chemosterilants provoke proliferation of the follicular epithelium when applied to females 0-18-hours-old; the juvenile hormone analogues induce such changes only when administered to mature larvae or newly formed pupae. The nuclei of the follicular epithelium proliferate, forming multi-layered epithelium, and proliferation always spreads to the second and third egg chambers. Chemosterilants often provoke proliferation in the first egg chamber only; follicular epithelium of the second and third egg chambers rarely proliferates.

It may be concluded that the analogues of the juvenile hormone used as chemosterilants certainly deserve our attention in the search for new methods of insect control.

DISCUSSION

E. BOLLER: What is the toxicity of these compounds to mammals?

B. BENNETT-ŘEŽÁBOVÁ: I do not have any exact data but preliminary results indicate high toxicity in some cases.

H. LEVINSON: Can the chemosterilizing effect of the juvenoid be reversed by natural juvenile hormone given in excess?

B. BENNETT-ŘEŽÁBOVÁ: The application of some juvenoids causes the ovaries to degenerate and an excess of natural juvenile hormone has no effect at all in such cases.

B.S. FLETCHER: Does this compound have a juvenile hormone effect on Musca domestica and, in particular, will it promote maturation of the ovaries in allatectomized females?

B. BENNETT-ŘEŽÁBOVÁ: We have not tried this compound on allatectomized females of the house fly.

B.S. FLETCHER: Have you tried similar experiments with Altosid® and, if so, did you find any effects?

B. BENNETT-ŘEŽÁBOVÁ: None of the tested compounds was tried on allatectomized females.

R. PAL: You have made a very important point that some of the juvenile hormones also act as sterilizing agents and therefore that advantage might be taken of this effect, the so-called "bonus effect", of insect growth regulators.

B. BENNETT-ŘEŽÁBOVÁ: Since analogues of the juvenile hormone can act as chemosterilants when given at the right time, the effect on population would be a sterilizing one as well.

D.W. WALKER: Would you elaborate on potential methods of delivering the juvenile hormone to the target organ, the reproductive system?

B. BENNETT-ŘEŽÁBOVÁ: The methods used to apply conventional chemosterilants could be employed after establishing the effect of the compound on other animals.

GENETIC MECHANISMS OF
INSECT CONTROL
(Session 6)

CONDITIONAL MUTATIONS FOR THE CONTROL OF INSECT POPULATIONS

M. FITZ-EARLE, D. T. SUZUKI
Department of Zoology,
The University of British Columbia,
Vancouver, B.C.,
Canada

Abstract

CONDITIONAL MUTATIONS FOR THE CONTROL OF INSECT POPULATIONS.
A genetic method for insect population control has been examined. The technique involves the replacement of a population of insects with standard chromosomes by those carrying compound autosomes, thus permitting the introduction of such controlling factors as conditional mutations. Strains of the fruitfly Drosophila melanogaster, bearing compound autosomes and temperature-sensitive (ts) lethal mutations, were competed against standards in cages over continuous generations at the permissive temperature of the ts mutant. When the compounds had successfully replaced the standards, the cages were then shifted to the restrictive temperature level of the ts lethal gene. Complete elimination of the caged populations was subsequently observed. A critical evaluation is given of conditional mutations suitable for insect control, in the light of these experiments and the environmental conditions to which released insects are likely to be subjected in the wild.

POPULATION REPLACEMENT

Following the successful application of the sterile insect technique to the control of the screw-worm Cochliomyia hominivorax (Coquerel) in the southern United States and northern Mexico [1], considerable attention has been focused upon more sophisticated genetic procedures for the suppression of pest insects. One of the potentially most powerful genetic methods for insect control involves the replacement, in a pest population, of individuals bearing standard chromosomes by those carrying chromosomal rearrangements termed compound autosomes which, in turn, may provide a vehicle for introducing controlling factors such as conditional mutations into the population [2,3,4].

Hybrids between certain types of chromosome rearrangements (e.g. translocations and compound autosomes) and lines carrying non-rearranged chromosomes often show reduced viability. In a mixed population of rearrangements and standards, in which there is no mating barrier, there is intrastrain fertility but a degree of interstrain sterility. If the numbers of the two chromosomal types are adjusted in a population, according to their relative fitnesses, a genetically unstable equilibrium can be established [5]. However, if the initial frequency of the rearrangements were to be increased beyond the equilibrium value, the standards would be driven out by the rearrangements. The rapidity of fixation of the chromosomally rearranged individuals would be a function of the relative fitnesses of the strains, their initial frequencies and whether generations were continuous (i.e. overlapping) or discrete. Throughout the population replacement, the fertility will be reduced due to the imposed genetic load. However, this does not necessarily suggest that the replaced population's size will be diminished since there may be density-dependent buffering of the population.

COMPOUND AUTOSOMES

     One type of chromosomal rearrangement suitable for population replacement
is the compound autosome, first synthesized in the late fifties for the fruit-
fly Drosophila melanogaster (Meigen) by E.B. Lewis, who also recognized its
potential for control programmes at that time. Compound autosomes differ from
standard chromosomes in that their homologous arms are attached to common
centromeres rather than to different centromeres and their formation is considered to be a translocation event (for a review, see [6]). The meiotic behaviour of a compound line is such that only 25% of all zygotes formed live
(c.f. 100% for standards), though these surviving zygotes are quite competitive. However, there is complete genetic isolation between compounds and
standards, a feature which makes compounds ideal for genetic replacement. At
this point in time, compounds have only been generated in D. melanogaster, although they are being actively sought in the Australian sheep blowfly, Lucilia
cuprina (Wiedemann) [7], the housefly Musca domestica [8] and the onion fly
Hylemya antiqua (Meigen) (A.S.Robinson, pers. comm.). The meiotic behaviour
of compounds from these species may not, however, parallel that of the fruit-
fly.
     By flooding a target pest population with a compound-bearing strain, it
would be possible to replace pest insects by those carrying factors that would
either eradicate the pest or render it innocuous. Many insect species provide
important links in the food chain and hence their eradication would be undesirable. The introduction of genes for the prevention of transmission of disease
into insects, such as mosquitoes where they are public health hazards, or of
mutants that render pests like biting flies less offensive to man, may be the
most satisfactory form of control for these kinds of pests. By contrast, many
pests that affect man's agriculture are exotics in the sense that they have
been introduced into an area along with man's crops and livestock. In these
instances, it would be acceptable to eradicate the insects, or at least to
reduce their numbers to economically tolerable levels, by the application of
conditional lethal mutations.

CONDITIONAL MUTATIONS

     One class of lethal factors that may be useful for suppressing certain
types of insects, are those affecting the diapause process which, if introduced into a population inhabiting a location for which diapause is obliga-
tory for survival, will cause the pest to be eradicated [9]. Such conditional lethality has been observed in a range of agricultural, forest, public
health and veterinary insect pests where the genetic control of diapause has
been investigated. A second class of conditional genes includes various
kinds of temperature-sensitive (ts) mutations. A ts mutant is one that ex-
presses the normal phenotype (survival) at permissive temperatures (e.g.,
22°C) but which either dies or is sterile at restrictive temperatures (e.g.
heat-sensitive 29°C, or cold-sensitive 17°C). In Drosophila, ts mutants
have been detected on all major chromosomes and have been extensively studied
(for a review, see [10]). Over 10% of those recessive lethal mutations induced by ethyl methanesulfonate (EMS) on the X and second chromosomes were
found to be temperature-sensitive [11,12]. Dominant heat-sensitive and cold-
sensitive lethals on chromosomes 2 and 3 were recovered in considerably lower
frequencies [13,14,15]. However, it is clear that wherever lethal mutations
are recoverable by treatment with mutagens such as EMS, a significant proportion will be temperature-sensitive.
     Phenotypes of ts mutants suitable for insect control fall into three
broad categories: (i) ts lethals, in which, following a temperature stress
during some stage of development (the temperature-sensitive period, TSP), the
insect dies at a later stage (the lethal phase, LP). Some ts lethals have

TABLE I. THE STRAINS OF Drosophila melanogaster USED IN THE
EXPERIMENTS

Strain	Mutations [a]
H28cn/H28cn	H28 – heat-sensitive egg lethal on left arm of chromosome 2
C(2L)VH2, lt; C(2R)P, px	
C(3L)VH2, ri; C(3R)VK1, e^s	
C(3L)VT1, se; C(3R)VK1, e^s	
C(2L)VFE1, H28; C(2R)P, px	
C(2L)VFE2, H28; C(2R)P, px	
C(2L)VFE3, H28; C(2R)P, px	
para^{ts}/para^{ts}; C(3L)SH2, +; C(3R)SH19, +	para^{ts} – X-linked, adult paralyzed at 29°C, normal at 22°C
C(2L)VFE1, H28; C(2R)VFE5, +	
C(2L)VFE2, H28; C(2R)VFE6, +	
$\frac{b/b}{e/e}$	

[a] Other than those described in Ref.[25].

relatively short TSPs, others will die subsequent to a temperature shock at
any time during development (indispensable). The TSP may not necessarily coincide with the LP but may be temporally separated from the expression of
lethality by several days. (ii) ts paralytics, in which prolonged exposure
of the adult individuals to restrictive levels of temperature initially causes
paralysis followed by lethality within relatively few hours. (iii) ts
steriles, in which exposure to a restrictive temperature during development
causes adult sterility (i.e. genetic death). To achieve eradication, the
rationale would be to replace, at permissive temperature, a native standard
pest population by a compound strain carrying a ts lethal, ts paralytic or
ts sterile gene. Then, if the temperature were either to increase or decrease to the restrictive level, lethality would occur. It is of interest to
note that in a study of mutations induced on chromosome 3 of D. melanogaster,
one third of all ts lethals were both heat- and cold-sensitive [16]. A strain
bearing such a conditional lethal gene, when introduced into the wild, would
be subjected to temperature stresses both in the spring and in the fall.
Temperature-sensitive lethal mutations have also been detected in the preda-
ceous wasp, Habrobracon serinopae [17,18] and the housefly M. domestica [19,
20], while they are being sought in the Australian sheep blowfly L. cuprina
[7] and the onion fly H. antiqua (M.Fitz-Earle, unpublished).
    Studies of cage populations of compound and standard strains of D.melano-
gaster under laboratory and field conditions have revealed that:
(i) Compounds can displace standards over discrete generations in bottles[2] and
     population cages [3]. It is also possible to displace standards in cage
     populations with continuous overlapping generations, a system that more
     accurately simulates the natural conditions [4]. Considerable variation
     exists, however, in the ability of compound strains to achieve replace-
     ment.

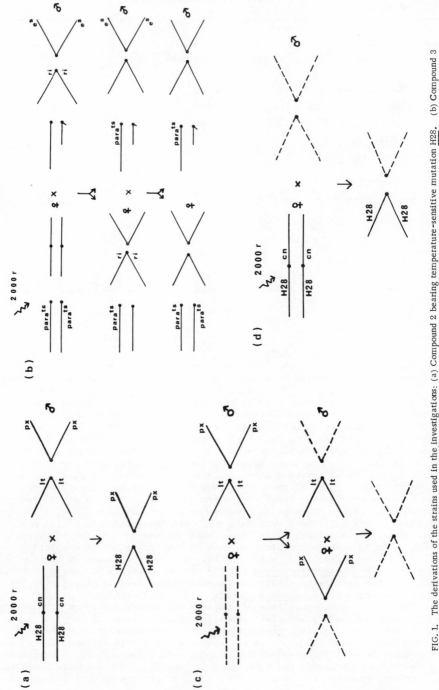

FIG. 1. The derivations of the strains used in the investigations: (a) Compound 2 bearing temperature-sensitive mutation H28.   (b) Compound 3 with paralytic temperature-sensitive mutant para^{ts} on the X chromosome.   (c) Native-derived compound 2.   (d) Native-derived compound 2 bearing mutation H28.   Solid lines = laboratory-derived material; broken lines = native-derived material.

(ii) The minimum initial ratio of compounds to standards required to subse-
      quently fix compounds in a population has been found to be close to the
      equilibrium ratio (4 compounds to 1 standard by Fitz-Earle [21], and
      5:1 by Cantelo and Childress [22]). These initial release ratios compared
      most favourably with data from other genetic control techniques.

(iii) When native-derived compound strains were tested in population cages
      under two different sets of field conditions, they were found to be ex-
      tremely successful in displacing standards [23]. Since the native-
      derived compounds were far superior to laboratory-derived compounds in
      their replacement ability, it was emphasized that for optimum success in
      control programmes, strains should be collected from the wild and mani-
      pulated genetically in the laboratory as rapidly as feasible prior to re-
      lease in the field.

The present paper describes the utilization of compound autosome lines of $\underline{D}$.
melanogaster carrying temperature-sensitive mutants, in a demonstration of the
principle of population replacement and the ancilliary principle of population
elimination.

MFTHODOLOGY

Strains

     The various lines of Drosophila melanogaster used in the experiments are
listed in Table I. One of the mutations is a heat-sensitive (29°C) lethal
that dies in the egg stage and is located on the left arm of chromosome 2
(H28). Another mutation, $para^{ts}$, is an X-linked recessive ts factor that
renders an adult immobile at 29°C but permits normal behaviour at 22°C [24].
Prolonged exposure of $para^{ts}$ flies to the restrictive temperature leads to
death. All other mutations used are described in some detail in Lindsley and
Grell [25]. The compound autosomes are reversed metacentrics and each unique
left or right arm is designated by location of synthesis (P-Pasadena, S-Storrs,
V-Vancouver), discoverer (FE-Fitz-Earle, H-Holm, K-Kiceniuk, T-Tabatabaie) and
a code number.

     Virgin females of the line bearing a ts lethal were collected, aged for
three days, treated with 2000r of γ-radiation and crossed to marked males
carrying compound autosomes (Fig. 1a). The progeny were screened for the
presence of putative ts compound male and female individuals, which were then
pair mated to yield discrete putative-ts compound lines. All non-segregation-
al progeny were discarded. The putative-ts compound strains were tested at
both 22°C (the permissive temperature) and 29°C (the restrictive temperature)
for the expression of the ts phenotype, and also for the presence of compound
left and right arms by crossing to known compound stocks. Nine ts compound
individuals were recovered from the 66 crosses (in bottles) between four
treated females bearing ts lethals on chromosome 2 and eight compound 2 males.
No ts compounds were obtained from comparable chromosome 3 matings, although
progeny from 20 bottles and 378 vials (one female and two males) were examined.
Of the four discrete lines established from the putative-ts compound 2s recovered, three proved to be temperature-sensitive ($C(2L)VFE1,H28; C(2R)P,px$,
$C(2L)VFE2,H28; C(2R)P,px$ and $C(2L)VFE3,H28; C(2R)P,px$) while the fourth was
non-temperature-sensitive.

     Females carrying the ts paralysis mutant $para^{ts}$ were collected as virgins,
aged for three days, irradiated with 2000r of γ-radiation and crossed to compound 3 males marked with $\underline{ri}$ and $\underline{e^s}$. Two generations later, a line of compounds associated with an adult paralysis gene ($para^{ts}/para^{ts}$ $C(3L)SH2,+;$
$C(3R)SH19,+$) was obtained (Fig. 1b).

     Finally, virgin females of two native standard strains (derived from
single females collected from an isolated fruit dump near Summerland, British
Columbia, Canada) were likewise aged, irradiated and crossed to marked males
carrying compound autosomes. Two unmarked native compound strains were thus

TABLE II. THE RESULTS OF CAGE COMPETITIONS BETWEEN
COMPOUND STRAINS BEARING TEMPERATURE-SENSITIVE (TS)
MUTATIONS AND STANDARD STRAINS AT VARIOUS RELEASE RATIOS:
REPLACEMENT AT THE PERMISSIBLE TEMPERATURE OF THE TS
MUTANT, ELIMINATION AT THE RESTRICTIVE TEMPERATURE

Compound strain	Standard strain	Ratio [a]	Days for replacement at permissive temperature	Days for elimination at restrictive temperature
C(2L)VFE1, H28; C(2R)P, px	±	25 : 1	74	33
		15 : 1	93	42
		12 : 1	32	38
		9 : 1	44	38
		7 : 1	105 [b]	-
		6 : 1	38	53
C(2L)VFE2, H28; C(2R)P, px	±	15 : 1	89	40
		10 : 1	56	45
		9 : 1	38	47
		7 : 1	105 [b]	-
		6 : 1	38	44
C(2L)VFE3, H28; C(2R)P, px	±	10 : 1	56	c
		9 : 1	124	c
		7 : 1	38	c
C(2L)VFE1, H28; C(2R)VFE5, ÷	b	30 : 1	49	48
		27 : 1	128	51
		7 : 1	112 [b]	-
C(2L)VFE2, H28; C(2R)VFE6, ÷	b	33 : 1	66	c
		25 : 1	60	c
		15 : 1	68	c
		7 : 1	96 [b]	-
para^{ts}/para^{ts};				
C(3L)SH2, ÷; C(3R)SH19, ÷	e	35 : 1	90 [b]	-
		24 : 1	96 [b]	-
		23 : 1	74 [b]	-
		11 : 1	72 [b]	-
		10 : 1	67 [b]	-
		8 : 1	42 [b]	-

[a] Compounds to standards.
[b] Compound failed to replace standard.
[c] Compound failed to be eliminated.

recovered two generations later (Fig. 1c). Virgin females of the original
line bearing the ts lethal H28 were similarly treated and crossed to males of
the newly derived native compounds. Their unmarked progeny putatively carried
the ts lethal gene and had genomes that were essentially 50% native and 50%
laboratory material (C(2L)VFE1,H28;C(2R)VFE5,+ and C(2L)VFE2,H28;C(2R)VFE6,+)
(Fig. 1d). The newly derived ts-bearing compound lines were also tested for
the presence of compound autosomes by crossing to known compound stocks, and
also tested for temperature-sensitivity by exposing them to both 22°C (permissive temperature) and 29°C (restrictive temperature).

Cages and growth conditions
     The cages used in the studies were Plexiglas® cubes of side 46cm, having
a screened vent on the top and a sleeve attached to the front to facilitate
access to flies and food dishes. Open dishes of food, containing 50cm³ of
standard cornmeal-agar medium, were introduced every three days until a maximum of seven were in the cage at any time. A covered dish of water was retained in the cage to ensure high humidity. All replacement phases of the experiments were conducted at 24.0+0.5°C (except for the para^{ts}/para^{ts};
C(3L)SH2,+;C(3R)SH19,+ tests which were at 21.0+0.5°C). The elimination
phases were performed at 29.0+0.5°C.

Experimental design and sampling regime
     Unmated compound and standard, males and females, were aged for 3-4 days
and then released into cages in the initial ratios listed in Table II. The
flies were permitted to increase in abundance over continuous generations.
Starting two to three weeks after releasing the compounds and standards into
a cage, and every three days until fixation, random samples of approximately
300 flies were removed with an aspirator. The flies were anaesthetized with
carbon dioxide, scored for a visible marker and then replaced into the cage,
to minimize disruption of the population composition. When three consecutive
three-day samples gave only standard or compound flies, fixation was considered to have been reached. The frequency of compound flies in each sample was
calculated and recorded. The number of days to achieve population replacement was subsequently deduced. Those cages in which compounds bearing ts
factors had gone to fixation were shifted to the restrictive temperature of
the ts mutants and the date was noted. The cages were retained at the restrictive temperature level until the populations were eliminated and the day when
this occurred was recorded. The number of days required for population elimination was then deduced.

POPULATION ELIMINATION

     With the exception of the strain carrying the paralysis factor, the com-
pounds bearing ts mutations went to fixation, when competed against standards
at the permissive temperature, over a wide range of initial ratios to a minimum of between 6:1 and 7:1 (Table II). The time period required to achieve
replacement ranged from 30 days to 130 days, with no obvious relationship
between initial ratio and time to fixation. The four unsuccessful replacement experiments were each initiated simultaneously in the ratio of 7:1. It
is speculated that their failure may be related to unknown environmental
factors following their establishment. Of the cages in which compounds became fixed at the permissive temperature (22°C), those containing flies of
the three genotypes C(2L)VFE1,H28;C(2R)P,px, C(2L)VFE2,H28;C(2R)P,px and
C(2L)VFE1,H28;C(2R)VFE5,+ were decimated following exposure to the restrictive temperature (29°C) (Table II). The period required for elimination was
in the order of 35-55 days. This is consistent with the fact that H28 is an
egg lethal and that any first instar larvae or later developmental stages
present in the cage at the time of the temperature shift would survive at the

restrictive temperature. The progeny of these survivors would, however, suc-
cumb to the temperature stress. The other two strains, C(2L)VFE3,H28;
C(2R)P,px and C(2L)VFE2,H28;C(2R)VFE6,+, having successfully displaced standards, failed to be eliminated at the restrictive temperature, suggesting that
at some stage during the experiment the H28 gene had been lost. Test matings
confirmed that the flies in the cages were still, in fact, compounds.

The line carrying the gene para^{ts} was unable to replace standards at the
ratios tested (Table II). Thus, the utility of an adult lethal in eliminating
populations could not be evaluated. Since each compound autosome is unique,
by virtue of its synthesis, it is possible that the failure of these experiments may be unrelated to the chosen paralytic factor but may be a function of
the compound arms or of the entire genome. To test this hypothesis, paralytic
mutations in combination with a variety of compound autosomes are to be synthesized and competed against standards.

The successful replacement and elimination experiments conducted in the
laboratory revealed no apparent differences between the abilities of entirely
laboratory material and partially native strains. However, it has been shown
elsewhere [23] that laboratory compounds fared less well than native-derived
stocks in their ability to displace standards in the wild. The implication
was that, in control programmes, it is important to collect native material
from the field and perform the necessary genetic manipulations, with the minimum of exposure to laboratory conditions, prior to re-introducing the insects
into the wild. In addition, the present experiments serve to emphasize the
need for a wide range of ts mutants and compound strains from which to choose
prior to release, and the value of preliminary cage trials with these strains
to test their ability to achieve both replacement and elimination of a population.

The present study also raises the question of the most advantageous
class of temperature-sensitive mutants to use in control programmes. None of
the strains tested under the constant temperature conditions of the laboratory
was eliminated in less than 70 days, partly because of the fact that the controlling factor chosen was an egg lethal. It is possible that in the wild
there may be insufficient time in which to achieve elimination if such a developmental ts lethal or sterile were used. Since a paralytic mutation, acting
as an adult lethal, gives an immediate response to a temperature stress, the
possibility arises that such mutants would be more advantageous in control
programmes. Although the strain bearing the para^{ts} gene was unable to replace standards in these experiments, it is likely that a compound line carrying one of the many other X-linked paralysis factors available [26] would be
more successful. It is also apparent that the restrictive temperature limit
of the mutant should be carefully chosen to take into consideration the fact
that in the wild the released insects may encounter significant diurnal temperature fluctuations. Thus, under certain conditions, the application of a
cold-sensitive mutant may be more effective than a heat-sensitive mutant,
while under different conditions a combined heat- and cold-sensitive gene
may be the most suitable.

ACKNOWLEDGEMENTS

This work benefited from the many fruitful discussions with Dr.D.G.
Holm. The authors wish to extend their appreciation to A.K.Junker, C.R.Reil-
koff and W.Rublee who provided technical assistance. The research was supported by Canada Council Grant S71-1687 to MFE and by NRC grant A1764 to DTS.
The senior author is a recipient of a Canada Council Killam Special Postdoc-
toral Research Scholarship.

REFERENCES

[1] BUSHLAND, R.C., "Sterility principle for insect control: Historical development and recent innovations", Sterility Principle for Insect Control or Eradication (Proc. Symp. Athens, 1970), IAEA, Vienna (1971) 3.

[2] FOSTER, G.G., WHITTEN, M.J., PROUT, T., GILL, R., Chromosome rearrangements for the control of insect pests, Science 176 (1972) 875.

[3] CHILDRESS, D., Changing population structure through the use of compound chromosomes, Genetics 72 (1972) 183.

[4] FITZ-EARLE, M., HOLM, D.G., SUZUKI, D.T., Genetic control of insect populations: Cage studies of chromosome replacement by compound autosomes in Drosophila melanogaster, Genetics 74 (1973) 461.

[5] LI, C.C., The stability of an equilibrium and the average fitness of a population, Am. Nat. 89 (1955) 281.

[6] HOLM, D.G., "Compound autosomes", Biology of Drosophila 1 (ASHBURNER, M., NOVITSKI, E., Eds), Academic Press, New York (in press).

[7] FOSTER, G.G., WHITTEN, M.J., "The development of genetic methods of controlling the Australian sheep blowfly, Lucilia cuprina", The Use of Genetics in Insect Control (PAL, R., WHITTEN, M.J., Eds), Elsevier, Amsterdam (in press).

[8] WAGONER, D.E., MCDONALD, I.C., CHILDRESS, D., "The present status of genetic control mechanisms in the housefly, Musca domestica L.", The Use of Genetics in Insect Control (PAL, R., WHITTEN, M.J., Eds), Elsevier, Amsterdam (in press).

[9] KLASSEN, W., KNIPLING, E.F., MOGUIRE, J.U., The potential for insect-population suppression by dominant conditional lethal traits, Ann. Entomol. Soc. Am. 63 (1970) 238.

[10] SUZUKI, D.T., Temperature-sensitive mutations in Drosophila melanogaster, Science 170 (1970) 695.

[11] SUZUKI, D.T., PITERNICK, L.K., HAYASHI, S., TARASOFF, M., BAILLIE, D., ERASMUS, U., Temperature-sensitive mutations in Drosophila melanogaster: I. Relative frequencies among γ-ray and chemically induced sex-linked recessive lethals and semilethals, Proc. Natl. Acad. Sci. U.S. 57 (1967) 907.

[12] BAILLIE, D., SUZUKI, D.T., TARASOFF, M., Temperature-sensitive mutations in Drosophila melanogaster: II. Frequency among second chromosome recessive lethals induced by ethyl methanesulfonate, Can. J. Genet. Cytol. 10 (1968) 412.

[13] SUZUKI, D.T., PROCUNIER, D., Temperature-sensitive mutations in Drosophila melanogaster: III. Dominant lethals and semilethals on chromosome 2, Proc. Natl. Acad. Sci. U.S. 62 (1969) 369.

[14] ROSENBLUTH, R., EZELL, D., SUZUKI, D.T., Temperature-sensitive mutations in Drosophila melanogaster: IX. Dominant cold-sensitive lethals on the autosomes, Genetics 70 (1972) 75.

[15] HOLDEN, J., SUZUKI, D.T., Temperature-sensitive mutations in Drosophila melanogaster: XII. The genetic and developmental effects of dominant lethals on chromosome 3, Genetics 73 (1973) 445.

[16] TASAKA, S.E., SUZUKI, D.T., Temperature-sensitive mutations in Drosophila melanogaster: XVII. Heat- and cold-sensitive lethals on chromosome 3, Genetics 74 (1973) 509.

[17] WHITING, P.W., Mutants in Habrobracon, Genetics 17 (1932) 1.

[18] SMITH, R.H., "Induced conditional lethal mutations for the control of insect populations", Sterility Principle for Insect Control or Eradication (Proc. Symp. Athens, 1970), IAEA, Vienna (1971) 453.

[19] MCDONALD, I.C., OVERLAND, D.E., Temperature-sensitive mutations in the housefly: the characterization of heat-sensitive recessive lethal factors on autosome III, J. Econ. Entomol. 65 (1972) 1364.

[20] MCDONALD, I.C., OVERLAND, D.E., Housefly genetics: II. Isolation of a heat-sensitive translocation homozygote, J. Hered. 64 (1973) 253.

[21] FITZ-EARLE, M., Minimum frequency of compound autosome in Drosophila melanogaster to achieve chromosomal replacement in cages, Genetica (in press).

[22] CANTELO, W.W., CHILDRESS, D., Laboratory and field studies with a compound strain of Drosophila melanogaster, Theor. Appl. Genet. (in press).

[23] FITZ-EARLE, M., HOLM, D.G., SUZUKI, D.T., Population control of caged native fruitflies in the field by compound autosomes and temperature-sensitive mutants, Theor. Appl. Genet. (in press).

[24] SUZUKI, D.T., GRIGLIATTI, T., WILLIAMSON, R., Temperature-sensitive mutations in Drosophila melanogaster: VII. A mutation (para^ts) causing reversible adult paralysis, Proc. Natl. Acad. Sci. U.S. 68 (1971) 890.

[25] LINDSLEY, D.L., GRELL, E.H., Genetic Variations of Drosophila melanogaster, Carnegie Inst. Wash. Publn No. 627 (1968).

[26] GRIGLIATTI, T.A., HALL, L., ROSENBLUTH, R., SUZUKI, D.T., Temperature-sensitive mutations in Drosophila melanogaster: XIV. A selection of immobile adults, Mol. Gen. Genet. 120 (1973) 107.

DISCUSSION

D.W. WALKER: As I interpret your work, you are causing homozygous gene locus change by radiation or other means which is debilitating as far as some aspect of environmental selection is concerned, either directly or indirectly, and, furthermore, the new homozygous mutant is dominant to the wild strain. This is a splendid concept for the laboratory where you have full control over the environment, but in the wild you do not have this degree of control over environmental stress, nor over the selective pressures of the environment. If you introduce a temperature-sensitive mutant into the field, climatic conditions could kill it off before it has had a chance to replace the natural population.

M. FITZ-EARLE: It is clear that if the restrictive temperature level were to be reached before the rearrangement carrying the temperature-sensitive mutant had been completely dispersed into the population, then only partial reduction of the target population would be achieved. To prevent this problem from arising, the most satisfactory strategy would be to conduct ecological studies in parallel with genetic ones. Thus information on such parameters as the best time to release the rearrangements, the period necessary to achieve replacement in the field and the time at which the restrictive temperature could be expected to be reached, would all be well established before a release. Under these circumstances the difficulty should not arise.

I think we should bear in mind that, in general, since all genetic control methods — whether they be the sterile insect technique or the application of chromosomal rearrangements — are essentially counter-evolutionary, our control efforts must be considered as no more than perturbations of the target pest populations.

PRELIMINARY RADIOBIOLOGICAL STUDIES
ON Hylemya antiqua Meigen AND
DATA ON THREE RADIATION-INDUCED
(0.5 krad) CHROMOSOMAL REARRANGEMENTS

A.S. ROBINSON
Association Euratom-Ital,
Wageningen

C. van HEEMERT
Department of Genetics,
Agricultural University,
Wageningen,
The Netherlands

Abstract

PRELIMINARY RADIOBIOLOGICAL STUDIES ON Hylemya antiqua Meigen AND DATA ON THREE RADIATION-INDUCED (0.5 krad) CHROMOSOMAL REARRANGEMENTS.

Using varying doses of X-rays (500-3000 rad), seven-day-old adult male onion flies, Hylemya antiqua, were irradiated and mated to virgin females. In eggs from individual females the percentage egg hatch and the percentage of late embryonic lethals was assessed. Late embryonic lethality could be observed by the brown appearance of the eggs. The sperm of the onion fly is relatively sensitive to the radiation induction of dominant lethal mutations. Three krad gave almost complete sterility. The percentage of late embryonic lethals showed an initial rise with dose to a peak at 1 krad, thereafter there was a decline. Arguments are put forward as to the merits of using a low radiation dose for the induction of chromosomal rearrangements for insect control. To test this males were irradiated with 500 rad and the F_1-progeny were screened for reduced fertility. Out of a total of 74 test-crossed F_1-males and females three rearrangements have been isolated and confirmed cytologically to the present time, two reciprocal translocations and one pericentric inversion. The duplication/deficiency gametes from such rearrangements following fusion with normal gametes lead to late embryonic lethality, and hence produce brown eggs. However, in the two translocation stocks viable duplication/deficiency larvae (7 - 9 days old) have been observed. The fertility and cytology of the three rearrangements are described. Only the female carriers of the inversion showed reduced fertility. The fertility of both translocations was in excess of 50% of the wild-type value. Preliminary inbreeding has been undertaken but as yet no homozygous stock has been established.

INTRODUCTION

Hylemya antiqua Meigen is the main insect pest attacking the onion crop in the Netherlands. In 1972, 7361 ha were planted with onion and the export value of the crop was $28.7 million. Mass rearing [1], ecology (Loosjes, unpublished data), radiobiology and cytology [2] and population modelling [3] of the insect have already been well studied, thus providing an excellent base for a possible genetic control method using radiation-induced chromosomal rearrangements.

The use of chromosomal rearrangements for insect control, specifically translocations, first proposed by Serebrovskii [4], was developed independently by Curtis [5], since when a host of publications both theoretical [6 - 9] and practical [10 - 13] have indicated the potential of the technique.

Chromosomal translocations can function in two interrelated ways in an insect control programme. Firstly, by subjecting the natural population to a high genetic load and secondly, as a transport/replacement system for the incorporation of conditional lethal factors or mutants which render the replacements innocuous.

The degree of genetic load necessary to produce an actual decrease in population numbers is influenced by many factors including density dependent regulation, immigration and emigration and the reproductive capacity of the natural population.

The work reported here is a continuation of the experiments begun by Wijnands-Stäb and van Heemert [2], and the short-term aim is the construction of a strain of Hylemya antiqua through the use of homozygous rearrangements which will generate a high genetic load when released either as a multiple homozygote or heterozygote into a native population.

RELEVANT LABORATORY DATA OF Hylemya antiqua

The generation interval in the laboratory is 4 - 6 weeks. Mated females can produce eggs over a period of 3 - 4 weeks, and up to 600 eggs have been recorded from one female. Following emergence there is a pre-oviposition period of about one week. In control matings the eggs from fertilized females, after incubation at 29°C and 80% humidity for three days, can be differentiated by three biological end points: (a) empty hatched eggs, (b) unhatched "brown" eggs, i. e. late embryonic lethals, and (c) unhatched white eggs, i. e. unfertilized eggs or early embryonic lethals. Following mass mating of control males and females, individually egged females gave the following percentages of eggs in the three categories: $93.4 \pm 3.9\%$, $3.0 \pm 1.3\%$ and $3.5 \pm 2.9\%$ respectively.

DOSE RESPONSE CURVES FOR DOMINANT LETHALITY AND LATE EMBRYONIC LETHALS INDUCED IN X-IRRADIATED SPERM

As mature sperm provides a homogeneous cell sample and is sensitive to the induction of chromosomal rearrangements, seven-day-old adult males were irradiated. The males were treated with varying doses of X-rays at a dose rate of 300 rad/min and mated in mass to an equal number of ten-day-old virgin females for three days. The males were then discarded and the females caged individually. In the eggs from these females the percentage egg hatch and the percentage brown eggs were measured. The results are shown in Fig. 1: the sperm of H. antiqua is relatively radiosensitive compared with other Diptera, i. e. a dose of 3 krad caused almost complete sterility. However, it was possible to rear and mate the few surviving F_1-progeny from the males given 3 krad. (The F_1-generation is defined as the progeny of irradiated males following outcrossing to control females.) The virtual straight line plot of the graph, log percentage egg hatch against dose, indicated that single-hit kinetics were involved for the majority of the dose range. The dose-response curve for the percentage of brown eggs, i. e. late embryonic lethals, showed an initial rise with increasing dose to a peak at 1 krad, thereafter there was a decline because at higher doses it became increasingly probable that an

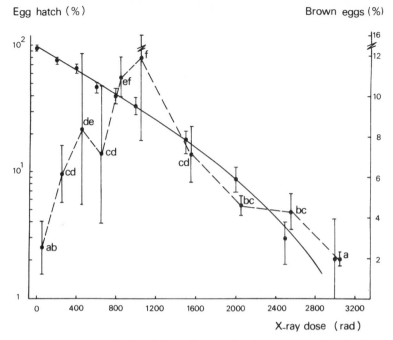

FIG.1.   Dose response curves for dominant lethality (●——●) and late embryonic lethality (●-----●) induced in mature sperm of <u>Hylemya</u> <u>antiqua</u>.   Values followed by the same letter are not significantly different from each other at the 5% level as determined by Duncan's multiple range test.

egg had at least one early acting dominant lethal, which could forestall the expression of later acting embryonic dominant lethals.   There were significant differences in the percentage of brown eggs for the different doses $(F = 5.71^{xxx}$ d.f. 9 and 40).   Von Borstel and Rekemeyer [14] recorded a similar dose response curve for late embryonic lethals following irradiation of <u>Habrobracon</u> sperm, although the peak was at 4 krad.   The brown appearance of the eggs is probably produced by the tyrosinase system which in <u>Drosophila</u> is active in embryos older than six hours [15].

THE CHOICE OF A RADIATION DOSE FOR THE INDUCTION OF CHROMOSOMAL REARRANGEMENTS

Radiobiological data indicate that the higher the dose of radiation given to the parental generation the higher the frequency of rearrangements recovered in the F_1-generation.   However, for insect control purposes, quality is of more importance than quantity and one of the most important aspects (for genetic control) of quality is the fitness of the rearrangement as a homozygote.   In most insect species so far studied a large proportion of induced translocations are either lethal when made homozygous or show severe fitness reductions, e.g. <u>Drosophila</u> [16, 17], <u>Aedes</u> <u>aegypti</u> [18]

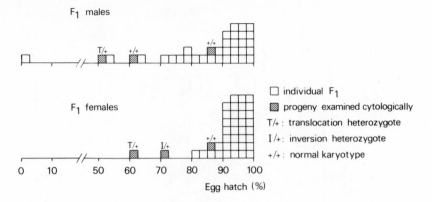

FIG.2. Fertility of test-crossed male and female Hylemya antiqua F_1's following radiation of parental males, with 500 rads of X-rays.

and Lucillia cuprina [19]. However, on the positive side there are reports in pest insects of viable translocation homozygotes [20 - 22]. The reduced viability of homozygous translocations can be due to position effects or damage at the translocation breakpoints, or to radiation-induced or naturally occurring recessive lethals. Sobels [17] has calculated the relative importance of these different aspects with translocations in Drosophila. It has been claimed that by a series of backcrosses, radiation-induced recessive lethals can be removed, but it is highly improbable that genetic engineering will remove the effect due to a true position effect or damage at the translocation breakpoint. Relevant to this argument is the observation that in maize [23] and Drosophila [24] crossing-over is reduced in the region of a translocation breakpoint, and it is conceivable that the absence of close pairing during meiosis in the region of the breakpoint would make extremely difficult the removal, by backcrossing, of recessive lethals within this particular chromosomal segment. The higher the dose of radiation used the higher is the probability that a recessive lethal would be included in the non-crossover region. Two other points indicated that, at least as far as the onion fly is concerned, the removal of recessive lethals by backcrossing would not be a worthwhile procedure.

1. In H. antiqua males there is no recombination, and in females there are as yet no data on the frequency of recombination.

2. With a generation time of six weeks, backcrossing is an extremely laborious and time-consuming process.

In a second experiment with these considerations in mind a very low radiation dose, 500 rads X-rays given to seven-day-old males, was used; it was considered that this dose would give a detectable frequency of translocated F_1-individuals, and that it was rather improbable that the same individuals would carry many induced recessive lethals. From Fig.1, it can be assessed that 500 rads would lead to about 50% egg hatch in the eggs fertilized by sperm from the irradiated males.

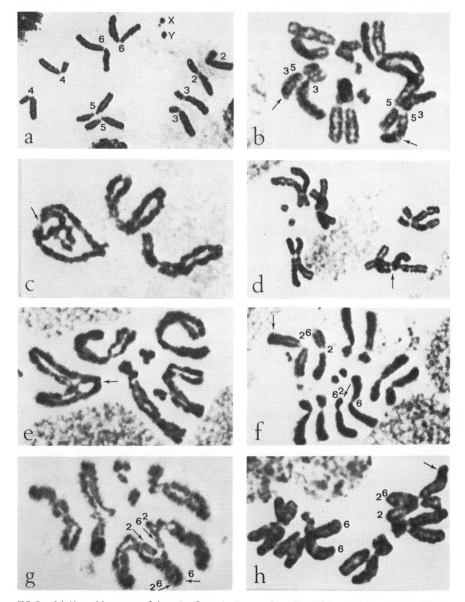

FIG.3. (a) Normal karyotype of the onion fly, mitotic metaphase (2n = 12) in a larval brain cell; (b) Mitotic
metaphase of a translocation heterozygous larva of T(3, 5) 6, arrows indicate the translocated arms;
(c) Mitotic prophase of an inversion heterozygote of In(6) 1 in a larval brain cell, arrow indicates position
of the centromeres of chromosome pair 6, cell incomplete: note the loop; (d) Mitotic metaphase of an
inversion heterozygous larva, arrow indicates the position of the two centromeres of chromosome pair 6,
note the disturbed somatic pairing in this pair (arms curved); (e) Diakinesis/prometaphase in the testes of
an inversion heterozygous male, arrow indicates the position of the centromeres of chromosome pair 6,
note the loop; (f) Spermatogonial metaphase of a translocation heterozygous male of T (2,6) 5, arrows indicate
the translocated arms; (g) Multivalent in a diakinesis/prometaphase stage of a translocation heterozygous
male of T (2, 6) 5, arrows indicate the paired arms involved in the translocation; (h) Mitotic metaphase of a
larva with a large duplication (see arrow) in chromosome 2.

TABLE I.   FERTILITY AND SEGREGATION RATIO OF THREE
RADIATION-INDUCED REARRANGEMENTS IN H. antiqua WHEN
TEST CROSSED TO CONTROL INSECTS

	Female		Male	
	Fertility (% Egg hatch)	Segregation ratio +/+ : HET	Fertility (% Egg hatch)	Segregation ratio +/+ : HET
T (3, 5) 6	62.2± 9.2	10 : 12	61.7 ± 7.1	11 : 6
T (2, 6) 5	81.9 ± 10.9	9 : 12	74.9 ± 5.6	11 : 13
In (6) 1	73.9± 6.7	30 : 31	97.4 ± 2.8	49 : 48
Control	96.7± 3.1			

RECOVERY OF REARRANGEMENTS IN THE F_1-GENERATION

There are no marker genes available in Hylemya antiqua, so initially
the presence of rearrangements is ascertained by reduced fertility in outcrosses of F_1-individuals to control insects.   It is also known that the
duplications and deficiencies from translocations can act as late embryonic
lethals and hence produce brown eggs [25].

Since single pair matings are not very successful in Hylemya antiqua,
the following mating techniques were used; F_1-females were confined in
mass with control males and subsequent to mating they were placed in
individual cages; F_1-males were placed individually in separate cages with
three control females.   Using these techniques it is possible that the
females do not always receive a full supply of sperm   (especially when
F_1-males are tested), and therefore a varying proportion of unfertilized
eggs may be laid.   Consequently, if the percentage egg hatch is calculated
from the total number of eggs laid, then large variations can occur.   To
reduce the variation the white unhatched eggs were deducted from the total
before calculating the egg hatchability.   Figure 2 shows the distribution
of the percentage egg hatch for 74 F_1's test-crossed to control insects.
Larvae from six stocks were examined cytologically for the presence
or absence of rearrangements in mitotic preparations.

As a routine procedure, larval brains of 7 - 9-day-old larvae were used
for cytological analysis.   Techniques were as described previously [2].
Somatic pairing makes it easy to identify homologous chromosomes (Fig.3a).
It also facilitates the detection of differences in length between the normal
and the rearranged chromosomes.   Mitotic prophases, because of their
pronounced telomeric pairing, were also used.   Meiotic pairing was studied
in diakinesis/prometaphase stages in the testes of young males in order
to obtain the critical evidence for the presence of rearrangements.   The
identification of rearrangements in mitotic preparations greatly increases
the efficiency of the screening process.   It would also be valuable in field
experiments.   As indicated in Fig.2, three rearrangements were identified,

two reciprocal translocations and one pericentric inversion. However, the progeny of one semi-sterile male showed no visible aberration. It is possible in this case that the exchanged segments were symmetrical and/or that the exchanged segments were very small. Both these conditions make it impossible for the translocations to be observed in mitotic preparations.

DATA ON THE THREE REARRANGEMENTS

The fertility and the segregation ratio of the three rearrangements are shown in Table I.

T (3, 5) 6

This was isolated in the progeny of an F_1-female. The long arm of chromosome 3 lost a large segment and gained a short segment from the short arm of chromosome 5 (Fig. 3b). The arm ratio in a normal chromosome 5 is 1.4, and the arm ratio of the translocated chromosome 5^3 is 1.0. The relatively small new chromosome 3 was used as a marker in an inbreeding programme for the isolation of homozygotes. In this stock duplication deficiency karyotypes (3 3 5^3 5) were regularly found in the larval stage.

There was no difference in the fertility of males and females carrying this stock when they were outcrossed to control insects. The fertility was in excess of 50% of the control value that would be expected following random alternate and adjacent segregation from a translocation heterozygote. It indicates that there was either preferential alternate segregation as found in Blatella germanica [26] and in males of Cochliomyia hominivorax [27], or survival to the larval stage of a large proportion of the duplication/deficiency karyotypes. In Glossina austeni [21] and Musca domestica [28] all translocations appear to have a fertility very close to the expected value of 50%. The reasons for these species differences are not clear.

Preliminary inbreeding work has been carried out in order to produce a homozygous stock. Following inbreeding in mass of a translocation stock it is expected that a proportion of the matings will be between heterozygotes and consequently translocation homozygotes can be generated. If the only viable zygotes from these matings are produced from genetically balanced sperm and eggs, and if the translocation homozygote is not egg-lethal, then the expected fertility of such crosses (from Table I) would be

$$\frac{62.2 \times 61.7}{100} = 38.4\%.$$

However, the preliminary data show a much higher fertility (50 - 55%) of such sib-crosses indicating that a proportion of unbalanced sperm and eggs complement each other's duplications and deficiencies and so produce viable genotypes [29]. Using the translocated 3rd chromosome as a marker, the translocation homozygote karyotype has been observed in the larval stage but as yet not in the adult stage. This fact together with the observation that the percentage pupation in such crosses is reduced might indicate larval lethality of the translocation homozygous genotype. Only a small number of sib-crosses have yet been tested and a much larger inbreeding programme is underway.

In (6) 1

This inversion was isolated in the progeny of an F_1-female showing a percentage egg hatch of 76%. It is a pericentric inversion, i.e. the centromere is included within the inverted segment.

In late mitotic prophases a ring configuration can be seen in chromosome pair 6 (Fig.3c). Mitotic metaphases showed the presence of a typical inversion configuration in chromosome pair 6 in that both arms of one of the pair always have a curve in about the middle (Fig.3d). An explanation of these observations can be given by considering the way somatic pairing is acting during the mitotic cycle. Centromeric pairing is of equal strength in prophase and metaphase. However, telomeric pairing is much stronger in the prophase. During the transition from prophase to metaphase the homologous areas around the centromeres proximal to the inversion break-points stay paired. Subsequently the chromosome ends distal to the break-point begin to loose their pairing because of the breakdown of telomeric attraction. In achiasmate male meiosis, pairing is rather complete over the total chromosome length, and as expected a loop is observed in the diakinesis/prometaphase stages (Fig.3e).

Two observations would indicate that the inversion breakpoints are equidistant from the centromere. Firstly, the arm ratio of the normal and inverted chromosome 6 is the same and, secondly, the centromere is positioned in the middle of the inversion loop (Fig.3c, e). Rough estimations indicate that the breakpoint positions are at the middle of the short arm and at one third of the length of the long arm from the centromere.

As indicated in Fig.3e, linear pairing during the meiotic sequence is achieved by the formation of an inversion loop. Crossing-over within the loop leads to the formation of duplication/deficiency gametes [30], and hence to a reduction in fertility. Such cross-over products have been cytologically observed in the eggs from test-crossed females of this inversion stock. Fertility reductions implicating inversions have also been observed in Aedes aegypti [31], Culex pipiens [32], Culex tritaeniorhynchus [33] and Musca domestica [34].

With In (6) 1, only the female carriers of the inversion exhibit reduced fertility, the male inversion heterozygotes having normal fertility (Table I). We conclude that this is strong evidence for the absence of crossing-over in the male of the onion fly: this is in agreement with data for other Cyclorrhaphid Diptera [35, 36], and confirms the previous observation of achiasmate male meiosis [37]. Both sexes showed the expected 1 : 1 segrega-tion of the inversion and wild-type gametes as determined by cytological analysis of the larvae from test-crosses. Inbreeding has been carried out with this stock in an attempt to obtain the inversion as a homozygote. However, there are three difficulties apparent. Firstly, as the inversion heterozygous male does not show reduced fertility it is impossible to differentiate the matings between the inversion females and wild-type males from those between inversion males and females. However, if the inversion homozygote is egg lethal then the fertility of the latter matings would be reduced by an additional 25%. Secondly, as the only gametes which are recovered from an inversion female are non-crossover types, it is impossible to remove radiation-induced or naturally occurring recessive lethals within the inverted segment. Thirdly, as indicated above, the inversion homozygote cannot be differentiated cytologically from the wild-type karyotype.

T (2, 6) 5

This was isolated in the progeny of an F_1-male, and the shortest and largest autosomes are involved. The short arm of chromosome 6 lost a large piece and received a small piece from the short arm of chromosome 2. Both of the translocated chromosomes can be easily differentiated from the other chromosomes in the karyotype (Fig.3f, g). Translocation homozygotes, if viable, will be very easy to discriminate from translocation heterozygotes and normal karyotypes by looking for the presence of two, one or none translocated chromosomes 6.

Further evidence for the asymmetry of the exchange can be obtained by the occurrence of duplication karyotypes $22^6 66$ in the larval stage (Fig.3h). Such individuals have a duplication for chromosome 6 and a very small deficiency for chromosome 2. The survival of such duplication types is perhaps the reason why the observed percentage egg hatch of this translocation is high (see Table I). Using the translocated chromosome 6 as a masker, this translocation could be observed in the late larval stage as a homozygote, but as yet not in the adult stage.

CONCLUSIONS

Several tentative conclusions can be drawn from these preliminary studies.

1. Chromosome rearrangements useful for genetic control can be induced by low doses of radiation. However, because of the small number of rearrangements so far tested for homozygous viability, a conclusion as to the merits of a low versus a high dose of radiation has still to be established.

2. Cytological observation of mitotic chromosomes in larval preparations greatly reduces the time involved in the isolation of rearrangements. In order to use this technique a translocation involving the exchange of asymmetrical pieces is necessary to differentiate between the translocation homozygote, the translocation heterozygote and the wild-type karyotype.

3. Because of the survival of duplication/deficiency karyotypes to the late larval stage, it is important for control of the onion fly that translocations are used which involve the exchange of large pieces of chromosome.

4. The use of inversions for exerting a genetic load in a wild population would appear to be limited as the males do not show reduced fertility because of the absence of crossing-over.

5. When inbreeding to produce homozygotes as large a number of individuals as possible should be used in order to maximize the size of the inevitable genetic bottleneck through which the homozygotes must pass.

ACKNOWLEDGEMENTS

We thank C. F. Curtis and J. Sybenga for their critical reading of the manuscript. The technical assistance of G. Schelling and W. van de Brink, and the photographic skill of K. Knoop, are also acknowledged.

REFERENCES

[1] TICHELER, J., "Rearing of the onion fly, Hylemya antiqua (Meigen), with a view to release of sterilized insects", Sterility Principle for Insect Control or Eradication (Proc. Symp. Athens, 1970), IAEA, Vienna (1971) 341.

[2] WIJNANDS-STÄB, K.J.A., van HEEMERT, C., Radiation induced semi-sterility for genetic control purposes in the onion fly Hylemya antiqua (Meigen): I. Isolation of semi-sterile stocks and their cytogenetical properties, Theor. Appl. Genet. 44 (1974) 111.

[3] WIJNANDS-STÄB, K.J.A., FRISSEL, M.J., "Computer simulation for genetic control of the onion fly Hylemya antiqua (Meigen)", Computer Models and Application of the Sterile-Male Technique (Proc. Panel, Vienna, 1971), IAEA, Vienna (1973) 95.

[4] SEREBROVSKII, A.S., On the possibility of a new method for the control of insect pests, Zool. Zh. 19 (1940) 618 (In Russian).

[5] CURTIS, C.F., Possible use of translocations to fix desirable genes in insect pest populations, Nature (London) 218 (1968) 368.

[6] CURTIS, C.F., HILL, W.G., Theoretical studies on the use of translocations for the control of tsetse flies and other disease vectors, Theor. Pop. Biol. 2 (1971) 71.

[7] CURTIS, C.F., ROBINSON, A.S., Computer simulation of the use of double translocations for pest control, Genetics 64 (1971) 97.

[8] McDONALD, P.T., RAI, K.S., Population control potential of heterozygous translocations as determined by computer simulations, Bull. World Health Organ. 44 (1971) 824.

[9] WHITTEN, M.J., Insect control by genetic manipulation of natural populations, Science 171 (1971) 682.

[10] LAVEN, H., COUSSERANS, J., GUILLE, G., Eradicating mosquitoes using translocations: a first field experiment, Nature (London) 236 (1972) 456.

[11] WAGONER, D.E., JOHNSON, O.A., NICKEL, C.A., Fertility reduced in a caged native house fly strain by the introduction of strains bearing heterozygous chromosomal translocations, Nature (London) 234 (1971) 473.

[12] ROBINSON, A.S., CURTIS, C.F., Controlled crosses and cage experiments with a translocation in Drosophila, Genetica 44 (1973) 591.

[13] RAI, K.S., GROVER, K.K., SUGUNA, S.G., Genetic manipulation of Aedes: incorporation and maintenance of a genetic marker and a chromosomal translocation in natural populations, Bull. World Health Organ. 48 (1973) 44.

[14] VON BORSTEL, R.C., REKEMEYER, M.L., Radiation induced and genetically contrived dominant lethality in Habrobracon and Drosophila, Genetics 44 (1959) 1053.

[15] HANLY, E.W., Lack of melanin formation in Drosophila embryo extract, Drosoph. Inf. Serv. 39 (1964) 100.

[16] PATTERSON, J.T., STONE, W., BEDICHEK, P.S., SUCHE, M., The production of translocations in Drosophila, Am. Nat. 68 (1934) 354.

[17] SOBELS, F.H., The viability of II-III translocations in homozygous condition, Drosoph. Inf. Serv. 48 (1972) 117.

[18] RAI, K.S., McDONALD, T.P., ASMAN Sr. M., Cytogenetics of two radiation induced, sex-linked translocations in the yellow fever mosquito Aedes aegypti, Genetics 66 (1970) 635.

[19] FOSTER, G.G., WHITTEN, M.J., "The development of genetic methods of controlling the Australian sheep blowfly, Lucilia cuprina", The Use of Genetics in Insect Control (PAL, R., WHITTEN, M.J., Eds.), Elsevier, Amsterdam (in press).

[20] LORIMER, N., HALLINAN, E., RAI, K.S., Translocation homozygotes in the yellow fever mosquito, Aedes aegypti, J. Hered. 63 (1971) 159.

[21] CURTIS, C.F., SOUTHERN, D.I., PELL, P.E., CRAIG-CAMERON, T.A., Chromosome translocations in Glossina austeni, Genet. Res. 20 (1972) 101.

[22] McDONALD, I.C., OVERLAND, D.E., House fly genetics: II. Isolation of a heat-sensitive translocation homozygote, J. Hered. 64 (1973) 253.

[23] BURNHAM, C.R., Chromosomal interchanges in maize: Reduction of crossing over and the association of non-homologous parts, Am. Nat. 68 (1934) 81.

[24] DOBZHANSKY, T., The decrease of crossing-over observed in translocations and its probable explanation, Am. Nat. 65 (1931) 214.

[25] HEEMERT, C. van, Isolation of a translocation homozygote in the onion fly Hylemya antiqua with a cytogenetic method in combination with the determination of the percentage of late embryonic lethals, Genen Phaenen 16 (1973) 17.

[26] ROSS, M.H., COCHRAN, D.G., German cockroach genetics and its possible use in control measures, Patna J. Med. 47 (1973) 325.

[27] LaCHANCE, L.E., RIEMANN, J.G., HOPKIN, D.E., A reciprocal translocation in Cochliomyia hominivorax: Genetical and cytogenetical evidence for preferential segregation in males, Genetics 49 (1966) 959.

[28] WAGONER, D.E., NICKEL, C.A., JOHNSON, O.A., Chromosomal translocation heterozygotes in the house fly, J. Hered. 60 (1969) 301.

[29] MULLER, H.J., SETTLES, F., The non-functioning of genes in spermatozoa, Z. Indukt. Abstamm.- u. VererbLehre 43 (1927) 285.

[30] ALEXANDER, M.L., The effect of two pericentric inversions upon crossing-over in Drosophila melanogaster, Univ. Texas Publ. 5204 (1952) 219.

[31] McGIVERN, J.J., RAI, K.S., A radiation induced paracentric inversion in Aedes aegypti (L.): Cytogenetic and interchromosomal effects, J. Hered. 63 (1972) 247.

[32] JOST, E., LAVEN, H., Meiosis in translocation heterozygotes in the mosquito Culex pipiens, Chromosoma 35 (1971) 184.

[33] BAKER, R., SAKAI, R.K., MIAN, A., Linkage group-chromosome correlation in a mosquito: Inversions in Culex tritaeniorhynchus, J. Hered. 62 (1971) 31.

[34] McDONALD, I.C., The isolation of cross-over suppressors on house fly autosomes and their possible use in isolating recessive genetic factors, Can. J. Genet. Cytol. 12 (1970) 860.

[35] LaCHANCE, L.E., DAWKINS, C., HOPKINS, D.E., Mutants and linkage groups of the screw worm fly, J. Econ. Entomol. 59 (1966) 1493.

[36] WAGONER, D.E., Linkage group-karyotype correlation in the house fly determined by cytological analysis of X-ray induced translocations, Genetics 57 (1967) 729.

[37] HEEMERT, C. van, Meiotic disjunction, sex determination and embryonic lethality in an X-linked 'simple' translocation in the onion fly, Chromosoma (in press).

DISCUSSION

R. PAL: Have you been able to obtain homozygous translocation lines and will you be looking into double translocation heterozygotes?

A.S. ROBINSON: We have not yet obtained a homozygous stock for a translocation. We have, however, observed homozygous individuals both as larvae and adults in several translocation lines.

L.E. LaCHANCE: Could you please explain how you got around the problem of poor mating performance in single pair matings in Hylemya?

A.S. ROBINSON: For test mating of females the F_1's were mated in mass with control males for two days. Subsequently the females were caged individually.

For test mating of males the F_1's were mated with three control females in individual cages.

G. DELRIO: Do you think it will be easy to obtain translocations to the homozygous state? In Ceratitis capitata one can obtain several heterozygous translocations but these do not seem viable in the homozygous state.

A.S. ROBINSON: I agree that obtaining translocations to the homozygous state is a crucial problem. We hope to obtain a higher proportion of such homozygous lines by the use of a low dose.

R. PAL: It is also our experience with Culex fatigans that higher radiation doses produce lethals and that it is difficult to obtain homozygous translocation lines. Exposure to lower levels of radiation is likely to be more productive.

CYTOLOGICAL STUDY OF GAMMA-IRRADIATED TESTES OF THE HOUSE FLY, Musca domestica L.

R. H. JAFRI, J. A. DAR
Zoology Department,
Panjab University,
Lahore, Pakistan

Presented by Z.A. Qureshi

Abstract

CYTOLOGICAL STUDY OF GAMMA-IRRADIATED TESTES OF THE HOUSE FLY, Musca domestica L.
House flies, Musca domestica L., were reared at 29° ± 2°C and 70 ± 5% r.h. Three dose levels, 2 krad, 2.5 krad and 3 krad, were used at 24 ± 6 hours before eclosion of pupae and 24 hours after emergence. The testes were examined by histological methods following hematoxylin staining, at hourly intervals for seven hours and 2, 4, 8, 10 and 12 days post irradiation. Cytological changes in all the irradiated groups showed a progressive pattern. At first, the nuclei of germ cells became pyknotic, then the vacuoles appeared in the nuclei. These lesions were followed by karyorrhexis of nuclei and degeneration of germ cells. Ultimately development of vacuoles in place of normal germ cells was thus a major phenomenon in the irradiated testes. Sperms became thick and twisty or rod-like 4-5 days after irradiation. The sperm bundles were also seen in a broken state almost at the same time. These changes coincided with dose and time. The radiosensitivity of germ cells was deduced from cytological changes observed at different time intervals. The early and the final stages of spermatogonia, spermatocytes and spermatids were not observed at one hour following exposure to all the three dose levels, whereas they were observed in the control group. These observations show that the early and the final stages were more radiosensitive than their respective intermediate stages. Moreover, the germ cells (spermatogonia, spermatocytes, spermatids and mature sperms) followed a decreasing pattern of radiosensitivity in that order. A dose of 3 krad was found to be most effective for producing sterility, particularly when the house flies were treated with this dose at the pupal stage. This dose was sufficient to suppress all the spermatogenic activity, at least up to 12 days after irradiation. Thus, for sterility purposes, the house flies should be irradiated with 3 krad of gamma rays at the pupal stage. The house flies recovered their fertility on 10th and 12th day following 2 krad and 2.5 krad of irradiation respectively. The immediate effect of the three dose levels of gamma rays was found to be stimulatory. This was evidenced by the fact that early stages of spermiogenesis were not observed at one hour following irradiation, and instead, the late stages of spermiogenesis were observed. This stimulatory effect was also observed in the mating behaviour of the house flies. The irradiated house flies began mating two to three days after emergence, whereas flies of the control group normally began mating four to five days after emergence.

1. INTRODUCTION

In a study of the sterilization of insects by irradiation, the fate of the spermatogonia has theoretical and practical importance. Since the primary spermatogonia are the stem cells of the testes, their complete destruction results in complete sterilization and the production of sperm ceases [1]. It has been worked out [1] that if some of the primary spermatogonia survive after treatment, they begin a new wave of sperm production. The fact that these cells can pass through the repeated mitotic and meiotic divisions necessary for sperm production, indicates that the recovery of fertility can be expected, as sperms derived from these surviving

spermatogonia become available for ejaculation. Such a recovery of
fertility has been observed in some species of Drosophila [2]. Thus, when
attempts are made to sterilize insect pests, a good deal of attention should
be paid to the fate of spermatogonia and sperms produced by it.

The most satisfactory way to check radiation-induced spermatogonial
death is the cytological examination of testes. Such observations have been
made for a limited number of insect species [2-7]. The purpose of this
study is to elucidate the cytological changes in the testes of the adult house
fly, Musca domestica L., exposed to different doses of gamma radiation.

2.   MATERIAL AND METHODS

The house fly, Musca domestica L., was collected from various
localities in Lahore, Pakistan. The larvae were reared on an artificial
medium [8], modified by the authors for the local conditions of Lahore.
The modified artificial medium for the larvae contained the following
ingredients:

Dried milk powder	45 g
Agar agar	20 g
Yeast powder	20 g
Beef extract	10 g
Molasses	2 g
Sodium benzoate	3 g
Distilled water	200 cm^3

The adults of the house fly, Musca domestica L., were fed on dried
milk powder, molasses and minced meat mixed in a ratio of 2:1:2. The
larvae were reared in sterilized test tubes and glass jars. The cages,
18 in × 18 in × 18 in, made of iron wire and cloth, were used for housing
the adult house flies. All the experiments were carried out at 29 ± 2°C
and 70 ± 5% r.h.

The pupae and adults of house fly were irradiated in the laboratories of
the Nuclear Institute of Agriculture and Biology, Atomic Energy Commission,
Lyallpur, Pakistan. The pupae and adults of the house fly were exposed
to doses of 2 krad, 2.5 krad and 3 krad of ^{60}Co gamma rays at a dose rate
of 4 krad per minute. The pupae were irradiated at 24 ± 6 hours before
eclosion. The adults were irradiated at 24 ± 6 hours after emergence.
Ten to fifteen testes were examined, using histological methods [9], at
intervals of one hour for seven hours and 2, 4, 8, 10 and 12 days post-
irradiation.

3.   RESULTS AND DISCUSSION

3.1. Control group

The present study shows that the cell types and the sequence of cell
proliferation in the anterior, middle and posterior region of the testes of
the house fly are similar to those described for Cochliomyia hominovorax [10]

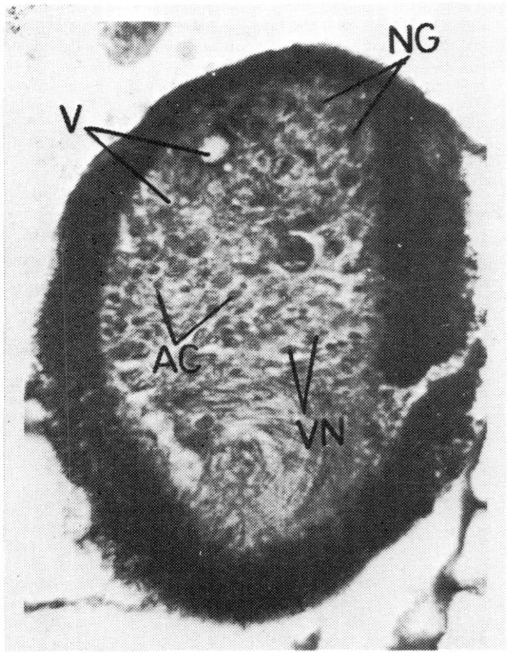

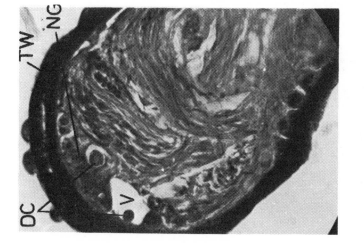

FIG.2. Longitudinal section of the testis of a 3-day-old house fly (4 days after 2 krad irradiation) showing cavities or vacuoles (V) beneath the testicular wall (TW), and degenerating germ cells (DC) and normal spermatogonia (NS) beneath the testicular wall (1000 ×).

FIG.1. Longitudinal section of the testis of a 7-day-old house fly (6 days after 2 krad irradiation) showing normal spermatogonia (NG), abnormal cells (AC) with vacuolated nuclei (VN) and vacuoles (V) (1000 ×).

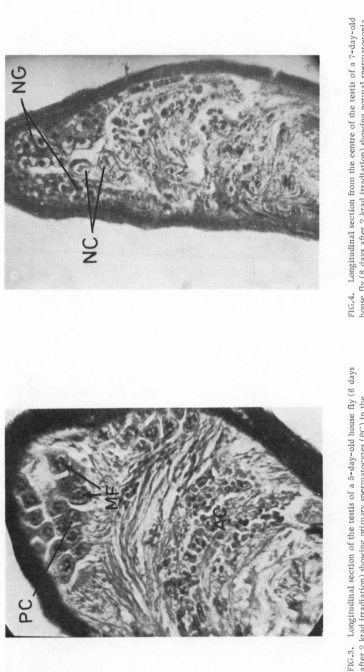

FIG.4. Longitudinal section from the centre of the testis of a 7-day-old house fly (8 days after 2 krad irradiation) showing normal spermatogonia (NG), and normal spermatocytes (NC) in the anterior region (1000 ×).

FIG.3. Longitudinal section of the testis of a 5-day-old house fly (6 days after 2 krad irradiation) showing primary spermatocytes (PC) in the anterior region, meiotic figures (MF) in primary spermatocytes, and abnormal cells (AC) of different size in the middle region (1000 ×).

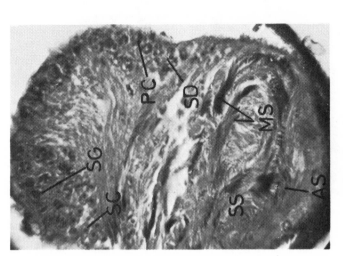

FIG.6.   Longitudinal section from the centre of an 11-day-old house fly (10 days after 2.5 krad irradiation) showing all the major stages of spermatogenesis: secondary spermatogonia (SG), primary spermatocytes (PC), secondary spermatocytes (SC), spermatids (SD) and mature sperms (MS); a few abnormal sperms (AS) are present in the sperm sac (SS) (1000 ×).

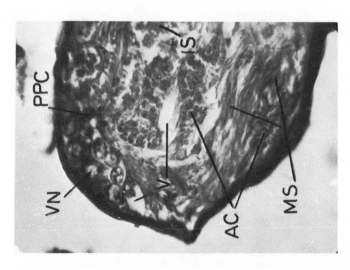

FIG.5.   Longitudinal section from the centre of the testis of a 24-hour-old house fly (one hour after 2.5 krad irradiation) showing primary spermatocytes with pyknotic (PPC) and vacuolated nuclei (VN) and vacuoles (V) or cavities beneath the testicular wall, and immature (IS) and mature sperms, vacuoles (V) and abnormal cells (AC) in the middle region (1000 ×).

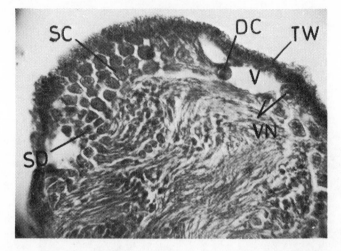

FIG. 7. Longitudinal section from the centre of the testis of a 24-hour-old house fly (48 hours after 2.5 krad irradiation) showing cavities or vacuoles (V) beneath the testicular wall (TW), degenerating cells (DC), primary spermatocytes with vacuolated nuclei (VN), secondary spermatocytes (SC) slightly larger in size, spermatids (SD) and immature sperms (1000 ×).

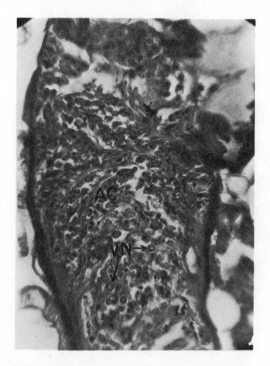

FIG. 8. Longitudinal section from the centre of the testis of a 3-day-old house fly (4 days after 2.5 krad irradiation) showing abnormal cells (AC) in the middle region; the nuclei of these cells are vacuolated (VN) (1000 ×).

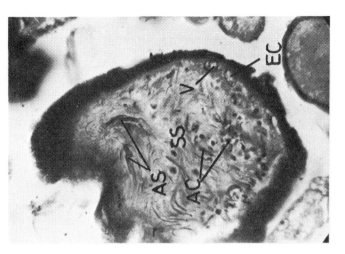

FIG.10. Longitudinal section from the lateral side of the testis of a 7-day-old house fly (8 days after 2.5 krad irradiation) showing abnormal cells (AC) and abnormal sperms (AS) intermixed with each other; and vacuoles (V) in the epithelial cells (EC) of the posterior lining of sperm sac (SS) (1000 ×).

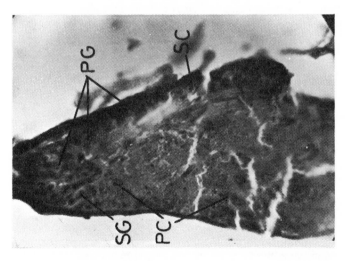

FIG.9. Longitudinal section from the centre of the testis of an 11-day-old house fly (12 days after 2.5 krad irradiation) showing primary spermatogonia (PG), secondary spermatogonia (SG), primary spermatocytes (PC) and secondary spermatocytes (SC) (1000 ×).

3.2. Treatment with 2 krad

When the house flies were irradiated as adults 24 hours after emergence, the spermatogonia appeared normal throughout the experiment. The spermatocytes showed some cytological changes on the second day following irradiation. These changes included pyknosis of the nuclei and vacuolization in the interstitial space. The sperm bundles were also in a broken state (Fig.1). These cytological changes were visible up to the 6th day following irradiation. After this period the testes appeared normal.

In cases where the house flies were irradiated as pupae, the gonial cells showed marked radiation damage 48 hours following irradiation. The cytological changes included vacuolization in nuclei, and liquefaction of cell mass resulting in cavities beneath the germinal epithelium (Fig.2). These changes were visible up to 4th day following irradiation. After this period the testis appeared normal with the exception of a few abnormal sperms present in the sperm sac (Figs 3 and 4).

3.3. Treatment with 2.5 krad

The spermatogonia showed no apparent cytological damage up to 24 hours in cases where the house flies were irradiated as adults. However, a few cytological lesions were visible 48 hours after irradiation. Spermatogonia were not visible on the 4th and 6th day post-irradiation. Instead of spermatogonia, cavities were visible beneath the germinal epithelium. Spermatogonia were again visible on the 10th day following irradiation. Spermatocytes showed abnormalities such as pyknosis, vacuolization and karyorrhexis of the nuclei at successive periods, starting from one hour post-irradiation to 6 days post-irradation (Fig.5). Normal spermatocytes were visible on the 10th day post-irradiation (Fig.6).

In cases where the house flies were irradiated as pupae, abnormal germ cells were present at all stages of observation. These abnormal cells could not be identified. It was difficult to ascertain whether they were spermatogonia or spermatocytes. Their cytological abnormalities included vacuolization in the nuclei, karyorrhexis, liquefication of cell mass and vacuolization beneath the germinal epithelium as well as in the interstitial space (Figs 7 and 8). On the 12th day post-irradiation as pupae, normal spermatogonia and spermatocytes were observed (Fig.9).

The sperms were also abnormal on the 2nd day post-irradiation as adults and pupae. The sperm bundles were seen in a broken state. The sperms were thick, curved or rod-like in appearance (Fig.10).

3.4. Treatment with 3 krad

The spermatogonia were partially present up to the 2nd day post-irradiation as adults. Cytological abnormalities were not visible in these spermatogonia. Spermatogonia were not visible in the testes from the 2nd to the 12th day post-irradiation as pupae. Instead of spermatogonia, cavities were seen beneath the germinal epithelium as well as in the interstitial space. During this period abnormal and degenerating germ cells of various dimensions with vacuolated nuclei were also visible.

Some spermatocytes were visible up to the 4th day post-irradiation as adults as well as pupae. They showed cytological abnormalities such

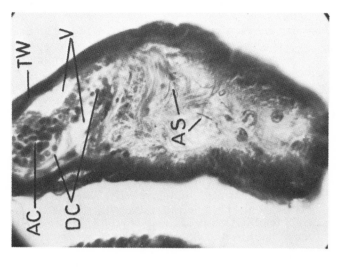

FIG.12.   Longitudinal section from the lateral side of the testis of an
11-day-old house fly (12 days after 3 krad irradiation) showing vacuoles (V)
or cavities beneath the testicular wall (TW), abnormal cells (AC),
degenerating germ cells (DC) in the anterior half and abnormal sperms (AS)
in the posterior half of the section (1000 ×).

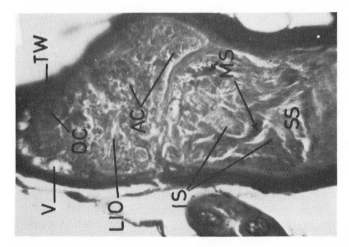

FIG.11.   Longitudinal section from the centre of the testis of a 5-day-old
house fly (6 days after 3 krad irradiation) showing a layer of vacuoles (V)
beneath the testicular wall (TW), a few degenerating cells (DC) in the
anterior region; abnormal cells (AC) are intermixed with various stages
of spermiogenesis (LIO) in the middle region, immature sperms (IS) and
mature sperms (MS) in the sperm sac (SS) (1000 ×).

as pyknosis of the nuclei, vacuolization in the nuclei, liquefaction of cell
mass and cavities in the interstitial space. These spermatocytes were not
visible after the 4th day post-irradiation as adults as well as pupae. Instead
of spermatocytes, many abnormal and degenerating germ cells and vacuoles
were observed (Fig.11).

Mature sperms also showed cytological abnormalities in both cases
when irradiated in the adult as well as in the pupal stage. The sperm
bundles were seen in a broken state. The mature sperms were thick, twisty
or rod-like in appearance (Fig.12).

The irradiated house flies began their mating two to three days after
emergence. The house flies of the control group began their mating four
to five days after emergence.

4. THE RECOVERY OF FERTILITY

The recovery of fertility is a common phenomenon observed after
irradiation. The word recovery used in the present work actually refers
to pseudo-recovery [11]. It has been demonstrated that the males of treated
insects recovered their fertility and became as fertile as the controls [1].

In the present study, all kinds of cytological abnormalities such as
pyknosis, vacuolization in the nuclei, karyorrhexis, liquefaction of cell
mass and cavities just beneath the germinal epithelium (Figs 1, 5, 7) were
observed up to the 7th day following irradiation doses of 2 krad and
2.5 krad. It appears that, because of these cytological abnormalities, the
reproductive power of the testes was suppressed up to the 7th day following
irradiation. The irradiated testes recovered their spermatogenic activity
afterwards. Thus, apparently normal spermatogonia and spermatocytes
were observed on the 10th day following irradiation as pupae (Fig.6).

5. STERILITY

From the evidence cited above and cytological lesions in the testes
(Figs 9,12) it is clear that 2 krad and 2.5 krad doses of gamma rays are
not sufficient to produce complete sterility in the house fly. This statement
is made on the basis of observations that 12 days after exposure of testes
to 2 krad and 2.5 krad, there was recovery of spermatogenic activity.

The recovery of spermatogonia was not visible up to 12 days following
3 krad irradiation (Fig.12). It therefore appears that the spermatogenic
activity of the testes was completely suppressed after exposure to 3 krad
of gamma rays. In other words, a complete sterility was achieved at least
up to 12 days after irradiation. Previously it has been suggested that a
dose of 2850 R of X-rays [12] or a dose of 2500 R of X-rays [13] would
sterilize the house fly.

It has already been pointed out [4] that post-irradiation changes
observed in spermatogonia, spermatocytes and spermatids are less readily
induced in fully differentiated spermatozoa. In the case of Drosophila, it
was observed [13] that after a moderate dose the males appeared fertile
for several days and then went through a sterile period. In another
investigation [14] mature sperms were less susceptible to radiation effects
than other stages of spermatogenesis. In the present study, a 24-hour-old

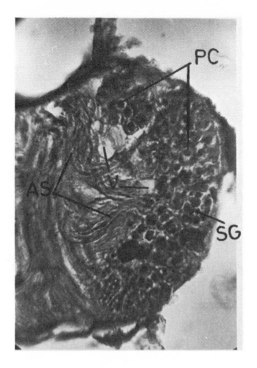

FIG. 13. Longitudinal section from the centre of the testis of a 3-day-old house fly (4 days after 2.5 krad irradiation) showing secondary spermatogonia (SG), and primary spermatocytes (PC) with compact nuclei, and vacuoles and abnormal sperms (AS) in the middle region (1000 ×).

irradiated adult house fly contained sufficient amount of mature sperms for ejaculation. There is a possibility that 2 to 3 days after irradiation with 2 krad, 2.5 krad and 3 krad as adults, when the adult house flies began mating, some unaffected mature sperms might be present in a sufficient amount for ejaculation (Fig.13). Thus, it appears that when the house flies were irradiated with 2 krad, 2.5 krad and 3 krad as adults, complete sterility was not achieved up to the 2nd day after irradiation. When the house flies were irradiated with the three doses at the pupal stage, the mature sperms showed radiation damage even when the house flies were 24 hours old. After a comparative study of post-irradiation effects on pupae and adults, it may be suggested that, for sterility, the house flies should be irradiated at the pupal stage rather than at the adult stage. A dose of 3 krad is the most effective for producing complete sterility.

6.   STIMULATION BY IRRADIATION

An important phenomenon but which has not been much studied is the stimulation by irradiation. It has been stated [11] that irradiation evidently impairs the ability of the cell to maintain metabolic equilibrium and processes

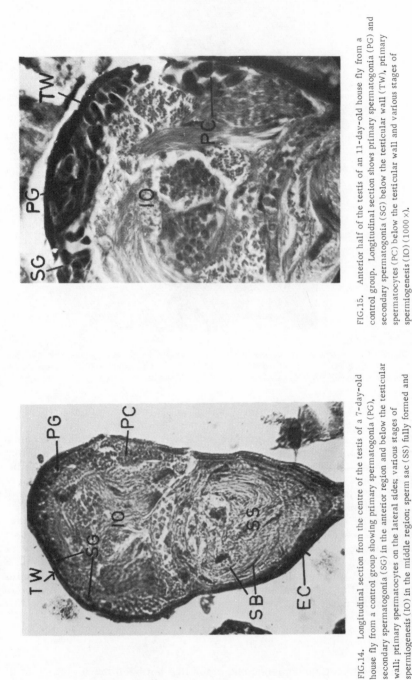

FIG.14. Longitudinal section from the centre of the testis of a 7-day-old house fly from a control group showing primary spermatogonia (PG), secondary spermatogonia (SG) in the anterior region and below the testicular wall; primary spermatocytes on the lateral sides; various stages of spermiogenesis (IO) in the middle region; sperm sac (SS) fully formed and filled with bundles of mature sperms (SB); epithelial cells (EC) lying in the basal portion of the sperm sac (400 ×).

FIG.15. Anterior half of the testis of an 11-day-old house fly from a control group. Longitudinal section shows primary spermatogonia (PG) and secondary spermatogonia (SG) below the testicular wall (TW), primary spermatocytes (PC) below the testicular wall and various stages of spermiogenesis (IO) (1000 ×).

within the cell so that the processes therefore proceed in part at an
increased tempo. To some extent the cell becomes prematurely aged.
This increase in tempo following irradiation is referred to as stimulation
by irradiation. Actually this cell stimulation is only a precipitate course
of the life function ending in cell death.

According to Ellinger [11], the relationship between dose and effect in
pharmacology is governed in a number of instances by the so-called
Arndt-Schulze law, which states that a small dose excercises a stimulating,
a medium dose a depressive and a large dose a destructive effect.

It was observed that in Rhodnius prolixus [14] irradiated males mated
more frequently than untreated ones and, as mating stimulates oviposition,
the quantity of eggs laid by the females also increased. In Musca autumnalis
[15] there was a slight increase in DNA synthesis two days after irradiation,
and an increase in the number of primary spermatocytes two to three days
after irradiation with small doses.

In the present work, mature sperms were observed in the middle region.
All other early stages of spermiogenesis were absent one hour after
irradiation as adults in all cases. Early, intermediate and late stages of
spermatids were characteristic of the middle region of the testes of the
control group (Figs 14 and 15). From the comparative study of irradiated
and control groups, it appears that, because of the initial stimulatory effect
of radiation, the early stages of spermiogenesis were transferred into the
late stages of spermiogenesis (immature and mature sperms). Perhaps
this was the reason for the disappearance of the early stages of spermiogenesis from the middle region at one hour post-irradiation as adult.

Mating behaviour of the adult house flies also supported the idea of
stimulation by irradiation. In the present study, it was discovered that
normally the control group of house flies began mating on the 4th or 5th
day after emergence. However, the flies of the irradiated groups began
mating on the 2nd and the 3rd day after emergence.

In short, it is evident from these observations that the immediate effect
of small doses (2 krad, 2.5 krad and 3 krad) of gamma rays is stimulatory.

ACKNOWLEDGEMENTS

We wish to thank Amir Mohammad, Director, Nuclear Institute of
Agriculture and Biology, Lyallpur, Pakistan, who provided facilities for
irradiating the experimental insects. We also wish to thank Zafar Qureshi,
Head of the Atomic Energy Entomology Division, Atomic Energy Agricultural
Research Centre, Tandojam, Pakistan, for his suggestions for the improvement of this work and for presenting this paper at the IAEA Symposium. We
are grateful to K.M. Aslam, Director (Training and Technical Assistance),
Pakistan Atomic Energy Commission, Islamabad, for giving permission for
this paper to be presented at the IAEA Symposium.

REFERENCES

[1] RIEMANN, J.G., THORSON, B.J., Comparison of effects of irradiation on the primary spermatogonia
    and mature sperms of three species of Diptera, Ann. Entomol. Soc. Am. 62 (1969) 613.
[2] KVELLAND, I., Radiosensitivity in different stages of spermiogenesis in Drosophila melanogaster,
    Hereditas 48 (1962) 221.

[3] SADO, T., Spermatogenesis of the silkworm and its bearing on radiation induced sterility: II,
     J. Fac. Agric. Kyushu Univ. 12 (1963) 387.

[4] MANDLE, A.M., The radiosensitivity of germ cells, Biol. Rev. 39 (1964) 288.

[5] RIEMANN, J.G. FLINT, H.M., Irradiation effects on midguts and testes of the adult boll weevil
     Anthonomous grandis, determined by histological and shielding studies, Ann. Entomol. Soc. Am.
     60 (1967) 289.

[6] OFFORI, E.D., Cytology of gamma irradiated gonads of Stomoxys calcitrans, Ann. Entomol. Soc. Am.
     63 (1970) 706.

[7] ASHRAFI, S.H., BROWER, H.J., TILTON, W.E., Gamma radiation effects on testes and mating success
     of the Indian Meal Moth, Plodia interpunctella, Ann. Entomol. Soc. Am. 65 (1972) 1144.

[8] HAMMEN, C.S., Nutrition of Musca domestica L., in single pair culture, Ann. Entomol. Soc. Am.
     49 (1956) 365.

[9] VAGO, C., AMARGIER, A., Coloration histologiques de virus d'insectes, Ann. Epiphyt. 14 (1963) 269.

[10] RIEMANN, J.G., A cytological study of radiation effects in testes of the Screw-Worm Fly, Cochliomyia
     hominivorax (Diptera: Calliphoridae), Ann. Entomol. Soc. Am. 60 (1967) 308.

[11] ELLINGER, F., "Fundamental biology of ionizing radiation", Ch. 7, Atomic Medicine (BEHRENS, C.F.,
     Eds), Williams and Wilkins Co., Baltimore (1959) 121.

[12] LaCHANCE, L.E., "The induction of dominant lethal mutations in insects by ionizing radiation and
     chemicals as related to the sterile male technique of insect control", Genetics of Insect Vectors of
     Disease (WRIGHT, J.W., PAL, R., Eds), Elsevier, Amsterdam (1967) 617.

[13] STRØMNAES, Ø., "Radiation sensitivity and repair of chromosomal damage", Isotopes and Radiation
     in Entomology (Proc. Symp. Vienna, 1967), IAEA, Vienna (1968) 341.

[14] GOMEZ-NUMEZ, J.C., GROSS, A., MACHADO, C., Gamma irradiation and the reproductive
     behaviour of male Rhodnius prolixus, Acta Cient. Venez. 15 (1964) 97.

[15] TUNG-PIERRE, SIK-CHUNG, Microspectrophotometric and cytological studies of the effect of Co-60
     irradiation on spermatogenesis and DNA synthesis in germ cells of face fly, Nuclear Science Abstracts
     26 (1971) No. 46958.

INDUCED STERILITY IN IRRADIATED
DIPTERA AND LEPIDOPTERA
Sperm transfer and
dominant lethal mutations

L. E. LaCHANCE
Metabolism and Radiation Research Laboratory,
Agricultural Research Service,
US Department of Agriculture,
Fargo, N. Dak.,
United States of America

Abstract

INDUCED STERILITY IN IRRADIATED DIPTERA AND LEPIDOPTERA: SPERM TRANSFER AND DOMINANT
LETHAL MUTATIONS.
    The ability of irradiated male insects to transfer sperm and the responses these males elicit from the
mated females differ in various species of insects. In the Diptera, sperm transfer, per se, plays a minor
role in changing the reproductive behaviour of the female; but the transfer of secretions from the ejaculatory
duct or accessory glands induces a monogamous response and stimulates oviposition. In the Lepidoptera,
eupyrene sperm must be transferred and incorporated in the spermathecae to elicit changes in mating behaviour
and stimulate oviposition; therefore, the ability of the irradiated and F_1 males to transfer eupyrene sperm
that contain dominant lethal mutations is most important for the competitiveness of lepidopteran males.
Likewise, the cytogenetic nature, mode of action and time of expression of dominant lethal mutations induced
in dipteran and lepidopteran sperm differ in the two orders of insects. At sterilizing doses, the lethal mutations
induced in dipteran sperm are expressed early in embryonic development; those induced in lepidopteran sperm
cause death of the embryo at a much later stage of development. Implications of these factors as they relate
to the application of the sterile-male technique are discussed.

INTRODUCTION

        Successful demonstrations of the sterile-insect release method (SIRM)
for population suppression or eradication have been largely limited to dipteran species. Extension of this technique to species in other orders of
insects requires specific knowledge about their radiation biology, ecology
and population dynamics. There are also important differences in reproductive
physiology and genetics to consider. As North and Holt [1] pointed out: "It
has become evident that duplication of the procedures used for sterile male
releases with dipteran species are not directly applicable to controlling
Lepidoptera. This does not imply, as one would suggest, that the technique
is of no value in controlling lepidopterous species. It means, however, that
researchers must develop modifications and apply techniques based on the
biological uniqueness of the Lepidoptera."

        Undoubtedly there will be critical differences between species, but
for the moment we can consider broader aspects that apply to many species
within the order.

        This paper deals with two aspects of radiation response. I attempt
to summarize the differences in reproductive physiology and expression of
dominant lethal mutations between dipteran and lepidopteran species that

occur as a result of radiation treatments. Hopefully, these observations, mostly from laboratory experiments, can be related to potential problems that would arise in field programs and point to possible modifications that might maximize the effectiveness of SIRM programs with lepidopteran pests.

SPERM TRANSFER AND FEMALE REPRODUCTIVE PHYSIOLOGY

During the copulatory process, the male insect transfers both sperm and semen containing a variety of secretions to the female. After receipt of these components, the female usually becomes sexually unreceptive (or much less so) and begins to oviposit. In the Diptera, the initiation of oviposition and the loss of sexual receptivity are directly related to the presence in the female of material from the male ejaculatory duct or the accessory gland. Hampton [2] showed that implanting house fly testes into females did not stimulate oviposition. Riemann and Thorson [3] found that female house flies are stimulated to oviposit and lose sexual receptivity as a result of the presence of accessory material from the male ejaculatory duct. Mating with castrated males with intact ejaculatory ducts stimulated oviposition and loss of sexual receptivity nearly as well as mating with normal males. Leahy and Craig [4] and Craig [5] showed that female monogamy in Aedes aegypti was induced by the male accessory gland material and that implantation of male accessory glands from A. aegypti and A. albopictus increased oviposition; it was not necessary for the material to reach the spermathecae. Work by Nelson et al. [6] with 3 dipteran species showed that male ejaculatory duct material could be extracted and was effective in inducing monogamy when injected into females. Leopold et al. [7] showed that the accessory secretion could be depleted by repeatedly mating male house flies. With each successive mating, there was a stepwise increase in the amount of time spent in copulo and in the number of females that remated. Thus for irradiated dipteran males to be competitive, they need only to locate females, mate, and transfer accessory material to the female. Sperm transfer plays a minor role.

In the Lepidoptera, the situation is far more complicated and is essentially reversed. Male Lepidoptera produce and transfer two types of sperm to the females. Both types are found in the testes as bundles of approximately 256 sperm. The nucleated eupyrene sperm move through the male reproductive tract and are deposited in the spermatophore as bundles that break down shortly after mating occurs; the anucleate apyrene sperm bundles break down before they are transferred to the females. Other features of morphogenesis were reported by Riemann [8].

Numerous studies have shown that in species of Lepidoptera the stimulus to oviposit and the loss of sexual receptivity are directly related to the presence of eupyrene sperm in the spermathecae of the female. Taylor [9] and Norris [10] showed that in Atteva punctella and Ephestia kuehniella a normal complement of sperm was required to initiate oviposition and produce long-term inhibition of female receptivity. Many other studies [11, 12, 13, 1, 14, 15] all showed that the presence of eupyrene sperm or an accompanying secretory material was directly related to initiation of the female ovipositional response For example, Karpenko and North [15] determined that the number of eggs deposited by virgin female Trichoplusia ni was no different from the number laid by females mating with males that transferred no sperm or only apyrene sperm. Indeed, they found a distinct increasing gradient in numbers of eggs produced by virgin females, by mated females without sperm, by mated females with only apyrene sperm to the high levels produced by females containing normal amounts of eupyrene sperm. Thus, the primary ovipositional stimulus was the presence of eupyrene sperm in the spermathecae of the female though there could be an effect of accessory fluid closely associated with these sperm. The manner in which these sperm exert their effect is unknown. Riddiford and Ashenhurst [16] found that in addition to the presence of sperm a factor in the brusa copulatrix of Cecropia moths was also essential in initiating oviposition in mated females.

In the Metabolism and Radiation Research Laboratory, many investigators working with Lepidoptera now use single-pair matings, then after the termination of the egging period the female is dissected and checked for number of spermatophores and presence of apyrene and eupyrene sperm in the spermathecae. The collective experience with at least 6 species of Lepidoptera and thousands of single-pair matings indicates that females with low ovipositional histories contain few or no eupyrene sperm in the spermathecae though they may have mated repeatedly.

Loss of sexual receptivity is admittedly more difficult to measure, but female pink bollworms, Pectinophora gossypiella, caged with males that did not transfer eupyrene sperm were found to mate significantly more often [17]. Also, Hofmann [14] showed that female Anagasta kuehniella lost receptivity only when eupyrene sperm reached the spermathecae; thus, if a secretion did promote nonreceptivity and oviposition, it apparently was specifically bound to the eupyrene sperm, and its action was dependent upon the presence of these sperm in the female spermathecae. Snow et al. [18] found that traps baited with a single female corn earworm, Heliothis zea, that had mated and received an apparently normal complement of sperm (both eupyrene and apyrene) did not attract males but that traps baited with a single female that had received immotile sperm attracted as many males as virgin females. Thus with the corn earworm, the female must have eupyrene sperm in the spermatheca, and the sperm must be motile if the female is to be unreceptive.

Development of more refined procedures for estimating sperm numbers [19] promises to be extremely useful in elucidating the various relationships between sperm content and female sexual receptivity.

SPERM TRANSFER BY IRRADIATED LEPIDOPTERAN MALES AND THE F_1 PROGENY

The relationship between the transfer of eupyrene sperm and female reproductive physiology in Lepidoptera would be purely academic except that in some species the irradiated males mate and transfer a spermatophore but eupyrene sperm often do not reach the spermathecae of the female. Some studies of this phenomenon are summarized in Table. I. The examples make plain that this situation is not found in all lepidopteran species. Irradiated males of Heliothis zea, Pectinophora gossypiella and Anagasta kuehniella transfer eupyrene sperm nearly as well as untreated males, at least under the conditions of these experiments. Holt and North [20] suggested that species with more complex reproductive systems may require a greater degree of precision and timing, especially in the inversion of the spermatophore bulb, and that they might therefore be more susceptible to radiation damage and to failure to inseminate females. In fact, they found that the inability of the irradiated cabbage looper males to transfer sperm to the spermathecae was caused by the inability of the males to incorporate sperm into the spermatophore bulb. Irradiated males required longer to mate, and the timing of the insertion of the spermatophore, the inflation of the spermatophore bulb, and the inclusion of sperm into the bulb were altered. It appears plausible that a generalized somatic effect of radiation treatments on the mating process may be responsible.

Nevertheless, recent studies indicate that the ability of irradiated male Lepidoptera to transfer sperm is not an all-or-none phenomenon and varies within a species. The dose of irradiation is an important factor as are

TABLE I. SPERM TRANSFER BY IRRADIATED LEPIDOPTERAN MALES [a]
MATING WITH UNTREATED FEMALES

Species	Dose (krad)	Percentage mated	Percentage ♀ inseminated	Reference
Trichoplusia ni	20	83	28	[13, 1]
	30	67	14	
Heliothis virescens	15.0	-	85	[35]
	22.5	-	90	
Heliothis virescens	25 - pupae	88.5	36.5	[19]
	25 - adult, 1-day-old	94.3	75.7	
	25 - adult, 2-day-old	85.9	84.4	
Heliothis zea	25	85	100	[36]
Anagasta kuehniella	0	100	93	LaChance & Richard (unpublished)
	50	100	92	
Pectinophora gossypiella	20	89	83	[37]
	30	95	83	

[a] Laboratory reared.

variables such as the age of the male and the experimental design. With
Heliothis virescens males, a dose of 25 krad administered to pupae 24 hours
before emergence or to 1-day-old adults significantly reduced sperm-transfer
ability; even when sperm were transferred, significantly fewer sperm were
found in the spermatheca. However, when 2-day-old males were treated, sperm
transfer and numbers were equal to those contributed by normal males [19].

Some preliminary results obtained at the Metabolism and Radiation
Research Laboratory with Anagasta kuehniella are reported in Table I. In
the case of doses of 50 krad to 3-day-old males, the treated males mated
and transferred eupyrene sperm when they were paired the same day they were
irradiated. However, when the males were kept separate from females for 1
or 2 days, the subsequent transfer of eupyrene sperm was considerably reduced.
Therefore, time between treatment of the male and copulation may play an
important role. These studies have not proceeded far enough to permit identification of the exact protocol required to guarantee sperm transfer, even in
this one species. Only further studies of the radiation biology and reproductive physiology of irradiated Lepidoptera will provide the information required
to improve this most important aspect of competitiveness in lepidopteran males.
Also, sperm transfer by released males may have far-reaching consequences in
the reproductive behavior of the females and thus affect the results of field
programs. However, a number of variables (age at irradiation, time between
irradiation and mating, and photoperiod) can be manipulated to maximize the
sperm-transfer ability of irradiated lepidopteran males.

TABLE II. SPERM TRANSFER BY THE UNTREATED F_1 PROGENY OF
IRRADIATED MALE LEPIDOPTERA

Species	Dose to P_1 ♂ (krad)	Percentage F_1 ♂ mated	Percentage ♀ inseminated	Reference
Heliothis virescens	0	92	88	[38]
	7.5	74	55	
	15	65	52	
	22.5	62	27	
Pectinophora gossypiella	0	94	90	[17]
	5.0	88	66	
	7.5	95	29	
	10.0	79	35	
	12.5	72	23	
	15	80	40	[39]
	20	58	46	
Ephestia cautella	0	94	100	[40]
	20	85	100	
	35	54	40	
	40	39	15	

When lepidopteran males are irradiated with a substerilizing dose of
radiation, a reduced number of F_1 progeny are produced. These progeny are
either partially or totally sterile, depending on the dose administered to
the male parent, but the F_1 males have higher levels of sterility than the
P_1 males, a phenomenon referred to as inherited partial sterility. This sterility of the F_1 progeny is not caused entirely by genetic damage; a substantial
part results from the failure of the F_1 males to inseminate the females with
eupyrene sperm. Table II summarizes the results of some experiments that demonstrated the reduced capability of the F_1 males to transfer sperm. It is interesting to note that in species such as Pectinophora gossypiella, irradiated P_1
males transferred eupyrene sperm to the females despite relatively high doses
of irradiation, but the F_1 progeny had a reduced sperm-transfer capability,
even at low doses to the P_1 males. Sugai and Suzuki [21] found that the F_1
progeny of male Bombyx mori treated with apholate produced few eupyrene sperm
though they produced a normal number of apyrene cells.

Although the deficiency in sperm transfer of the F_1 males may appear
identical to that of irradiated P_1 males, this is not the case. The end effect
may be the same, but the causative factors are different. The F_1 males are not
irradiated; they merely inherit an irradiated genome that contains a variety
of chromosomal aberrations. It is not known how chromosomal aberrations can
influence the processes of eupyrene sperm formation or transfer. Studies by
Riemann [22] showed gross ultrastructural abnormalities in the eupyrene sperm
of the F_1 males of Anagasta kuehniella.

Thus in suppression programs involving releases of partially sterilized lepidopteran males, it may be dangerous to expect the F_1 progeny to be fully competitive. However, no data have been obtained concerning sperm production or transfer by F_1 progeny reared in the field on natural hosts. We are just beginning to appreciate how lack of sperm transfer by the irradiated males and the F_1 progeny affects the reproductive behavior of the female. It therefore seems too early for either pessimism or optimism concerning the quality and competitiveness of the Lepidoptera that will eventually be tested in population suppression programs.

RADIATION-INDUCED DOMINANT LETHAL MUTATIONS IN INSECT SPERM

When an irradiated or chemosterilized male mates and transfers sperm and the sperm fertilize the eggs, the failure of the eggs to hatch and even post-hatch deaths are attributed to dominant lethal mutations induced in the treated cell. No basic information was available about dominant lethal mutations in the screwworm, Cochliomyia hominovorax, at the time of the successful screwworm eradication programs in Curacão and the Southeastern U. S., and the many SIRM field demonstrations with fruit flies (Tephretidae) were likewise done without such knowledge though studies of other dipteran species had been published (for review see LaChance [23]). However, fundamental information concerning the cytogenetic nature and time of expression of dominant lethal mutations in lepidopteran insects may prove to be invaluable in the application of the sterility method to these insects. Because transfer of eupyrene sperm to the female and the induction of dominant lethal mutations in the sperm are the key events in lepidopteran sterility, it seems logical that these biological events should be understood. This knowledge might provide a means of improving the poor performance of irradiated insects, permit the use of lower doses of radiation in producing sterility, and perhaps provide information leading to the induction of sterility without irradiation. Little is known about dominant lethal mutations in lepidopteran species. Cytogenetic studies of induced dominant lethal mutations in insects have been largely limited to Diptera [24, 25, 26] and Hymenoptera [27, 28, 29], i.e. to species with a single kinetochore per chromosome (monokinetic). In Diptera and Hymenoptera, most dominant lethal mutations are usually expressed during the early cleavage divisions. Cytological examination of eggs fertilized by irradiated sperm has revealed that embryonic development often stops after only a few cleavage divisions and usually before blastoderm formation; the eggs contain chromosome bridges and fragments between dividing cleavage nuclei that lead to mitotic inhibition, formation of polyploid cleavage nuclei and cessation of embryonic development. Thus, mechanical difficulties in the chromosome distribution to cleavage nuclei and in the orderly separation of cleavage nuclei appear to initiate a series of events that are lethal.

In contrast, in lepidopteran and hemipteran species, the kinetic activity is diffused along the entire chromosome (holokinetic) (see LaChance and Riemann [30] for references), and the radioresistance of lepidopteran species was originally thought to be directly related to this type of chromosome structure [30, 31, 12]. However, hemipteran species with holokinetic chromosomes can be sterilized with fairly low doses of radiation. Thus, there is still no satisfactory explanation for the fact that dominant lethal mutations can be induced in the sperm of dipteran and hymenopteran species with doses below 10 krad and in hemipteran species with doses of 5-15 krad; yet the majority of lepidopteran species must be treated with doses between 25 and 90 krad to induce dominant lethal mutations in all sperm. Indeed, one would expect that embryonic development would not proceed very far after treatment with such large doses of ionizing radiation; in fact, the reverse is true. Observations of unfixed and unstained eggs of both Lepidoptera and Hemiptera [1, 12, 13, 32,

33] indicate that when they are fertilized by an irradiated sperm, embryonic development proceeds to a very late stage before lethality occurs. Also, studies of fixed and stained eggs have so far corroborated these observations. LaChance and Riemann [30] studied embryonic development in Oncopeltus eggs fertilized with irradiated (10 krad) sperm. Most eggs showed a decreased rate of development, but at least 70% reached the blastoderm stage. Thereafter, developmental abnormalities occurred, and death ensued. When chromosomes in the dividing cells of these embryos were counted, many cells with more or less than the normal 16 chromosomes were found, and different cells in the same embryo often had a wide range of chromosome numbers. Thus, in Oncopeltus the eggs fertilized with irradiated sperm were able to proceed to a post-blastoderm stage though many of the cells in the developing embryo had grossly abnormal chromosome numbers.

In a similar study of the cabbage looper, Trichoplusia ni [34], males were irradiated with 20 krad and crossed with untreated females. Many of the eggs showed no sign of embryonic development. Thus, a major cause of nonhatching eggs may be lack of fertilization though at this dose, the males usually inseminate the females. When the eggs did show embryonic development, it was much slower in the treated group than in eggs from females crossed with untreated males (no chromosomal abnormalities were present, but chromosome counts in individual cells were not possible). Virtually all embryos that did begin to develop reached the late, fully differentiated, pre-hatch larval stage; since 20 krad is not a fully sterilizing dose for cabbage loopers, some of the larvae hatched. Nevertheless, many eggs from the treated group contained well-formed larvae that did not hatch, an indication that when sperm bearing dominant lethal mutations fertilized an egg, development proceeded to a very late stage. More recent laboratory studies of lepidopteran species that are able to transfer eupyrene sperm despite fully sterilizing doses (Pectinophora gossypiella and Anagasta kuehniella) indicate that a few dominant lethal mutations are expressed fairly early in development after doses of 30 krad and 50 krad, respectively, but that a significant number of eggs fertilized with irradiated sperm are able to complete embryonic development despite these doses.

Table III compares some points concerning chromosome structure, radiosensitivity, and dominant lethal mutations in the sperm of species representing three orders. The assumption that both hemipteran and lepidopteran chromosomes are holokinetic and therefore should respond similarly to radiation treatments may be in error. However, the following hypothesis emerges: 1) possession of holokinetic chromosomes does not per se confer radioresistance to sperm but only permits the escape from mechanical chromosomal abnormalities (bridges and

TABLE III. RADIOSENSITIVITY OF SPERM, CHROMOSOME STRUCTURE AND EXPRESSION OF DOMINANT LETHAL MUTATIONS OF THREE SPECIES

	House fly	Milkweed bug	Cabbage looper
Kinetochore	Monokinetic	Holokinetic	Holokinetic
99% Dominant lethal dose (krad)	3	10	30
Embryonic death	Early cleavage	Post-blastoderm	Pre-hatching larvae

fragments) during early cleavage divisions; 2) in Oncopeltus, the blastoderm stage is reached, but other developmental abnormalities originating during cleavage such as mitotic spindle disorientation and chromosome imbalance lead to death in the post-blastoderm stages; 3) the lepidopteran species are able to escape most of the developmental crises and post-blastoderm deaths so embryonic development usually reached the pre-hatch larval stage.

Further development of hypotheses concerning the cytogenetics, mode of action and time of expression of dominant lethal mutations in Lepidoptera must await answers to the following questions: (1) Do embryonic deaths in Lepidoptera resemble those resulting from fertilization with sperm containing chromosome duplications and deficiencies? In Diptera and Hymenoptera [29], crosses of males heterozygous for a chromosome translocation mating with a normal female produce embryos that hatch (fertilized by chromosomally balanced sperm) and embryos that die relatively late in development (fertilized by unbalanced sperm). (2) Is there some mechanism that regulates chromosome orientation and separation during the cleavage divisions in lepidopteran embryos? Such a mechanism would favor orderly distribution of the complete genome to all cleavage nuclei (despite radiation-induced chromosome fragmentation and translocations), and most lethal mutations would be associated with transcription and point events and expressed late in embryonic development. (3) Are many of the Lepidoptera polyploids? If they are, the time of death due to chromosome imbalance would be greatly delayed. (4) When do the genomes contributed by the male and female begin to control embryonic development, i.e. when does active RNA synthesis begin in lepidopteran embryos? Obviously an irradiated sperm containing a lethal mutation cannot produce the lethal event until the damaged genetic material normally begins to function.

The information would be extremely useful in furthering our understanding of dominant lethal mutations in lepidopteran species and in maximizing the probability of success of the sterility approach for control of these species.

REFERENCES

[1] NORTH, D.T., HOLT, G.G., Population control of Lepidoptera: The genetic and physiological basis, Manitoba Entomol. 4 (1970) 53.

[2] HAMPTON, U.M., Reproduction in the housefly (Musca domestica L.), Proc. R. Entomol. Soc. Lond. 27 A (1952) 29.

[3] RIEMANN, J.G., THORSON, B.J., Effect of male accessory material on oviposition and mating by female house flies, Ann. Entomol. Soc. Am. 62 (1969) 828.

[4] LEAHY, M.G., CRAIG, G.B., Accessory gland substance as a stimulator for oviposition in Aedes aegypti and A. albopictus, Mosq. News 24 (1965) 448.

[5] CRAIG, G.B., Mosquitoes: Female monogamy induced by male accessory gland material, Science 156 (1967) 1499.

[6] NELSON, D.R., ADAMS, T.S., POMONIS, J.G., Initial studies on the extraction of the active substance inducing monocoitic behaviour in house flies, black blow flies and screwworm flies, J. Econ. Entomol. 62 (1969) 634.

[7] LEOPOLD, R.A., TERRANOVA, A.C., SWILLEY, E.M., Mating refusal in Musca domestica: Effects of repeated mating and decerebration upon frequency and duration of copulation, J. Exp. Zool. 176 (1971) 353.

[8] RIEMANN, J.G., "Metamorphosis of sperm of the cabbage looper, Trichoplusia ni, during passage from the testes to the female spermatheca", Comparative Spermatology (BACCETTI, B., Ed.), (Proc. Symp. Rome-Siena, 1969), Accademia Nazionale Dei Lincei, Quaderno N. 137 (1970) 321.

[9] TAYLOR, O.R., Relationship of multiple mating to fertility in Atteva punctella (Lepidoptera: Yponomeutidae), Ann. Entomol. Soc. Am. 60 (1967) 583.

[10] NORRIS, M.J., Contributions toward the study of insect fertility: 2. Experiments on the factors influencing fertility in Ephestia kühniella Z. (Lepidoptera: Phycitidae), Proc. Zool. Soc. Lond. 4 (1933) 903.

[11] FLINT, H.M., KRESSIN, E.L., Transfer of sperm by irradiated Heliothis virescens (Lepidoptera: Noctuidae) and relationship to fecundity, Can. Entomol. 101 (1969) 500.

[12] NORTH, D.T., HOLT, G.G., Inherited sterility in progeny of irradiated male cabbage loopers, J. Econ. Entomol. 61 (1968) 928.

[13] NORTH, D.T., HOLT, G.G., "Genetic and cytogenetic basis of radiation-induced sterility in the adult cabbage looper, Trichoplusia ni", Isotopes and Radiation in Entomology (Proc. Symp. Vienna, 1967), IAEA, Vienna (1968) 391.

[14] HOFMANN, H.C., Physiological factors controlling mating and oviposition in Anagasta kuehniella. PhD dissertation, North Dakota State Univ., 1972.

[15] KARPENKO, C.P., NORTH, D.T., Ovipositional response elicited by normal, irradiated, F$_1$ male progeny or castrated male Trichoplusia ni (Lepidoptera: Noctuidae), Ann. Entomol. Soc. Am. 66 (1973) 1278.

[16] RIDDIFORD, L.M., ASHENHURST, J., The switchover from virgin to mated behavior in female Cecropia moths: the role of the bursa copulatrix, Biol. Bull. 144 (1973) 162.

[17] LaCHANCE, L.E., BELL, R.A., RICHARD, R.D., Effect of low doses of gamma irradiaton on reproduction of male pink bollworms and their F$_1$ progeny, Environ. Entomol. 2 (1973) 653.

[18] SNOW, J.W., JONES, R.L., NORTH, D.T., HOLT, G.G., Effects of irradiation on ability of adult male corn earworms to transfer sperm, and field attractiveness of females mated to irradiated males, J. Econ. Entomol. 65 (1972) 906.

[19] RAULSTON, J.R., GRAHAM, H.M., Determination and quantitative sperm transfer by male tobacco budworms irradiated at different ages, J. Econ. Entomol. (in press).

[20] HOLT, G.G., NORTH, D.T., Effects of gamma irradiation on the mechnisms of sperm transfer in Trichoplusia ni, J. Insect Physiol. 16 (1970) 2211.

[21] SUGAI, S., SUZUKI, S., Male sterility in progeny of the male silkworm, Bombyx mori L., orally administered apholate (Lepidoptera: Bombycidae), Appl. Entomol. Zool. 6 (1971) 126.

[22] RIEMANN, J.G., Ultrastructure of sperm in F$_1$ progeny of irradiated males of the Mediterranean flour moth, Anagasta kuehniella, Ann. Entomol. Soc. Am. 66 (1973) 147.

[23] LaCHANCE, L.E., "The induction of dominant lethal mutations in insects by ionizing radiation and chemicals — as related to the sterile-male technique of insect control", Genetics of Insect Vectors of Disease (WRIGHT, J.W., PAL, R., Eds), Elsevier Press, Amsterdam (1967) 617.

[24] FAHMY, O.G., FAHMY, M.J., Cytogenetic analysis of the action of carcinogens and tumor inhibitors in Drosophila melanogaster: II. The mechanism of induction of dominant lethals by 2:4:6-tri (ethyleneimino)-1:3:5-triazine, J. Genet. 52 (1954) 603.

[25] LaCHANCE, L.E., RIEMANN, J.G., Cytogenetic investigations on radiation and chemically induced dominant lethal mutations in oocytes and sperm of the screwworm fly, Mutat. Res. 1 (1964) 318.

[26] LaCHANCE, L.E., LEOPOLD, R.A., Cytogenetic effect of chemosterilants in house fly sperm: Incidence of polyspermy and expression of dominant lethal mutations in early cleavage divisions, Can. J. Genet. Cytol. 11 (1969) 648.

[27] WHITING, A.R., Effect of X-rays on hatchability and on chromosomes of Habrobracon eggs treated in first meiotic prophase and metaphase, Am. Nat. 79 (1945) 193.

[28] WHITING, A.R., Dominant lethality and correlated chromosome effects in Habrobracon eggs X-rayed in diplotene and in late metaphase I, Biol. Bull. 89 (1945) 61.

[29] von BORSTEL, R.C., REKEMEYER, M.L., Radiation-induced and genetically contrived dominant lethality in Habrobracon, Genetics 44 (1959) 1053.

[30] LaCHANCE, L.E., RIEMANN, J.G., Dominant lethal mutations in insects with holokinetic chromosomes: 1. Irradiation of Oncopeltus (Hemiptera: Lygaeidae) sperm and oocytes, Ann. Entomol. Soc. Am. 66 (1973) 813.

[31] LaCHANCE, L.E., SCHMIDT, C.H., BUSHLAND, R.C., "Radiation-induced sterilization", Pest Control: Biological, Physical, and Selected Chemical Methods (KILGORE, W.W., DOUTT, R.L., Eds), Academic Press, New York (1967) 147.

[32] BAUER, H., Die kinetische Organisation der Lepidopteran-Chromosomen, Chromosoma 22 (1967) 102.

[33] LaCHANCE, L.E., DEGRUGILLIER, M., LEVERICH, A.P., Cytogenetics of inherited partial sterility in three generations of the large milkweed bug as related to holokinetic chromosomes, Chromosoma 29 (1970) 20.

[34] LaCHANCE, L.E., Dominant lethal mutations in insects with holokinetic chromosomes: 2. Irradiation of sperm of cabbage looper, Ann. Entomol. Soc. Am. 67 (1974) 35.

[35] PROSHOLD, F.I., BARTELL, J.A., Inherited sterility and postembryonic survival of two generations of tobacco budworms, Heliothis virescens (Lepidoptera: Noctuidae), from partially sterile males, Can. Entomol. 104 (1972) 221.

[36] NORTH, D.T., HOLT, G.G., "Radiation studies of sperm transfer in relation to competitiveness and oviposition in the cabbage looper and corn earworm", Application of Induced Sterility for Control of Lepidopterous Populations (Proc. Panel Vienna, 1970), IAEA, Vienna (1971) 87.

[37] LaCHANCE, L.E., RICHARD, R.D., Radiation response in the pink bollworm: A comparative study of sperm bundle production, sperm transfer, and oviposition by wild and laboratory-reared males (in preparation).

[38] PROSHOLD, F.I., BARTELL, J.A., Inherited sterility in progeny of irradiated male tobacco budworms: Effects of reproduction, developmental time, and sex ratio, J. Econ. Entomol. 63 (1970) 280.

[39] CHENG, W.Y., NORTH, D.T., Inherited sterility in the F_1 progeny of irradiated male pink bollworms, J. Econ. Entomol. 65 (1972) 1273.

[40] GONEN, M., CALDERON, M., Effects of gamma radiation on Ephestia cautella (Wlk.) (Lepidoptera, Phycitidae): II. Effects on the progeny of irradiated males, J. Stored Prod. Res. 7 (1971) 91.

DISCUSSION

K. SYED: You have shown in your data that there was a progressive decrease in sperm transfer with increased doses of radiation. At the same time there was also a progressive decrease in the percentage of males that mated. Does this mean that these increasing doses in some way affect the mating ability of the treated males?

L.E. LaCHANCE: The sperm transfer data were based only on females that had a spermatophone. In this manner we eliminated all unmated females. This shows that some males mate and yet their eupyrene sperm do not reach the spermathecae. However, please note that in many lepidopteran species the P_1 males can transfer sperm at high doses but even at lower doses their F_1 sons do not transfer sperm.

D.W. WALKER: Do you interpret all the debilitation beyond the F_1 adult in Lepidoptera to be genetic (i.e. chromosomal)? If so, one would expect all the F_2 eggs to be embryonated.

L.E. LaCHANCE: I did not discuss any data on the F_2 generation. What I did stress was that many F_1 males did not transfer eupyrene sperm to normal females. This undoubtedly has a genetic-physiological basis.

D.W. WALKER: Male Lepidoptera respond much less to fractionated dose treatment than the females. Can you explain this?

L.E. LaCHANCE: No. Irradiation of female Lepidoptera is quite a different matter. For example, some irradiated females lay eggs that do not hatch depending on dose, but their F_1 progeny are usually quite fertile.

S. LOAHARANU: Is it possible to distinguish eupyrene and apyrene sperms in the spermatid stage?

L.E. LaCHANCE: At the spermatid stage both apyrene and eupyrene sperm occur in bundles of circa 256 sperm. I am sure one can differentiate between them at the electron microscope level, but I am not sure if that is true at the light microscope level. Also, these types of sperm are formed at different times in the larval and pupal stages. Therefore, when many apyrene bundles in the spermatid stage are present in the testes, there would be very few eupyrene sperm in that stage.

B. AMOAKA-ATTA: Can you tell me what effect double mating of normal female Lepidoptera with (1) a sterilized male and (2) a normal male has on the movement of eupyrene sperm from the bursa copulatrix into the spermatheca of the female? I am asking this because there is possibly a stimulatory factor that causes movement of the eupyrene sperm into the spermatheca which may be lacking in the sterilized male but which may be introduced into the female by the normal male during the second mating.

L.E. LaCHANCE: I cannot answer that question in detail. Recent work by F.I. Proshold in our laboratory suggests that when eupyrene sperm leave the spermatophore and do not reach the spermatheca, they would not remain in the bursa for a very long period. It appears unlikely that a second mating would stimulate these sperm to move to the spermatheca.

G.W. RAHALKAR: If you allow an F_1 male, which has been derived from an irradiated P_1 male parent, to mate successively with virgin normal females, and then analyse the sperms transferred in these females, what is the proportion of eupyrene and apyrene sperms?

L.E. LaCHANCE: I am afraid I do not have any data on multiple matings of F_1 males but this is a most interesting idea.

J.L. MONTY: Is there a genetic barrier to the occurrence of F_1 sterility in irradiated Diptera?

L.E. LaCHANCE: No. Semi-sterile progeny are found after irradiation of fruit flies, house flies, mosquitoes, onion flies, etc. The trouble is that the high levels of dominant lethals induced at relatively low doses drastically reduce the radiation dose one can use for recovery of F_1 sterile progeny.

PRELIMINARY REPORT ON MATING STUDIES OF THREE VARIETIES OF Cadra (Ephestia) cautella Walker

M. S. H. AHMED, S. B. LAMOOZA,
N. A. OUDA, I. A. ALHASSANY
Nuclear Research Institute,
Atomic Energy Commission,
Tuwaitha, Baghdad,
Iraq

Abstract

PRELIMINARY REPORT ON MATING STUDIES OF THREE VARIETIES OF Cadra (Ephestia) cautella Walker.
Three morphologically distinguishable strains (B, C and D) of Cadra (Ephestia) cautella Walker were started from moths collected in Baghdad, and have been bred for more than 15 generations. Results of some crosses between these strains revealed a kind of cytoplasmic incompatibility when males of either strain D or C were mated with females from strain B. Males of strain C caused a significant decrease in egg hatch when placed with males and females of strain B. Therefore, the phenomenon of incompatibility might be useful in suppressing certain populations of this insect. Field surveys carried out for a short period have shown that heterozygous individuals are prevalent which might suggest the existence of a kind of balanced polymorphism. However, extensive surveys are required to establish the actual situation in the Cautella populations. The availability of strains with easily distinguishable colours might be a useful tool in measuring various biological and ecological aspects of C. cautella populations.

1. INTRODUCTION

In a previous study [1] a unidirectional cytoplasmic incompatibility was revealed in crosses between two strains of the fig moth, Cadra (Ephestia) cautella Walker, originating from two different regions. These two strains were arbitrarily designated in our laboratory as strain A and strain B. Strain A was obtained from the US Army Quartermaster Research and Development Command, Natick, Mass., United States of America. The second strain (B) was started from two pairs of identical fig moths caught in a date store in Baghdad, Iraq. The study was elaborated further and included in another report [2].

In the present report the results of some crosses of two Cadra (Ephestia) cautella varieties (arbitrarily designated as strain C and strain D) with strain (variety) B are recorded. In addition, some crosses with strain A have been carried out for comparison.

Also, a preliminary assessment of the proportions of C. cautella moths with various forewing colours in nature was conducted for a short period in a limited area in Baghdad.

2. MATERIALS AND METHODS

The two new strains (C and D) have been cultured from two morphologically distinguishable dark females captured while mating with two males of lighter

413

a. Strain B

b. Strain C

c. Strain D

FIG. 1. Variations in forewing coloration in three pure forms of Cadra (Ephestia) cautella Walker.

colours in a date store in Baghdad City on the same day. B, C and D strains have so far been reared pure for more than 15 generations in the laboratory (except that during the first and second generations of strains C and D, some insects appeared with different colours of the forewings which were dis-carded without further investigation). Adults of the three varieties differ significantly as they have quite distinct wing colours and patterns (Fig.1). Moths belonging to strain C are the darkest of all (melanic form), the forewings of which are transversely unbanded (or slightly banded) and mostly with pale brown longitudinal stripes (Fig.1b). Their dark brownish colour resembles in general dry-date colour of some date varieties. Strain D moths (Fig.1c) are intermediate in colour with dark spots on the clearly banded forewings (similar to strain A [2]) whereas strain B adults (Fig.1a) have more or less the same features as D or A but are much paler (usually the width of the band in these strains is very variable). Thus any contamination of the culture could be easily avoided, keeping the stocks pure.

Larvae and adults of strains B, C and D have been determined as belonging to the species Ephestia cautella Walker (Pyralidae) by the Commonwealth Institute of Entomology, British Museum (Natural History).

The present tests were conducted at about 25°C and 40-60% r.h. Larvae were reared on culture medium consisting of ground wheat, to which about 12% glycerine was added. Pupae were collected, sexed and placed in cotton-stoppered shell vials for emergence. All moths used in the present tests were less than 24-h old when each cross was started.

Fecundity, fertility (% egg hatch), longevity and sex ratio were measured for some possible crosses between the four strains. For this purpose adults of the fig moth were allowed to mate (single-paired mating) in small vials (25 × 75 mm) provided with a drop of 10% sugar solution on a small piece of cottonwool. Eggs were usually collected daily throughout the whole life of almost every female adult, counted and some of them transferred to a piece of black filter paper spread in a petri dish to measure the hatchability. Some of these eggs (or hatched larvae) were placed directly in beakers with the appropriate amount of the medium to allow them to develop into adults in order to calculate the sex ratio and percentages of malformation among the emerged moths. Because of many insurmountable difficulties the replicates of many crosses have been carried out at different times.

Mating frequency was determined by dissecting females and counting the number of spermatophores present in the bursa copulatrix [3]. Since spermatophores might be transferred even by castrated males [4], spermatheca of at least one female from each type of the parental crosses was examined, and the presence of moving eupyrene sperm was preliminarily determined.

The suppressing effect of strain C males has been measured by following the method described in Ref.[2].

Field catches were carried out by making daily visits to a locality where dry dates were stored. All adult moths, whether field-collected or progeny of different crosses, were arbitrarily arranged in several colour groups as far as could be ascertained with the naked eye. Such a classification system was useful in estimating the colour of another moth [5]. The distinguishable colours of the forewings of the three strains (A or D, B and C), have served as standards of comparison.

TABLE I.   RESULTS OF SOME CROSSES BETWEEN FOUR DIFFERENT STRAINS OF Cadra (Ephestia) cautella Walker

Crosses Female × male	No. pairs (replicates)	Average eggs per female ± S.D.	No. eggs incubated	Percentage hatch	Average in days		Average spermatophores per female [a]
					Female	Male	
B × B	8	181.3 ± 78.4	1450	74.6 ± 33.7	7.00 ± 3.91	5.63 ± 2.72	1.0 ± 0
D × D	4	409.7 ± 39.2	200	78.0 ± 10.7	11.35 ± 3.50	9.25 ± 1.70	1.50 ± 0.57
C × C	10	310.3 ± 123.0	1811	88.9 ± 9.7	9.10 ± 2.02	6.50 ± 1.50	1.0 ± 0
A × A	14	172.4 ± 113.4	2149	92.1 ± 30.2	9.0 ± 3.58	7.92 ± 1.44	1.15 ± 0.37
D × B	15	285.0 ± 109.6	2648	84.7 ± 32.6	7.41 ± 3.11	5.69 ± 2.15	1.0 ± 0 [b]
B × D	16	145.1 ± 68.0	2295	0.13 ± 0.63	6.31 ± 2.38	6.25 ± 2.2	1.33 ± 0.77 [c]
C × B	7	192.3 ± 122.1	1108	78.3 ± 23.2	6.13 ± 2.03	6.13 ± 2.23	1.14 ± 0.38
B × C	12	171.7 ± 120.3	1820	0.66 ± 1.4	7.09 ± 2.02	5.54 ± 1.75	1.20 ± 0.56
C × A	12	279.1 ± 89.1	2624	82.1 ± 28.5	9.58 ± 2.39	8.09 ± 2.42	1.27 ± 0.64
A × C	16	193.4 ± 113.1	2327	75.3 ± 38.3	8.71 ± 2.37	6.25 ± 1.73	1.27 ± 0.45
D × A	10	254.5 ± 106.7	1881	67.4 ± 30.3	10.2 ± 2.48	8.5 ± 2.46	1.0 ± 0
A × D	6	208.3 ± 125.5	825	92.5 ± 34.7	9.5 ± 3.61	7.17 ± 2.31	1.2 ± 0.44

a   Actively moving large numbers of apparently eupyrene sperm were found in spermatheca of at least one female from every cross (for crosses B × C and B × D, 3 and 2 females were dissected respectively).
b   Only 5 females were dissected.
c   Only 9 females were dissected.

TABLE II. SEX RATIO AND OTHER BIOLOGICAL DATA OF THE PROGENY OF SOME CROSSES BETWEEN DIFFERENT STRAINS OF Cadra (Ephestia) cautella Walker

Crosses Female × male	No. eggs incubated	No. larvae	Percentage pupation	No. adults emerged	Percentage malformed moths	Percentage males [a]
A × A	3864	–	47.28	1199	11.84	49.04
B × B	2743	–	43.37	706	18.98	51.70
C × C	2019	–	32.64	290	15.0	54.83
D × D	2564	–	44.19	646	11.45	52.63
C × B	141	–	58.87	74	9.64	43.24
B × C	95	–	0.0	0	–	–
D × B	527	–	69.67	499	2.73	53.11
	–	462				
B × D	350	–	0.0	0	–	–
D × A	568	–	35.21	127	19.68	49.61
A × D	425	–	61.65	139	11.51	43.17
$F_1^b \times F_1^b$	221	158	36.20	39	15.38	64.10
(Unknown)	–	–	–	335 [c]	0.30	49.55

a In all cases the male : female ratio does not significantly deviate from 1:1 as determined by X^2 test at the 5% level of probability.

b F_1 moths obtained from crossing D♀♀ × B♂♂ (see column 7).

c Field-collected moths of different colours (March–April 1974) in a small room used as a store (containing about 100 kg of dry dates, Zahdi variety), in a house situated near a dry-date store in Baghdad.

TABLE III. THE EFFECT ON HATCHING PERCENTAGE OF CONFINING ONE MALE OF Cadra (Ephestia) cautella STRAIN C WITH A PAIR OF STRAIN B (1 B♀ × 1 B♂)

Crosses Female : male : male	No. of replicates (pairs)	No. of eggs/female	Percentage hatch	Reduction in egg hatch (%)	
				Theoretical	Observed
1 B♀ : 1 B♂ : O	8	181.25	74.62		
1 B♀ : 1 B♂ : 1 C♂	8	290.63	34.66	50	53.55

3. RESULTS AND DISCUSSION

Results of some crosses between the four different <u>Cadra</u> (<u>Ephestia</u>) <u>cautella</u> varieties (strains) are summarized in Table I. Any cross appeared to be normally fertile when the male was of B strain, whereas the recipro- cal crosses (e.g. B ♀ × C ♂ or B ♀ × D ♂) were only slightly fertile (0.66% and 0.13% hatch respectively), and gave no progeny when 95 and 350 eggs of both crosses were separately incubated on appropriate medium (Table II). All other types of inter- or intrastrain crosses shown in Table I exhibited normal fertility. It is worthwhile to mention that more than 50% of the eggs laid by females of strains C and D mated with B males developed to adulthood where even the sex ratio was not distorted on a basis of 1 ♀ : 1 ♂ ratio (Table II). However, the low percentage of egg hatch was also noticed when F_1 males (males of the first generation of the crosses C female × B male, and D female × B male) were crossed with females of strain B. Further information concerning such a slight fertility in both parental crosses and backcrosses is withheld pending the collection of more data. If this low fertility is due to a real development of B female eggs inseminated by spermatozoa of C or D males (and not to other factors), then the underlying mechanisms of incompatibility between the different strains (or varieties) are probably controlled by both cytoplasmic factors (for further details see Ref.[6]) and chromosomal genes. This premature conclusion needs to be experimentally confirmed. However, it seems that there is no absolute barrier to gene exchange between these three Iraqi forms. It is also thought that this cytoplasmic incompatibility probably acts as a kind of an isolating mechanism that maintains to a certain degree the different forms (varieties) in a single population.

It is interesting to note that F_1 individuals of one of the intervarietal crosses were fertile when intercrossed, and gave offspring with more males but the sex ratio was not significantly different from 1 : 1 (Table II). From this Table it is also seen that malformed moths have been frequently encountered in the progeny of inter- or intrastrain crosses as well as among the moths captured in the field.

The classification of 335 moths captured in the field during a period of less than one month showed that 79 (23.6%) of them resembled strain B, 3 (0.9%) were similar to strain D or A, and only 1 male (0.3%) resembled the dark brown adult of strain C. The remainder (75.2%) presumably resembled the different heterozygotes of various crosses between the 3 Iraqi varieties (28.26% were similar to the progeny of O and A female × B male [2]. One female moth was obviously paler than strain B colour.

It is, therefore, concluded, as stated above, that there are no intrinsic crossing barriers between these varieties since hybrids have been found. About eight intermediate colour shades were observed. This variation in the colour of the forewings seems to be multifactorial, a well-known fact in other moths [7]. Some of the colours of the heterozygotes may have selective advantages in different habitats where there are different varieties of dates and other dry foodstuffs. In order to prove this notion and to collect information about the speciation and geographical distribution of different forms or races of <u>C. cautella</u>, extensive surveys are urgently needed in the different regions of Mesopotamia as well as other countries, especially neighbouring ones. Such surveys will certainly throw more light on the nature and kind of the polymorphism discussed in this report. However, the present data indicate that heterozygotes are at an advantage if mere numbers are considered which might suggest the existence of a kind of balanced polymorphism as defined by Ford [8]. Here it is necessary to recall that a kind of hybrid vigour (heterosis) did appear in F_1 moths [1,2] of the fertile interstrain cross (A♀ × B♂).

As previously mentioned, C. cautella heterozygotes were the most common forms found in nature, and at the same time mated pairs caught in the field have shown that heterogamic matings are much more frequent than homogamic, a fact that confirms the occurrence of gene flow between the different varieties.

Turning now to the problem of insect control, it is evident that males of strain C (or D) could be successfully used to suppress the population of strain B. Data of eight replicates (Table III) indicated that when one male of strain C was placed in a vial with one male and one female of strain B a significant (at 1% level of probability) reduction in egg hatch occurred. It is also possible that F_1 males of the fertile crosses between these strains (C or D female × B male) can play the same role as C males.

Finally, it is obvious that the phenomenon of incompatibility, whatever the underlying mechanisms, may be exploited in the suppression of certain populations (varieties or forms) of this insect of economic importance [9,10]. Apparently it is also possible to suppress a population consisting of several (sympatric) forms through combining simultaneously both the incompatibility phenomenon and the principles of radiation-induced sterility.

Also, moths of strains having different colours from wild populations can be easily used in tests requiring marked moths instead of using chemical dyes or radioisotopes in order to measure population density and dispersal, or to study insect behaviour as indispensable preparatory steps for sterile-insect-release programmes and other pest-control purposes.

ACKNOWLEDGEMENTS

The authors wish to thank the following staff of this Institute for their field and laboratory assistance: J. Jassim, K. Dawood and H. Jassim. Special thanks are also extended to M. Abdullah for his help in photography.

REFERENCES

[1] AHMED, M.S.H., ALHASSANY, I.A., LAMOOZA, S.B., OUDA, N.A., Crossing experiments with two Ephestia cautella strains (unpublished).

[2] AHMED, M.S.H., ALHASSANY, I.A., LAMOOZA, S.B., OUDA, N.A., "Cross-mating studies on two strains of the fig moth, Ephestia cautella (Walker)", submitted to the Third FAO Technical Conference on the Improvement of Date Production, Processing and Marketing.

[3] NORTH, D.T., HOLT, G.G., "Genetic and cytogenetic basis of radiation-induced sterility in the adult male cabbage looper Trichoplusia ni", Isotopes and Radiation in Entomology (Proc. Symp. Vienna, 1967), IAEA, Vienna (1968) 391.

[4] KARPENKO, C.P., NORTH, D.T., "Ovipositional response elicited by normal, irradiated, F_1 male progeny, or castrated male Trichoplusia ni (Lepidoptera: Noctuidae), Ann. Entomol. Soc. Am. 66 (1973) 1278.

[5] FORD, E.B., Genetic research in the Lepidoptera, Ann. Eugen. 10 (1940) 227.

[6] LAVEN, H., "Speciation and evolution in Culex pipiens", Genetics of Insect Vectors of Disease (WRIGHT, J.W., PAL, R., Eds), Elsevier, Amsterdam (1967) 251.

[7] FORD, E.B., Moths, 3rd Edn, Collins, London (1972).

[8] FORD, E.B., The genetics of polymorphism in the Lepidoptera, Adv. Genet. 5 (1953) 43.

[9] AHMED, M.S.H., AL-HAKKAK, Z.S., AL-SAQUR, A.M., "Inherited sterility in the fig moth,
     Cadra (Ephestia) cautella Walker", Int. Conf. peaceful Uses atom. Energy (Proc. Conf. Geneva,
     1971) 12, UN, IAEA, Vienna (1972) 383.

[10] AHMED, M.S.H., OUDA, N.A., LAMOOZA, S.B., ALHASSANY, I.A., Effect of gamma radiation
     on some embryonic stages of two stored-date insect species, Acta Aliment. (in press).

DISCUSSION

K. SYED: It appears from your data that you are dealing with biologically distinct entities regardless of whether or not they are considered to be of the same species taxonomically. This sort of situation should be examined more carefully from the ecological point of view, as I feel there are likely to be significant differences which would be of very great practical importance in controlling these insects.

It looks as though these strains are in the process of evolution and this should be ascertained before attempting any kind of genetic manipulation, because it is possible that the course of action adopted might simply serve to accelerate the natural process of evolution and have no useful effect whatever.

H. LEVINSON: The malformed phycitids you have shown resemble the malformed moths produced by omitting polyenic fatty acids from the larval diet. Have you established whether the linoleate or linolenate metabolism of the strains used is defective?

M.S.H. AHMED: No investigations of that kind have been conducted.

H. LEVINSON: Have you any idea whether the females of your different strains display the usual calling behaviour before copulation?

M.S.H. AHMED: As far as I know, yes, they do.

B.S. FLETCHER: In your mating experiments with a multiple choice situation do you find evidence of a "strange male" effect similar to that reported by workers on Drosophila?

M.S.H. AHMED: Our male-choice tests reported in Ref.[2] revealed that B males, when placed with equal numbers of A and B females for two days, mated more frequently with A females, and thus exhibited a significant degree of negative assortative mating. As for males of strain A, the significance test indicated no deviation from random mating, since they mated as often with B as with A females.

STERILITE HEREDITAIRE CHEZ Gonocerus acuteangulatus Goeze (Rhynchote, Coreidae)*

G. DELRIO
Istituto di Entomologia Agraria,
Università degli Studi,
Padoue

R. CAVALLORO
Groupe de biologie,
Direction générale Recherche, science et éducation,
CCE, Centre commun de recherche,
Ispra (Varese),
Italie

Abstract—Résumé

INHERITED STERILITY IN Gonocerus acuteangulatus Goeze (Rhynchota, Coreidae).
Researches have been made on sterility induced in Gonocerus acuteangulatus Goeze by ionizing radiations. The males are particularly radiosensitive and lose their mating ability at doses greater than 8 krad; on the other hand, the sexual behaviour of females remains unaltered. Examination of the effects of substerilizing doses for male and female lines on the following generations reveals an inherited sterility. In the F_1 generation the sterility was greater than that of the irradiated parents and was reduced in the following generations F_2 and F_3. In addition, mortality of larvae from eggs laid by substerile insects and a distortion of the sex ratio in favour of males were revealed. Since this species is particularly harmful to Corylus avellana L. and is responsible for two harmful diseases in hazel nuts, possible applications using the substerile male technique taking into account the greater competitiveness of F_1 males rather than that of directly irradiated insects are discussed.

STÉRILITÉ HÉRÉDITAIRE CHEZ Gonocerus acuteangulatus Goeze (Rhynchote, Coreidae).
Des recherches ont été effectuées sur la stérilité de Gonocerus acuteangulatus Goeze irradié. Les mâles sont particulièrement sensibles et perdent la capacité de s'accoupler aux doses supérieures à 8 krad; les femelles au contraire ne modifient pas leur comportement sexuel. L'examen des effets de doses substériles pour les lignes masculine et féminine sur les générations successives a révélé une stérilité héréditaire. On a trouvé que dans la génération F_1 la stérilité est supérieure à celle des parents irradiés et qu'elle diminue successivement dans F_2 et encore plus dans F_3. On a aussi mis en évidence un certain pourcentage de mortalité des larves écloses d'œufs pondus par les insectes substériles et une modification du rapport des sexes à l'avantage des mâles. Comme cette espèce est particulièrement nuisible au Corylus avellana L. et qu'elle cause deux graves altérations, le «vide traumatique» et le «punaisé» des noisettes, on envisage des applications pratiques éventuelles, dans le cadre d'une lutte intégrée, au moyen de mâles substériles, compte tenu de la compétitivité plus élevée des mâles F_1 par rapport aux insectes directement irradiés.

INTRODUCTION

L'Hétéroptère Gonocerus acuteangulatus Goeze s'est révélé l'espèce la la plus nuisible au noisetier dans presque toute l'Italie, où il cause deux des plus graves altérations aux fruits: le «vide traumatique» et le «punaisé» des noisettes [1, 2].

---

* Contribution du Programme de biologie, Direction générale XII de la Commission des Communautés européennes.

423

Le dégât dû à cette espèce acquiert une importance économique considérable, car il dépasse souvent la limite de tolérance commerciale.

La lutte contre cet insecte est particulièrement difficile; l'emploi actuel de produits organochlorés peut conduire à de graves déséquilibres de la biocœnose complexe du noisetier [3].

C'est pour cette dernière raison qu'on est en train d'étudier la possibilité d'une lutte intégrée par l'utilisation de parasites oophages et de la technique du mâle stérile [4].

Jusqu'à présent, on n'a étudié que chez quelques Rhynchotes la possibilité d'induire des mutations létales dominantes au moyen de rayonnements ionisants: par exemple chez Rhodnius prolixus (Stål) [5], Perkinsiella saccharicida (Kirk) [6], Dysdercus peruvianus G. [7], Oncopeltus fasciatus (Dallas) [8], Lygus hesperus Knight [9], Circulifer tenellus Baker [10]. Les Rhynchotes possèdent des chromosomes holocinétiques [11] et offrent une radiorésistance inférieure à celle de l'autre ordre d'insectes à chromosomes holocinétiques, les Lépidoptères [12].

De même que chez plusieurs Lépidoptères [13-18] on a observé chez un Hétéroptère, Oncopeltus fasciatus Dallas, la propriété de transmission de la stérilité aux générations successives, due à des fragmentations et à des réarrangements chromosomiques complexes à la suite de l'irradiation des parents à des doses substériles [8, 19].

Cette étude a porté en particulier, pour le Gonocerus acuteangulatus, sur l'effet des rayonnements gamma sur le comportement sexuel des adultes, sur l'effet de doses substériles différentes sur les mâles et les femelles de trois générations successives, sur le développement embryonnaire des œufs pondus par des femelles normales accouplées avec des mâles substériles, et sur la compétitivité sexuelle et le taux d'éclosion des œufs pondus par des femelles accouplées alternativement avec des mâles normaux ou stériles.

MATERIEL ET METHODES

Les adultes de Gonocerus acuteangulatus ont été récoltés dans des noiseraies de Sicile et élevés au laboratoire à 28°C de température, 75% d'humidité relative et 12 heures de photopériode sur de petites plantes de Buxus sempervirens L. et des akènes de Castanea sativa Mill. [20].

Les essais ont été conduits pendant deux ans. L'appareil utilisé pour les irradiations était un Gammacell 220, avec une source de ^{60}Co et un débit initial de dose égal à 1866 R/min ± 2,5%.

Lors des irradiations effectuées dans les mêmes conditions de géométrie à température ambiante (22 ± 1°C), on a toujours tenu compte de la décroissance physique de la source et les doses ont été calculées sur la base du temps d'exposition des insectes.

Les adultes irradiés ont été accouplés avec des individus vierges du sexe opposé, et chaque couple a été tenu isolé dans de petites cages. Les œufs pondus ont été ramassés tous les 5 à 7 jours et gardés en boîtes de Pétri à 27°C et 75% HR, pour le contrôle des éclosions. Les œufs ont été classés en quatre groupes: les œufs éclos; ceux qui présentaient un chorion recroquevillé; les œufs non éclos où le développement embryonnaire s'était arrêté au stade 1 ou au stade 2.

Les œufs qui n'avaient pas été fécondés présentaient donc un chorion recroquevillé sans forme subsphérique; les œufs au stade 1, fécondés, avaient une forme normale, mais sans aucune trace de développement embryonnaire; les œufs au stade 2 présentaient un embryon développé, où l'on pouvait entrevoir les yeux et les segments abdominaux.

Les larves obtenues de tous les couples irradiés à la même dose ont été tenues dans la même cage; le nombre d'adultes a été contrôlé. Les adultes F_1 et F_2 ont été accouplés entre eux ou avec des insectes normaux, selon toutes les combinaisons possibles, et tenus en couples séparés, tandis que les larves, pour le même type d'accouplement, ont été tenues dans la même cage.

Les essais de compétitivité sexuelle ont été faits en plaçant dans la même cage une femelle normale avec un mâle stérile et un mâle normal, respectivement marqué et non marqué avec des poudres fluorescentes (Day-Glo A-17, Switzer Brothers Inc., Cleveland, Etats-Unis); les accouplements ont été contrôlés ainsi que l'éclosion des œufs pondus. D'autres essais ont été effectués, en faisant accoupler la femelle avec un mâle normal et laissant ensuite la femelle avec un mâle stérile et vice versa, et en contrôlant l'éclosion des œufs.

RESULTATS ET DISCUSSION

Transmission de la stérilité à trois générations successives

La stérilité augmente chez les deux sexes en fonction de la dose (fig. 1), mais s'il est possible d'obtenir la stérilisation complète de la femelle, elle n'est pas atteinte chez les mâles, puisqu'ils perdent leur capacité de s'accoupler déjà à partir d'une dose de 8 krad (avec 79,6% seulement de mutations létales dominantes).

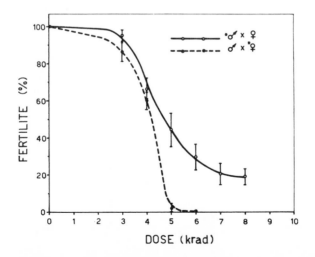

FIG. 1. Effet de l'irradiation gamma sur la fertilité des mâles et des femelles de Gonocerus acuteangulatus Goeze.

TABLEAU I. Gonocerus acuteangulatus Goeze: EFFET RETARDE DE LA STERILITE DANS LES GENERATIONS SUCCESSIVES (LIGNE MASCULINE) DES MALES SOUMIS A UNE IRRADIATION GAMMA

Génération	Nombre de couples étudiés	Nombre d'œufs	Eclosions (%)	Intervalle (% d'éclosions)	Nombre de couples par groupes d'éclosions				
					0 à 20 éclosions	21 à 40 éclosions	41 à 60 éclosions	61 à 80 éclosions	81 à 100 éclosions
Dose: 4 krad									
P♂ (4 krad) × N♀	10	243	66,66	43,8 - 86,5	-	-	2	7	1
F₁♂ × N♀	8	167	24,5	0 - 83,3	5	2	-	-	1
N♂ × F₁♀	5	125	42,4	4,6 - 73,7	1	1	2	1	1
F₁♂ × F₁♀	9	655	18,1	0 - 77,6	5	3	-	1	-
F₂♂ × N♀	4	306	88,2	81,2 - 100,0	-	-	-	2	2
N♂ × F₂♀	6	138	82,6	75,4 - 100,0	-	-	1	1	4
F₂♂ × F₂♀	3	96	57,2	43,5 - 100,0	-	-	2	-	1
F₃♂ × N♀	4	119	96,6	80,0 - 100,0	-	-	-	1	3
N♂ × F₃♀	5	150	87,3	0,0 - 100,0	1	-	-	1	3
F₃♂ × F₃♀	3	147	72,1	59,0 - 100,0	-	-	1	1	1
Dose: 5 krad									
P♂ (5 krad) × N♀	10	379	44,5	21,5 - 70,2	-	3	5	2	-
F₁♂ × N♀	10	180	30,0	0 - 100,0	4	4	-	-	2
N♂ × F₁♀	5	116	48,2	23,4 - 64,5	-	1	3	1	-
F₁♂ × F₁♀	8	271	31,7	0 - 68,3	3	2	1	2	-
F₂♂ × N♀	5	168	33,9	24,9 - 100,0	-	3	1	-	1
N♂ × F₂♀	3	75	70,6	31,3 - 100,0	-	1	-	-	2
F₂♂ × F₂♀	5	128	56,2	6,7 - 88,5	1	-	2	1	1
F₃♂ × N♀	5	146	59,5	15,3 - 100,0	1	2	-	-	2
N♂ × F₃♀	2	48	52,0	23,5 - 67,7	-	1	-	1	-
F₃♂ × F₃♀	4	51	43,1	12,1 - 79,6	2	-	1	1	-

Dose: 6 krad

P ♂ (6 krad) × N ♀	12	342	26,9	17,5 – 38,6	4	8	–	–	–
F₁ ♂ × N ♀	7	186	21,5	0 – 83,4	4	2	–	–	1
N ♂ × F₁ ♀	6	172	36,6	0 – 79,5	3	2	1	1	–
F₁ ♂ × F₁ ♀	9	249	18,4	0 – 43,2	6	2	1	–	–
F₂ ♂ × N ♀	6	187	65,2	33,6 – 100,0	–	3	1	–	2
N ♂ × F₂ ♀	6	210	47,1	18,5 – 100,0	1	2	–	1	2
F₂ ♂ × F₂ ♀	5	143	27,2	0 – 65,2	2	1	1	1	–
F₃ ♂ × N ♀	6	184	90,2	32,5 – 100,0	–	1	1	–	4
N ♂ × F₃ ♀	6	196	73,4	47,6 – 100,0	–	–	2	2	2
F₃ ♂ × F₃ ♀	6	285	59,6	0 – 79,8	2	1	1	2	–

Dose: 7 krad

P ♂ (7 krad) × N ♀	15	457	20,5	8,1 – 41,4	10	4	1	–	–
F₁ ♂ × N ♀	6	205	37,0	0 – 79,4	2	2	1	1	–
N ♂ × F₁ ♀	6	294	30,2	15,2 – 44,5	2	3	1	1	–
F₁ ♂ × F₁ ♀	4	378	5,8	0 – 22,1	3	1	–	–	–
F₂ ♂ × N ♀	6	196	70,9	0 – 100,0	1	–	1	2	2
N ♂ × F₂ ♀	5	295	58,9	18,7 – 97,8	1	1	–	1	2
F₂ ♂ × F₂ ♀	5	231	36,3	0 – 51,8	2	2	1	1	–
F₃ ♂ × N ♀	4	180	83,3	25,0 – 100,0	–	1	1	1	2
N ♂ × F₃ ♀	3	102	70,5	53,5 – 100,0	–	–	1	–	2
F₃ ♂ × F₃ ♀	10	326	47,8	0 – 66,6	1	3	5	1	–

TABLEAU I (suite)

Génération	Nombre de couples étudiés	Nombre d'œufs	Eclosions (%)	Intervalle (% d'éclosions)	Nombre de couples par groupes d'éclosions				
					0 à 20 éclosions	21 à 40 éclosions	41 à 60 éclosions	61 à 80 éclosions	81 à 100 éclosions
				Dose: 8 krad					
P ♂ (8 krad) × N ♀	13	415	20,4	15,3 - 37,6	11	2	-	-	-
F$_1$ ♂ × N ♀	10	352	5,3	0 - 33,3	9	1	-	-	-
N ♂ × F$_1$ ♀	6	175	9,8	0 - 56,4	4	1	1	-	-
F$_1$ ♂ × F$_1$ ♀	9	194	4,1	0 - 18,2	9	-	-	-	-
F$_2$ ♂ × N ♀	4	156	51,2	28,4 - 100,0	-	1	2	-	1
N ♂ × F$_2$ ♀	3	113	46,9	23,2 - 75,7	-	2	-	1	-
F$_2$ ♂ × F$_2$ ♀	3	86	25,5	6,6 - 37,9	1	2	-	-	-
F$_3$ ♂ × N ♀	6	215	93,9	51,8 - 100,0	-	-	1	-	5
N ♂ × F$_3$ ♀	4	198	68,1	47,3 - 100,0	-	-	2	-	2
F$_3$ ♂ × F$_3$ ♀	3	172	83,1	25,5 - 100,0	-	1	-	-	2

On a remarqué un comportement analogue avec incapacité d'accouplement chez le Pentatomidae Nezara viridula L. irradié à 10 krad [21].

Les femelles s'accouplent normalement jusqu'à la dose de 30 krad, mais elles interrompent la ponte des œufs à 6 krad, tandis qu'à 5 krad elles pondent environ la moité moins d'œufs que les témoins. La durée de vie des insectes irradiés aux doses considérées ne présente pas de différences significatives par rapport à celle des témoins; des doses supérieures à 15 krad sont nécessaires pour relever une augmentation de mortalité.

On sait que quelques Rhynchotes offrent une baisse de vitalité des individus irradiés aux doses stérilisantes, par exemple l'Omoptère Cicadellidae Circulifer tenellus (Baker) [10] ou l'Hétéroptère Rhodnius prolixus Stål [22].

Les mâles F_1 accouplés avec des femelles normales ont présenté une stérilité supérieure à celle des couples parentaux (tableau I). Ce n'est que quand le mâle de la génération parentale a été irradié à 7 krad que l'éclosion des œufs de la génération F_1 a été, inexplicablement, plus élevée. On n'a remarqué aucune corrélation entre le degré de stérilité de la génération parentale, qui augmente proportionnellement à la dose, et la stérilité de la génération F_1.

Aux doses de 4, 5, 6 et 7 krad, F_1 présente une stérilité moyenne entre 78,5% et 63%; ce n'est qu'à la dose de 8 krad que la stérilité de F_1 est sensiblement plus élevée (jusqu'à 94,7%).

Les femelles F_1 accouplées à des mâles normaux ont une stérilité presque toujours plus élevée que celle des mâles F_1; l'éclosion des œufs, par contre, est beaucoup plus réduite après les accouplements entre mâles F_1 et femelles F_1.

Dans les générations F_2 et F_3 en ligne masculine, la stérilité est considérablement réduite, mais les éclosions restent toujours inférieures à celles des témoins.

L'analyse de la fertilité de chaque couple des différentes générations suivant la génération parentale et lors de tous les accouplements possibles démontre que cette fertilité est très variable dans la plupart des cas (entre 0 et 100).

Par exemple, de tous les mâles F_1 obtenus de parents irradiés à n'importe quelle dose, plus de la moitié (58%) ne dépasse pas 20% de fertilité, et 27% environ des autres ne dépasse pas 40%.

Pour les doses de 7 et 8 krad, on a contrôlé quelques couples des générations F_2, F_3 et F_4, descendant en ligne féminine de femelles F_1 accouplées avec mâles normaux. D'après les données obtenues (tableau II) il n'est pas possible de tirer de conclusions générales, sauf que la stérilité héréditaire peut se manifester même en ligne féminine jusqu'à la quatrième génération.

Effets sur le développement embryonnaire et post-embryonnaire

L'étude du stade d'arrêt du développement embryonnaire peut donner des indications sur l'importance des dommages génétiques induits dans les gamètes des mâles [23].

Nous avons effectué cette recherche sur Gonocerus acuteangulatus, sur les œufs pondus par les insectes des différentes générations F_1, F_2, F_3 descendant de mâles irradiés à 7 krad.

TABLEAU II. Gonocerus Acuteangulatus Goeze: EFFET RETARDE DE LA STERILITE DANS LES GENERATIONS SUCCESSIVES (LIGNE FEMININE) DES MALES SOUMIS A UNE IRRADIATION GAMMA

Génération	Nombre de couples étudiés	Nombre d'œufs	Eclosions (%)	Intervalle (% d'éclosions)	Nombre de couples par groupes d'éclosions				
					0 à 20 éclosions	21 à 40 éclosions	41 à 60 éclosions	61 à 80 éclosions	81 à 100 éclosions
Dose: 7 krad									
P♂ (7 krad) × N♀	15	457	20,5	8,1 – 41,4	10	4	1	-	-
N♂ × F₁♀	6	294	30,2	15,2 – 44,5	2	3	1	-	-
F₂♂ × N♀	1	55	100,0	100,0	-	-	-	-	1
N♂ × F₂♀	4	109	66,0	0 – 100,0	1	1	2	-	1
F₂♂ × F₂♀	1	28	25,0	-	-	1	-	-	-
F₃♂ × N♀	2	37	35,1	27,3 – 100,0	-	1	-	-	1
N♂ × F₃♀	2	85	51,7	47,5 – 100,0	-	1	1	-	1
F₃♂ × F₃♀	7	351	42,7	35,1 – 100,0	-	2	3	-	2
F₄♂ × N♀	2	52	75,0	25,4 – 100,0	-	1	-	-	1
N♂ × F₄♀	-	-	-	-	-	-	-	-	-
F₄♂ × F₄♀	1	3	33,3	33,3	-	1	-	-	-
Dose: 8 krad									
P♂ (8 krad) × N♀	13	415	20,4	15,3 – 37,6	11	2	-	-	-
N♂ × F₁♀	6	175	9,8	0 – 56,4	4	1	1	-	-
F₂♂ × N♀	3	91	38,4	0 – 55,4	1	-	2	-	-
N♂ × F₂♀	3	68	10,2	0 – 45,8	1	1	1	-	-
F₂♂ × F₂♀	1	21	14,2	14,2	1	-	-	-	-

Les œufs dont le développement s'arrête à une phase avancée représentent dans la génération parentale la moitié environ des œufs non éclos; cette proportion augmente toujours plus dans les générations suivantes (tableau III). On peut expliquer cela en supposant que les altérations aux gamètes dans F_1, F_2, F_3 sont inférieures à celles de la génération parentale. Dans les croisements où les deux sexes dérivent d'un parent irradié, il semble que le développement des œufs s'arrête à un stade plus précoce, et cela peut-être à cause de l'accumulation des effets létaux des deux gamètes.

Le pourcentage d'adultes obtenus des larves est inférieur à celui des témoins dans la génération parentale et surtout dans les générations F_1 et F_2 (tableau III).

Il semble donc logique de supposer que les effets létaux se manifestent non seulement au stade embryonnaire, mais aussi au stade post-embryonnaire.

Par contre on n'a relevé aucune différence dans le nombre d'adultes F_1 obtenus de larves dérivant de mâles irradiés aux différentes doses. Le rapport entre les sexes, qui chez les insectes normaux est de $1:1$ environ, s'est élevé pour F_1 à $1,4:1$ en faveur des mâles. Ce fait a déjà été observé, par exemple dans le cas des Lépidoptères Pectinophora gossypiella (Saunders) [18] et Heliothis virescens (F.) [16].

Compétitivité sexuelle des mâles

Le comportement sexuel des mâles et des femelles se révèle polygame: on relève un nombre élevé d'accouplements successifs, de longue durée, suivis de pontes régulières.

L'irradiation abaisse fortement la compétitivité sexuelle des mâles. Des essais préliminaires ont été effectués à la dose de 7 krad. Quand les mâles irradiés ont été mis en présence de couples normaux, $1/10$ seulement des accouplements des femelles a eu lieu avec le mâle stérile.

Le nombre d'accouplements du mâle irradié est inférieur à celui du mâle normal.

Accouplements moins fréquents et compétitivité sexuelle réduite ont été aussi relevés chez les mâles stériles de Rhodnius prolixus Stål [22]. Les mâles F_1 de Gonocerus acuteangulatus, au contraire, présentent une compétitivité de $0,9$, presque égale donc à celle des insectes normaux; elle a été calculée sur la base de l'éclosion des œufs en population mixte de mâles stériles et couples normaux [24].

Le nombre moyen d'accouplements, en outre, est égal à celui des témoins. Quand des accouplements successifs de femelles non irradiées ont lieu la première fois avec des mâles normaux et la deuxième fois avec des mâles substériles, l'éclosion moyenne des œufs a été de 86,6%; par contre, quand on a interverti l'ordre des accouplements le taux d'éclosion a été de 68,5%.

Evidemment l'effet de l'irradiation se manifeste non seulement sur le comportement sexuel des mâles, mais aussi au moment de la fécondation de l'œuf.

CONCLUSIONS

En ce qui concerne la production de mutations létales dominantes chez Gonocerus acuteangulatus et d'autres Rhynchotes, les doses de rayonnements

TABLEAU III. Gonocerus acuteangulatus Goeze: ARRET DU DEVELOPPEMENT EMBRYONNAIRE ET POURCENTAGE D'ADULTES OBTENUS DANS LES GENERATIONS SUCCESSIVES (LIGNE MASCULINE) DES MALES SOUMIS A UNE IRRADIATION GAMMA DE 7 krad

Génération	Nombre de couples étudiés	Nombre d'œufs pondus	Œufs stériles (%)	Stade d'arrêt du développement (%)			Nombre d'adultes obtenus sur 100 larves
				0^a	1^b	2^c	
Témoins	16	524	0,2	0,2	0	0	65,8
P ♂(7 krad) × N ♀	15	457	79,5	1,1	52,8	46,1	60,3
F_1 ♂ × N ♀	6	205	63,0	0,8	51,1	48,1	50,4
N ♂ × F_1 ♀	6	294	69,8	0,5	34,0	65,5	47,2
F_1 ♂ × F_1 ♀	4	378	94,2	0,7	40,2	59,1	38,0
F_2 ♂ × N ♀	6	196	29,1	0,5	16,1	83,4	53,3
N ♂ × F_2 ♀	5	295	41,1	0,2	28,1	71,7	50,8
F_2 ♂ × F_2 ♀	5	231	63,7	1,3	35,7	63,0	33,3
F_3 ♂ × N ♀	4	180	16,7	0,7	12,9	86,4	66,3
N ♂ × F_3 ♀	3	102	29,5	0,6	23,4	76,0	35,4
F_3 ♂ × F_3 ♀	10	326	52,2	0,8	50,5	48,7	55,1

[a] Œufs recroquevillés.
[b] Développement embryonnaire initial.
[c] Développement embryonnaire avancé.

gamma sont notablement inférieures à celles nécessaires pour produire le même effet chez les Lépidoptères, et cela malgré la présence de chromosomes holocinétiques dans les deux ordres. Le comportement sexuel des mâles de Gonocerus acuteangulatus est gravement troublé par l'irradiation; en effet à 9 krad déjà, le mâle devient incapable de s'accoupler, tout en gardant une longévité tout à fait égale à celle de l'insecte non traité. Cet aspect rend impossible la stérilisation complète de l'adulte, puisque à 8 krad on n'obtient que 79,6% de stérilité chez le mâle.

En outre les mâles irradiés ont une compétitivité sexuelle réduite, qui se manifeste aussi par la diminution de la fréquence des accouplements. Les mâles F_1 descendant de parents substériles ont présenté une stérilité supérieure à celle de leurs parents et on n'a relevé ni réduction de compétitivité sexuelle ni diminution dans la fréquence d'accouplements.

De toute façon la stérilité baisse rapidement déjà dans la deuxième génération et encore plus dans la troisième.

Comme on sait, la transmission de la stérilité aux générations suivant la génération parentale irradiée a été déjà relevée chez l'Hétéroptère Oncopeltus fasciatus et mise en rapport avec la transmission de fragments et de réarrangements chromosomiques aux différentes générations [8, 19]. Chez Oncopeltus fasciatus la fertilité de F_1 est plus élevée que dans la génération parentale irradiée, tandis que chez Gonocerus acuteangulatus elle est inférieure; cela peut s'expliquer par le fait que dans cette dernière espèce on n'a pas atteint chez les parents un degré de stérilité aussi élevé que dans l'autre.

Pour ce qui concerne la stérilité héréditaire, le comportement similaire relevé dans les deux espèces fait supposer un comportement égal même au point de vue cytogénétique.

Chez les Lépidoptères, autre ordre d'insectes qui présente une stérilité héréditaire dans les générations successives [25], il existe une corrélation directe entre la dose délivrée à la génération parentale, la mortalité larvaire et le degré de stérilité de F_1, par exemple chez Laspeyresia pomonella L. [26] et Pectinophora gossypiella (Saunders) [18].

Ce phénomène n'est pas confirmé chez Gonocerus acuteangulatus, où la mortalité larvaire et la stérilité de F_1 semblent indépendantes de la dose délivrée aux parents.

En outre, on a observé chez Gonocerus acuteangulatus que la mortalité des œufs pondus par des femelles normales accouplées avec des mâles irradiés se manifeste tard dans le développement embryonnaire.

Ce fait est connu pour les Lépidoptères [25] et l'Hétéroptère Oncopeltus fasciatus, où la mortalité embryonnaire intervient généralement après la formation du blastoderme [23].

Une autre observation particulièrement importante est la modification à l'avantage des mâles du rapport entre les sexes dans la F_1 de Gonocerus acuteangulatus, comme cela a déjà été signalé pour les Lépidoptères. Pour ce qui est d'une application pratique éventuelle, il est important d'avoir constaté le phénomène de la stérilité héréditaire de Gonocerus acuteangulatus puisque les insectes directement irradiés se révèlent très sensibles aux rayonnements et peu compétitifs, ce qui fait supposer que la technique des lâchers de mâles stériles serait d'application difficile.

En outre, la lutte au moyen des mâles stériles pourrait être insérée dans le cadre d'une lutte intégrée, même avec l'emploi de parasites oophages, dont l'un, l'Hyménoptère Chalcididæ Anastatus bifasciatus Fonsc.,

se développe normalement (Genduso, communication personnelle) sur les
œufs stériles à embryon au stade 2.

Naturellement, d'autres recherches seront nécessaires pour mieux
éclaircir les différents aspects de la biologie et du comportement de
l'insecte normal et stérile en laboratoire et en plein champ.

REFERENCES

[1]   BOSELLI, F.B., Studio biologico degli Emitteri che attaccano le nocciuole in Sicilia, Boll. Lab. Zool.
      Gen. Agr. Portici 26 (1932) 142-309.
[2]   GENDUSO, P., Esperimenti di lotta contro gli Eterotteri del Nocciuolo eseguiti in Sicilia negli anni
      1954 e 1955, Boll. Ist. Ent. Agr. Palermo 2 (1956) 57-85.
[3]   VIGGIANI, G., «Nematodi, acari e insetti dannosi al nocciuolo», C.R. 3e J. Phyt. Phytoph. Circum-
      méditer. Sassari (1971) 390-98.
[4]   GENDUSO, G., «Gli artropodi dannosi al nocciuolo in Sicilia e relativi metodi di lotta», C.R. 3e
      J. Phyt. Phytoph. Circum-méditer. Sassari (1971) 339.
[5]   BALDWIN, W.F., SHAVER, E.L., Radiation induced sterility in the insect Rhodnius prolixus, Can. J.
      Zool. 41 (1963) 637-48.
[6]   SHIPP, E., OSBORN, A.W., HUTCHINSON, P.B., Radiation sterilization of the sugar cane leafhopper,
      Perkinsiella saccharicida, of the family Deltocephalidae, Nature (London) 2111 (1966) 98-99.
[7]   SIMON, J.E., «Cría masal de Dysdercus peruvianus G. y su esterilización mediante rayos gamma»,
      Isotopes and Radiation in Entomology (Proc. Symp. Vienna, 1967), IAEA, Vienna (1968) 287-300.
[8]   LaCHANCE, L.E., DEGRUGILLIER, M., Chromosomal fragments transmitted through three generations
      in Oncopeltus (Hemiptera), Science 166 (1969) 235-36.
[9]   STRONG, F.E., SHELDAHL, J.A., HUGHES, P.R., HUSSEIN, E.M.K., Reproductive biology of
      Lygus hesperus Knight, Hilgardia 40 (1970) 105-47.
[10] AMERESEKERE, R.V.W.E., GEORGHIOU, G.P., Sterilization of the beet leafhopper: Induction of
      sterility and evaluation of biotic effects with a model sterilant (OM-53139) and ^{60}Co irradiation, J. Econ.
      Entomol. 64 (1971) 1074-80.
[11] HUGHES-SCHRADER, S., SCHRADER, F., The Kinetochore of the Hemiptera, Chromosoma 12 (1961)
      327-50.
[12] BAUER, H., Die kinetische Organisation der Lepidopteren Chromosomen, Chromosoma 22 (1967) 102-25.
[13] COGBURN, R.T., TILTON, W., BURKHOLDER, W.E., Gross effects of gamma radiation on the Indian-
      meal moth and Augoumois grain moth, J. Econ. Entomol. 57 (1966) 682-85.
[14] NORTH, D.T., HOLT, G.G., Inherited sterility in progeny of irradiated male cabbage loopers,
      J. Econ. Entomol. 61 (1968) 928-31.
[15] WALKER, D.W., QUINTANA, V., Inherited partial sterility among survivors from irradiation-eradication
      experiment, J. Econ. Entomol. 61 (1968) 318-19.
[16] PROSHOLD, F.J., BARTELL, J.A., Inherited sterility in progeny of irradiated male tobacco budworms:
      effects on reproduction, developmental time, and sex ratio, J. Econ. Entomol. 63 (1970) 280-85.
[17] ASHRAFI, S.H., TILTON, E.W., BROWER, J.H., Inheritance of radiation induced partial sterility in
      the Indian meal moth, J. Econ. Entomol. 65 (1972) 1265-68.
[18] CHENG, W., NORTH, D.T., Inherited sterility in the F_1 progeny of irradiated male pink bollworms,
      J. Econ. Entomol. 65 (1972) 1272-75.
[19] LaCHANCE, L.E., DEGRUGILLIER, M., LEVERICH, A.P., Cytogenetics of inherited partial sterility in
      three generations of the large milkweed bug as related to holokinetic chromosomes, Chromosoma 29
      (1970) 20-41.
[20] GENDUSO, F., MINEO, G., Difesa del nocciolo dagli artropodi dannosi, I. Possibilità di allevamento
      permanente in laboratorio del Gonocerus acuteangulatus Goeze, Boll. Ist. Ent. Agr. Oss. Fitopat.
      Palermo 8 (1972) 29-35.
[21] MAU, R., MITCHELL, W.C., ANWAR, M., Preliminary studies on the effects of gamma irradiation
      of eggs and adults of the Southern green stink bug, Nezara viridula (L.), Proc. Hawaiian Ent. Soc. 19
      (1967) 415-17.
[22] GOMEZ-NUNEZ, J.C., «Sterility and Chagas' disease vector control», Sterility Principle for Insect
      Control or Eradication (Proc. Symp. Athens, 1970), IAEA, Vienna (1971) 157-70.
[23] LaCHANCE, L.E., RIEMANN, J.G., Dominant lethal mutations in insects with holokinetic chromo-
      somes, I. Irradiation of Oncopeltus (Hemiptera : Lygaeidae) sperm and oocytes, Ann. Entomol. Soc.
      Am. 66 (1973) 813-19.

[24] DELRIO, G. , ANSELMI, L. , CAVALLORO, R. , Valutazione della competitività degli insetti a diversi
       livelli di sterilità, Boll. Lab. Ent. Agr. Portici, sous presse.
[25] NORTH, D. T. , HOLT, G. G. , Population control of Lepidoptera: The genetic and physiological basis,
       Manit. Entomol. 4 (1970) 53-69.
[26] FOSSATI, A. , STAHL, J. , GRANGES, J. , « Effect of gamma irradiation dose on the reproductive
       performance of the P and F₁ generations in the codling moth, Laspeyresia pomonella L. », Application
       of Induced Sterility for Control of Lepidopterous Populations (Proc. Panel Vienna, 1970), IAEA, Vienna
       (1971) 41-47.

DISCUSSION

G.W. RAHALKAR: In all your experiments you have used a very small number of pairs and I feel that it is rather hazardous to draw the conclusions which you have drawn on such a basis.

G. DELRIO: Your remark would be valid if one were simply to consider the data relating to a single dose but here one needs to take into account the whole range of doses applied. It should not be forgotten that Gonocerus acuteangulatus is a large insect and altogether we checked the fertility of about 400 pairs for all the doses employed. Besides, these were single pairs, which provide a better basis for analysis.

J. TICHELER: What is the area of hazelnut-growing land and what is the value of the crop?

G. DELRIO: I do not know the area involved but the insect certainly causes tremendous damage to the crops. In recent years the damage, especially that due to the disease caused by the insects which leaves a bitter taste in the nuts, has far exceeded the internationally agreed level.

A. FELDMANN: You say that you observed differences in radio-sensitivity in the frequency of induction of dominant lethality and delayed sterility after male or female irradiation. The stage at which the sperm and oocytes are irradiated in relation to meiosis is very important in understanding the differences in radiosensitivity. Could you tell me which stages of sperm and oocytes were irradiated?

G. DELRIO: Our investigations mainly involved irradiating adults aged from 24 to 48 hours, and no special study was made of the meiosis of the spermatozoids and oocytes. At that age the males possess mature spermatozoids and the females have fully formed eggs.

L.E. LaCHANCE: I would like to comment on Mr. Feldmann's question. I know nothing about this species but in Oncopeltus and other Hemiptera the eggs are laid while the oocyte nucleus is in the first meiotic stage. Therefore, as Mr. Delrio says that the females had mature eggs when they were irradiated, I suspect they were pre-meiotic or early MI stages.

B. NA'ISA: It would appear that sterility in males can only be measured up to 8 krad, since with higher doses no matings take place. Can you tell us why doses greater than 8 krad lead to failure of the males to mate? Is it due to damage to the gonads or some somatic cells?

G. DELRIO: It is difficult to explain this phenomenon in any species of insect and, in the particular case of the Hemiptera, very little is known about their sexual behaviour. It could be that complex reactions are triggered off in the male as a result of irradiation.

F.M. EVENS: Did doses of less than 8 krad cause any other changes in the behaviour or morphology of the insects, and were the F_1, F_2 and F_3 generations affected in any way?

G. DELRIO: We did not observe any morphological changes in the irradiated insects but the sexual behaviour of the directly irradiated males was always disturbed, even at doses less than 8 krad. The treatment had no effect at all on the sexual behaviour of the F_1, F_2 and F_3 generations.

INDUCTION OF STRUCTURAL CHROMOSOME MUTATIONS IN MALES AND FEMALES OF Tetranychus urticae Koch (Acarina:Tetranychidae)

A.M. FELDMANN
Association Euratom-Ital,
Wageningen,
The Netherlands

Abstract

INDUCTION OF STRUCTURAL CHROMOSOME MUTATIONS IN MALES AND FEMALES OF Tetranychus urticae Koch (Acarina: Tetranychidae).

Percentages of structural chromosome mutations in the F_1-progeny of Tetranychus urticae Koch were determined after irradiation of spermatids and prophase I oöcytes with 1.5 MeV neutrons and 250 kV X-rays. X-rays were more effective than neutrons at doses between 1.0 and 2.0 krad for the induction of heritable structural chromosome mutations in spermatids. At doses lower than 0.5 krad, neutrons were as effective as X-rays and at higher doses neutrons were more effective. Irradiation of prophase I oöcytes did not demonstrate differences in relative effectiveness of neutrons and X-rays at doses lower than 4 krad. At higher doses, X-rays were more effective. In general, the relative effectiveness of neutrons and X-rays for the induction of structural chromosome mutations is dose dependent. The association of structural chromosome mutations with recessive lethals in spermatids, is positively correlated with the irradiation dose. This observation is important from the point of view of the practical application of heritable structural chromosome mutations in genetic control, because the best probability to isolate a structural chromosome mutation in a viable homozygous condition, exists when the spermatids were irradiated at a low dose (0.2 - 0.5 krad X-rays). The holokinetic nature of the chromosomes of T. urticae is substantiated by the correlation between the radiobiological properties of T. urticae with those of species with holokinetic chromosomes.

INTRODUCTION

In the past, Tetranychus urticae (Koch) offered serious problems in chemical control in Dutch greenhouses, particularly where roses were grown. The species built up numerous resistances to organophosphorous compounds and specific acaricides [1]. However, since the introduction, seven years ago, of pentac into chemical control, the two-spotted spider mite can be controlled satisfactorily in greenhouses. Since it was experimentally shown in the United States of America [2] that resistance to pentac can develop in populations of Tetranychus urticae, alternative control measures must be seriously considered. There is a very important aspect which favours a genetical method of control. This is the special isolation properties offered by a greenhouse. This isolation is ecological, genetical and physical. The ecological isolation is mainly caused by the different climatic conditions in autumn and winter present in the greenhouses, by which diapausing immigrants are not reactivated in the spring [3].

The genetical isolation of greenhouse populations of T. urticae is based on the frequent incompatibility between different populations, by which F_1-hybrids are rendered partially sterile [4].

437

Another favourable factor for genetic control in this species is the easy mass-rearing technique; in addition the biological and genetical properties of the species have been thoroughly studied [5, 6].

The sterile-male technique, the delayed-sterility technique and the translocation method of control require knowledge of the radiobiological properties of the species. This study deals with one aspect, namely the induction by different radiation dosages and radiation sources of heritable sterility factors, e.g. translocations, inversions and fragments. The various chromosome rearrangements are referred to structural chromosome mutations (SCM).

ASSESSMENT OF STRUCTURAL CHROMOSOME MUTATIONS IN Tetranychus urticae

Until now it has been very difficult to study SCM by conventional cytological methods in T. urticae. The chromosomes are very tiny and they differ only by 10% in length from each other [7]. Also genetical methods, such as the pseudo-linkage technique [8], cannot be used, because of the high recombination index and the scarcity of marker genes. Because of these restrictions the presence of SCM was determined by fertility measurement.

T. urticae reproduces by arrhenotokous parthenogenesis, mated females produce both fertilized (diploid) eggs, which give rise to females, and unfertilized eggs, which develop parthenogenetically into males (with the haploid number of chromosomes). Virgin females produce only a haploid progeny, e.g. males. This method of parthenogenesis is also characteristic for Hymenoptera, on which much radiobiological work has been done, particularly on Habrobracon juglandis (Ashmead) [9]. In Habrobracon the arrhenotokous parthenogenesis was exploited in determining with what frequency unmated F_1-females, progenerated by irradiated parents, are heterozygous for recessive lethals [10, 11]. However, some complications are associated with this method [9] because translocations and inversions also lead to a certain level of mortality in the progeny, when passing through meiosis of a heterozygous gonial cell [8]. In the method here presented, these complications are overcome by mating the F_1-females to untreated males, after these females have produced a sample of unfertilized (haploid) eggs. When the F_1-female is heterozygous for a structural chromosome mutation with the segregation property of a translocation or inversion, she will produce a haploid and a diploid progeny exhibiting the same reduced survival. However, an F_1-female, heterozygous for a recessive lethal, produces a haploid progeny with reduced survival, but a diploid progeny with full survival. By establishing the mortalities of the haploid and diploid progeny of single F_1-females from parents of which the male or female is irradiated, the frequency of F_1-females heterozygous for a SCM or a recessive lethal or both, was determined.

From the point of view of the practical applicability of homozygous SCM in genetic control purposes, it is also important to know what proportion of SCM is closely associated with recessive lethals, by which the homozygous situation is inviable. For that reason, the frequency of SCM associated with a recessive lethal was also determined.

It is assumed that F_1-females, heterozygous for a SCM associated with a recessive lethal, produce a haploid progeny of which the surviving fraction is free of the recessive lethal but also partly free of the SCM, depending on the linkage between the recessive lethal and the SCM. The survival of the diploid progeny should be higher than the survival of the haploid progeny because of complementation of the recessive lethal by the allele, derived from the fertilizing gamete. In the case of one recessive lethal and random segregation of the chromosomes, it is expected that the survival of the diploid progeny is about 50% and the viability of the haploid progeny is about 25%.

MATERIALS AND TECHNIQUES

Mites were reared on bean leaf punches (7.1 cm^2) on wet cotton wool, at 28°C and 50-70% r.h. Under these conditions, the developing time from egg to adult is 10 days.

Fifty males or females were irradiated with different doses of X-rays (250 kV, dose rate = 100 rad/min) or fast neutrons (mean energy = 1.5 meV, dose rate = 100 rad/min) as one-day-old adult virgins. Immediately after the radiation treatment the irradiated mites were mass mated for 1 hour. The males were then removed and the females were allowed to produce eggs. The diploid F_1-progeny studied was derived from the egg sample produced in the 24-48 h period after the radiation treatment. By such standardization of the sampling method the induction of SCMs in spermatids and prophase I (diplotene) oöcytes was studied [12].

From the F_1-progeny, about 100 female teleiochrysalids were individually isolated, to ensure virginity, on bean leaf punches (7.1 cm^2). The virgins were allowed to produce haploid eggs for three days on which survival was determined. Then the females were transferred individually to a fresh bean-leaf punch where they were mated to an untreated male. For three days the fertilized F_1-females produced a haploid-diploid progeny, on which the survival and the sex ratio was determined. The number of haploid eggs produced in the haploid-diploid sample was calculated by correcting the number of adult F_2-males by their survival frequency established in the F_2-progeny of the same virgin F_1-female.

By subtraction, the number of diploid eggs was calculated and then compared with the observed number of diploid females to assess their survival.

The number of both sterile or infecundant F_1-females was subtracted from the total number of F_1-females in the sample. From the resulting figure the percentages of the different classes were calculated.

RESULTS

The frequency of F_1-females, heterozygous for at least one SCM, induced at different doses of X-rays or neutron radiation of spermatids and prophase I oöcytes, is shown in Fig. 1, A-D.

On the horizontal axis are the different radiation doses in krad and on the vertical axis is the percentage of the tested F_1-females heterozygous for a SCM. The height of the column gives the percentage of the F_1-females heterozygous for at least one SCM; the height of the shaded areas give the percentage of F_1-females heterozygous for at least one structural chromosome mutation associated with at least one recessive lethal. At the top of each

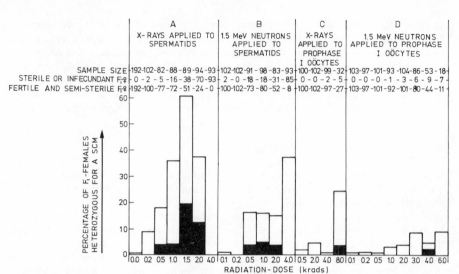

FIG. 1.   Percentages of structural chromosome mutations in the F_1-progeny of <u>Tetranychus urticae</u> Koch after irradiation of spermatids and prophase I oöcytes with 1.5 MeV neutrons and 250 kV X-rays.   On the horizontal axis are the different radiation doses in krad and on the vertical axis is the percentage of the tested F_1-females, heterozygous for at least one structural chromosome mutation.   The total height of the column shows the percentage of the F_1-females heterozygous for SCM.   The shaded area shows the percentage of F_1-females heterozygous for at least one structural chromosome mutation associated with at least one recessive lethal.   At the top of each column is given the number of tested F_1-females, the number of either sterile or infecund F_1-females: and by subtraction the number of normal and semi-sterile F_1-females.

column is given the number of tested F_1-females from which the number of both sterile and infecundant females is subtracted, giving the number of fecundant or semi-sterile F_1-females.

Irradiation of spermatids with X-rays

The induction of chromosome mutations by X-rays in spermatids is shown in Fig. 1, A.   It shows that the maximum yield of SCM is 60.8% at 1500 rad X-rays.   However 19.6% of all tested F_1-females have a SCM associated with a recessive lethal, which corresponds to 32.2% of all the F_1-females heterozygous for a SCM.

At doses between 0 and 1500 rad the percentage of F_1-females heterozygous for at least one SCM increases from 0.5% (control value) up to 60.8%. At higher doses the percentage is reduced, and at 4.0 krad no F_1-females can be tested because they are all either sterile or infecund.   The frequency of SCM associated with a recessive lethal is also dose dependent.   At 200 rad no F_1-females are observed which are heterozygous for both a SCM and a recessive lethal.

At doses between 500 rad and 2000 rad, the frequency of SCM associated with a recessive lethal fluctuates with a maximum of 19.6% at 1500 rad.   The relative frequency of SCM associated with a recessive lethal, related to all SCM, increased from 21.4% at 500 rad to 33.3% at 2000 rad.

Irradiation of spermatids with fast neutrons

The results obtained by neutron radiation of spermatids are given in Fig. 1, B.

After doses of 100 and 200 rads of neutrons there was no increase in the observable percentages of SCM. A sudden rise to 16% is observed at 500 rads of neutrons, and this level stays constant up to 2000 rad. Also the percentage of F_1-females heterozygous for a SCM associated with a recessive lethal is constant between 500 and 2000 rad at about 4.5%. This percentage related to all SCM is about 31%. However at 4000 rad all F_1-females heterozygous for a SCM did not have a recessive lethal.

Irradiation of prophase I eggs with X-rays

The dose response for the percentage of F_1-females heterozygous for a SCM induced in prophase I oöcytes by X-rays is presented in Fig. 1, C.

At 0.5 krad and 4.0 krad no significant induction of SCM was observed. At 2.0 krad X-rays, 4.9% of the F_1-females were heterozygous for a structural chromosome mutation which is a significant increase from the control ($X_1^2 = 6.3 : P = 0.01$). At 8.0 krad 22.2% of the F_1-females were heterozygous for a SCM of which 16.6% have a SCM associated with a recessive lethal.

Irradiation of prophase I eggs with fast neutrons

After neutron irradiation of prophase I eggs, the percentage of F_1-females heterozygous for a SCM shows a steady increase from 0.5 krad up to 3 krad neutrons (see Fig. 1, D). At doses lower than 0.5 krad neutrons no significant induction compared with the control was observed ($X_1^2 = 0.2 : 0.70 > P > 0.50$).

After 1.0 krad neutrons the induction of SCM is raised significantly compared with the control ($X_1^2 = 3.36 : 0.10 > P > 0.05$). At doses higher than 3.0 krad neutrons, the percentage of F_1-females heterozygous for a SCM did not increase further. The decrease at 4 krad is not significantly different from the frequency at 3.0 krad ($X_1^2 = 0.75 : 0.50 > P > 0.30$).

Almost no SCM associated with a recessive lethal were observed, except at 4.0 krad, where 2.3% of the F_1-females heterozygous for both a SCM and a recessive lethal were observed.

DISCUSSION

Because of the technique used in this study only a proportion of the total number of induced SCM could be measured. Those causing either complete sterility or infecundity or no reduction in fertility in heterozygous condition could not be observed.

These limitations in measuring the real frequency of all physically induced SCM are also encountered in most work published on other organisms, but the differentiation of limitations will depend on the detection method used and on the species studied. Therefore the results presented in this report can only partly be compared with those published elsewhere on other species.

Cytological investigation of the meiotic process reveals that during anaphase I and anaphase II the chromosomes move as straight orientated rods to the poles, and radiation-induced chromosomal fragments do not lag but

move independently to the poles. Some fragments can also pass through a
number of cleavage divisions [12].

Both the high radio-resistance of sperm (32 krad [13]) and oöcytes, and
the behaviour of the chromosomes and fragments during meiosis and cleavage
divisions, make it most probable that the chromosomes of T. urticae are
holokinetic [14, 15]. Another parallel with species with holokinetic chromosomes is the observation that sub-sterilized males of T. urticae mated with
untreated females produce F_1-females that are totally sterile, semi-sterile,
infecund or normal. The ratio of the different groups of F_1-females is
dose dependent (see Table I).

Table I shows that the progeny of irradiated males inherit a large degree
of sterility, which is characteristic of species having holokinetic chromosomes.
This phenomenon of delayed sterility offers possibilities for control of this
species [14], i.e. male irradiation with 4 krad X-rays induces a totally
sterile female F_1-progeny.

In a species with holokinetic chromosomes, random segregation of
fragments during meiosis leads to the production of about 50% duplication-
deficiency gametes [14]. This means that in T. urticae the heritable sterility
factors or structural chromosome mutations are most probably translocations,
inversions and fragments.

From the results presented in Fig. 1 it was deduced that both X-rays and
fast neutrons induce more SCM in mature sperm than in oöcytes. However,
these cell types are irradiated at different developmental stages in relation
to meiosis: sperm as spermatids and oöcytes in prophase I. The differences
described in radio sensitivity in males and females to the induction of
heritable sterility factors or SCM seems to be a general phenomenon, especially
in those orders having holokinetic chromosomes, e.g. Lepidoptera, Hemiptera
and Homoptera [15-17]. In Oncopeltus fasciatus Dallas [14] and in Siteroptes
graminum Reuter [18], irradiation of metaphase I eggs results in unequal
segregation of chromosomes during the meiotic divisions thus leading to the
production of duplication-deficiency gametes, causing dominant lethality
expressed during the embryological or larval developmental stages of the
F_1-generation.

TABLE I.   THE EFFECT OF X-RAYS ON THE F_1-PROGENY OF
IRRADIATED MALES OF Tetranychus urticae Koch MATED WITH
UNTREATED VIRGIN FEMALES

Dose X-rays (rads)	Number of F_1-females analysed	Fully fertile F_1-females		Semi-sterile F_1-females		Sterile F_1-females		Infecund F_1-females	
		Number	%	Number	%	Number	%	Number	%
0	192	190	99.0	2	1.0	0	0.0	0	0.0
500	82	59	72.0	18	22.0	1	1.2	4	4.8
1500	89	4	4.5	47	52.8	17	19.1	21	23.6
4000	93	0	0.0	0	0.0	39	41.9	54	58.1

In this way, the meiotic process acts as a sieve by which most of the SCM are sieved out, and results in a certain degree of dominant lethality. Therefore irradiation of pre-meiotic oöcytes of species with holokinetic chromosomes results in a higher level of dominant lethality than does irradiation of post-meiotic sperm. This difference in the induction of dominant lethality in irradiated oöcytes and mature sperm has been observed in Lepidoptera [16, 17] and Hemiptera [19]. The meiosis-disturbing chromosome rearrangements induced in mature sperm are, according to this theory, sieved out by the meiosis taking place in the F_1-individuals. These F_1-individuals are rendered sterile, semi-sterile or perhaps infecund by this process. The F_1-adults, derived from pre-meiotic irradiated eggs, will largely be free of those meiosis-disturbing factors and thus be more fertile than the F_1-progeny from irradiated parental males.

Comparison of the relative biological effectiveness (RBE) of 1.5 MeV neutrons versus 250 kV X-rays is complex and dose dependent. The dose-response relationship of SCM induced by neutrons in spermatids is totally different in kinetics from the dose response of SCM induced in spermatids by X-rays. At 0.5 krad X-rays and 0.5 krad neutrons, the same percentage of F_1-females, heterozygous for a SCM, is observed. X-rays are more effective than neutrons at doses between 1 krad and 2 krad, but at 4 krad X-rays all female F_1-progeny is sterile.

At 4.0 krad neutron irradiation of spermatids, 81.7% of the female F_1-progeny is infecund or sterile (9.6%), but of the remaining F_1-females, 37.5% is heterozygous for a SCM. In conclusion, at doses lower than 0.5 krad, 1.5 MeV neutrons are as effective as X-rays; at intermediate dosages (1-2 krad) X-rays are more effective than neutrons, but at higher dosages (4 krad), neutrons are more effective than X-rays.

Both for X-rays and for neutrons a significant induction of SCM is observed at 2 krad after irradiation of oöcytes (Fig. 1). However, both levels do not differ significantly from each other. At 4 krad irradiation of oöcytes, X-rays induce a low percentage (1.0%) of SCM, and this percentage does not differ significantly from the percentage (4.6%) observed at 4 krad neutron irradiation ($X_1^2 = 1.8 : 0.20 > P > 0.10$). It is argued therefore that for doses up to 4 krad, neutrons are as effective as X-rays in inducing SCM in prophase I oöcytes. From the assumption that saturation of induction of SCM in oöcytes is reached at 3 krad neutrons, it is concluded that at doses higher than 4 krad, X-rays are more effective (Fig. 1).

In the literature, the RBE values for the induction of lethals and translocations in sperm are contradictory. Some authors report lower yields of lethals and translocations in neutron-irradiated sperm than in X-ray irradiated sperm. Other authors present opposite results (for a discussion see Sobels and co-workers [20]). It is shown that in the experiments resulting in RBE values higher than 1, the sampling technique is often imperfect with regard to the homogeneity of the sampled developmental sperm stages. Differential radiosensitivity of different developmental sperm stages give spurious results when admixture occurs. In the experiments presented here, the samples are homogeneous, and probably for this reason agreement exists with those results derived from well-defined gamete samples [20]. However, the mean energy of the neutrons used in these experiments on T. urticae is 1.5 MeV whereas Sobels [20] used 15 MeV neutrons. This difference in the neutron energy used is important, since 1.5 MeV neutrons are 1.25 - 2 times more effective in inducing lethals and translocations than 15 MeV neutrons [20,21].

The values for total production of SCM in T. urticae after irradiation of spermatids correspond rather well with the results obtained from Culex pipiens L. [22,23]. Laven [22,23] estimated the frequency of SCM with the same technique as is used in T. urticae, i.e. the frequency of semi-sterile F_1-animals after irradiation of mature sperm. With dosages between 0.5 and 3.0 krad, 10 - 20% of the surviving F_1-individuals of C. pipiens were semi-sterile, but at dosages up to 5.0 krad this percentage increases up to 50%. Those results contrast with the maximum yield of translocations (17.9%) in Drosophila measured by the pseudo-linkage technique after irradiation of males with 4.452 krad X-rays [24]. In contrast, using cytological screening, about 40% of the F_1-larvae have been shown to possess translocations [25].

Only the yield of translocations in C. pipiens corresponds with the yield of SCM in T. urticae, but not the corresponding irradiation doses. At 4 krad X-rays, all F_1-females in T. urticae were sterile or infecund (Table I), but at 4 krad X-rays on sperm of C. pipiens, the frequency of semi-sterile F_1-individuals is 40%. It is very probable that the decline in the yield of the SCMs in T. urticae after sperm irradiation at higher dosages of X-rays is due to the holokinetic nature of the chromosomes; most of the surviving fraction of the F_1-animals of arthropods with holokinetic chromosomes will be rendered sterile [14].

From the point of view of the practical application in genetic control of structural chromosome mutations according to the translocation method [26,27] the viability of the SCM in the homozygous condition is a very important parameter. It is shown in Fig. 1 that a certain percentage of SCM is associated with a recessive lethal effect. The possibility of getting those SCM in a viable homozygous condition depends on the linkage between the SCM and the recessive lethal. Repeated backcrossing may succeed in disposing of this recessive lethal, but on the basis of the results presented in this report it seems most probable that the SCM induced at low radiation doses are most promising. Especially after X-ray irradiation of spermatids of parental males of T. urticae, a positive correlation is observed between the height of the radiation dose and the frequency of SCM associated with recessive lethals.

ACKNOWLEDGEMENTS

I wish to thank W. Helle, A.S. Robinson and D. Snieder for their critical reading of the manuscript. The technical assistance of J. Popma, R. Plaxton and Mrs. H.E. Broecks-Doorn is much appreciated.

REFERENCES

[1] HELLE, W., ZON, van, A.Q., Rates of spontaneous mutation in certain genes of an arrhenotokous mite, Tetranychus pacificus, Entomologia Exp. Appl. 10 (1967) 189.

[2] McENROE, W.D., LAKOCY, A., The development of pentac resistance in an outcrossing swarm of the two spotted spider mite, J. Econ. Entomol. 62 (1969) 283.

[3] HELLE, W., Genetic variability of photoperiodic response in an arrhenotokous mite (Tetranychus urticae), Ent. exp. appl. 11 (1968) 101.

[4] HELLE, W., PIETERSE, A.H., Genetic affinities between adjacent populations of spider mites, Ent. exp. appl. 8 (1965) 305.

[5] BOUDREAUX, H.B., Biological aspects of some phytophagous mites, Annu. Rev. Entomol. 8 (1963) 137.

[6] HELLE, W., OVERMEER, W.P.J., Variability in tetranychid mites, Annu. Rev. Entomol. 18 (1973) 97.

[7] PIJNACKER, L.P., Personal communication.

[8] MULLER, H.J., ALTENBURG, E., The frequency of translocations produced by X-rays in Drosophila, Genetics 15 (1930) 283.

[9] WHITING, A.R., Genetics of Habrobracon, Adv. Genet. 10 (1961) 295.

[10] HEIDENTHAL, G., X-ray induced recessive lethals in Habrobracon, Genetics 37 (1952) 590.

[11] ATWOOD, K.C., BORSTEL, von, R.C., WHITING, A.R., An influence of ploidy on the time of expression of dominant lethal mutations in Habrobracon, Genetics 41 (1956) 804.

[12] PIJNACKER, L.P., Personal communication.

[13] HENNEBERRY, T.J., Effects of γ-radiation on the fertility of the two spotted mite and its progeny, J.Econ. Entomol. 57 (1964) 672.

[14] LaCHANCE, L.E., DEGRUGILLIER, M., LEVERICH, A.P., Cytogenetics of inherited partial sterility in three generations of the Large Milkweed Bug as related to holokinetic chromosomes, Chromosoma 29 (1970) 20.

[15] NORTH, D.T., HOLT, G.G., "Inherited sterility and its use in population suppression of Lepidoptera", Application of Induced Sterility for Control of Lepidopterous Populations (Proc.Panel Vienna, 1970), IAEA, Vienna (1971) 99.

[16] SNIEDER, D., TER VELDE, H.J., Delayed sterility and chromosomal aberrations in succeeding generations after irradiation of Adoxophyes orana F.R. (Lepidoptera, Tortricidae) with substerilizing dosages of X-rays and fast neutrons (in press).

[17] PROSHOLD, F.I., BARTELL, J.A., Fertility and survival of tobacco budworms, Heliothis virescens (Lepidoptera: Noctuidae), from gamma-irradiated females, Can.Entomol. 105 (1973) 377.

[18] COOPER, R.S., Experimental demonstration of holokinetic chromosomes, and of differential radio-sensitivity during oögenesis, in the Grass Mite, Siteroptes graminum (Reuter), J.Exp.Zool. 182 (1972) 69.

[19] LaCHANCE, L.E., RIEMANN J.G., Dominant lethal mutations in insects with holokinetic chromosomes: I. Irradiation of Oncopeltus (Hemiptera: Lygaeidae) sperm in oöcytes, Ann.Entomol.Soc.Am. 66 (1973) 813.

[20] SOBELS, F.H., BROERSE, J.J., R.B.E. values of 15-MeV neutrons for recessive lethals and translocations in mature spermatozoa and late spermatids of Drosophila, Mutat.Res. 9 (1970) 395.

[21] EDINGTON, C.W., RANDOLPH, M.L., A comparison of the relative effectiveness of radiations of different average linear energy transfer on the induction of dominant and recessive lethals in Drosophila, Genetics 43 (1958) 715.

[22] LAVEN, H., JOST, E., Inherited semisterility for control of harmful insects: I. Production of semisterility due to translocation in the mosquito, Culex pipiens L., by X-rays, Experientia 27 (1971) 471.

[23] LAVEN, H., JOST, E., MEYER, H., SELINGER, R., "Semisterility for insect control", Sterility Principle for Insect Control or Eradication (Proc.Symp.Athens, 1970), IAEA, Vienna (1971) 415.

[24] PATTERSON, J.T., STONE W.S., BEDICHEK, S., SUCHE, M., The production of translocations in Drosophila, Am.Nat. 68 (1934) 359.

[25] BAKER, W.K., The production of chromosome interchanges in Drosophila virilis, Genetics 34 (1949) 167.

[26] CURTIS, C.F., HILL, W.G., Theoretical studies on the use of translocations for the control of tsetse flies and other disease vectors, Theor.Pop.Biol. 2 (1971) 71.

[27] LAVEN, H., Genetische Methoden zur Schädlingsbekämpfung, Anz.Schädlingsk. 41 (1968) 1.

DISCUSSION

G.W. RAHALKAR: No doubt the data available on the relative biological effectiveness (RBE) of fast neutrons and X-rays in inducing lethal mutations or translocations in insects are conflicting. However, in your experiments the RBE values are different at different dose levels. Can you comment on the probable reasons for this?

A.M. FELDMANN: It has to be stressed that the only structural chromosome mutations (SCM) that can be observed are those which do not induce, or are not associated with, total sterility or total infecundity, when the SCM is in a heterozygous condition. When higher doses of radiation are applied to spermatids and oöcytes, the induced SCM are associated with higher levels

of F_1-sterility.   This implies that at higher radiation doses a certain proportion of SCM cannot be observed because they are obscured by the associated sterility or infecundity of the F_1-females.   In Fig. 1 it is shown that at
higher doses the proportion of F_1-females, which are fully sterile or infecund
is higher after X-ray irradiation than after fast neutron irradiation of sperm.
Because of this phenomenon the RBE of SCM is lower at higher radiation dose

CONTROL OF BLOOD-SUCKING INSECTS
BY THE STERILE-MALE TECHNIQUE
(Session 7)

POTENTIAL OF THE STERILE-MALE TECHNIQUE FOR THE CONTROL OR ERADICATION OF STABLE FLIES, Stomoxys calcitrans Linnaeus

G.C. LaBRECQUE, D.W. MEIFERT, D.E. WEIDHAAS
Insects Affecting Man Research Laboratory,
Agricultural Research Service,
US Department of Agriculture,
Gainesville, Fla.,
United States of America

Abstract

POTENTIAL OF THE STERILE-MALE TECHNIQUE FOR THE CONTROL OR ERADICATION OF STABLE FLIES, Stomoxys calcitrans Linnaeus.
   Stable flies, Stomoxys calcitrans Linnaeus, can be mass reared inexpensively and with relative ease; they can be sterilized with chemicals or gamma irradiation with little loss in viability, and their specific breeding sites and reproductive capacity limit density. The rates of increase indicate that it should be feasible to check population growth by introducing sterile insects at low ratios. Genetic manipulation other than dominant lethality appears a likely method of control. Studies with mathematical models of populations indicate that the fly can be controlled at a cost of US $0.01 to US $0.19 per acre by the sterile-male technique.

The stable fly, Stomoxys calcitrans Linnaeus, sometimes called the dog fly, is a world-wide pest of man and animals. The preferred feeding hosts are domestic animals, but if they are absent, this fly readily attacks man. Light to moderate infestations may be tolerated because insecticidal control is often costly, difficult and short-lived. However, environmental conditions such as variations in wind direction, untended commercial wastes, improper agricultural practices or concentrations of aquatic grasses can help to create concentrations of stable flies that will cause serious economic losses, for example, in agriculture and in the tourist trade [1-4].
   The technique of sterile-male releases could be a part of an integrated pest management scheme for the control of stable flies. However, the successful development of the technique depends upon many factors including methods of mass producing the target insect, the ability to sterilize effectively without adverse effects on behaviour, the ability to evaluate results, and an understanding of the relative and absolute densities of the species, its growth rate and its migration. Over the past years, we have been investigating these factors as they relate to the feasibility of using the sterile-male technique against stable flies. The present study summarizes the pertinent published and unpublished data on this subject obtained in our laboratory. It is not a complete documentation of methods and results but an evaluation of existing data important to the development and potential of the sterility method against stable flies.

REARING TECHNOLOGY

The stable fly can be reared in large numbers by using the following procedure: A measured amount of stable fly eggs is placed on larval medium (3 parts wheat bran, 1 part pelletized sugar can bagasse, and 5 parts tap water). The 11th to 14th day after the eggs are placed in the medium (depending on the rearing temperature), we isolate the pupae by water flotation, and then dry them in a forced air drier [5]. To maintain the strain and to ensure an adequate supply of eggs, we confine lots of about 500 pupae in separate cages (0.15 m^3). When the adults eclose, we supply them daily with citrated bovine blood on cotton pads which also serve as the oviposition medium. Once the adults are 5-days-old, the eggs laid on the blood pads are collected each day, separated from the pads by washing with water, and measured in a tapered graduated centrifuge tube (1 mlitre averages 14 400 eggs). Thus, exclusive of the costs of buildings and utilities, the first million stable flies can currently be reared for less than US $1000 (which includes the cost of basic equipment), and subsequent flies from the same facilities will cost US $150/million. However, the numbers of adults emerging from known numbers of eggs in our colony at the time of these cost estimates showed only 52% survival from egg to emerging adult. If survival can be increased so that it is comparative with that obtained in rearing small numbers of flies, and if procedures can be further automated, it should be possible to reduce production costs.

SEGREGATION OF SEXES

Since both sexes of the stable fly are obligate blood feeders, the number of released insects must be kept below the thresholds of annoyance. A method of limiting or eliminating the females from the insects to be released would help, but we know of no method now available that will readily and economically accomplish a separation of the sexes in either the immature or adult stages. The procedure worked out by Bailey and co-workers [6], a modification of the forced air method developed by Henneberry and co-workers [7], will successfully sex adult stable flies if the flies have had no blood meal for at least 18 hours and are at least 4 days old. (Aging is essential since the accuracy of the sexing is dependent upon the weight and volume differential between the ovarian and testicular development of sexually mature adults.) However, the method can only be used when small numbers of adults are required or when adequate holding space is available so the insects can reach sexual maturity without undue harm from crowding.

Strains of insects that produce a distorted sex ratio with a high pre-ponderance of males [8, 9] might be an alternative to sexing since geneticists agree that such a meiotic drive factor can be isolated. However, with stable flies, this mechanism is still theoretical and cannot be isolated in less than one year. Another interesting possibility is a sexual differentiation involving a conditional lethal mutation of the type described in house flies, Musca domestica L., by McDonald [10]. The genetic makeup of this house fly strain is such that when the larval stage is reared at 25° ± 2°C, both sexes reach maturity; however, when the temperature of the larval medium is raised to 33° ± 2°C, all females die in the late larval or pupal stage, and only males emerge. In any case, until the problem of sex segregation is resolved, sterile-male releases of stable flies will have to be conducted when field populations are low or have been reduced by other control measures.

METHODS OF STERILIZATION AND EFFECT ON BEHAVIOUR

Sterility can be induced in the stable fly by either chemicals or irradiation. In the case of chemosterilization immersion of pupae is currently preferred over other methods [11-13]. Both males and females are completely sterilized when pupae are immersed in a 5.0% solution of either tris(1-aziridinyl)-phosphine sulphide (thiotepa) or bis(1-aziridinyl) ethylphosphine sulphide for 1 hour; also, the males show little or no debilitating effect such as reduced competition with normal males for females. The residues remaining on the insect can be monitored with existing chromatographic methods to nanogram amounts. Chemosterilized males have been used successfully in preliminary field studies made to evaluate the sterile-male approach to control of this species [14].

Stable flies can also be sterilized with little or no loss in longevity and mating competitiveness by exposing pupae that are more than 2 days old or adults to 2 krad of gamma irradiation [15, 16]. With radiosterilization, there is no possibility of environmental contamination such as could occur with chemosterilants. Also, females exposed to 2 krad or more will not oviposit; some chemosterilized females will. (A secondary advantage of no oviposition in irradiated females is that they can sometimes be used as markers.) Our field cage trials indicate that the mating competitiveness of radiosterilized male stable flies is not seriously affected by the exposure, but this may not be true of flies released in the field. (Further information on competitiveness is given later in the section on field trials.) Our data indicate that male stable flies must be exposed to 4 krad to obtain 100% sterility, but such a dose is usually accompanied by some loss in sexual competitiveness. However, we have also found (unpublished data) that 98-100% sterility in males and 100% sterility in females can be obtained with 2 krad with no loss in vigour and longevity. This dose would therefore be the recommended dose until the field population is suppressed to low levels. Then, the insects should be sterilized with 4 krad.

METHODS OF ASSESSMENT AND EVALUATION

Developments in the sterile-male approach depend upon actual field trials, but effective trials cannot be achieved unless techniques are available to assess the sterility and density of the natural population before and during treatment. For assays of sterility, we use sweep-net collections of adult flies around favoured aggregation sites. In release programmes, the collected native females can be separated from released irradiated females by the use of markers, or the females can be held for an extended period to determine lack of oviposition and/or ovarian development that would indicate radiosterilized females. At present, we assess field sterility by allowing captured females to oviposit in the laboratory; we then place a selected sample of eggs in larval medium, and the number of eggs that hatch and reach the pupal stage from a known number of eggs seeded initially is then used to determine sterility of the field populations.

Methods of assaying the density of stable fly populations are available, but none are completely satisfactory. Currently, we derive quantitative estimates primarily by the mark-release-recapture method, that is, we release a specific number of laboratory-reared flies into the environment and collect a sample of flies from the field one day later, either by hand netting or by using sticky traps [4, 14, 17]. The ratio of marked to unmarked flies captured and the number of marked flies released provide sufficient data so we can achieve a rough estimate of the native population. These sticky traps are essentially panels of wood or translucent plastic treated with an adhesive and mounted on poles or suspended from beams or ceilings. They give a fairly accurate qualitative index of population trends, but the captured insects are dead or so covered with adhesive as to make them use- less for sterility evaluations, and sometimes it is not even possible to distinguish between the colours of the dye markers. A second method that we use to obtain quantitative estimates of populations involves counting the number of flies per animal from a random sample of animals in a herd. When the average number of flies per animal is applied to the total number of animals in the herd, we can derive the average number of flies feeding or resting on animals in the environment. In our field cage studies (unpublished data), we found that each fly resting or feeding on an animal represents 55 in the environment. This factor is used with the number of flies on the animals to produce a rough estimate of the field population.

RELATIVE DENSITIES AND GROWTH RATES

Hoffman [1] in a well-documented review concerning the distribution and abundance of stable flies indicated that populations are normally low. In Florida, Williams (personal communication) estimates averages of 5 to 10 flies per dairy animal when flies are abundant, but he has some- times observed as many as 200 per cow. In our studies in central Florida [17], an average maximum of 9.3 stable flies per dairy animal was observed throughout a year, but the numbers ranged from 0.1 to 0.9 from October through December; from 0.5 to 4.0 from January through March; from 4.3 to 9.3 from April through June; and from 0.8 to 8.4 from July to September. With these data, we were able to calculate changes in density from one three-week period to the next, the interval required to complete one generation when the flies are reared in the laboratory at 25°-35°C. Our calculations showed the following ranges for the rates of change every three weeks.

September through December   — 0.2 - 2.0

January through March   — 1.4 - 2.5

April through June   — 0.6 - 1.7

July through September   — 0.2 - 2.0

Thus, we found no larger increases in density over time (the greatest was 2.5-fold) when the population was subjected to normal environmental stress. Also, LaBrecque et al. [14] reported similar rates of increase for a popula- tion of stable flies that were subjected to the additional stress caused by the release of sterile males. In other words, the normal low rates of

increase were not changed despite a reduction in density that allowed for
a greater carrying capacity. This finding and the levels of increase are
particularly encouraging for further development of the sterile-male release
technique if these preliminary data are subsequently confirmed. However,
the results of similar studies with house flies and a comparison of the biology
of the two species support our conclusion. Weidhaas and LaBrecque [18]
showed that a population of house flies on Grand Turk Island attained rates
of increase of 2 to 10X when the population was reduced below the carrying
capacity by treatment with a chemosterilant bait; thus, the rates were
higher than for stable flies. However, over the same relative life span
(LD-50 — longevity in days — about 20 days) our laboratory and field
studies of house flies and stable flies have shown that house fly females:
(1) oviposit sooner (3.5 versus 7.8 days); (2) produce more total eggs
(1195.3 versus 292.4); and (3) have a lower natural sterility (17.3 versus
22.9%). Also, more immature house flies survive (83.5 versus 52.0%)
(unpublished data). Thus, the lower reproductive rate of stable fly populations compared with house fly populations seems reasonable.

MIGRATION

Field releases of marked flies have shown that the stable fly is capable
of flights of as much as 15 miles and can cover 5 miles within 2 h [19, 20].
Also, in laboratory studies with flight mills, Bailey and co-workers [21]
observed one 3-day-old female that flew 29.11 km in 24 h. Thus, an extensive flight potential has been demonstrated. On the other hand, we have
found that released sterile or marked flies can be recaptured near the
release site for extended periods — as much as 14 days. The extensive
flight potential indicates, of course, that rapid reinfestation of areas by
stable flies can occur. Thus, the sterile-male technique will be useful
only in isolated areas or over extremely large areas.
    Unfortunately little is known of the relationship between dispersal and
mating.

RELEASE EXPERIMENTS

As early as 1970, we conducted field experiments at dairy installations
in central Florida to investigate the potential of the sterile-male technique
against stable flies. In the first study, LaBrecque and co-workers [14]
released 7-day-old adults chemosterilized by immersing 4 to 5-day-old
pupae in a water solution of 5.0% metepa or methiotepa for 60 min. The
flies (mixed sexes) were then held for 7 days after eclosion to ensure:
(1) the loss of most of the chemosterilant residue and (2) the insemination
by the sterile males of most of the females which were only partially
sterilized by the treatment. Thereafter, releases involving about 8000 to
14 000 flies per release were made every 3-4 days. The ratio of released
flies to field flies approached 1:1 and was adequate to introduce sterility
reaching 77.3% in the natural population. Thus, (1) most of the released
flies remained in the vicinity of the release site if adequate sources of food
were present; (2) the insect was relatively long lived; (3) the native populations could be reduced by sterile-male releases; and (4) and most important,

TABLE I. STERILITY INTRODUCED INTO A FIELD POPULATION OF STABLE FLIES BY THE INTRODUCTION OF RADIOSTERILIZED INSECTS OF BOTH SEXES
(M = marked or released, UM = unmarked)

Generation[a]	Treatment area							Control area	
	No. flies released (× 1000)	No. flies/release (× 1000)	Sterility (%)	Density counts[b]			Est. no. field flies (× 1000)	Sterility (%)	Density counts[b]
				Ratio		Total			
				M	UM				
1			24						
2	323	22	55	1	1		22	37	35
3	334	22	86	c				25	56
4	740	49	86	6	1	61	7	19	37
5	845	56	100	95	1	96	0.6	17	44
6	310	62	73	14	1	52	4		32
6			94	7	1	95		56	59

[a] Releases initiated at start of second generation and terminated during 1st week of 6th.
[b] Number of adults netted per 0.5 h. Average of 4 samplings weekly.
[c] Unmarked flies released during this period.

only a low ratio of sterile to native flies was necessary to obtain significant reductions in the native population. When similar studies were conducted at the same sites in 1971 and 1972 with stable flies irradiated as 3 to 4-day-old pupae with 2 krad and allowed to eclose near grazing cattle (unpublished data), the ratio of sterile to native flies never exceeded 1.4 to 1, but the sterility introduced into the populations once again equalled the theoretically expected sterility.

Considerable loss of young adult flies through predators and adverse meteorological conditions occurred in the 1971 and 1972 tests when pupae were released. We therefore conducted another sterile-male release programme at the same site in the summer of 1973. In this test, we used flies irradiated as adults (less than 24-h old) with 2 krad from a ^{60}Co source. The newly emerged flies were collected each day at the laboratory, immobilized by chilling, measured volumetrically to give us an estimate of numbers, placed in lots of 5000 into 10 × 25 cm polyvinylchloride cylinders, marked with a fluorescent dye and irradiated. Shortly after irradiation, the containers with the flies still in a chilled, immobilized state were transported in plastic foam chests containing ice to the field where they were distributed at protected sites in the vicinity of concentrations of cattle. (Since high temperatures, particularly at noon, are highly detrimental to survival of young released adults, the releases were made at dusk, Monday through Friday.)

At the time of the initial release, the field population was estimated (by the mark-release-recapture method previously described) at about 22 000. The first of the daily releases averaged 22 000 flies. Subsequently the number was gradually increased to slightly more than 50 000 (Table I). By one generation (20 days) after the start of the releases, the sterility induced in the native population was higher than 85%. Then as the ratio of released to field flies increased from 1:1 to 95:1 (determined by weekly samplings with ten sticky traps), the sterility of the captured (netted) flies increased to 100%, and the estimated field population was reduced by over 97% from 22 000 to about 600 (Table I). In the latter part of the study, flies were so scarce that in two instances we were unable to collect any for evaluation of sterility. Meanwhile, sterility in the untreated control averaged 24% (range of 17 to 56). Sterility remained high in the treated population for two weeks after the last release, but re-infestation was rapid: the ratio of released to native flies decreased from 14:1 to 7:1. Also, releases averaged 50 000 daily during the latter part of the programme, but the recaptures following marked releases indicated that only 147 000 released flies per day remained in the environment. Thus, there was a daily loss of 35% of the released flies calculated over a two-week survival period. However, concurrent field cage studies with tethered cattle were showing a daily death rate of 20% when conditions were similar to those in the field (unpublished data). This natural mortality coupled with a rate of emigration of 15% could easily account for the 35% daily loss. Our data satisfied the hypothesis that these flies tend to remain in the environment (supported by the released-field ratios) and that they do disperse (indicated by observations and by recaptures 12.8 km (8 miles) away). The released insects appeared to be sexually competitive with the field strain. As observed in the 2nd generation, that is, through days 20-40 of the test, a 1:1 ratio (sterile:fertile) produced 40% sterility so the mating competitiveness ratio was about 2:1; therefore approximately two irradiated males equated one field male.

DISCUSSION

From the information available to date, the stable fly can be mass reared inexpensively and with relative ease, and can be sterilized with chemicals or gamma irradiation. At present, both sexes must be reared and released, but the isolation of a male determinant factor is within reason and should remedy this disadvantage. The specific breeding sites and limited reproductive capacity of the stable fly keep density and rate of increase at low levels, particularly during seasonal extremes. For example, most stable fly breeding, particularly during times of seasonal extremes, was observed to be restricted largely to areas where agriculture was confined to cattle and horses though we have observed it in poultry and pig-rearing installations and on range cattle. Thus, waste matter can accumulate that is conducive to year-round fly breeding as opposed to vegetable wastes or grass accumulations that are seasonal, but initiate and support short-lived but unbearable outbreaks. Although the insect has great mobility, preliminary studies indicate that populations can be held in check or reduced by releasing sterile males at a low ratio of sterile to fertile insects. However, the marked effect of such low ratios also indicates that genetic mechanisms such as translocations, meiotic drive and other factors, and dominant lethality, have potential for population control.

PROJECTIONS

To obtain an estimate on the absolute number of stable flies present within an extensive area, we asked the Florida county extension directors for the number of domestic animals in each county. Reports from 60 of the 67 directors showed that the state had an estimated 2.5 million cattle, horses, and hogs but the approximately 360 000 cattle and horses that were confined in about 6000 installations would be the principal source of year-round breeding. This total was therefore used to determine that in each of Florida's 58 560 square miles there are about six animals that provide potential breeding sites for the stable fly. Since an observed fly density of 1 per animal normally indicates about 50-60 flies in the environment, a count of 1 fly per animal and six animals per square mile would indicate 300-360 flies per square mile. With the calculated densities and the parameters established as the result of field testing, a model can be constructed to derive estimates for the numbers of insects needed for and the costs of a sterile-male release programme. (Isolation determines whether such a programme will be designed to eradicate or control. If infiltration from the outside sources can be prevented or eliminated, the final goal should be elimination because many of the costs of both goals are identical.) For the model, we assume that a 10 000 square mile area (a 100 mile square block, which is roughly the total area of Maryland or Massachusetts) is to be treated and that the programme must begin when fly densities are low or reduced to manageable levels by other measures. We assume that 10 flies per animal represents a high density and that 1 fly per animal represents a low density (1 fly per animal would also represent 3.0×10^6 flies in the total area). The average daily mortality rate is taken to be 25%, which means that the number of new adult flies emerging into the area each day would be the same as the number of flies lost each day (8.5×10^5) if the population is stable (1-fold

TABLE II. PARAMETERS USED TO ESTABLISH A MODEL FOR
DETERMINING THE FEASIBILITY OF A STERILE-MALE RELEASE
PROGRAMME INVOLVING STABLE FLIES S. calcitrans

Test site	10 000 square miles
Number of animals/square mile (Fla. average)	6
Number flies/animal	1
Number flies in environment/animal	50
Number flies/square mile	300
Number flies in test site	3.0×10^6
Average daily mortality rate	25%
Number flies emerging daily	
1-fold rate of increase	0.75×10^6
5-fold rate of increase	3.75×10^6
Costs per 1×10^6 flies	
Rearing	$160
Sterilizing and packaging	$150
Duration of releases − 5 generations	105 days

TABLE III. TENTATIVE COST OF STERILE-MALE RELEASE
PROGRAMME TO ERADICATE OR CONTROL THE STABLE FLY
S. calcitrans

Initial sterile to fertile ratio	Cost (dollars) per			
	10 000 square miles		Acre	
	1 ×	5 ×	1 ×	5 ×
10 : 1	244 125	1 220 625	0.04	0.19
5 : 1	122 063	610 313	0.02	0.10
2.5 : 1	61 031		0.01	

rate of increase). If the population is increasing 5-fold each generation,
five times more (3.75×10^6) new flies would be emerging each day. (We
chose these two rates of increase for our model because stable fly populations often have low rates of increase around 1×, but the high rate is not
as high as 5-fold.) Then release ratios could be 10 : 1, 5 : 1 or 2.5 : 1,
depending on the growth rates and the objective, and we have included

calculations for these three levels in our model. The assumptions made
are summarized in Tables II and III as are the calculated costs for rearing,
sterilizing and packaging (not including the cost of distribution). Also, the
estimated costs assume a programme covering 105 days (five generations)
of releases, though theoretically high levels of control could be obtained
in three generations. (The five-generation span provides a margin of safety
for the calculations.)

 The actual numbers of stable flies that were released in a given situation
such as that described by the model would have to be determined by further
research. However, we feel that the calculations indicate the method to be
economically feasible in large unisolated areas or in smaller, isolated
situations, whether long-term control or a temporary reduction is desired.

NOTE: Mention of a pesticide in this paper does not constitute a recommendation of this product by the
US Department of Agriculture.

REFERENCES

[1] HOFFMAN, R.A., The Stable Fly Stomoxys calcitrans (Linnaeus): Biology and Behavior Studies,
 Unpublished PhD dissertation, Oklahoma State University, Stillwater, Okla. 1968.
[2] JAMES, M.T., Stomoxys calcitrans, in The Flies that cause Myiasis in Man, USDA Misc. Publ.
 No.631 (1947).
[3] UNITED STATES DEPARTMENT OF AGRICULTURE, Losses in Agriculture, Agricultural Handbook
 No.291, US Government Printing Office, Washington, D.C. (1965).
[4] WILLIAMS, D.F., "Sticky traps for sampling populations of adult stable flies Stomoxys calcitrans (L.)",
 Joint meeting Entomol. Soc. Can., Entomol. Soc. Quebec and Entomol. Soc. Am., Montreal,
 Canada, 26-30 Nov. 1972.
[5] BAILEY, D.L., Forced air for separating pupae of house flies from rearing medium, J. Econ. Entomol.
 63 1 (1970) 331.
[6] BAILEY, D.L., LaBRECQUE, G.C., WHITFIELD, T.L., A forced air column for sex separation of
 adult house flies, J. Econ. Entomol. 63 5 (1970) 1451.
[7] HENNEBERRY, T.J., McGOVERN, W.L., YOEMANS, A.H., MASON, H.C., Sexing large numbers
 of Drosophila melanogaster adults by size differential, J. Econ. Entomol. 57 5 (1964) 769.
[8] HICKEY, W.A., "Genetic distortion of sex ratios in Aedes aegypti", Proc. XII Int. Cong. Entomol.,
 London, England (1965) 226.
[9] SWEENEY, T.L., Sex Ratio Distortion caused by Meiotic Drive in Culex pipiens L., Unpublished
 PhD dissertation, University of California, Los Angeles, Calif., 1972.
[10] McDONALD, I.C., A male-producing strain of the house fly, Science 172 (1971) 489.
[11] CASTRO, UMANA, J.J., Chemical Sterilization of the Stable Fly Stomoxys calcitrans (Linne.) with
 Metepa and Hempa, Unpublished PhD dissertation, University of Florida, Gainesville, Fla., 1967.
[12] HARRIS R.L., Chemical induction of sterility in the stable fly, J. Econ. Entomol. 55 6 (1962) 882.
[13] LaBRECQUE, G.C., FYE, R.L., MORGAN, J., Jr., Induction of sterility in adult house flies and
 stable flies by chemosterilization of pupae, J. Econ. Entomol. 65 3 (1972) 751.
[14] LaBRECQUE, G.C., MEIFERT, D.W., RYE, J., Jr., Experimental control of stable flies Stomoxys
 calcitrans (Diptera: Muscidae) by releases of chemosterilized adults, Can. Entomol. 104 (1972) 885.
[15] WHITE, S.A., The Effect of Ionizing Radiation on the Stable Fly Stomoxys calcitrans L., Unpublished
 PhD dissertation, University of Florida, Gainesville, Fla., 1971.
[16] OFFORI, E.D., The Effects of Gamma Irradiation on the Reproduction and Longevity of the Stable
 Fly Stomoxys calcitrans (L.), Unpublished PhD dissertation, University of North Dakota, Fargo, 1969.
[17] LaBRECQUE, G.C., MEIFERT, D.W., WEIDHAAS, D.E., Dynamics of house fly and stable fly
 populations, Fla. Entomol. 55 2 (1972) 101.
[18] WEIDHAAS, D.E., LaBRECQUE, G.C., Studies on the population dynamics of the house fly Musca
 domestica L., Bull. World Health Organ. 43 (1970) 721.

[19] KING, M.V., LENERT, L.G., Outbreaks of Stomoxys calcitrans L. ("dog flies") along Florida's northwest coast, Fla. Entomol. 19 (1936) 33.
[20] EDDY, G.W., ROTH, A.R., PLAPP, F.W., Studies on the flight habits of some marked insects, J. Econ. Entomol. 55 5 (1962) 603.
[21] BAILEY, D.L., WHITFIELD, T.L., SMITTLE, B.J., Flight and dispersal of the stable fly, J. Econ. Entomol. 66 2 (1973) 410.

DISCUSSION

J. L. MONTY: What precautions do you intend to take to prevent excessive damage to people and animals during a release programme aimed at suppression?

D. E. WEIDHAAS: We are planning several precautions to prevent excessive damage or annoyance to people and animals. Firstly, the sterile-male technique will probably not be useful without prior suppression of population density. This could be accomplished in two ways, i.e. by the use of other control methods or by carrying out releases when densities are very low. In either case the number released will have to be below the damage or annoyance threshold. We hope that genetic methods of eliminating females during rearing will be developed which would reduce the number released by half. I should also point out that, when the stable fly population is low, the insects will be concentrated around animals and farms rather than congregations of people.

D. W. WALKER: In your oral presentation you described work on mosquitoes in El Salvador. Do you have any estimates of the per capita cost of malaria suppression or eradication by SIRM in El Salvador?

D. E. WEIDHAAS: Not at the moment. We hope that the coming pilot test in El Salvador will enable us to make such estimates as it will give us practical experience of working in a large area.

R. PAL: It is gratifying, as we draw near the end of this symposium, to hear a few optimistic reports on the genetic control of insects — Mr. D. Cuisance on tsetse flies and Mr. D. E. Weidhaas on stable flies — even though the experiments involved have been carried out in relatively small, isolated areas. The WHO/ICMR unit on the genetic control of mosquitoes in New Delhi has carried out a series of experiments on Culex fatigans in villages and has achieved an even higher degree of sterility but, because of massive immigration, a parallel decrease in the natural population was not achieved. Since immigration is somewhat less of a problem with Aedes aegypti, we are now conducting experiments on this species in a township. Increasing emphasis is also being placed on Anopheles stephensi at this unit.

D. E. WEIDHAAS: I hope that I did not leave a mistaken impression by the way I presented my data; I was trying to emphasize both the effect of the release of sterile males and the dynamics of the population. In the experiment on Anopheles albimanus in El Salvador we achieved 100% sterility at most collection sites, and the overall sterility approached 100%. In such experiments there is a problem in obtaining sufficient females for meaningful values at the end of the experiment when populations have been reduced to extremely low levels. Thus, it is difficult to demonstrate 100% sterility. It was interesting to note in our experiment that about the time we stopped releases we were unable to sample any mosquitoes by three independent methods for two weeks. This, of course, does not mean that there were no mosquitoes in the area, but it indicates that the number was very, very low.

L. C. MADUBUNYI (Chairman): While we are expressing optimism about eradicating the tsetse fly and other haematophagous insects by the sterile-male release technique, we should bear in mind two things; firstly, our experience in West Africa (Nigeria in particular) has revealed that certain species of the tsetse fly are now peridomestic and this seems to be a relatively recent behavioural change. We should now start considering how to deal with the peridomestic tsetse fly after getting rid of the wild tsetse, i. e. now is the time to initiate work on the ecology of the peridomestic tsetse fly; secondly, intensive efforts are now being made to preserve game in Africa. The tsetse fly happens to be an important inhabitant of game reserves and the preservation of game will help the tsetse fly to consolidate its ground. I imagine the sterile-male technique will be very valuable against the tsetse fly in such game reserve ecosystems.

MALES STERILES DE Glossina
tachinoides West. LACHES DANS DEUX GITES
NATURELS: RESULTATS, PERSPECTIVES

D. CUISANCE*, J. ITARD**
Institut d'élevage et de médecine
vétérinaire des pays tropicaux

Abstract—Résumé

RELEASES OF Glossina tachinoides West. MALES AT TWO NATURAL BREEDING SITES: RESULTS AND PROSPECTS.

Sterile-male releases of Glossina tachinoides were made during the dry seasons in 1972 and 1973, at breeding sites along the Chari and Logone rivers. The high density of wild flies at the first site makes it impossible to determine the effect of the releases (7000 males sterilized with 15 000 rad). With regard to the batches irradiated to a greater or lesser extent, it can be said that the factors of "rearing" and "travel" reduce the flying ability of the insects and disrupt their natural rhythm during the first few days: their average lifetime is reduced to 4.8 d; and the "irradiation" factor reduces the maximum lifetime to 12 d — less than half the value for unirradiated males (28 d). The sterile males disperse over the same parts of the breeding ground and seek out the same ecological conditions for survival, but do so at a much slower pace. Their resting places during the day and night are the same, as is the elevation of the refuge points. The sterile males acquire feeding habits similar to those of the wild insects, though not until the sixth day after release.

Releases carried out at the second site (4625 males sterilized with 10 000 rad) confirm the behavioural characteristics already discussed. Better performance (longevity) is obtained by making a number of releases at the most effectively protected spots and by reducing the radiation dose. Deterioration of the control site makes it impossible, by comparison, to measure the precise effect of the sterile males at the experimental breeding site, where the population is decreasing. That the sterile males do exert an influence is evident from the high percentage of sterility found in the females under observation. In view of the limitations observed during the first experiment, certain modifications were made in the second test in order to cut down as far as possible the phase required for adaptation of the released insects. The results suggest certain comments as to possible solutions and indications for use of the sterile-male technique as applied to G. tachinoides under natural conditions.

MALES STERILES DE Glossina tachinoides West. LACHES DANS DEUX GITES NATURELS: RESULTATS, PERSPECTIVES.

Des lâchers de mâles stériles de Glossina tachinoides ont été effectués au cours des saisons sèches de 1972 et 1973 dans des gîtes riverains des fleuves Chari et Logone. La densité élevée de glossines sauvages dans le premier gîte ne permet pas d'observer un effet des lâchers effectués (7000 mâles stérilisés à 15000 rad). Par comparaison avec des lots plus ou moins irradiés, on note que les facteurs «élevage, voyage» abaissent leur capacité d'envol et perturbent leur rythme d'activité durant les premiers jours: ils raccourcissent la durée de survie moyenne (4,8 j). Le facteur «irradiation» diminue la survie maximale (12 j): elle est plus de deux fois inférieure à celle des mâles sauvages (28 j). Les mâles stériles se dispersent dans les mêmes zones géographiques du gîte et recherchent les mêmes conditions écologiques pour leur survie, mais le font de façon beaucoup plus lente. Les lieux de repos diurnes et nocturnes sont identiques et les hauteurs de ces points-refuges sont égales. Les mâles stériles acquièrent un comportement alimentaire comparable à celui des mâles sauvages, mais seulement vers le sixième jour après le lâcher.

Les lâchers effectués dans le deuxième gîte (4 625 mâles stérilisés à 10 000 rad) confirment les aspects du comportement déjà signalés. De meilleures performances (longévité) sont obtenues en fractionnant les lâchers aux lieux les mieux protégés et en abaissant la dose d'irradiation. Des dégradations dans le gîte témoin ne permettent pas de mesurer par comparaison l'influence exacte des mâles stériles dans le gîte d'expérience dont la population baisse. Un effet des mâles stériles est certain, si on considère le haut pourcentage de stérilité rencontré chez les femelles mises en observation. Compte tenu des contraintes

* Laboratoire de recherches vétérinaires de Farcha, N'Djaména, Tchad.
** Maisons-Alfort, France.

observées au cours de la première expérience, certains palliatifs ont été utilisés dans la seconde afin d'écourter au maximum la phase d'adaptation de l'insecte lâché. Les résultats notés suggèrent quelques remarques sur les solutions à envisager et sur les indications de l'utilisation de la méthode de lâcher de mâles stériles de G. tachinoides dans les conditions naturelles.

INTRODUCTION

Depuis plusieurs années, des études régulières ont permis d'approfondir l'écologie de Glossina techinoides dans la région du Bas-Chari et du Bas-Logone [1, 2]. Dans un certain nombre de gîtes riverains, la dynamique des populations a pu ainsi être appréciée, en particulier la distribution de ces dernières et leurs fluctuations saisonnières. Cette région du Tchad est caractérisée par une pluviométrie réduite pendant trois mois et par une saison sèche très longue. Cette dernière favorise le regroupement des glossines qui se concentrent dans des îlots de végétation dense que l'on appelle gîtes, où les populations sont isolées pendant les mois chauds.

Au laboratoire, de grands progrès ont été réalisés en matière d'élevage en masse de ces insectes [3, 4]; de nombreuses données ont été fournies sur la stérilisation des mâles de Gl. tachinoides en particulier [5-7], et sur leur compétitivité [5]. Ces données demandaient à être confirmées sur le terrain. Mettant à profit l'isolement naturel de populations de Gl. tachinoides bien connues, des lâchers ont été effectués régulièrement pendant la saison sèche de 1972 dans un gîte riverain du fleuve Chari ainsi qu'au cours de la saison sèche de 1973 dans un gîte bordant le fleuve Logone.

Les résultats de ces deux séries d'observations sont ici présentés ainsi que quelques conclusions sur les perspectives d'utilisation de la méthode du lâcher de mâles stériles, compte tenu des données acquises.

1. RESULTATS

1.1. Observations faites au cours de la saison sèche de 1972 [8-10]

1.1.1. Lieux d'expérience

Les lâchers ont eu lieu dans le parc national de Kalamaloué à une vingtaine de kilomètres de N'Djaména (ex Fort-Lamy) où deux gîtes particulièrement isolés et d'accès facile ont été retenus.

— Situation

Ces deux îlots de végétation dense s'étirent au bord du Chari (950 m × 100 m) pour le premier, au bord du Serbewel, son défluent (1200 m × 100 m), pour le second. Distants d'environ 4 km, l'un sert de gîte d'expérience, l'autre de gîte témoin.

— Dynamique des populations dans ces gîtes

La population s'installe en saison sèche et fraîche. Avec le début de la saison sèche et chaude, les glossines, dispersées dans la réserve,

TABLEAU I. BILAN AU MOMENT DU LACHER (1972)

Comportement des glossines au lâcher	Glossines irradiées à 15 000 rad (%)	Glossines sauvages (%)
Envol	80,7	86,4
Tombées au sol	9,3	2,3
Mortes	8,9	10,0

se regroupent dans ces gîtes où elles trouvent dans la végétation ripicole dense une protection contre les rigueurs climatiques de cette saison. La population croît rapidement en densité pour atteindre un maximum en avril et décroître lentement jusqu'aux pluies.

Dans ces deux gîtes, la densité en glossines est élevée: 6000 glossines en moyenne dans le gîte d'expérience, 8000 glossines dans le gîte témoin (méthode de marquage-recapture appliquée toutes les semaines).

1.1.2. Comportement au lâcher

— Irradiation

Agés de 2 à 10 jours, les mâles issus de l'élevage de Maisons-Alfort [3,4] sont soumis au rayonnement gamma d'une bombe au cobalt à la dose de 15 000 rad.

— Conditions de transport

Placés dans des boîtes isothermes, ils sont expédiés par voie aérienne et arrivent au Tchad dans les 10 heures qui suivent. Ils sont acheminés dans les gîtes en voiture puis en bateau (durée: 1 h).

— Alimentation

Dès l'arrivée, les glossines sont nourries sur lapin avant d'être marquées puis lâchées.

— Marquage

Chaque mâle reçoit sur la partie supérieure du thorax une petite tache de gouache acrylique dont la couleur change à chaque séance de lâcher.

— Bilan au moment du lâcher

9572 mâles ont été expédiés de Maisons-Alfort. Ils étaient tous irradiés à 15 000 rad à l'exception d'un lot de 197 individus irradiés à 6000 rad et 317 non irradiés. Chaque lâcher de glossines d'Alfort s'accompagnait d'un lâcher de glossines sauvages capturées dans le gîte la veille du lâcher, afin de comparer le comportement respectif de chaque lot. Douze séances de

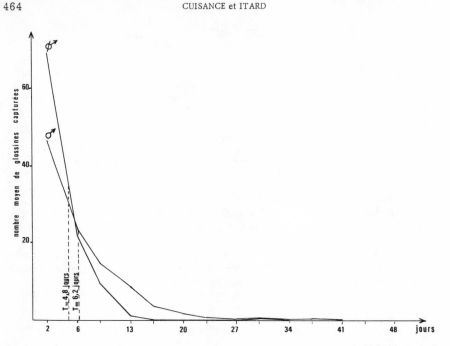

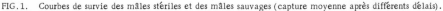

FIG.1.　Courbes de survie des mâles stériles et des mâles sauvages (capture moyenne après différents délais).

lâchers ont ainsi eu lieu en un seul point du gîte; 300 à 500 mâles ont été libérés toutes les semaines de février à mai.

Les facteurs «élevage, voyage» abaissent le pourcentage d'envol (moindre vigueur, ailes anormales ou abîmées); ils sont sans effet sur le taux de mortalité comparativement aux glossines sauvages (tableau I).

Par comparaison entre eux des lots irradiés à des doses variables, on observe que le facteur «irradiation» abaisse le pourcentage d'envol (moindre vigueur) mais n'entraîne aucune mortalité anormale.

Ces trois facteurs agissent en commun pour handicaper légèrement, au lâcher, les glossines importées et en particulier les mâles stériles.

1.1.3.　Comportement dans le gîte

— Rythme journalier d'activité

Des captures régulières ont permis de calculer le pourcentage horaire de glossines lâchées retrouvées, c'est-à-dire, d'apprécier la façon dont les glossines se mobilisent au long de la journée pour venir piquer les captureurs.

Le transport et les manipulations (contention, marquage, etc.) perturbent légèrement le rythme d'activité des glossines lâchées (saison sèche et fraîche), au moins pendant les 48 premières heures. Le facteur «élevage» semble responsable d'un optimum thermique d'activité plus bas chez les mâles stériles (27°C) que chez les mâles sauvages (31°C); la température des salles d'élevage est en effet de 25°C.

TABLEAU II. DISTANCES MOYENNES PARCOURUES A PARTIR DU
POINT DU LACHER
(en mètres)

	Délai de recapture			
	2 j	6 j	9 j	13 j
Mâles stériles	61,75	104,54	144,96	188,00
Mâles sauvages	99,47	121,35	114,22	116,37

La comparaison des lots plus ou moins irradiés n'a pas révélé de
différences entre eux; le facteur «irradiation» est sans effet.

— Longévité

Après la vigueur sexuelle, la longévité conditionne en grande partie
la compétitivité des mâles stériles. Des recaptures régulières sont faites
après chaque lâcher au bout de 2, 6, 9, 13, 16, 20, 23 ... jours.

Durée de survie moyenne ou longévité de groupe. Si on appelle «période»
le temps au bout duquel la population a décru de moitié, on note que les
mâles stériles ont une longévité moyenne de 4,8 j, inférieure à celle des
mâles sauvages (6,2 j) lâchés en même temps qu'eux. La décroissance est
rapide pour les mâles stériles alors qu'elle est lente pour les mâles
sauvages (fig. 1).
Les facteurs «élevage, voyage» sont responsables de cette longévité
moyenne inférieure à celle des mâles sauvages, le facteur «irradiation»
étant sans effet par comparaison avec les autres groupes.

Durée de survie maximale ou longévité individuelle. Par les recaptures
régulières on obtient une évaluation assez bonne de la durée maximale après
les lâchers au bout de laquelle il est possible de capturer encore un individu
marqué. Il s'agit des individus qui se sont le mieux adaptés et qui ont vécu
le plus longtemps: mâles stériles: 12,1 j (1 cas à 30 j); mâles sauvages:
28,4 j (1 cas à 41 j).
Les mâles stériles ont une longévité maximale plus de deux fois
inférieure à celle des mâles sauvages et celle-ci baisse lorsque le taux
d'irradiation augmente.
Avec la saison chaude, les longévités baissent nettement chez les mâles
stériles; elles restent constantes chez les mâles sauvages.

— Dispersion

Vitesse et étendue de la dispersion (tableau II). Les mâles stériles
se dispersent moins vite que les mâles sauvages pendant les deux premiers
jours; la dispersion devient semblable vers le sixième jour. Cette capacité
moindre à se diluer dans le gîte est due au manque de vigueur déjà signalé
ainsi qu'à l'état des ailes, l'élevage et le transport en cage provoquant
l'usure prématurée de celles-ci.

TABLEAU III. HAUTEURS MOYENNES DE REPOS DIURNE SUR LES
TRONCS DE Morelia senegalensis
(en centimètres)

	15 h	16 h
Mâles stériles	50,92 ± 5,16	35,28 ± 5,00
Mâles sauvages	46,46 ± 6,70	38,22 ± 8,95

Zones de dispersion. Cinq zones ont été définies dans le gîte et on
a comparé les fractions de glossines d'élevage et de glossines sauvages
qui y étaient respectivement capturées. La distribution géographique est
identique et les deux catégories de glossines peuplent en même proportion
les mêmes parties du gîte, qui sont celles les mieux protégées. Bien que
numériquement réduits les lâchers de glossines plus ou moins irradiées
autorisent les mêmes conclusions.

Variation de la dispersion dans le gîte. Avec l'élévation de température
au cours de la saison sèche et les passages d'animaux sauvages (éléphants),
la végétation s'éclaircit dans certaines zones du gîte et les mâles stériles,
comme les mâles sauvages, respectent les déplacements saisonniers en
se réfugiant dans les lieux les mieux protégés (végétation dense des zones
les plus basses).

— Lieux et hauteurs de repos

Ces observations se sont déroulées en espace limité dans une grande
cage de 6 m de haut et 30 m de périmètre entourant un bosquet de Morelia
senegalensis dans un biotope riche en G. tachinoides. Un lot de 420 glossines sauvages est prélevé dans le gîte puis marqué (gouache acrylique
fluorescente) et lâché en même temps qu'un lot de mâles irradiés également
marqués, au nombre de 859. Tous sont nourris avant le lâcher qui a lieu
aux heures très chaudes (14 h).

Observations diurnes. Du fait des fortes températures, les glossines
se réfugient sur les troncs de Morelia. Le fait est habituellement observé
pour les glossines sauvages. Les mâles stériles, réagissant à ces conditions climatiques, adoptent des lieux de repos identiques.
Deux mesures des hauteurs de repos sont faites à 15 h et à 16 h
(tableau III). Il n'y a aucune différence significative entre les hauteurs
enregistrées. Le comportement des mâles stériles et des mâles sauvages
est le même dans le choix et la hauteur des lieux de repos aux heures
chaudes.

Observations nocturnes. La détection de nuit des glossines marquées
est relativement aisée grâce à l'usage d'une lampe portative à rayonnement
ultra-violet qui rend particulièrement visible la petite tache de gouache.
Au crépuscule, mâles stériles et mâles sauvages cherchent en fort
pourcentage à sortir de l'espace limité de la cage et se retrouvent sur le

TABLEAU IV. HAUTEURS MOYENNES DE REPOS NOCTURNE
(en centimètres)

	Sur Morelia	Sur grillage
Mâles stériles	195,00 ± 28,00	82,49 ± 8,38
Mâles sauvages	216,66 ± 48,00	132,20 ± 42,83

grillage de celle-ci tandis que le restant gagne les parties supérieures de
la végétation (Morelia).

La mesure des hauteurs de repos nocturne (tableau IV) montre qu'il
n'existe aucune différence significative entre les mâles stériles et les mâles
sauvages. Par contre, les hauteurs d'arrêt sur le grillage sont plus faibles
chez les mâles stériles que chez les mâles sauvages (moindre vigueur,
usure des ailes, etc.).

— Comportement alimentaire

Etat de réplétion des glossines au repos. 62% environ des mâles stériles
au repos sont gorgés, contre 50% des mâles sauvages. Peu dispersés les
premiers jours, les mâles stériles se nourrissent plus volontiers sur
l'équipe de captureurs qui pénètre dans le gîte et vient à leur portée.

Choix des hôtes nourriciers. Les méthodes sérologiques permettant
d'identifier l'origine du repas de sang chez la glossine montrent que le
choix des hôtes nourriciers des mâles stériles est très différent de celui
des mâles sauvages pendant les 48 premières heures. Il s'en rapproche
lentement dans les jours suivants et devient identique vers le sixième jour;
il est alors le suivant: bovidés (Guib harnaché): 69,23%; primates (homme):
15,38%; suidés (phacochère): 7,69%; reptiles (varans): 7,69%.

Du fait d'une faible dispersion les premières heures, les mâles stériles
sollicitent surtout les captureurs qui pénètrent dans le gîte. Mieux répartis
ensuite dans le gîte, leur source principale de nourriture devient la faune
sauvage (fig. 2).

1.1.4. Efficacité

Cette dernière résulte de la compétitivité que les mâles stériles
peuvent soutenir vis-à-vis des mâles sauvages. Elle est la conséquence
de leur vigueur mais aussi de leur nombre relatif.

Après douze lâchers, aucune modification de la densité et de la composition de la population du gîte receveur n'a pu être observée; ceci est
attribuable en grande partie aux quantitiés insuffisantes de mâles stériles
lâchés dans une population naturelle de très forte densité où ils se retrouvent
dans un rapport de 0,1 ♀/1♂ .

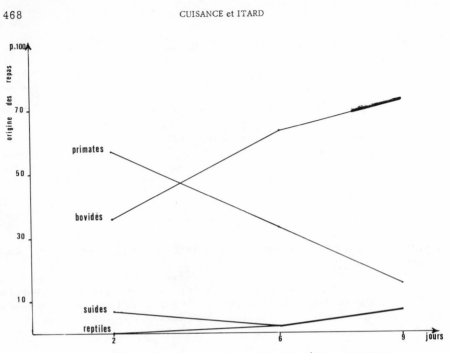

FIG.2.   Evolution de l'origine des repas de sang des mâles stériles avec le temps.

1.2. Observations faites au cours de la saison sèche de 1973 [11]

1.2.1. Lieux d'expérience

Deux gîtes ont été retenus sur les berges du fleuve Logone à 30 km
au sud de N'Djaména, l'un servant de gîte d'expérience, l'autre de gîte
témoin. Leur aspect et leur composition végétale sont identiques à ceux
précédemment décrits; ils s'étirent en fuseau de 600 m × 100 m pour le
premier, 300 m × 70 m pour le second, et sont bien isolés.

La dynamique des populations naturelles qu'ils abritent est semblable
à celle décrite, mais on souligne ici que la sécheresse anormale a préci-
pité cette évolution (concentration plus précoce des glossines) et a surtout
provoqué une chute de densité importante. Les conditions climatiques
exeptionnelles ont mis à notre disposition deux populations de G. tachinoides
isolées, accessibles et de faible densité (elle est estimée à 300 glossines
environ pour chaque gîte).

1.2.2. Comportement au lâcher

— Irradiation

A la suite d'une exposition de 30 minutes au rayonnement gamma, la
dose reçue par les mâles, âgés de 1 à 6 jours, est de 7600 à 10 000 rad;
elle assure une stérilité à 95%.

TABLEAU V. BILAN AU MOMENT DU LACHER (1973)

Comportement des glossines au lâcher	Glossines irradiées à 10 000 rad (%)
Envol	68,7
Tombées au sol	14,9
Mortes	16,3

— Conditions de transport, d'alimentation et de marquage

Elles sont identiques à celles de la précédente observation.

— Protocole de lâcher

Les lâchers n'ont plus lieu en un seul point mais en trois, correspondant aux endroits les mieux protégés et les plus denses en glossines sauvages.
Seize lâchers s'étalent sur quatre mois et demi à un rythme hebdomadaire à peu près respecté, sauf au mois de mars (grèves).
Sur 6727 mâles reçus, 4625 prennent leur envol normalement.

— Bilan au moment du lâcher (tableau V)

Le taux d'envol (68,7%), plus bas qu'en 1972, est attribuable à la défection des transports aériens entraînant des acheminements prolongés et des lâchers aux heures chaudes.

1.2.3. Comportement dans le gîte

Les résultats de la première observation se confirment ici. On notera en particulier que les mâles stériles peuplent les mêmes zones géographiques du gîte que les mâles sauvages et recherchent les mêmes conditions écologiques pour leur survie en fonction des rigueurs climatiques.

— Longévité

La durée de survie moyenne est supérieure de 2 jours à celle observée en 1972: mâles stériles: 6,5 j; mâles sauvages: 8 j. Ceci est attribuable à la dispersion rapide résultant de lâchers en plusieurs points du gîte qui permettent aux mâles de trouver rapidement un refuge favorable.
La durée de survie maximale est nettement supérieure elle aussi: mâles stériles: 20,3 j (1 cas à 26 j); mâles sauvages: 28,7 j (1 cas à 44 j).
Ces meilleures performances sont dues au fractionnement des lâchers dans les lieux les mieux protégés ainsi qu'à l'usage d'une irradiation ménagée.

TABLEAU VI. EVOLUTION DU TAUX DE MALES STERILES DANS LE
GITE EN FONCTION DES QUANTITES LACHEES ET DU RYTHME DES
LACHERS

Mois	Nombre de mâles stériles lâchés	Nombre de séances de lâchers	Rapport mâles stériles/ mâles fertiles au moment du lâcher	Rapport mâles stériles/ mâles fertiles dans les jours suivants
février	1053	4	2,09/1	1,45/1
mars	601	1	11,55/1	0,07/1
avril	862	5	6,95/1	2,84/1
mai	1376	4	12,07/1	8,09/1
juin	733	2	11,63/1	12,15/1

1.2.4. Efficacité

L'objet de cette observation portait principalement sur l'appréciation
de l'efficacité des lâchers dans une population dont la densité est devenue
naturellement faible sous l'action de la sécheresse.

— Rapport mâles stériles/mâles sauvages

Sur quatre mois et demi qui correspondent à la période des lâchers
on trouve en moyenne, lors des captures, 2,8 mâles stériles pour 1 mâle
sauvage (0,1 mâle stérile pour 1 mâle sauvage en 1972).

— Evolution du taux de mâles stériles en fonction des lâchers (tableau VI)

Le taux de mâles stériles dans le gîte est important lorsque les
fractions lâchées sont très supérieures à la population sauvage. Un rapport
au lâcher de 10/1 semble souhaitable.
Il se maintient à un niveau valable si les lâchers sont réguliers,
rapprochés (tous les 7 j environ), et demeurent quantitativement importants.
Ces conditions sont impératives et l'inobservation d'une seule d'entre
elles fait vite tomber le rapport mâles stériles/mâles sauvages.

— Evolution de la densité de population

A un maximum atteint en janvier fait suite une décroissance rapide
dans les mois suivants. Des dégradations imprévues dans le gîte témoin
(sécheresse) n'ont pas permis de mesurer l'effet exact des mâles stériles
sur la baisse de densité de la population du gîte d'expérience qui résulte
en partie de l'effet de conditions climatiques sévères mais certainement
aussi de l'action des mâles stériles si on considère le taux de fertilité
des femelles sauvages capturées.

— Contrôle de la fertilité des femelles

Sur 8 femelles capturées en fin d'expérience dans le gîte témoin,
toutes ont pondu entre le quatrième et le huitième jour suivant leur mise
en cage.

Sur 13 femelles du gîte d'expérience, 9 sont demeurées stériles.

Ces chiffres, bien que réduits, indiquent une forte perturbation de la fertilité des femelles sauvages de cette population.

2. PERSPECTIVES

Ces deux séries d'observations sur G. tachinoides permettent de noter un certain nombre de contraintes dans l'utilisation de la méthode du lâcher de mâles stériles. En fonction des données de la première expérience, quelques palliatifs ont été utilisés dans la seconde. L'ensemble permet de proposer quelques solutions et de mieux définir les indications de cette technique particulière.

2.1. Contraintes observées

— L'éloignement du centre de production par rapport au lieu du lâcher semble un inconvénient majeur. Le facteur «voyage» abaisse la vigueur générale des mouches et compromet leur compétitivité.

— La glossine d'élevage paraît se différencier quelque peu de l'insecte sauvage:

— le maintien en cage occasionne une usure prématurée des ailes, voire des anomalies (ailes non dépliées ou atrophiées);

— certains auteurs ont montré qu'une insuffisance de la musculature alaire pouvait se manifester [12] (processus de «domestication»?) [13];

— dans nos observations, l'optimum thermique d'activité des glossines lâchées semble se rapprocher de celui enregistré dans les salles d'élevage et diffère de celui des glossines sauvages.

— Les facteurs «élevage» et «voyage» semblent les plus néfastes: baisse des capacités d'envol, longévité moins grande, dispersion plus faible les premiers jours. Le facteur «irradiation» n'aurait qu'une action de second ordre en abaissant surtout la durée de survie maximale et pour une moindre part la vigueur de l'insecte.

Tous ces facteurs sont responsables d'une phase d'adaptation des mâles stériles qui abaisse leur compétitivité vis-à-vis des mâles sauvages et les rendent vulnérables.

— La méthode exige un minimum de connaissances écologiques de la zone à «traiter» (densité de la population sauvage, dispersion saisonnière, lieux de plus grande concentration, etc.), donc un certain nombre d'investigations [14, 15].

— Dans le cadre de ces observations sur G. tachinoides un lâcher toutes les semaines paraît indispensable; ce rythme doit être ininterrompu.

2.2. Palliatifs utilisés

Il convient d'écourter le plus possible la durée de la phase d'adaptation en facilitant aux mâles stériles leur insertion dans la population sauvage [16].

— Les lâchers se font au niveau des zones du gîte les mieux protégées.

— On choisit de préférence les heures où les conditions climatiques (température, humidité relative, lumière) se rapprochent le plus de celles des salles d'élevage (matin et soir en saison chaude).

— On favorise leur dispersion en fractionnant le total à lâcher sur toute la surface du gîte.

— Les mâles stériles sont nourris avant le lâcher.

— L'abaissement du taux d'irradiation (10 000 rad) des adultes entraîne une longévité meilleure tout en assurant une stérilité correcte.

— On a choisi un gîte de faible densité afin de compenser par la quantité lâchée la qualité inférieure du mâle stérile (nécessité d'avoir un rapport 10/1

2.3. Solutions à envisager

Il est bien sûr souhaitable d'arriver à une production massive d'insectes dans les élevages, mais il reste à améliorer surtout la qualité de l'insecte que l'on va lâcher. Ces observations suggèrent les quelques remarques suivantes.

— On soignera au maximum la transition entre les conditions de l'insectarium et celles du gîte où l'on intervient. Dans la perspective de lâchers d'adultes, le transport sera le plus court possible et se fera en enceintes climatisées. On pourrait même envisager de maintenir les mâles pendant les heures précédant le lâcher dans des conditions climatiques intermédiaires entre celles des salles d'élevage et celles des gîtes afin d'atténuer les effets néfastes des brusques variations thermiques, hygro-métriques et lumineuses.

— La remise dans le milieu naturel au stade pupal quelques jours, voire quelques heures avant l'éclosion semble être une solution favorable [17]. L'adulte ne subit pas les effets d'un transport traumatisant et l'adaptation aux conditions naturelles est en grande partie facilitée. De ce fait, on peut espérer abaisser le rapport σ/σ, les mâles stériles s'adaptant plus vite et étant plus compétitifs. Une meilleure «qualité» du mâle stérile permet de réduire les quantités à lâcher. La stérilisation aurait lieu alors dans les tout derniers jours précédant l'éclosion, à la fin du stade pupal.

2.4. Indications de l'utilisation de la méthode chez G. tachinoides

Si la méthode du mâle stérile reste séduisante, il demeure qu'elle exige que de nombreuses conditions soient remplies pour son application pratique sur le terrain.

— Le biotope

Elle n'est pas la solution à n'importe quelle situation. Elle conviendra aux gîtes de surface délimitée, à végétation très dense, inaccessibles en partie aux techniques de pulvérisation classiques. Cette méthode semble, de ce fait, indiquée dans le cas des espèces riveraines et en particulier pour G. tachinoides.

— La population

Les mâles stériles seront lâchés dans des populations naturelles de faible densité et le plus isolées possible.

— La saison d'intervention

La saison sèche au Tchad provoque un repli et une concentration
saisonnière des glossines dans les zones bien boisées du bord de l'eau.
Cette situation est très favorable à des lâchers de mâles stériles, la
population-cible étant alors répartie sur une faible surface.

— L'objectif visé

Il n'est pas dans ses objectifs d'obtenir l'éradication immédiate d'une
population de glossines. La méthode du mâle stérile reste une méthode
à moyen ou à long terme et ne saurait être utilisée en cas d'épidémies ou
d'épizooties graves.
A partir des éléments d'observation rassemblés, il apparaît que la
technique du mâle stérile pourrait rendre de grands services, en particulier
dans l'éradication de populations de basse densité (traitements aux insecticides) ou résiduelles (sécheresse, déforestation), inexpugnables avec les
moyens classiques habituels.

CONCLUSION

Ces deux séries d'observations permettent une meilleure connaissance
du comportement d'insectes de laboratoire lâchés dans le milieu naturel.
Elles montrent que le mâle stérile de G. tachinoides soumis à plusieurs
manipulations présente une phase d'adaptation non négligeable pour atteindre
les performances des mâles sauvages. Pour remédier à cette moindre
compétitivité générale, il devient nécessaire d'établir dans le gîte à
«traiter» un rapport de nombre très en faveur des mâles stériles. Les
efforts de recherche doivent permettre d'obtenir des élevages de masse
toujours plus productifs, mais si la quantité est indispensable, il convient
de rester soucieux de la qualité de l'insecte élevé.
La durée de cette expérimentation, conditionnée par les facteurs
saisonniers, reste courte, mais elle permet de noter des résultats
encourageants. Ces premiers essais, bien que limités, laissent entrevoir
l'utilité certaine de la méthode dans la perspective d'une lutte intégrée.

REFERENCES

[1] GRUVEL, J., Contribution à l'étude écologique de Glossina tachinoides Westwood, 1850 (Diptera,
Muscidae) dans la réserve de Kalamaloué, vallée du Bas-Chari), Thèse Doct. ès Sciences, Paris, 1974.
[2] GRUVEL, J., CUISANCE, D., «Données récentes sur l'écologie de G. tachinoides, Application aux
méthodes du mâle stérile», Colloque sur les moyens de lutte contre les trypanosomes et leurs vecteurs, Paris,
12 - 15 mars 1974.
[3] ITARD, J., «Techniques d'élevage des glossines, Perspectives offertes pour l'utilisation de la méthode
de lutte par lâchers de mâles stériles», Conseil scient. int. de recherches sur la trypanosomiase (ISCTR),
13e réunion, Lagos, Publ. n° 105 (1971) 243-48.
[4] ITARD, J., «Situation actuelle des élevages de glossines à Maisons-Alfort», Colloque sur les moyens de
lutte contre les trypanosomes et leurs vecteurs, Paris, 12 - 15 mars 1974.
[5] ITARD, J., Stérilisation des mâles de Glossina tachinoides West. par irradiation aux rayons gamma,
Rev. Elev. Méd. Vét. Pays Trop. 21 4 (1968) 479-91.

[6] ITARD, J., Elevage, cytogénétique et spermatogénèse des insectes du genre Glossina, Stérilisation des
      mâles par irradiation gamma, Ann. Parasit, Hum. Comp. 46 3bis (1971) 35-63.

[7] ITARD, J., «Sterilization by gamma irradiation of adult male Glossinae, Low dosage irradiation
      (4000 to 6000 rads) of adult male G. tachinoides», Conseil scient. int. de recherches sur la trypano-
      somiase (ISCTR), 13ᵉ réunion, Lagos, Publ. n° 105 (1971) 321-25.

[8] CUISANCE, D., ITARD, J., Comportement de mâles steriles de Glossina tachinoides West. lâchés
      dans les conditions naturelles, environs de Fort-Lamy (Tchad), I. Transport, lâchers, rythme d'activité,
      action sur la population sauvage, Rev. Elev. Méd. Vét. Pays Trop. 26 1 (1973) 55-76.

[9] CUISANCE, D., ITARD, J., Comportement de mâles steriles de Glossina tachinoides West. lâchés
      dans les conditions naturelles, environs de Fort-Lamy (Tchad), II. Longévité, dispersion, Rev. Elev.
      Méd. Vét. Pays Trop. 26 2 (1973) 169-86.

[10] CUISANCE, D., ITARD, J., Comportement de mâles stériles de Glossina tachinoides West. lâchés
      dans les conditions naturelles, environs de Fort-Lamy (Tchad), III. Lieux et hauteurs de repos,
      Comportement alimentaire, Rev. Elev. Méd. Vét. Pays Trop. 26 3 (1973) 323-38.

[11] CUISANCE, D., ITARD, J., Lâchers de mâles stériles de Glossina tachinoides West. dans un gîte
      naturel de faible densité (Bas-Logone, Cameroun) sous presse.

[12] DAME, D.A., BIRKENMEYER, D.R., BURSELL, E., Development of the thoracic muscle and flight
      behaviour of Glossina morsitans orientalis Vanderplank, Bull. Ent. Res. 55 (1969) 345-50.

[13] BOLLER, E., Behavioral aspects of mass-rearing of insects, Entomophaga 17 1(1972) 9-25.

[14] GLOWER, P.E., Importance of ecological studies in tsetse fly control, Bull. World Health Org. 37
      4 (1967) 581-614.

[15] BILIOTTI, E., L'écologie, fondement et support de la lutte biologique, Ann. Parasit. Hum. Comp.
      46 3bis (1971) 5-10.

[16] CUISANCE, D., «Quelques aspects du comportement des mâles stériles de G. tachinoides W. lâchés
      dans les conditions naturelles; leur incidence sur l'utilisation de la méthode du mâle stérile»,
      Organisation commune de lutte contre les endémies en Afrique Centrale, 8ᵉ Conf. tech., Yaoundé
      (Cameroun) 28 février-3 mars 1973.

[17] DAME, D.A., SCHMIDT, C.H., The sterile-male technique against tsetse flies, Glossina spp.,
      Bull. Ent. Soc. Am. 16 1 (1970) 24-30.

DISCUSSION

B. NA'ISA: Your paper has merely served to emphasize the difficulty
of applying the sterile-male technique to the tsetse fly, i.e. the difficulty
of reproducing large numbers of flies in the laboratory and the lack of a
method of accurately determining the number of flies in the wild population.
Although the correct season (dry season) was chosen in which to release
the flies, you have suggested that, to achieve better results, more frequent
releases need to be made, i.e. weekly (because the wild flies outnumbered
the sterilized by 10:1). I would suggest that non-persistent insecticide
should have been used to reduce this ratio, in order to give the sterilized
flies a chance to mate with the wild population.

D. CUISANCE: In the first experiment our aim was to determine the
behaviour of laboratory insects released into the natural environment. The
aim of the second experiment was to measure the efficiency of sterile-male
releases. The wild population was isolated and its density low because of
the climatic conditions (dryness). Therefore the use of insecticides was
not called for. If we had been dealing with a high-density population
preliminary insecticide treatment would of course have been desirable.
We took advantage here of the favourable conditions for experimental
observation created by the extreme dryness in that region of Chad.

J.F. COZ: How did you measure changes in the population density in the experimental and control areas?

D. CUISANCE: We used two conventional methods simultaneously, the first being the marking and recapture method with a recapture time of 48 hours. It was employed once a week to obtain an approximate indication of the true density. This method seemed particularly suitable, as it was being applied to an isolated population during a period when the climatic conditions remained stable and did not give rise to any violent change in the population density of the area. Moreover, the marked flies we released seemed to mix well with the wild ones in the course of 48 hours. The other method used was the fly-round method which consists in monitoring a fixed circuit for five hours and recording the number of flies caught hourly per catcher. This provides what we call the apparent density. The population density curves obtained with the two methods are in satisfactory agreement.

P.R. FINELLE: What is the effect of the "irradiation" factor on the fly behaviour or, in other words, have you tried releasing both normal and sterilized males and, if so, were there any differences in behaviour?

D. CUISANCE: In our first experiment three batches of flies were released, the first two having been irradiated with 15 krad and 6 krad respectively and the third being non-irradiated. Observations showed that the flight capacity decreases with increasing dose and that irradiation also reduces the longevity of the released flies.

A.A. AMODU: How do you calculate longevity? Is your calculation based on recapture data alone?

D. CUISANCE: The calculation of longevity is based on the results of the frequent and regular capturing operations following each release. The large numbers of marked insects recaptured enabled us to obtain fairly reliable data.

A.A. AMODU: Why were the irradiated released flies attracted to feeding on men?

D. CUISANCE: The sterile males are not particularly attracted by man. However they do tend to bite the fly-catchers, because the latter come within their range as they move around the area. When the sterile males are well dispersed — by about the sixth day — the number of natural hosts (wild animals) bitten is high and the harnessed antelope is then a welcome prey.

F.M. EVENS: To what do you attribute the change in behaviour of the irradiated flies at around the sixth day? Is it due to the elimination of the weak specimens or is it due to the flies adapting to their new environment?

D. CUISANCE: It is without doubt due to the elimination of the insects which were too weak to survive, for I think that by the sixth day the rest would have adapted to the natural environment. This seems to be confirmed by their choice of host which is the same as that of males of the wild strain.

F.M. EVENS: The presence of flies with abnormal wings cannot be due to their transportation. In our experience this is caused by sub-optimal rearing conditions.

D. CUISANCE: The existence of anomalies in wing structure has, of course, something to do with the rearing but transport does cause premature wear of the wings.

F.M. EVENS: You are proposing to overcome transport difficulties by putting out irradiated pupae to eclose in the field. I should like to point out that you would thereby run the risk of increasing the trypanosomic infestation rate. I would propose an intermediate solution whereby you would irradiate and transport the pupae, allow them to eclose and give the first feed in the laboratory and then release the adults.

D. CUISANCE: Yes, that is a good idea. There is no doubt that transporting the pupae is less traumatic for the insect which is to be released, and it avoids wear of the wings due to agitation of the adult flies in their cage.

IAEA-SM-186/32

A TECHNIQUE FOR DETECTING ABORTIONS
IN WILD POPULATIONS OF
Glossina SPECIES*

L.C. MADUBUNYI
Department of Veterinary Pathology,
University of Nigeria,
Nsukka, Nigeria

Abstract

A TECHNIQUE FOR DETECTING ABORTIONS IN WILD POPULATIONS OF Glossina SPECIES.

Female Glossina morsitans morsitans collected from the field in February 1973 were dissected in the laboratory. Records of ovarian configuration, number of ovulations, stage of pregnancy, mean spermathecal value, size of the intra-uterine content and egg in the follicle next in the ovulation sequence (FNOS) were kept for each fly dissected. Because of the significant curvilinear relationship between the FNOS and stage of pregnancy, the probable uterine content (stage of pregnancy) was defined within 95% fiducial limits on the basis of size of the egg in the FNOS. By comparing the size of fully developed eggs with the size of the eggs in the FNOS it was possible to determine whether or not a given female could have larviposited a viable and full-grown 3rd instar larva within the 24 h preceding its capture. By combining these considerations, it was determined that 8.72% of the 195 parous flies dissected had aborted. It was also possible to predict the stage of pregnancy and ovulation cycle at which the abortions occurred.

INTRODUCTION

The genus Glossina is among the pest genera which the IAEA has earmarked for suppression through the sterile-male technique. Such a suppressive measure involving interference with the tsetse's reproductive performance requires, among other things, accurate knowledge of the actual rate of increase (r) of tsetse populations in nature. In order to obtain this statistic, two parameters need to be known, namely the birth-rate (m_x) and survival rate (l_x) [1].

Following success in establishing laboratory colonies of the tsetse and subsequent advances in colonization techniques, progressively improved estimates of l_x and m_x have been reported for various Glossina species under laboratory conditions [2-5]. Saunders [6] has outlined a method of deriving l_x for a natural population of the tsetse based on the calculated probable age-composition of a sample of non-teneral female Glossina palpalis palpalis Robineau-Desvoidy. Similar information on the m_x of wild tsetse populations is totally lacking. Consequently the estimates of r_m attempted by Glasgow [7] and of r by Saunders [6] had to be based on cautiously assumed m_x values.

Determination of m_x of wild tsetse populations is complicated by reproductive losses whose frequencies have yet to be established. The main sources of reproductive loss are copulation without insemination, follicular degeneration, egg retention in virgins and abortions [8-11].

---

* This work was carried out by the author at the National Council for Scientific Research, P.O. Box 49, Chilanga, via Lusaka, Zambia.

Whereas these reproductive abnormalities may occur frequently in laboratory colonies, they are probably rare among wild tsetse populations [6, 9].
Apparently, however, abortions are suspected to be not infrequent in wild
tsetse populations particularly towards the terminal portion of a female's
reproductive life [7]. Unlike other reproductive abnormalities, abortion
leaves no trace in the female genital tract, and consequently it is the most
elusive source of reproductive loss to detect in wild tsetse populations.
Nevertheless, a fair knowledge of abortion rates in natural populations is
crucial to any valid estimates of the tsetse's m_x.

In the laboratory, abortions have actually been observed through
recoveries of under-sized non-viable larvae prematurely ejected by the
pregnant female [10, 12] or inferred from gaps in larviposition [12 - 14].
For obvious reasons it is impractical to make similar observations in the
field. However, using empty uteri as criterion for abortions, Jordan [15]
concluded that an adverse factor, probably inability of females to obtain
a blood meal at just the right stage of gestation, was causing a high rate
of abortions in a natural population of G. palpalis which he sampled in
midwestern Nigeria. Harley [9] in Uganda reported that one out of 2666
wild Glossina brevipalpis Newstead and two out of 2477 wild Glossina
palpalis fuscipes Newstead which were uniseminated had evidence of
ovulation but empty uteri. He believed that they must have aborted infertile
eggs before capture, and concluded that abortions are rare in nature.

In the course of studies of survival and age-specific pregnancy rates
of a wild population of Glossina morsitans morsitans Westwood in Central
Zambia, a fairly precise technique was devised not only for detecting the
elusive abortion but also for predicting the stage of pregnancy at which
it occurred. Since a method for detecting abortions in wild tsetse has so
far never been described, details of the technique and associated observations
are presented below.

MATERIALS AND METHODS

In February 1973, all female Glossina morsitans morsitans captured
during routine daily transect fly rounds using human bait at our field station
(15°19' S; 29°16' E; elevation 3000 - 4000 m) were immediately preserved
in Machado's fluid (10 parts C_2H_5OH, 4 parts H_2O, 1 part CH_3COOH and
1 part $CH_2OHCHOHCH\,OH$), and subsequently transported to the laboratory.
Each fly was dissected in 0.9% saline under a binocular microscope fitted
with a calibrated micrometer eyepiece.

The ovaries of each fly were examined for configuration and evidence
of ovulations, i.e. presence of follicular relics. The uterus of each fly
was also examined, the stage of pregnancy noted and linear measurement
of the uterine content taken where applicable. Then the fly was assigned
to an age category after Saunders [16] and Challier [17]. Irrespective of
whether or not a fly was parous, the length of its largest egg follicle next
in the ovulation sequence (FNOS) and its spermathecal value [2] were
determined.

The size of the FNOS was regressed on the stage of pregnancy. Four
stages of pregnancy (1 - 4) were recognized, each corresponding to egg,
1st, 2nd and 3rd larval instars in utero respectively. The nature of the

regression was statistically analysed. Fiducial limits (95%) were attached
to the mean size of the FNOS at each stage of pregnancy for predictive
purposes.

RESULTS

All age groups from 0 - 80 days old and all stages of pregnancy were represented in the dissections of the 420 female Glossina morsitans morsitans captured
in February 1973 (Table I). Approximately 5% of all flies dissected had
empty spermathecae. Of these apparently uniseminated females, 16 were
nullipars (5 in the ovarian category Oa and 11 in Ob) and 4 were parous
(2 each in ovarian categories 2 and 4 + 4n respectively). Among the inseminated females mean spermathecal values were high, approaching 2 in all
age groups except ovarian category 1 probably because of the small sample
size.

Approximately 54% of the 420 flies dissected were nullipars and 46% bore
clear evidence of successful ovulations before capture. Of the parous flies,
91.28% were pregnant with either an egg or larva in utero and 8.72% had
empty uteri. No instances of follicular degeneration or egg retention were
encountered.

Full-grown intra-uterine eggs measured 1.387 mm ± 0.016 (n = 84)
(unless otherwise stated all means are accompanied by standard errors).
The larvae measured 1.490 ± 0.037, 2.720 ± 0.133, 3.960 ± 0.095 mm during
their 1st (n = 57), 2nd (n = 23) and 3rd (n = 13) instars respectively. The
FNOS of nulliparous flies in the Oa ovarian category (age group 0 - 4 days)
measured 0.581 mm ± 0.014; n = 17, and those of Ob flies (age group
4 - 10 days) measured 1.026 ± 0.015 (n = 208). About 9% of the Ob nullipars
carried eggs measuring 1.368 - 1.748 mm in their inside right ovarioles.
Since this was within the range of the size of full-grown eggs found within
the uterus of parous flies, it appears that those 19 Ob nullipars were
moments away from their first ovulation at the time of their capture.

As pregnancy progressed the size of the FNOS increased at variable
rates (Fig. 1). Its growth was gradual up to the 2nd stage of pregnancy and
became rapid thereafter. The regression of FNOS length on the stage of
pregnancy ($\hat{Y}$ = 0.8326 - 0.0525X + 0.0318X^2) was significant (P < 0.05; df = 174).
Apparently most of the growth of the FNOS occurred during the stage of
pregnancy when the uterus contained a 2nd instar larva. As a consequence
of this significant relationship of the FNOS to pregnancy stage, the size of
the FNOS was defined for each stage of pregnancy with 95% fiducial limits
(Table II).

The probable uterine content (stage of pregnancy) of each parous fly
with an empty uterus (Table III) was determined by fitting the size of its
FNOS into the calculated confidence intervals. Of the 17 parous but non-
pregnant females, 11 had their FNOS size in the interval 0.970 - 1.240 mm
indicating that there should have been a 3rd instar larva in their respective
uteri. However 7 of these 11 flies could also have contained a 2nd instar
larva in utero since the sizes of their FNOS were within the interval
0.917 - 1.100 mm. Of the remaining 6 flies, 3 each had an FNOS measuring
0.912 mm which was above the upper limit for the 2nd stage of pregnancy
(i.e. 1st instar larva in utero) and below the lower limit for the 3rd stage
of pregnancy (i.e. 2nd instar larva in utero). The FNOS in each of the

TABLE I. AGE COMPOSITION, AGE-SPECIFIC PREGNANCY STAGES AND MEAN SPERMATHECAL VALUES OF FEMALE Glossina morsitans morsitans FROM A NATURAL POPULATION

	Oa Ob	1	2	3	4 + 4n	5 + 4n	6 + 4n	7 + 4n
	a b c d e	a b c d e	a b c d e	a b c d e	a b c d e	a b c d e	a b c d e	a b c d e
No.	17 208	1 0 0 0 1	4 0 0 0 2	5 1 8 10 8	26 39 20 11 6	0 1 0 0 0	1 3 1 0 0	1 7 1 2 0
Sub totals	225	2	6	68	102	1	5	11
% of total	53.7	0.48	1.43	16.19	24.28	0.24	1.19	2.62
% of parous flies only	-	1.03	3.08	34.87	52.31	0.51	2.56	5.64
Mean spermathecal	1.77	1.25	2.00	1.62	1.60	2.00*	1.75	1.64
value ± S.E.	±0.03	±0.25	±0.00	±0.06	±0.04		±0.19	±0.13

* Single observation.

N. B.   Oa = nullipars with inside right follicle $\leqq$ 0.65 mm.
     Ob = nullipars with inside right follicle $\geqq$ 0.65 mm.
     a = parous female with egg in utero.
     b = parous female with 1st instar larva in utero.
     c = parous female with 2nd instar larva in utero.
     d = parous female with 3rd instar larva in utero.
     e = parous but non-pregnant female.

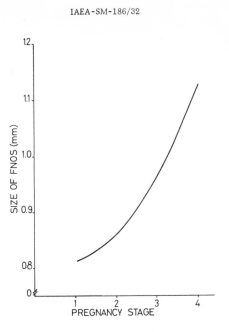

FIG.1.   Mean length (mm) of the egg follicles next in the ovulation sequence (FNOS) at various stages of
pregnancy in Glossina morsitans morsitans under natural conditions.

TABLE II.   SIZE (mm) OF EGG FOLLICLES NEXT IN THE OVULATION
SEQUENCE (FNOS) AT DIFFERENT STAGES OF PREGNANCY IN A
NATURAL POPULATION OF Glossina morsitans morsitans

Pregnancy stage[a]	Sample size	Mean length of FNOS	±Standard error	±95% fiducial limit
1	84	0.816	0.018	0.035
2	57	0.837	0.021	0.042
3	23	1.009	0.044	0.091
4	13	1.105	0.062	0.135

[a] 1 = egg in utero.
     2 = first instar larva in utero.
     3 = second instar larva in utero.
     4 = third instar larva in utero.

TABLE III. LENGTH (mm) OF THE EGG FOLLICLE NEXT IN THE
OVULATION SEQUENCE (FNOS) IN PAROUS BUT NON-PREGNANT
FEMALES FROM A WILD POPULATION OF Glossina morsitans
morsitans

Ovarian category	Individual FNOS measurements
1	0.912
2	1.064; 1.178
3	0.988; 0.836; 0.912; 1.064; 0.760; 0.912; 0.988; 1.216
4 + 4n	1.224; 0.988; 1.216; 0.684; 1.026; 0.988

FNOS when egg should be in utero 0.781 - 0.851 mm.
FNOS when 1st instar larva should be in utero 0.795 - 0.879 mm.
FNOS when 2nd instar larva should be in utero 0.917 - 1.100 mm.
FNOS when 3rd instar larva should be in utero 0.970 - 1.240 mm.

remaining parous but non-pregnant females fell either within the interval
0.781 - 0.851 mm or below the lower limit in which case each should have
had at least an egg in utero.

DISCUSSION

An empty uterus in a parous female tsetse is indicative of either
successful parturition of a mature 3rd instar larva or abortion of any of
the developing stages. Since ovulations occur with regularity within 24 h
of each parturition [18, 19] and, according to Foster [20], within 1 h of
parturition in G. austeni and G. morsitans respectively, the period during
which the uterus may remain empty in a normally reproducing parous tsetse
is very limited. However, before parturition the ovariole next in the
ovulation sequence (FNOS) contains a fully developed egg [11]. For the
purposes of the present study, the size of the intra-uterine egg is considered
a logical yard-stick of the size of a fully-developed egg. Consequently,
if a fly from the population sampled in this study had a normal parturition
but had not yet ovulated at the time of its capture, the length of its FNOS
should be 1.387 mm (±0.032 95% fiducial limits). Conversely, if the fly
had aborted within 1 - 24 h before its capture, its FNOS would not have
attained the maximum size. If, on the other hand, normal parturition
occurred and was followed by ovulation before the fly was captured, not
only should its uterus contain an egg or larva but also the size of its FNOS
would be correlated with the stage of pregnancy.

Since the length of the FNOS in each of the parous but non-pregnant
flies was below the lower limit of the calculated confidence interval for a

normal parturition (1.355 - 1.419 mm), none of them should have larviposited a viable 3rd instar larva within the 24 h preceding their capture.
It can therefore be safely concluded that 8.72% of the 195 parous flies dissected had aborted. Apparently, 17.64% of these abortions occurred at the
egg and 1st larval instar respectively, and 41.18% and 23.52% were aborted
at the 2nd and 3rd larval instar stages of intra-uterine development
respectively. These observations are in agreement with the suggestion
of Saunders & Phelps [21] that abortion may occur at any stage of the
gestation period. It is also self-evident that abortions were not limited
to any particular ovulation cycle or age group (Table I). Thus abortion was
observed among flies which had ovulated from 1 - 4 or more times. Because
few of the captured flies were in the (5 + 4n) - (7 + 4n) ovulation cycles, the
absence of aborters among them cannot be taken as contradicting the prediction of Glasgow [7] that abortions might be more frequent among very
old females approaching the end of their reproductive lives. Nevertheless,
it is also not uncommon among relatively younger flies (minimum age
10 - 50 days). With a sample more representative of all age groups,
ovulation cycles and stages of pregnancy, it should now be possible to attempt
realistic estimates of m_x for natural tsetse populations by incorporating the
necessary corrections for abortions.

Abortion rates have been reported for two other species of Glossina
in the wild, namely G. brevipalpis (0.04%), Glossina palpalis fuscipes
(0.08%) by Harley [9]. There is nothing from literature to suggest that
certain species of the tsetse are less prone to abortion than others, thus
these different rates cannot be ascribed to interspecies variations. Nevertheless poor nutrition is a recognized abortion-inducing environmental
stress [10] which could account for the widely varying rates reported.
It is probable that Harley's populations of G. brevipalpis and G. palpalis
fuscipes were better nourished than the population of G. morsitans morsitans
sampled in the present study. It has already been reported that nutritional
stress was evident between January and February 1973 in the area from
which the flies dissected in this study were collected [22]. Abortion rates
can therefore be used to complement other indices of nutritional stress in
natural populations of the tsetse.

Only the 16 nulliparous females with empty spermathecae can be
categorically branded true virgins. The two females which had ovulated
twice but had empty spermathecae were probably either never inseminated
or were only partially so initially. If they were initially uninseminated
their pregnancies might have been prematurely terminated as a result of
egg infertility [9]. If they were partially inseminated the quantity of sperm
they received during copulation must have been so little as to be completely
depleted after fertilizing only two ova. The other two parous flies in the
ovarian category 4 + 4n which had empty spermathecae were most likely
females whose sperm store had depleted with time. Nevertheless about
93% of all nullipars and 98% of all parous flies had mean spermathecal
values above zero. Furthermore, over 50% of the flies in the 7 + 4n age
group had mean spermathecal values of 2 indicating that their sperm
complement had probably been recharged through multiple matings. It has
already been demonstrated in the laboratory that multiple matings can occur
in tsetse populations [2, 23 - 25]. Nevertheless the data indicate that
copulation without insemination was uncommon and that the mating machinery
was efficient in the population of G. morsitans morsitans investigated.

ACKNOWLEDGEMENTS

The assistance of G. Masumba in the dissections and L. M. Didee in statistical analyses is gratefully appreciated.

REFERENCES

[1] ANDREWARTHA, H.G., BIRCH, L.C., The Distribution and Abundance of Animals, University of Chicago Press, Chicago (1954).

[2] NASH, T.A.M., The fertilisation of Glossina palpalis in captivity, Bull. Entomol. Res. 46 (1955) 357.

[3] NASH, T.A.M., JORDAN, A.M., BOYLE, J.A., A promising method for rearing Glossina austeni (Newst.) on a small scale, based on the use of rabbits' ears for feeding, Trans. R. Soc. Trop. Med. Hyg. 60 (1966) 183.

[4] JORDAN, A.M., NASH, T.A.M., BOYLE, J.A., The rearing of Glossina austeni Newst. with lop-eared rabbits as hosts. 1: Efficacy of the method, Ann. Trop. Med. Parasitol. 61 (1967) 182.

[5] JORDAN, A.M., CURTIS, C.F., Productivity of Glossina austeni Newst. maintained on lop-eared rabbits, Bull. Entomol. Res. 58 (1968) 399.

[6] SAUNDERS, D.S., Survival and reproduction in a natural population of the tsetse fly Glossina palpalis palpalis (Robineau-Desvoidy). Proc. R. Entomol. Soc., London, Series A 42 (1967) 7.

[7] GLASGOW, J.P., The Distribution and Abundance of the Tsetse, Pergamon Press, Oxford (1963).

[8] BUXTON, P.A., LEWIS, D.J., Climate and tsetse flies: Laboratory studies upon Glossina submorsitans and tachinoides, Philos. Trans. R. Soc. London, Ser. B 224 (1934) 175.

[9] HARLEY, J.M.B., Studies on the age and trypanosome infection rate in females of Glossina pallidipes Aust., G. palpalis fuscipes Newst. and G. brevipalpis Newst. in Uganda, Bull. Entomol. Res. 57 (1966) 23.

[10] MELLANBY, H., Experimental work on reproduction in the tsetse fly, Glossina palpalis, Parasitology 29 (1937) 131.

[11] SAUNDERS, D.S., The ovulation cycle in Glossina morsitans Westwood (Diptera : Muscidae) and a possible method of age determination for female tsetse flies by examination of their ovaries, Trans. R. Entomol. Soc., London, Series A 112 (1960) 221.

[12] SAUNDERS, D.S., The effect of starvation on the length of the interlarval period in the tsetse fly Glossina morsitans orientalis Vanderplank, J. Entomol. (A) 46 (1972) 197.

[13] HARLEY, J.M.B., The influence of temperatures on reproduction and development in four species of Glossina (Diptera, Muscidae), Proc. R. Entomol. Soc., London, Series A 43 (1968) 170.

[14] SAUNDERS, D.S., Reproductive abnormalities in the tsetse fly, Glossina morsitans orientalis Vanderplank, caused by a maternally acting toxicant in rabbit food, Bull. Entomol. Res. 60 (1971) 431.

[15] JORDAN, A.M., The pregnancy rate in Glossina palpalis (R. - D.) in Southern Nigeria, Bull. Entomol. Res. 53 (1962) 387.

[16] SAUNDERS, D.S., Age determination for female tsetse flies and the age compositions of samples of Glossina pallidipes Aust., G. palpalis fuscipes Newst. and G. brevipalis Newst., Bull. Entomol. Res. 53 (1962) 579.

[17] CHALLIER, A., Amélioration de la méthode de détermination de l'âge physiologique des Glossines, Bull. Soc. Path. Exot. 58 (1965) 250.

[18] BUXTON, P.A., The Natural History of Tsetse Flies, Mem. London Sch. Hyg. Trop. Med. 10, Lewis, London (1955).

[19] SAUNDERS, D.S., DODD, C.W.H., Mating, insemination and ovulation in the tsetse fly, Glossina morsitans, J. Insect Physiol. 18 (1972) 187.

[20] FOSTER, W.A., "Influence of medial neurosecretory cells on reproduction in female G. austeni", Second Symposium on Tsetse Fly Breeding in the Laboratory and its Practical Applications, Trans. R. Soc. Trop. Med. Hyg. 66 (1972) 322.

[21] SAUNDERS, D.S., PHELPS, R.J., "Reproduction in Glossina: breeding sites", The African Trypanosomiases (MULLIGAN, H.W., Ed.), George Allen & Unwin, London (1970).

[22] MADUBUNYI, L.C., Morphometric indices of nutritional state in a natural population of Glossina morsitans morsitans Westwood., J. Anim. Ecol. 43 (1974) 469.

[23] JORDAN, A.M., The mating behaviour of females of Glossina palpalis (R. - D.) in captivity, Bull. Ent. Res. 49 (1958) 35.

[24] CURTIS, C.F., Radiation sterilisation and effect of multiple mating of females in Glossina austeni,
      J. Insect Physiol. 14 (1968) 1365.
[25] DAME, D.A., FORD, H.R., Multiple mating of G. morsitans Westw. and its potential effect on the
      sterile male technique, Bull. Entomol. Res. 58 (1968) 213.

DISCUSSION

R. GALUN: Do you think that the trapping procedure could cause abortion?

L.C. MADUBUNYI: It is possible and I plan to investigate this question in an effort to refine the technique further.

F.M. EVENS: Your figure of 8.72% abortions of parous flies corresponds approximately to our observations during laboratory rearing of this insect.

Can you explain why, as shown in Table I, you found so few flies of the first and second larval stages and so many multiparous (3 - 4 larvae)?

L.C. MADUBUNYI: I attribute it to sampling error. When one looks at the numbers of nulliparous and older parous flies, the proportion of flies in the first and second ovarian categories seems out of phase. I am not certain that the sample gives the best representation of all age groups.

TECHNIQUES FOR REARING
Glossina tachinoides Westw.

E.D. OFFORI, P.A. DORNER
Division of Research and Laboratories,
International Atomic Energy Agency,
Vienna

Presented by H.W. Wetzel

Abstract

TECHNIQUES FOR REARING Glossina tachinoides Westw.

Techniques developed at the Seibersdorf Laboratory for rearing Glossina tachinoides Westw. are here described and the performance of a colony held for 12 months is discussed. Flies were held in a room 3.6 × 3.2 × 2.4 m high, maintained at 24.5° - 26°C and 72 - 75% relative humidity. They were fed six times a week on rabbits' ears. Seven-day-old males were mated with females 2 - 4 days old and females held for 108 days during which period pupa production averaged 8.5 per female. For separation of sexes at emergence and after mating, nitrogen gas was used to immobilize flies for about 2 - 3 minutes. The technique facilitated handling and had no adverse effect on flies. We estimated that our fly room would hold conveniently a maximum of 20000 flies, made up of 15000 producing females, 3000 young females and about 2000 males for mating to stock females. At the rate of production achieved in our laboratory, 15000 females would yield 3000 - 4000 surplus males each week for use in a sterile-male release programme.

INTRODUCTION

Glossina tachinoides Westw. is an important transmitter of human and animal trypanosomiasis in several parts of west and west-central Africa. This species tends to be distributed along rivers and streams, especially during the dry season. Therefore, at certain times of the year, isolated populations of G. tachinoides exist in these riverine habitats, thus providing an ideal condition for eradication by the sterile-male technique. This method requires the production of large numbers of flies in order to obtain sufficient surplus males for sterilization and release. Rearing of G. tachinoides was first attempted by Buxton and Lewis [1], who fed flies on sheep. However, the first successful laboratory colony was initiated by Itard and Maillot [2], who fed their flies six times a week, initially on guinea-pigs and chickens. Later, Itard and his co-workers [3], following techniques developed by Nash and co-workers [4] for maintaining G. austeni, fed G. tachinoides on rabbits with remarkable success.

Our studies at the Seibersdorf Laboratory of the International Atomic Energy Agency were carried out primarily to investigate methods of mass-rearing and to obtain material for irradiation studies.

COLONY INITIATION AND REARING PROCEDURE

Several batches of pupae were obtained from a laboratory-adapted stock in Maisons-Alfort, France (courtesy of J. Itard) between May and October 1972. Eclosion averaged 91%, yielding over 700 females, of which 600 were mated and used to begin the colony.

Insectary conditions

During the first five months of our study, flies were held in the same insectary as used for holding G. morsitans Westw. Temperatures in this room averaged 24.5°C - 26°C, and relative humidity about 65%. We therefore provided additional humidity for pupae and also adult males by holding pupal containers and male cages in trays lined with damp towelling or plastic sponge (Spontex®). Later, the colony of G. tachinoides was removed to a room maintained at 25°C - 26°C and 72 - 75% relative humidity. Dim lighting (12-hour photophase) was provided by a rheostatically controlled 40-watt tungsten bulb.

Fly handling procedure

Adults

Emerged adults were separated by sex and held in round (12.5 cm dia.) or oblong (16 cm × 7 cm) PVC-netting cages. To facilitate handling, flies were immobilized, initially by cooling at 4°C. Although this treatment appeared to have had no adverse effect on females, we noted that over 70% of males died a few days after cold treatment. We therefore resorted to the use of nitrogen gas, which when passed into the emergence cage for about two or three minutes quickly immobilized adults. Flies recovered conscious-ness within three minutes of removal from the nitrogen atmosphere; subsequent experiments showed that this treatment had no adverse effect on either males or females.

Flies were fed six times a week on rabbits' ears, following the procedure described by Mews and co-workers [5]. Cages were left on the animals' ears for about 7 - 10 minutes, which was sufficient time for full engorgement.

For mating, cages containing separate sexes were connected by means of a tube, the male cage covered with black cloth and the female cage lighted, whereupon males moved into the female cage. After mating, sexes were separated by immobilizing with nitrogen gas. Males, at least 7-days-old, were mated to 2-4-day-old females, and the sexes (equal numbers) kept together for 24 - 48 hours or 72 hours if a week-end intervened. Flies were fed six hours before mating and also daily thereafter, except on Sunday.

Mated females were held 10 per cage in PVC-netting cages (12.5 cm dia.) and three cages of 30 females constituted one unit. Experiments indicated that this size of cage could hold up to 15 females without resulting in unduly high mortality or lowered productivity. Cages were supported on a metal rack placed in a plastic tray 40 cm × 28 cm × 2 cm deep. This size tray could hold six cages, thus the maximum capacity of each tray is 90 females. Each tray was lined with one or two layers of paper towelling which provided pupation sites for larvae. Females were kept for 108 days and then dis-carded. Samples of surviving females were dissected and their spermathecae examined to determine the degree of insemination. Males were usually dis-carded after having been used once or twice for mating.

Pupae

Pupae were collected daily, except on Sunday. During the initial stages of our investigation, pupae were weighed daily; as the colony increased in

size, individual and daily weighing was curtailed, but samples of about
100 pupae were weighed in bulk once a week.

Pupae were placed in open plastic dishes without sand and maintained in
this state for about two weeks before being placed in an emergence cage.
Additional humidity was provided by placing the emergence cage on a tray
lines with damp towelling or plastic sponge (Spontex ®).

PERFORMANCE OF THE COLONY

At the end of November 1972 the colony contained 525 mated females,
most of which were parental flies. No more flies were added from outside.
By mid-March, only F_1 and subsequent generations were present. The colony
increased rapidly and was limited to about 1000 mated females by the end of
April 1973, and was held at this level until the investigation was terminated
six months later.

The first larviposition occurred on day 16 or 17 of adult life, i.e.
14 - 15 days after insemination. Data on 60 mated females kept individually
showed that the inter-larval period varied from 6 days to 11 days (the majority
about 8 and 9 days). Production data for the 12 months covered by this
study are shown in Table I and Fig. 1. Eclosion ranged from 85% to 97%, with

TABLE I. PERFORMANCE OF A COLONY OF Glossina tachinoides Westw.
HELD FOR 12 MONTHS AND FED ON RABBITS 6 TIMES A WEEK:
FOR SEXING AT EMERGENCE AND SEPARATION OF SEXES AFTER
MATING, FLIES WERE IMMOBILIZED FOR 3 - 5 MINUTES
USING NITROGEN GAS

Month	Number of females in colony	Daily mortality (%)	Mean number of pupae[a] per female per month	Pupal eclosion (%)
November (1972)	525	0.30	1.57	85.8
December	638	0.22	1.75	90.6
January (1973)	664	0.40	1.91	95.4
February	699	0.18	1.80	92.1
March	750	0.34	2.03	96.5
April	908	0.59	2.61	94.8
May	1000	0.33	2.40	96.4
June	999	0.35	2.73	97.2
July	1036	0.50	3.08	94.7
August	1035	0.36	3.11	95.7
September	1031	0.37	3.09	97.3
October	934	0.69	3.14	95.2

a Mean pupal weight = 18.5 mg (range 14.8 - 22.4 mg).

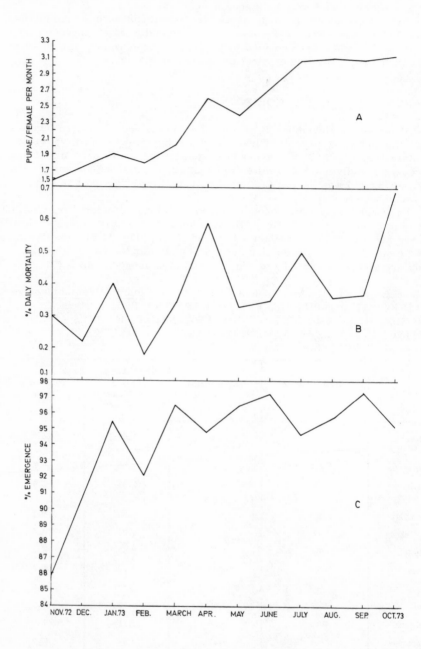

FIG.1. Monthly record of (A) pupal eclosion, (B) adult mortality, and (C) productivity of a colony of
<u>Glossina tachinoides</u> Westw. fed on rabbits six times a week. Flies were immobilized by nitrogen gas for
sexing at emergence and after mating.

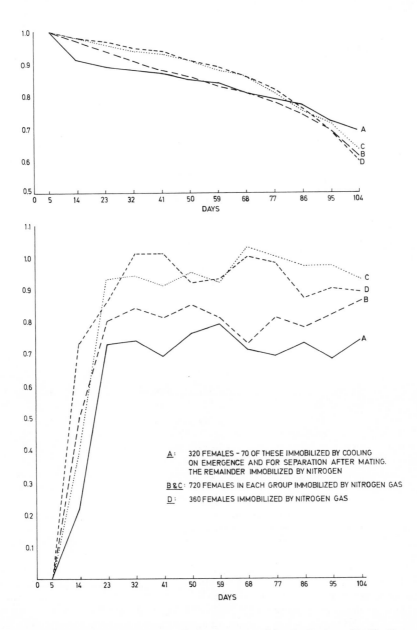

FIG. 2. Performance (longevity and fecundity) of successive groups of female <u>Glossina</u> <u>tachinoides</u> Westw. fed six times a week on rabbits. Flies were immobilized by cooling or by nitrogen gas for sexing at emergence and after mating.

equal sex ratio; mean daily mortality was 0.36% (range 0.18 - 0.69), and
pupae per female per month increased from 1.57 in November 1972 to 3.0 in
July 1973, and remained at this level during the last four months of the
investigation. Pupal weight averaged 18.5 mg (range 14.8 - 22.4).

Upon the assumption that the interlarval period was 8 - 9 days, the
expected number of pupae per female during the 108 days that they were kept
would be 10 or 11. Actual production from our colony from mid-March to
mid-October 1973 was 8.5 pupae per female.

We followed the trend in performance of the colony by computing at
four-monthly intervals the fecundity and longevity of females that had completed 108 days of life. This is shown graphically in Fig. 2 and includes
also data for the first group of 320 females, some of which were subjected to
cooling for separation at emergence and after mating. All other females
were treated with nitrogen gas. It is obvious that nitrogen treatment did not
adversely affect fly survival, pupa production or adult emergence (Fig. 1 and
Table I). In all groups, over 50% of the females were alive and still productive at the time they were discarded.

Availability of surplus males

At the steady level of 1000 females in the colony (May-October) production
averaged 95 - 100 pupae daily. At the eclosion rate of 95%, and 1:1 sex ratio,
over 250 males were obtained each week. About 20% of these were retained
for mating with colony females; thus approximately 200 surplus males were
available each week for experiments.

SPACE UTILIZATION

Flies were held in a room measuring 3.6 m × 3.2 m × 2.4 m high. Since
our primary objective was to develop methods of mass rearing, we
investigated also ways of making the best use of available space. Trays
holding fly cages were supported on shelves constructed in the form of a
trolley, each 120 cm × 40 cm × 180 cm high as shown in Fig. 3, and consisting
of eight shelves. A maximum of ten shelves was possible, and each shelf
accommodated four plastic trays with capacity for a maximum of six cages
per tray. The trolley system was convenient for wheeling cages out of the
holding room into the feeding area.

Stocking at the rate of ten females per cage, the capacity of each trolley
was 2400 flies. Our fly-holding room could conveniently accommodate five
trolleys together with fixed wall shelving to hold male cages, pupa containers
and emergence cages.

Thus we estimated that the five trolleys could accommodate a minimum
of 12 000 females if cages were stocked with ten flies each. Stocking at
15 flies per cage, the maximum number would be 18 000, including about
15 000 producing females which would yield 3000 - 4000 surplus males weekly
for use in a sterile-male release programme. Assuming that 2000 males
were retained for mating with colony females, the total size of the colony
to be held in a room of the size indicated would be 20 000 flies. An additional
room 5 m × 4 m would be required for feeding purposes.

We estimated that 100 - 120 G. tachinoides could be applied to each
rabbit daily for feeding. Therefore, a colony of 20 000 flies would require

FIG.3. Trolley rack system used at Seibersdorf Laboratory to hold tsetse fly cages. The arrangement facilitates moving cages from the holding room to the feeding and handling area.

about 165 - 200 rabbits. Allowing about 0.21 m³ of space per rabbit (i. e. 65 cm × 65 cm × 50 cm), the maximum space needed for 200 rabbits would be 42 m³. Therefore an additional room 6 m × 4 m would be required for housing rabbits.

ACKNOWLEDGEMENTS

Grateful acknowledgements go to J. Itard (Maisons-Alfort, France) for providing us with pupae and advising us on several aspects of the rearing, and to H. Kraus and F. Ivantschitz of our laboratory who were responsible for the feeding and maintenance of the colony. B. A. Butt and D. J. Nadel critically read the manuscript.

REFERENCES

[1] BUXTON, P.A., LEWIS, D.J., Climate and tsetse flies: Laboratory studies upon Glossina submorsitans and tachinoides, Philos.Trans.R.Soc.London, Ser.B. 224 (1934) 175.
[2] ITARD, J., MAILLOT, L., Notes sur un élevage de Glossines (Diptera-Muscidae) entrepus, à partir de pupes expédiées d'Afrique, à Maisons-Alfort, France, Rev.Elev.Méd.Vét.Pays Trop. 19 1 (1966) 29.

[3] ITARD, J., MAILLOT, L., BRUNET, J., GIRET, M., Observations sur un élevage de Glossina tachinoides West. après adoption du lapin comme animal-hôte, Rev.Elev.Méd.Vét.Pays Trop. 21 3 (1968) 387.

[4] NASH, T.A.M., JORDAN, A.M., BOYLE, J.A., A promising method for rearing Glossina austeni (Newst.) on a small scale, based on use of rabbits' ears for feeding, Trans.R.Soc.Trop.Med.Hyg. 60 2 (1966) 183.

[5] MEWS, A.R., OFFORI, E.D., BAUMGARTNER, H., LUGER, D., "Techniques used at the IAEA Laboratory for rearing the tsetse fly Glossina morsitans Westw.", Proc. 13th Meeting ISCTR, Lagos (1971) 243.

EFFECT OF BACTERIA ON TSETSE FLIES
FED THROUGH MEMBRANES*

B. BAUER, H.W. WETZEL
Joint FAO/IAEA Division of Atomic Energy
in Food and Agriculture,
International Atomic Energy Agency,
Vienna

Abstract

EFFECT OF BACTERIA ON TSETSE FLIES FED THROUGH MEMBRANES.

The bacterial flora in colonies of G. morsitans fed through membranes and exhibiting the "black abdomen" symptom was investigated. Aeromonas spp., Hafnia spp., Flavobacterium spp., Micrococcus spp., Pseudomonas aeruginosa and some bacilli were isolated from the guts of these flies, whereas the flies from the rabbit stock colony were not infected. Gut contents of dead flies and isolates of the most common bacteria, P. aeruginosa and Aeromonas spp., were offered to non-infected flies. Most of these infected flies died with the "black abdomen" symptom within one week following infection. Sources of infection were investigated, and several precautionary measures against bacterial contamination were tested.

INTRODUCTION

Among colonies of Glossina morsitans Westwood fed through an agar-parafilm membrane system, 95% of the moribund and dead flies exhibit "black abdomen" symptoms. In flies of this species maintained on rabbits only, this occurrence is less than 2%. Microscopic examinations reveal abundant bacteria in the gut of moribund flies from membrane-fed colonies, whereas such infections are very rare among rabbit-fed flies.

The isolation and identification of these bacteria are described. Experiments to determine the pathogenicity of the two bacteria most frequently associated with "black abdomen", sources of bacterial contamination in the agar-membrane system and preventive measures are also described.

METHODS

The flies were fed six times a week on defibrinated horse blood through membranes [1]. Moribund and dead flies were collected daily. These were dissected under aseptic conditions, the gut contents plated on agar/blood medium, and incubated at 37°C for 18 to 24 hours. The different bacterial isolates and their Gram stains were sent for identification to the Microbiological Institute[1] of the Veterinary School, Vienna.

* This research was supported in part by the Federal Republic of Germany, Gesellschaft für Strahlen- und Umweltforschung, and the Agricultural Research Service, US Department of Agriculture. Co-operative Agreement No. 12-14-100-11,170 (50).

[1] The help of H. Tiemann is gratefully acknowledged.

Experiment 1

The gut contents of dead flies were mixed with aseptically collected, bacteria-free and defibrinated horse blood. The infected blood was fed to teneral females on two successive days. Thereafter, the flies were fed aseptic defibrinated horse blood through membranes. The flies, maintained in groups of 10 to 15 females, were observed during their entire life. Control groups were treated likewise, except for the feeding of infected blood, and were observed for 25 days. The abdomens of the dead flies were opened and the contents plated as above for re-isolation and identification of the bacteria.

Experiment 2

Pseudomonas aeruginosa and Aeromonas spp. , the bacteria most frequen in dead flies (see below), were suspended in saline, and 0.1 ml of each suspension was mixed with 5 ml of defibrinated blood. Each of the infective mixtures was fed once through membranes to a group of flies. A control group was treated and kept similarly but without infection. All dead flies were examined microscopically for bacterioses.

Experiment 3

Teneral and 1- to 6-day-old flies from a rabbit colony were offered once 0.1 ml of the above suspensions diluted 10^{-8} with saline and then mixed with 5 ml of defibrinated horse blood.

To confirm the presence of viable bacteria in the inocula of Experiments 2 and 3, a loopful of each diluted suspension was streaked on blood/agar medium and incubated at 37°C for 16 hours.

Experiment 4

This experiment aimed at clarifying sources of bacterial infection and prophylactic measures required.

Swabs were made from untreated glass plates, membrane surfaces and the hands of laboratory personnel handling the membranes.

The following preventive measures were tested:
  (a) The agar for the membranes was sterilized by autoclaving.
  (b) Membranes were prepared and handled aseptically.
  (c) The grooved glass plates (which contain the blood and support the membranes) were washed, dried overnight, and cleaned with 75% ethanol shortly before use.
  (d) The glass plates were flamed with a gas burner.

RESULTS AND DISCUSSION

Experiment 1

All the flies fed infected blood died within 9 days, each with the "black abdomen" symptom. There was no mortality among the control flies during the observation period of 25 days.

The following bacteria were isolated from the guts of the dead flies: Aeromonas spp. , Flavobacterium spp. , Hafnia spp. , Micrococcus spp. , Pseudomonas aeruginosa. Some bacilli were isolated as well. Among these bacteria, Pseudomonas spp. and various Bacillaceae also have been found by other workers in tsetse flies fed through membranes [2-4]. In our experiments, P. aeruginosa and Aeromonas spp. were the bacteria most common in membrane-fed colonies.

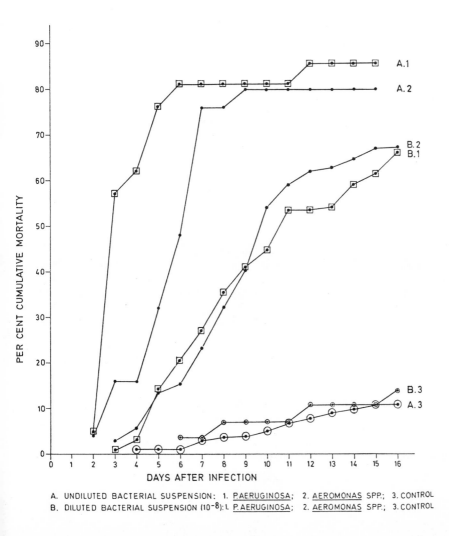

A. UNDILUTED BACTERIAL SUSPENSION: 1. P.AERUGINOSA; 2. AEROMONAS SPP.; 3. CONTROL
B. DILUTED BACTERIAL SUSPENSION (10⁻⁸): 1. P.AERUGINOSA; 2. AEROMONAS SPP.; 3. CONTROL

FIG.1. Cumulative mortality of female Glossina morsitans fed bacteria-infected blood on day 0 (flies 1-5-days-old).

Experiment 2

Trends of cumulative mortality for flies fed undiluted suspensions of
P. aeruginosa and Aeromonas spp. are illustrated in Fig. 1.   About 75% to 80%
of the flies were dead within 6 to 7 days following infection.   Post mortem,
the same bacteria (P. aeruginosa and Aeromonas spp.) were isolated from
these flies, thus confirming that these organisms were pathogenic for
G. morsitans.

Experiment 3

The bacterial suspensions diluted, mixed with blood and plated as described
yielded 8 colonies (range 6-12) on the average for both P. aeruginosa and
Aeromonas spp. This constitutes further proof that viable bacteria were
present in the inocula of Experiments 2 and 3.

As shown in Fig. 1, mortality was low during the first 4 days, most
likely because in this case, the inoculum being small, it took longer for the
bacteria to overcome the resistance of the flies.   Likewise, the maximum
mortality was also lower (65%).   As Fig. 1 also shows, with the undiluted
inocula (Experiment 2), 50% of the flies had died within 3 days as opposed to
9 to 10 days with the 10^{-8} dilutions (Experiment 3).   However, no conclusion
can be drawn regarding the relative pathogenicity of P. aeruginosa and
Aeromonas spp. since there could be no control over the amount of inoculum
taken in by the flies.

Experiment 4

Swabs taken from the untreated glass plates and agar membranes yielded
the following bacteria: Aeromonas spp., Flavobacterium spp., Micrococcus
spp., P. aeruginosa, and likewise from the hands of the personnel.

Cleaning the glass plates with 75% alcohol proved inefficient.   Swabs
from these plates yielded the same bacteria as those isolated before, although
in small numbers.   However, except for spores, the most effective method
of decontaminating the plates was flaming.

The agar membranes, sterile when freshly prepared, can subsequently
become contaminated particularly during manipulation. This can be prevented
by repeated cleaning of the hands with an antiseptic.   However, replacing
agar membranes by a recently developed synthetic membrane offers less
chances of bacterial contamination [5].   The use of such membranes and
clean surroundings should decrease contamination by microorganisms.

Airborne contamination of blood is much more difficult to prevent.
This is currently being tested by feeding the flies on a laminar air flow bench.

All these measures should be taken and maintained from the first days
of emergence, and not when an increase in mortality is noted.   Another
possibility is the use of antibacterial additives to the blood.

REFERENCES

[1] LANGLEY, P.A., Further experiments on rearing tsetse flies in the absence of the living hosts, Trans.R.
       Soc'.Trop.Med.Hyg. 66 (1972) 310.
[2] AZEVEDO, J.F. de, dos SANTOS, A.M.T., da COSTA PINHÃO, R., "New data on the feeding membrane
       and blood meals type of Glossina morsitans", Proc.Int.Symp.Tsetse Fly Breeding under Laboratory Conditions
       and its Practical Application, Lisbon (1970) 199.

[3] NOBRE, G., dos SANTOS, A.M.T., "The use of antibiotics in artificial breeding of Glossina morsitans —
    a bacteriological study", Proc. Int. Symp. Tsetse Fly Breeding under Laboratory Conditions and its Practical
    Application, Lisbon (1970) 335.
[4] OLIVEIRA, J.C., NOBRE, G., "Bacteriological and mycological study of a laboratory-bred Glossina colony",
    Proc. Int. Symp. Tsetse Fly Breeding under Laboratory Conditions and its Practical Application, Lisbon
    (1970) 341.
[5] BAUER, B., WETZEL, H.W., Unpublished data.

DISCUSSION

A.A. AMODU: What is the mechanism of pathogenesis of the bacteria infecting the flies?

B. BAUER: Once infection has occurred, reproduction of the bacteria begins in the gut of the flies, finally reaching the state of septicaemia.

A.A. AMODU: Have you studied the infection process and have you noticed any inability to feed before the death of the insect?

B. BAUER: The infection process depends on the amount of bacteria taken in by the flies. Infected flies did not refuse to feed and in a number of experiments infected flies fed one hour before death.

B. NA'ISA: What was the control group treated with in experiment 2?

B. BAUER: It was fed with aseptic defibrinated horse blood six times a week and handled the same way as the test group.

B. NA'ISA: Your flies were fed on horse blood and yet in nature the horse is not a preferred host of G. morsitans. Is it not possible that the infection occurring in the flies could have come from the horse blood, which may have been collected during the bacteremic stage, or from air-borne contamination or contamination carried by laboratory attendants.

B. BAUER: The horse blood proved to be bacteria-free in a number of tests. The blood is kept in 100 ml-bottles which are used only once.

P.R. FINELLE: Could you tell us more about the various bacteria involved?

B. BAUER: The bacteria found in our flies are ubiquitous and, in the case of P. aeruginosa, difficult to eliminate.

P.R. FINELLE: How long does the blood-feed last and how often is the membrane changed?

B. BAUER: It takes ten minutes to feed each group of flies. The same blood is used for different groups of flies for slightly over half an hour, after which the membrane is replaced.

P.R. FINELLE: Have you considered sterilizing your feeding equipment and the blood with ultra-violet rays.

B. BAUER: Yes, but it is not practicable with membrane feeding.

INFLUENCE OF ANTIBIOTICS
ON Glossina morsitans Westwood*

H.W. WETZEL, B. BAUER
Joint FAO/IAEA Division of Atomic Energy
in Food and Agriculture,
International Atomic Energy Agency,
Vienna

Abstract

INFLUENCE OF ANTIBIOTICS ON Glossina morsitans Westwood.
    Certain bacterial infections of tsetse flies cause high mortality. Oxytetracycline was tested as an antibiotic mixed with the blood fed to tsetse flies through membranes. Feeding the antibiotic to young females in this manner drastically reduced fecundity. Therefore oxytetracycline cannot be recommended for controlling bacterial infections in tsetse fly colonies.

INTRODUCTION

In a paper[1] already presented in this Symposium, the problem of rapid and high mortality in Glossina morsitans Westwood, contaminated with certain bacteria, was reported. Since this problem occurs primarily in tsetse fly colonies fed via membranes, and since one of the primary objectives of our tsetse fly programme is to develop membrane feeding, the solution of this problem is important. Two solutions appear feasible. One is essentially aseptic tsetse fly rearing; the second is to cure the bacterial infections when they occur. However, even under essentially aseptic conditions, bacterial infections in membrane-fed tsetse fly colonies are possible. Therefore, the development of therapeutic methods is very important.

METHODS

Antibiotics were tested for effectiveness against bacteria isolated from dead or moribund tsetse flies exhibiting the "black abdomen" symptom. The compounds tested were neotetracycline and oxytetracycline; both were effective and the latter was selected for further testing.

A colony of G. morsitans fed five days weekly defibrinated bovine blood via membranes and one day weekly on rabbit ears had a very high mortality rate. To solve this problem, the entire colony of 650 producing females was fed on two successive days defibrinated blood containing 25 ppm oxytetracycline. Immediately after this treatment, "black abdomen" mortality

---

* This research was supported in part by the Agricultural Research Service, US Department of Agriculture, Co-operative Agreement No. 12-14-100-11,170 (50), and by the Federal Republic of Germany, Gesellschaft für Strahlen- und Umweltforschung.

[1] BAUER, B., WETZEL, H.W.,"Effect of bacteria on tsetse flies fed through membranes", these Proceedings, IAEA-SM-186/54.

dropped to nearly zero. However, the fecundity of the colony also fell drastically. We therefore investigated whether oxytetracycline could be used to control "black abdomen" death in membrane-fed G. morsitans without reducing fecundity.

In these tests, 25 ppm oxytetracycline in defibrinated horse blood was used since this was the minimum concentration effective in vitro against tsetse fly bacteria. The oxytetracycline blood mixture was never more than three days old, and the commercial solution was used immediately upon opening a bottle. All antibiotic-treated flies were fed this mixture through a silicone membrane and it was ensured that each fly fed on the treated blood.

Only female flies were used. They were mated on the third day after emergence. In group-cage holdings, 8-12 flies were kept together. The observation period was 90 days (10 age groups). Mortality was recorded, and pupae collected and weighed daily except on Sundays. Aborted pupae were recorded separately. The productivity was calculated as normal pupae produced per female obtained by dividing the total number of pupae produced after treatment by the total number of flies per group at the beginning of the 90-day test.

In Experiment 1, 10-12 females were kept in each cage. They were fed six days weekly through a silicone membrane on antibiotic-free defibrinated horse blood. This blood was always fed within six days of its collection. As each fly deposited her first larva, the fly was removed from the group-cage and held separately. The treatment groups in this experiment were:

(1) Ten flies treated on day 1 after emergence, another group on day 2 after emergence, another on day 3 after emergence, etc., and the last group on day 23 after emergence.

(2) Individual groups of 10 flies each treated on two successive days, i.e. on the 1st and 2nd day after emergence, 2nd and 3rd day after emergence, etc., until the 22nd and 23rd day after emergence.

(3) Individual groups of 10 flies each treated on three successive days, i.e. the 1st, 2nd and 3rd day after emergence until the 14th, 15th and 16th day after emergence.

In Experiment 2, flies fed five times weekly on rabbit ears were used. The antibiotic was administered on two successive days immediately after the 1st or 2nd larval deposition. After treatment the flies were returned to the rabbit regime.

In Experiment 3, the flies were fed as in Experiment 1, but were from a larger colony. These flies were individually selected within 16 hours after their first larval deposition and 10 females were kept in each cage. The tetracycline was applied on the 7th, 8th or 9th day after first larval deposition.

Control groups of flies were kept for each experiment, and except for the oxytetracycline treatment, were handled exactly as the test group.

RESULTS

In Experiment 1, there was no difference between the 1, 2 or 3 treatment therefore the data were pooled. In addition, data were pooled for treatments on days 1 through 4, 5 through 7, 8 through 16 and 17 through 23. These pooled data were very similar.

TABLE I. EFFECT OF OXYTETRACYCLINE TREATMENT ON PUPAL PRODUCTION OF FEMALE G. morsitans; 25 ppm OXYTETRACYCLINE ADMINISTERED IN DEFIBRINATED HORSE BLOOD THROUGH A SILICONE MEMBRANE

Age when treated (days)	No. of flies	Mean number of pupae per female during 90-day test period excluding pupae produced before treatment		Reduction in fecundity (%)
EXPERIMENT 1 (Membrane-fed colony)				
1 - 4	67	0.37	(0.1 - 0.7)[a]	90.3
5 - 7	73	1.45	(0.4 - 1.8)	62.2
8 - 16	197	2.13	(0.9 - 3.6)	45.4
17 - 23	130	2.74	(2.2 - 3.6)	29.3
Control	65	3.83	(3.4 - 4.6)	-
EXPERIMENT 2 (Rabbit-fed colony)				
22 + 23	45	2.8	(2.6 - 2.9)	52.6
Control	86	5.9	(5.7 - 6.2)	-
31 + 32	45	3.0	(2.9 - 3.1)	32.9
Control	86	4.4	(4.2 - 4.8)	-
EXPERIMENT 3 (Membrane-fed colony)				
26	35	1.9	(1.7 - 2.0)	42.5
27	40	1.7	(1.3 - 2.1)	48.5
28	42	1.8	(1.6 - 2.4)	45.5
Control	60	3.3	(3.1 - 3.5)	-

[a] Figures in brackets = range.

All oxytetracycline treatments (Experiments 1-3) resulted in reduced pupal production (Table I).

Mortality in these experiments ranged from 50% to 70% during the 90-day test period.

DISCUSSION

Nobre and Santos [1] recommended the use of 25 ppm tetracycline in blood for preventing bacterial infections in tsetse flies fed through membranes. This treatment reduced mortality considerably since most of the bacteria infecting the blood were sensitive to tetracycline.

Azevedo and co-workers [2] recorded that flies fed a tetracycline/blood mixture for 25 days starting from emergence died without producing pupae. From this they concluded that no ovulation occurred.

In our experiments, the greatest reduction in fecundity was observed during the seven days following emergence. This finding should be taken into account together with the sequence of the egg development cycle in tsetse flie [3-8]. In the teneral female, three follicles are present of which two are very small and one is close to one quarter the mature size. When ovulation occurs (i.e. about seven days after emergence), the 4th follicle appears. Our data suggest that the reduction in fecundity following administration of oxytetracycline is due to inhibition of follicle development by this antibiotic. However, the drug does not affect any follicle larger than about one quarter the mature size; such a follicle will continue its development and ovulate. Likewise, the treatment does not affect a larva in the uterus. Our records also show that the reduction in fecundity was not caused by an increase in the rate of abortions.

No visible effect on the symbionts in the intestine cells of the tsetse flies could be detected.[2]

Flies treated with oxytetracycline within seven days of emergence can thus be expected to produce one larva at most. Even on day 4 after emergence, the most developed follicle very seldom exceeds one quarter the mature size, hence the 90% reduction in fecundity when oxytetracycline is administered during that period (Table I, Experiment 1). As soon as a follicle is formed in each of the four ovarioles, the treatment will have an effect during the next cycle. The degeneration of the ovaries is irreversible.

As shown in Table I (Experiments 1-3), the younger the fly at the time of treatment, the more drastic the reduction in pupal production.

The relatively slight reduction in fecundity following treatment between days 17 and 23 (Experiment 1) compared with the data of Experiments 2 and 3 is caused by the high mortality of flies kept individually in tubes and fed through membranes.

The highest mortality resulting from bacterial infection affects flies 20-30 days old. Therefore, to decrease the incidence of such bacterial disease, oxytetracycline should be administered when the flies are 10-to-20 days old. As shown in Table I (Experiment 1), the reduction in fecundity when flies in this age group are thus treated is about 45%. However, when oxytetracycline is administered therapeutically and not as a prophylactic measure, i.e. when the antibiotic is given after the first diseased or dead flies are discovered (flies 20-30 days old), fecundity drops by some 52% (Table I, Experiment 2).

As shown in this investigation, the considerable reduction in pupal production following the administration of oxytetracycline restricts the use of this drug in tsetse fly colonies to situations where no other antibiotic is effective.

REFERENCES

[1]   NOBRE, G., dos SANTOS, A.M.T., "The use of antibiotics in artificial breeding of Glossina morsitans — a bacteriological study", Proc. Int. Symp. Tsetse Fly Breeding under Laboratory Conditions and its Practical Application, Lisbon (1970) 335.

[2]   AZEVEDO, J.F. de, dos SANTOS, A.M.T., da COSTA PINHÃO, R., "New data on the feeding membrane and blood meals type of Glossina morsitans", Proc. Int. Symp. Tsetse Fly Breeding under Laboratory Conditions and its Practical Application, Lisbon (1970) 199.

---

[2]   The study on the symbionts of tsetse flies has been carried out by G. Nogge, Bonn University, Bonn, Federal Republic of Germany.

[3] SAUNDERS, D.S., The ovulation cycle in Glossina morsitans West. (Diptera-Muscidae) and a possible method of age determination for female tsetse flies by the examination of their ovaries, Trans. R. Ent. Soc. 112 (1960) 221.

[4] SAUNDERS, D.S., Ovaries of Glossina morsitans, Nature (London) 185 (1960) 121.

[5] SAUNDERS, D.S., Studies on ovarian development in tsetse flies (Glossina, Diptera), Parasitology 51 (1961) 545.

[6] CHALLIER, A., Observations sur l'ovulation chez Glossina palpalis gambiensis Vanderplank, Bull. Soc. Path. Exot. 57 (1964) 985.

[7] CHALLIER, A., Amélioration de la méthode de détermination de l'âge physiologique des glossines, Bull. Soc. Path. Exot. 58 (1965) 250.

[8] ITARD, J., «Oogénèse chez Glossina tachinoides West. élevée au laboratoire», Control of Livestock Insect Pests by the Sterile-Male Technique (Proc. Panel Vienna, 1967), IAEA, Vienna (1968) 65.

DISCUSSION

F.M. EVENS: Do the insects treated with oxytetracycline survive as long as the controls?

H.W. WETZEL: Yes, or they may live even longer.

F.M. EVENS: Then, perhaps it would be possible to develop a "sterile-female" technique. As we know, males can mate successfully only five or six times. Therefore, if we introduced a surplus of oxytetracycline-treated females, we could exhaust the males and achieve a reduction in the population.

H.W. WETZEL: If the females are not needed to maintain a rearing colony, they could be released after tetracycline administration. But, if the rearing unit is to be kept operating at the same level, only the release of surplus males can be considered.

H. LEVINSON: The inhibitory effect of various antibiotics on (nucleo)protein biosynthesis has been known for some time, and it is for this reason that the addition of antibiotics to (synthetic) insect diets has been abandoned by a number of insect nutritionists. Instead of antibiotics it might be worthwhile trying an antibacterial compound such as p-aminobenzene sulfonamide which seems harmless to insects but not to bacteria.

H.W. WETZEL: The use of tetracycline to cure the alarmingly high mortality in our tsetse fly colony was dictated by the fact that all the bacteria isolated were very susceptible to this antibiotic. Some other antibiotics tested in the meantime have not had the same effect on the ovaries and these are now available for curing infections, if required. The search for blood-additives to cure or prevent bacterial contamination or fly infection is continuing.

R. PAL: What method is being used for the mass rearing of tsetse flies for the Tanga project?

H.W. WETZEL: The tsetse flies (G. morsitans) in Tanga are being mass-reared by feeding six times a week on goats, and a small trial colony is being fed on rabbits. Good results have been obtained in both cases.

STUDIES ON THE ERADICATION OF
Anopheles pharoensis Theobald BY THE STERILE-MALE TECHNIQUE USING [60]Co
XI. Release-recapture experiments for flight range and dispersion

A. F.-M. WAKID
Radiobiology Department,
Atomic Energy Department,
Cairo

A. A. ABDEL-MALEK
Entomology Department,
Faculty of Science, Cairo University, Giza

A. O. TANTAWY
Genetics Department,
Faculty of Agriculture,
Alexandria University,
Alexandria,
Egypt

Abstract

STUDIES ON THE ERADICATION OF Anopheles pharoensis Theobald BY THE STERILE-MALE TECHNIQUE USING [60]Co: XI. RELEASE-RECAPTURE EXPERIMENTS FOR FLIGHT RANGE AND DISPERSION.

Anopheles pharoensis Theobald is a mosquito vector of malaria in Egypt. Efforts to control it by using the sterile-insect technique are being made. An account is given of experiments in a field study on the dispersion, flight range and survival of the mosquito in the experimental area. Mosquitoes were labelled with ^{32}P during the larval stage and released from a release centre. Recapture of the released mosquitoes was achieved by scattering outlet window traps fixed in 10 surrounding collecting stations. A total of 65020 labelled mosquitoes were released and 103 recaptured. Of the recaptured mosquitoes 64 were females, giving a recapture rate of about 0.2%, and 0.12% of the males were recovered. The distribution of the recaptured mosquitoes from the release centre was highly irregular. The mean flight range of the adult mosquito was estimated to be 1087 m from the release centre with a maximum range of 2500 m. One female mosquito was recaptured 23 days from the beginning of the release. The oldest male mosquito caught was recaptured 22 days after the end of the release period.

INTRODUCTION

Since time immemorial malaria has been endemic in tropical countries, sapping the energy of the inhabitants, reducing their productivity and seriously affecting food production. Its importance in Egypt has been growing steadily in recent years, and considerable attention has been focused on its control.

The high hopes for mosquito eradication by means of residual insecticides were soon belied by the discovery of resistance in vector

mosquitoes [1]. This discovery has once again emphasized the value of
the use of other biological methods for control or eradication.

Studies on the possibility of controlling Anopheles pharoensis using
the sterile-insect technique have been carried out successfully in Egypt [2-7]
The population dynamics of mosquitoes present in a semi-isolated area of
Fayoum Province, Egypt, were studied for the years 1965-67 inclusive [8, 9].
Uptake, distribution, persistence, decay and the biological effects of ^{32}P
have been studied in our laboratory [10, 11]. All these studies have given
promising results that show the possibility of using the sterile-insect release
method against Anopheles pharoensis.

The present work is a field study on the dispersion, flight range and
survival of mosquitoes labelled with ^{32}P and released in a semi-isolated
area in Fayoum Province, Egypt.

METHODS AND TECHNIQUE

The experimental area is known as Manshaat Sinnoris, an area of
Fayoum Province 90 km southwest of Cairo. The area covers about 18 km^2
and is bounded by desert and Lake Karoun to the north and by agricultural
land in the other directions. The whole of Fayoum Province, however, can
be considered an isolated area.

Ten adult mosquito traps were fixed to ten scattered mosquito collecting
stations. The traps are of the outlet window type having a wooden frame
30 × 30 × 30 cm with a wooden bottom and wire-screened top. Three of its
sides are of glass and the fourth side has a muslin conical sleeve extending
to the inside of the trap with its narrow opening kept in position by four
stretched strings fixed in the frame of the opposite side of the trap. Each
trap was firmly positioned in a bedroom or stable window from the
outside (Fig. 1), leaving the other windows open all night to attract mosquitoes
(mostly females) into the room. Before the sunlight appeared the windows
were closed. After sunrise, most of the mosquitoes present in the room
were attracted to light coming through the trap. They entered the trap
opening in an attempt to escape and were trapped inside.

The traps were distributed in a T-shaped arrangement throughout the
area (Fig. 2). The T-stem extends to the south about 3 km from the junction
of the Cairo-Fayoum main road and the road leading to Lake Karoun. Each
of the two arms of the T extends about 3 km east and west of the junction.
All the stations as well as the release centre were located on the main roads.
Traps number 1 and 2 were located on the Cairo-Fayoum main road (the
eastern arm of the T) 2500 and 1500 m from the release centre respectively,
and traps 4 and 5 were located on the road leading to Lake Karoun (the
western arm of the T) 1500 and 2250 m from the release centre respectively.
Traps number 3, 6, 7, 8, 9 and 10 were located on the southern part of Cairo-
Fayoum main road (the T-stem) with distances of 250, 1250, 1350, 1750,
2500 and 2750 m from the release centre.

Experimental mosquitoes were obtained by collecting larvae from the
rice water in the surrounding rice fields. The larvae were taken to the
breeding laboratory in Fayoum where they were reared to the pupal stage
in enamel dishes with radioactive Nile water containing 0.2 μCi/mlitre of
^{32}P [10]. The radioactive pupae were collected daily from the rearing
dishes, and transferred into pure water in a large enamel dish. The dish

FIG.1. An outlet window trap fixed in a stable window.

with the radioactive pupae was placed in an open release cage on the roof
of the Sinnoris Malaria Hospital which was used as the release centre,
located near the junction of the Cairo-Fayoum main road and the road leading
to Lake Karoun. The cage was also provided with a 10% sugar solution as
food for the emerging adults.

Releases were made once every month from August 1969 until October
1969 and from May 1970 until October 1970. In each release the crop of
pupae in seven successive days was used. This made an interval between
each two successive releases of not less than 23 days.

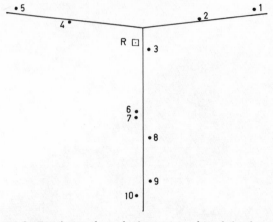

FIG.2. Distribution of traps for the recapture of <u>Anopheles</u> <u>pharoensis</u>
R = Release centre
1-10 = Trap numbers

Daily recapture of mosquitoes from the ten stations was done by spray-catching, cotton sheets being spread over all horizontal surfaces. and the room or stable then sprayed with 0. 2% pyrethrin in kerosene. Mosquitoes found inside each collecting station were added to those collected in the trap, and were then sorted in a field laboratory in Fayoum and the radioactive ones identified as to sex. Radioactivity was traced by a laboratory monitor.

RESULTS AND DISCUSSION

Recovery of released mosquitoes

Table I shows the recovery of the released labelled mosquitoes through-out the nine releases made in this experiment. The recovery rate ranged between a minimum of 0.03% to a maximum of 0.37% with an average of 0.16%. Of a total of 65020 labelled adult mosquitoes released, 103 were recaptured. Of the recaptured mosquitoes 64 were females, giving a recapture rate of about 0.2% (for a 1:1 sex ratio among the released mosquitoes), whereas 0.12% of the males were recaptured.

Distribution of recaptures

Study of the distribution of recaptures (Table II) showed that they did not follow a regular pattern. The distribution from the release centre was highly irregular, 51% being caught 1250 m from the release point (trap number 6), and half of this percentage was caught 250 m from the release point (trap number 3). Only three individuals (two females and one male) were caught from trap number 5 which was located 2250 m from the release centre. One female and one male were caught 2500 m from the release centre (trap number 9).

TABLE I.   RECOVERY RATE OF THE RELEASED LABELLED Anopheles
pharoensis THROUGHOUT THE RELEASES IN FAYOUM (1969 AND 1970)

Release number	Number of released mosquitoes	Number of recaptured mosquitoes			Percentage recaptured
		Labelled		Unlabelled	
		♂	♀		
1	7013	1	1	1583	0.03
2	6978	1	1	5733	0.03
3	5104	10	9	6602	0.37
4	5531	3	2	3047	0.09
5	7865	8	16	10357	0.31
6	9029	7	11	7904	0.20
7	9276	3	9	1943	0.13
8	7100	4	12	5890	0.23
9	7124	2	3	2363	0.07
Total	65020	39	64	45422	Av. =0.16

TABLE II.   DISTRIBUTION AND FLIGHT RANGE OF LABELLED
Anopheles pharoensis IN ALL RELEASES

Trap number	Distance from release centre (m)	Number of recovered labelled mosquitoes	
		♂	♀
3	250	10	18
5	2250	1	2
6	1250	18	35
7	1350	3	4
8	1750	6	4
9	2500	1	1

In general these results confirmed the opinion of public health workers
that dispersion in well-watered inhabited areas is over a fairly short range.
Gillies arrived at the same conclusion concerning Anopheles gambiae [12].
Conversely, in the drier and more open parts of the tropics it is possible
that a completely different pattern may emerge, as has been found by Russell
and co-workers (1944) on Anopheles culicifacies [13].

As early as 1949, Bugher and Taylor reported that released Aedes
aegypti labelled with ^{32}P and ^{89}Sr could be recaptured up to a maximum
distance of 1.2 km [14]. Gillies applied the same technique to Anopheles
gambiae, and found that the mean flight range of females was 1 km and of

males 0.8 km. Individuals of both sexes were caught at the maximum range
of 3.6 km [12]. Both sexes of Anopheles stephensi were capable of flying
a distance of 4.3 km and could fly 1.8 km overnight from the point of release
[15, 16]. With Anopheles sergenti in Siwa Oasis, it was reported that the
males dispersed up to 3.4 km and the females up to 2.5 km from the point
of release [17].

Age of recaptures

As stated above, each release lasted for a period of one week. For
this reason it was impossible to make an accurate estimate of the survival
rate of the released mosquitoes. However, about 29% of the recoveries was
made during the first week, 49% during the second week, 19% during the
third week, 12% during the fourth week, and less than 1% during the fifth
week represented by one male recaptured at 29 days from the beginning of
the release. This age distribution of recaptures was surprising as the
maximum recovery was made during the second and not the first week as
was expected. One female mosquito was collected 23 days from the beginning
of the release, i.e. 16 days after the last day of the release.

Gillies arrived at similar results with regard to Anopheles gambiae,
and reported that no females were recovered more than 23 days after release
and that the two oldest mosquitoes caught were both males, recaptured at
26 and 30 days after release. However, he mentioned that over 70% of the
recaptured mosquitoes were caught during the first week [12].

REFERENCES

[1] RIBEIRO, H., MEXIA, J.T., Mosqu. News 30 4 (1970).
[2] ABDEL-MALEK, A.A., TANTAWY, A.O., WAKID, A.M., J. Econ. Entomol. 59 2 (1966) 272.
[3] TANTAWY, A.O., ABDEL-MALEK, A.A., WAKID, A.M., J. Econ. Entomol. 59 6 (1966) 1392.
[4] ABDEL-MALEK, A.A., TANTAWY, A.O., WAKID, A.M., J. Econ. Entomol. 60 1 (1967) 20.
[5] TANTAWY, A.O., ABDEL-MALEK, A.A., WAKID, A.M., J. Econ. Entomol. 60 1 (1967) 23.
[6] TANTAWY, A.O., ABDEL-MALEK, A.A., WAKID, A.M., J. Econ. Entomol. 60 3 (1967) 696.
[7] ABDEL-MALEK, A.A., TANTAWY, A.O., WAKID, A.M., J. Econ. Entomol. 60 5 (1967) 1300.
[8] ABDEL-MALEK, A.A., TANTAWY, A.O., WAKID, A.M., J. Econ. Entomol. 62 2 (1969) 348.
[9] TANTAWY, A.O., ABDEL-MALEK, A.A., WAKID, A.M., Alex.J. Agric. Res. 18 (1970) 155.
[10] WAKID, A.M., ABDEL-MALEK, A.A., TANTAWY, A.O., Isot. Radiat. Res. 2 (1969) 49.
[11] WAKID, A.M., TANTAWY, A.O., ABDEL-MALEK, A.A., Isot. Radiat. Res. 3 (1970) 35.
[12] GILLIES, M.T., Bull. Entomol. Res. 52 (1961) 99.
[13] RUSSELL, P.F., KNIPE, F.W., RAMACHANDRA RAO, T., PUNTAM, P., Exp. Zool. 97 (1944) 135.
[14] BUGHER, J.C., TAYLOR, M., Science 110 (1949) 146.
[15] QURAISHI, M.S., in The Role of Science in the Development of Natural Resources with Particular
     Reference to Pakistan, Iran and Turkey; a Symposium held under the auspices of GENTO Scientific
     Council, Lahore, Pergamon, New York (1964) 425.
[16] QURAISHI, M.S., J. Econ. Entomol. 58 (1965) 821.
[17] ABDEL-MALEK, A.A., ABDEL-AAL, M.A., Bull. World Health Organ. 35 (1966) 968.

DISCUSSION

B.S. FLETCHER: Why did you arrange your traps in a T-formation
rather than a cross.

A.F.-M.WAKID: The traps were placed in villages located along the road
as shown in Fig.2 of the paper. Villages not on the road would have been
more difficult to reach.

R. GALUN: Could the distribution of mosquitoes not have been affected by the traffic along the road?

A.F.-M.WAKID: Yes, it might have been to some extent.

R. PAL: The analysis of data on insect dispersal is rather difficult unless the experiment has been prepared in such a way as to facilitate such analysis and I feel that the advice of an ecologist would be very valuable for such purposes. A thorough knowledge of the ecology of the vector is absolutely essential for the genetic control of insects and I think it might be appropriate to invite ecological experts to these symposia in the future.

CONTROL OF Lepidoptera BY THE
STERILE-MALE TECHNIQUE
(Session 8)

ASPECTOS DE LA BIOLOGIA
Y COMPORTAMIENTO DE ADULTOS
DE Heliothis virescens Fabricius
AL IRRADIAR PUPAS DIAPAUSICAS*

D. ENKERLIN S. , R. A. MANCILLA BERGANZA,
C. E. PEÑA ANDELIZ
Instituto Tecnológico y de Estudios Superiores
de Monterrey, Mexico

Abstract—Resumen

THE BIOLOGICAL ASPECTS AND BEHAVIOUR OF Heliothis virescens Fabricius ADULTS FROM IRRADIATED DIAPAUSAL PUPAE.

The purpose of this work was to study the behaviour of Heliothis virescens Fabricius by treating diapausal pupae 2-5 days after formation with 2.5 krad gamma rays, storing them at 18°C for a photophase of 10 h of light and 14 h of darkness and interrupting the diapause after different periods of storage: immediately and after 6 to 8, 14 to 15, 15, 30 and 60 days. In inducing diapause, bioclimatic chambers with adjustable photoperiod and temperature were used. The diapausal pupae (2-5 days after formation) were sexed, divided into two equal parts and sent to the laboratory of the US Department of Agriculture, Animal and Plant Health Special Service, Plant Protection and Quarantine Program, in Monterrey, Nueva Leon State, Mexico, where one part was irradiated with a ^{60}Co bomb. The other part was not irradiated but used for pairing with the irradiated diapausal pupae. The pupae were placed in an environment of 26°C with 14 h of light and 10 h of darkness to break the diapause. After emergence the adults were paired as follows: Irradiated male with normal female, irradiated female with normal male, irradiated male with irradiated female and, for control purposes, normal male with normal female. The pairs were placed in cylindrical plastic cages 18.5 cm in height and 7.5 cm in diameter, two pairs (two males with two females) per cage. The cages were provided with two oviposition areas — a cloth cover in the upper part and a paper filter in the lower part. A 25% water-honey solution was used as adult feed. The pairings yielded data on the fertility of eggs, transference and/or reception of spermatophores and adult longevity for both generations P_1 and F_1. The data are discussed in the paper. The most important result was that the dose of 2.5 krad led to sub-sterility, younger pupae being the most susceptible. The time of storage did not have a marked influence on the parameters studied. The sterility was transmitted with varying intensity to generation F_1.

ASPECTOS DE LA BIOLOGIA Y COMPORTAMIENTO DE ADULTOS DE Heliothis virescens Fabricius AL IRRADIAR PUPAS DIAPAUSICAS.

El objetivo de esta serie de trabajos fue estudiar el comportamiento del gusano de la yema del tabaco Heliothis virescens Fabricius, irradiando pupas diapáusicas a los 2 a 5 días de haberse formado con 2,5 krad de rayos gamma y almacenándolas a 18°C y un fotoperíodo de 10 h de luz y 14 h de obscuridad, rompiendo la diapausa a diferentes edades de almacenamiento, inmediatamente, a los 6 a 8, 14 a 15, 15, 30 y 60 días. En la inducción de la diapausa se utilizaron cámaras bioclimáticas manipulando el fotoperíodo y la temperatura. Las pupas diapáusicas que se formaron de 2 a 5 días se sexaban y se dividían en número de dos partes iguales, llevándose posteriormente al laboratorio del U.S. Department of Agriculture, Animal and Plant Health Special Service, Plant Protection and Quarantine Program, localizado en Monterrey, N. L. , México, en donde una parte se irradiaba con una bomba de cobalto-60. La otra parte no se irradiaba porque se trataba de los individuos diapáusicos que se usaban para cruzarlos con los diapáusicos irradiados. Para romper la diapausa, las pupas se colocaron bajo condiciones de 26°C y 14 h de luz y 10 h de obscuridad. Emergidos los adultos se efectuaban las cruzas de macho irradiado con hembra normal, hembra irradiada con macho normal, macho irradiado con hembra irradiada y macho normal con hembra normal como testigo. Las parejas se colocaron en jaulas de plástico cilíndricas de 18,5 cm de altura por 7,5 cm de diámetro, haciendo cruzas dobles (dos machos con dos hembras). Estas jaulas estaban provistas de dos áreas de

---

* Este trabajo ha sido financiado en parte con un contrato de investigación del OIEA.

oviposición: manta de cielo en la parte superior y papel filtro en la parte inferior. Para la alimentación de los adultos se utilizó una solución de aguamiel al 25%. De las cruzas efectuadas se tomaron datos de fertilidad de los huevecillos, transferencia y/o recepción de espermatóforos y longevidad de los adultos, tanto para la generación P_1 como para la F_1. Los datos resultantes se discuten en el trabajo. Como resultado más importante se obtuvo una subesterilidad con 2, 5 krad, siendo más susceptibles las pupas de menor edad. El tiempo de almacenamiento no tuvo una influencia muy marcada sobre los parámetros estudiados. La esterilidad en diversa intensidad fue transmitida a la generación F_1.

Como es conocido, dosis por debajo de 22, 5 krad causan subesterilidad cuando se tratan pupas no diapáusicas de H.virescens, mientras que Flores [1], al tratar pupas en diapausa, encontró que dosis de 5 krad o más afectan en un alto grado a esta especie; el mismo autor inició trabajos para determinar el efecto de 2, 5 krad, para causar subesterilidad en pupas diapáusicas. Sánchez [2], llevó a cabo un estudio almacenando las pupas diapáusicas en frío (5°C) y completa obscuridad, irradiando las pupas antes de romper la diapausa. Ninguno de los dos autores pudieron llegar a conclusiones definitivas.

Como continuación de los mencionados trabajos de Flores y Sánchez, en otra serie experimental se estudiaron únicamente los efectos de la dosis de 2, 5 krad de rayos gamma sobre adultos al irradiar pupas previamente inducidas a diapausa del mismo H.virescens. En este caso, en tres fechas distintas se indujeron a diapausa un total de 2 539 larvas, manipulando el fotoperíodo y la temperatura de cámaras bioclimáticas según el método de Roach y Adkisson [3]. La alimentación de las larvas consistió en una dieta artificial formulada originalmente por Vanderzant y modificada por Raulston como lo cita Flores [1].

Del total de larvas inducidas 50% llegaron al estadio de pupa y de éstas 86, 6% estaban en diapausa (tabla I).

Una vez obtenidas las pupas en diapausa se sexaban para hacer los tratamientos por separado. Las pupas de cada sexo se dividían para los diferentes tratamientos y se llevaban a una bomba de cobalto-60, del laboratorio del U.S. Department of Agriculture, Animal and Plant Health Special Service, Plant Protection and Quarantine Program, localizado en la ciudad de Monterrey, N.L., México. Las pupas testigo recibieron la misma manipulación, con excepción de la irradiación.

Las pupas de las tres fechas de inducción a diapausa, ya sexadas, se irradiaron a las edades de 2 a 3, 6 a 8 y 14 a 15 días de edad. Una vez emergidos los adultos se hicieron las cruzas siguientes: hembra irradiada (HI) con macho no irradiado (Mn); macho irradiado (MI) con hembra no irradiada (Hn); ambos sexos irradiados (HI × MI) y las cruzas de adultos no irradiados (HN × MN) de las mismas pupas diapáusicas.

Cronológicamente se iniciaron los estudios con adultos de pupas diapáusicas irradiadas a la edad de 6 a 8 días, haciendo cruzas simples (una hembra con un macho). Las cruzas de adultos obtenidos de pupas diapáusicas irradiadas a la edad de 2 a 3 y de 14 a 15 días se hicieron dobles (dos hembras con dos machos) y cronológicamente casi coincidieron.

Las parejas se colocaron en jaulas de plástico de 18, 5 cm de alto por 7, 5 cm de diámetro, con dos áreas de oviposición, consistentes una, de manta de cielo (parte superior) y la otra, de papel filtro (parte inferior). La alimentación de los adultos consistió de aguamiel en la proporción de tres partes de agua por una de miel.

TABLA I. NUMERO INICIAL DE LARVAS DE H.virescens DE PRIMER
ESTADIO INDUCIDAS A DIAPAUSA EN TRES FECHAS, PORCENTAJE DE
PUPAS OBTENIDAS Y PORCENTAJE DE PUPAS DIAPAUSICAS
(Monterrey, N.L., 1972)

Fecha de inducción	N° de larvas	% de pupas obtenidas	% de pupas diapáusicas
12-4-72	600	65,50	83,28
2-7-72	789	43,00	88,99
8-7-72	1,150	41,31	87,68
Promedio		49,93	86,65

TABLA II. NUMERO PROMEDIO DE ESPERMATOFOROS POR HEMBRA
DE H.virescens, OBTENIDOS DE PUPAS DIAPAUSICAS IRRADIADAS CON
2,5 krad A DIFERENTES EDADES (Monterrey, N.L., 1972)

Cruzas	2 - 3 días		6 - 8 días		14 - 15 días	
	N° de parejas	Promedio de espermatóforos × hembra	N° de parejas	Promedio de espermatóforos × hembra	N° de parejas	Promedio de espermatóforos × hembra
Fi × Mn	16	1,5	34	2,8	14	1,9
Fn × Mi	28	1,7	38	2,1	26	1,3
Fi × Mi	10	1,1	30	1,6	16	1,1
Fn × Mn	24	2,1	20	2,9	20	1,6

Nota: Según la prueba de X^2 no hubo diferencia en transferencia de espermatóforos según la edad de irradiación.

TABLA III. PORCENTAJE DE ESTERILIDAD DE HUEVECILLOS DE
H.virescens (CORREGIDO POR ABBOT), OBTENIDOS DE ADULTOS
CUYAS PUPAS EN ESTADIO DE DIAPAUSA FUERON IRRADIADAS A
DIFERENTES EDADES (Monterrey, N.L., 1972)

Cruzas	2 - 3 días		6 - 8 días		14 - 15 días	
	N° de parejas	% de esterilidad	N° de parejas	% de esterilidad	N° de parejas	% de esterilidad
Fi × Mn	16	43,8	34	43,5	14	2,5
Fn × Mi	28	59,8	28	47,0	26	15,2
Fi × Mi	10	100,0	30	77,3	16	45,9

TABLA IV. NUMERO PROMEDIO DE ESPERMATOFOROS POR HEMBRA F_1 DE H.virescens, PROVENIENTES DE PUPAS DIAPAUSICAS (P_1) IRRADIADAS A DIFERENTES EDADES CON 2,5 krad (Monterrey, N.L., 1972)

Cruzas[a]		N° promedio de espermatóforos según fecha de irradiación de pupas		
P_1	F_1	2 - 3 días	6 - 8 días	14 - 15 días
Fi × Mn	$\underline{F}$ × Mn	2,0	2,4	1,6
	Fn × $\underline{M}$	2,1	1,5	1,5
Fn × Mi	$\underline{F}$ × Mn	1,1	2,1	1,1
	Fn × $\underline{M}$	1,2	1,6	2,2
Fi × Mi	$\underline{F}$ × Mn	-[b]	1,3	0,9
	$\underline{F}$ × $\underline{M}$	-[b]	1,5	1,5
Testigo		1,9	2,1	1,9

[a] 18 a 25 parejas por cruza.
[b] No se obtuvo generación F_1.

TABLA V. PORCENTAJE DE ESTERILIDAD DE HUEVECILLOS PUESTOS POR HEMBRAS F_1 DE H.virescens DESPUES DE IRRADIAR PUPAS DIAPAUSICAS P_1 A DIFERENTES EDADES CON 2,5 krad (Monterrey, N.L., 1972)

Cruzas[a]		% de esterilidad de huevecillos según fecha de irradiación de pupas[b]		
P_1	F_1	2 - 3 días	6 - 8 días	14 - 15 días
Fi × Mn	$\underline{F}$ × Mn	27,6	32,4	18,3
	Fn × $\underline{M}$	25,7	36,8	24,6
Fn × Mi	$\underline{F}$ × Mn	100,0	29,4	39,8
	Fn × $\underline{M}$	96,2	30,9	64,6
Fi × Mi	$\underline{F}$ × Mn	-[c]	98,1	49,9
	Fn × $\underline{M}$	-[c]	92,6	32,8

[a] 18 a 25 parejas por cruza.
[b] Corregido por la fórmula de Abbot.
[c] No se obtuvo generación F_1.

De dichas cruzas se tomaron datos de transferencia o recepción de espermatóforos y fertilidad de los huevecillos, para lo cual se tomó una muestra al azar de la tela (aproximadamente 4 cm^2), número de larvas F_1 criadas, mortalidad de larvas, cantidad de pupas obtenidas y relación de sexo.

En casi todas las cruzas se obtuvo progenie; ésta se crió bajo condiciones normales y los adultos obtenidos se cruzaron de nuevo con individuos del sexo opuesto no irradiado pero de pupas diapáusicas y se efectuaron las mismas observaciones que las de los progenitores.

Las hembras adultas provenientes de pupas diapáusicas irradiadas a la edad de 6 a 8 y 14 a 15 días mostraron estar menos afectadas por las radiaciones para recibir espermatóforos que a la edad de 2 a 3 días. En las tres edades a que fueron irradiadas las pupas, al cruzar adultos tratados de ambos sexos, se manifestó una incapacidad mayor relativa a la transferencia y/o recepción de espermatóforos de machos y hembras, respectivamente (tabla II). En la mayoría de las cruzas el promedio de espermatóforos transferidos por los machos tratados fue más bajo que el promedio transferido por los machos no tratados.

Se puso de manifiesto que la fertilidad en los padres tratados resultó más afectada que la de los individuos no tratados y, dentro de las cruzas, fue el macho el que mostró ser más susceptible (tabla III). Cuando se irradiaron pupas diapáusicas a la edad de 2 a 3 días, las cruzas de los adultos de este origen presentaron un 100% de esterilidad; los adultos de pupas diapáusicas que se irradiaron a las edades de 6 a 8 y de 14 a 15 días presentaron fertilidad, pero más reducida que las cruzas con individuos provenientes de pupas en que se irradió sólo uno de los sexos.

En lo que se refiere a la progenie el porcentaje de pupas obtenido fue variable, no presentando ninguna influencia la edad de la irradiación de las pupas (P_1) padres, por lo que se supone que esta variación se debió a efectos de manipulación en el laboratorio. En relación al sexo en las pupas F_1, tal como lo señalan Proshold y Bartell [4], hay una dominancia de los machos con respecto a las hembras a causa de las radiaciones.

En lo que respecta a la compatibilidad sexual para las cruzas F_1 se aprecia que hembras y machos, cuyas madres P_1 provenían de pupas diapáusicas que se irradiaron a la edad de 2 a 3 días, mostraron una recuperación en la función de aceptar y/o transferir espermatóforos en comparación con sus padres (tabla IV). Para las edades de 6 a 8 y 14 a 15 días a las cuales se irradiaron las pupas diapáusicas, la actividad de los adultos fué variable, con una recuperación (6 a 8 días) y reducción (14 a 15 días) en comportamiento sexual comparado con los padres.

En lo que se refiere a la esterilidad heredada se observó que las hembras (F_1), provenientes de padres en donde el macho se irradió en estadio de pupa diapáusica de 2 a 3 días de edad, transmiten a su descendencia una esterilidad de 100% (tabla V); se comprobó también que el macho proveniente de los mismos padres estaba muy afectado (96,15% de esterilidad). No hubo descendencia de la cruza cuando ambos padres fueron irradiados en el estado de pupa diapáusica. A la edad de 14 a 15 días a la que se irradiaron las pupas diapáusicas (P_1), se manifestó una variabilidad, pero siempre con reducción en la fertilidad comparada con las cruzas cuyos individuos no se irradiaron.

En lo que respecta al desarrollo de la progenie F_2, se observó un efecto variable de las larvas a pupas, o sea que la susceptibilidad de los individuos a las radiaciones no se manifestó en forma dependiente, lo cual hace suponer que ello se debió más bien a las condiciones de manipulación que a la irradiación. La relación de sexo en esta progenie siempre fue dominante para los machos, concordando los resultados con Proshold y Bartell [4].

El objetivo de un trabajo más reciente fue medir el mismo efecto de radiaciones gamma sobre pupas diapáusicas a una intensidad de 2,5 krad,

TABLA VI. NUMERO DE LARVAS DE PRIMER ESTADIO DE H.virescens INDUCIDAS A DIAPAUSA EN TRES FECHAS, PORCENTAJE DE MORTALIDAD Y PORCENTAJE DE PUPAS DIAPAUSICAS (Monterrey, N.L., 1973)

Fechas de inducción	N° de larvas inducidas	% de larvas muertas	% de pupas diapáusicas
1	2060	34,80	88,99
2	2048	40,57	70,25
3	2036	48,68	90,04
Promedio		42,35	83,09

TABLA VII. PORCENTAJE DE EMERGENCIA NORMAL DE ADULTOS DE H.virescens AL ALMACENAR PUPAS DIAPAUSICAS SIN IRRADIAR E IRRADIADAS A LA EDAD DE 2 a 5 DIAS CON 2,5 krad, DURANTE 15, 30 Y 60 DIAS A 18°C, 10 h DE LUZ Y 14 h DE OBSCURIDAD (Monterrey, N.L., 1973)

Días de almacenamiento	Tratamiento	N° de pupas	% de adultos[a]
15	MT	112	67,8
	MN	112	75,8
	HT	96	67,6
	HN	96	76,8
30	MT	112	57,5
	MN	112	61,9
	HT	96	59,3
	HN	96	62,8
60	MT	112	47,3
	MN	112	51,9
	HT	96	48,2
	HN	96	53,6

[a] Adultos emergidos con apariencia normal.

MT = Macho tratado   HT = Hembra tratada
MN = Macho normal   HN = Hembra normal

pero almacenando esta fase a 18°C y un fotoperíodo de 10 h de luz y 14 h de obscuridad, y rompiendo la diapausa 15, 30 y 60 días después del almacenamiento. Para este estudio el número de larvas de primer estadio inducidas a diapausa fué de 7 044, las que se alimentaron según el método ya mencionado. Como resultado de la inducción a diapausa se obtuvieron 1 260 pupas hasta los cinco días a partir de la iniciación de la formación de este estadio, correspondiendo este número al 83,09% del total de pupas (tabla VI)

TABLA VIII.   FERTILIDAD DE H.virescens AL CRUZAR INDIVIDUOS
DIAPAUSICOS SIN IRRADIAR E IRRADIADOS EN ESTADIO DE PUPA A LA
EDAD DE 2 A 5 DIAS CON 2,5 krad Y ALMACENARLAS DURANTE 15,
30 Y 60 DIAS A 18°C, 10 h DE LUZ Y 14 h DE OBSCURIDAD
(Monterrey, N.L., 1973)

Días de almacenamiento	Cruzas[a]	N° de huevos fértiles	% de fertilidad	% de esterilidad[b]
	MT × HN	376	17,54	76,60
	MN × HT	431	24,21	67,78
15	MT × HT	239	13,03	82,66
	MN × HN	1698	75,13	——
	MT × HN	225	16,12	77,66
	MN × HT	353	21,31	69,06
30	MT × HT	172	9,41	86,34
	MN × HN	1350	68,87	——
	MT × HN	171	11,85	81,76
	MN × HT	211	15,50	76,14
60	MT × HT	87	6,01	90,75
	MN × HN	1148	64,96	——

[a] Dos parejas por jaula; 20 a 34 por cruza.
[b] Corregida por la fórmula de Abbot.

TABLA IX.   COMPATIBILIDAD SEXUAL DE H.virescens AL IRRADIAR
PUPAS DIAPAUSICAS A LA EDAD DE 2 A 5 DIAS CON 2,5 krad Y
ALMACENARLAS DURANTE 15, 30 Y 60 DIAS A 18°C, 10 h DE LUZ Y
14 h DE OBSCURIDAD (Monterrey, N.L., 1973)

Días de almacenamiento	Cruzas	N° de parejas[a]	N° de hembras con espermatóforos	N° total de espermatóforos	Media por hembra
	MT × HN	34	26	76	2,23
	MN × HT	30	27	66	2,20
15	MT × HT	30	22	48	1,60
	MN × HN	34	32	120	3,53
Promedio					2,39
	MT × HN	28	22	55	1,96
	MN × HT	24	18	45	1,87
30	MT × HT	28	20	36	1,29
	MN × HN	28	28	75	2,68
Promedio					1,95
	MT × HN	22	16	39	1,77
	MN × HT	20	14	31	1,55
60	MT × HT	22	18	27	1,23
	MN × HN	24	22	68	2,83
Promedio					1,82

[a] Dos parejas por jaula.

obtenidas a esa edad (1511) y que corresponde a algo más del 50% de las larvas que sobrevivieron (42, 35% murieron por manipulación y hongos). Las pupas diapáusicas se sexaban dividiéndose en partes iguales para proceder al tratamiento mediante radiación gamma, en la ya mencionada unidad.

Para esta serie de pruebas, las pupas diapáusicas irradiadas y sin tratar se dividían en lotes para tres períodos de almacenamiento: 15, 30 y 60 días, a 18°C, con 10 h de luz y 14 h de obscuridad.

Para romper la diapausa, las pupas se colocaron en estas condiciones: 26°C, 14 h de luz y 10 h de obscuridad. Emergidos los adultos se efectuaban las cruzas de macho irradiado (MT) con hembra normal (HN), hembra irradiada (HT) con macho normal (MN), macho irradiado (MT) con hembra irradiada (HT) y macho normal (MN) con hembra normal (HN) como testigo.

Las parejas se colocaron en las jaulas de plástico cilíndricas descritas antes, haciendo cruzas dobles (dos machos con dos hembras).

De las cruzas efectuadas se tomaron datos de fertilidad de los huevecillos, transferencia y/o recepción de espermatóforos y longevidad de los adultos.

El efecto del almacenamiento sobre las pupas se evaluó por la emergencia de los adultos, encontrándose estadísticamente que a medida que aumentaban los días de almacenamiento la emergencia decrecía significativamente, pero no se encontró diferencia entre los tratamientos de irradiación (tabla VII). La fertilidad de los adultos provenientes de pupas diapáusicas almacenadas durante tres períodos de tiempo se evaluó por el número de larvas emergidas de los huevecillos (tabla VIII). Se notó que el almacenamiento afecta dicha fertilidad, la cual se redujo en relación directa al tiempo de almacenaje tanto en individuos irradiados como en los no irradiados. El efecto de las radiaciones gamma fue más marcado que el del almacenamiento, causando una subesterilidad en los individuos irradiados, la cual se transmitió a la progenie F_1. A su vez esta esterilidad fue más marcada al irradiarse machos cruzados con hembras normales y desde luego en cruzas en que ambos padres provenían de pupas irradiadas.

La transferencia y/o recepción de espermatóforos (tabla IX) se vio afectada por el almacenamiento y las radiaciones gamma, siendo el efecto significativamente mayor (prueba «t») a los 30 y 60 días de almacenamiento; los machos irradiados se vieron ligeramente más afectados en transferir espermatóforos que las hembras irradiadas en aceptarlos. Las cruzas entre adultos en que ambos sexos provenían de pupas irradiadas fueron las más afectadas, ya que las hembras presentaron el menor número de espermatóforos. Las pruebas de «t» arrojaron cifras significativas, excepto para la comparación de machos tratados con hembras tratadas.

La longevidad de los adultos se vio más afectada por las radiaciones gamma que por el almacenamiento (tabla X). La prueba de «t» indica que individuos tratados de ambos sexos vivieron menos que los normales.

En todas las cruzas de los progenitores se obtuvo progenie, la cual se crio bajo condiciones normales y se hicieron observaciones de larvas que llegaron al estadio de pupa y adultos emergidos. Se observó que los mayores porcentajes de larvas que llegaron al estadio de pupa y de adultos emergidos con apariencia normal corresponden a individuos provenientes de padres hembras y machos no irradiados, almacenados en estado de pupa diapáusica.

TABLA X. LONGEVIDAD DE ADULTOS DE AMBOS SEXOS DE
H. virescens PROVENIENTES DE PUPAS DIAPAUSICAS SIN IRRADIAR
E IRRADIADAS A LA EDAD DE 2 A 5 DIAS CON LA DOSIS DE 2,5 krad
Y ALMACENADAS DURANTE 15, 30 Y 60 DIAS A 18°C, 10 h DE LUZ Y
14 h DE OBSCURIDAD (Monterrey, N.L., 1973)

Días de almacenamiento	Tratamiento	N° de individuos	Longevidad de los machos $\bar{x}$ Calculada	Tratamiento	N° de individuos	Longevidad de las hembras $\bar{x}$ Calculada
15	MT	64	10,1	HT	60	11,3
	MN	64	12,2	HN	68	13,4
30	MT	56	11,4	HT	52	10,5
	MN	52	13,8	HN	56	14,3
60	MT	44	9,3	HT	42	10,9
	MN	44	12,6	HN	46	13,5

TABLA XI. FERTILIDAD DE LA PROGENIE F_1 DE H. virescens AL
CRUZAR INDIVIDUOS PROVENIENTES DE PADRES DIAPAUSICOS SIN
IRRADIAR E IRRADIADOS A LA EDAD DE 2 A 5 DIAS CON 2,5 krad Y
ALMACENADOS DURANTE 15, 30 Y 60 DIAS A 18°C, 10 h DE LUZ Y
14 h DE OBSCURIDAD (Monterrey, N.L., 1973)

Días de almacenamiento	Padres (P_1)	Generación (F_1)[a]	% de fertilidad	% de esterilidad corregido[b]
15	MT × HN	M × HN	5,82	92,09
		H × MN	3,82	94,81
	MN × HT	M × HN	51,17	30,49
		H × MN	47,73	35,16
	MT × HT[c]	MT × HT	1,95	97,35
	MN × HN	MN × HN	73,62	———
30	MT × HN	M × HN	4.93	93,20
		H × MN	1,62	97,76
	MN × HT	M × HN	46,01	36,55
		H × MN	47,23	34,87
	MT × HT[c]	MT × HT	00,00	100,00
	MN × HN	MN × HN	72,52	———
60	MT × HN	M × HN	3,27	95,26
		H × MN	3,49	94,94
	MN × HT	M × HN	42,19	38,89
		H × MN	37,47	45,73
	MT × HT[c]	MT × HT	00,00	100,00
	MN × HN	MN × HN	69,05	———

[a] Dos parejas por jaula; de 14 a 30 por cruza.
[b] Corregido por la fórmula de Abbot.
[c] Faltaron las cruzas de P_1 F_1
      MT × HT   M × HN
                H × MN

TABLA XII. LONGEVIDAD EN DIAS DE LA PROGENIE F_1 CUYOS PADRES FUERON IRRADIADOS EN ESTADIO DE PUPAS DIAPAUSICAS A LA EDAD DE 2 A 5 DIAS CON 2,5 krad Y ALMACENADOS DURANTE 15, 30 Y 60 DIAS A 18°C, 10 h DE LUZ Y 14 h DE OBSCURIDAD (Monterrey, N.L., 1973)

Padres	Generación	Media calculada por tratamiento		
P_1	$F_1{}^a$	15 días	30 días	60 días
MT × HN	M	10,5	10,9	11,9
	H	11,3	10,0	10,4
MN × HT	M	12,1	12,4	11,8
	H	11,7	11,4	12,3
MT × HT	M	9,8	11,1	8,9
	H	8,2	10,9	10,8
MN × HN	M	13,6	14,2	12,8
	H	14,5	12,6	13,1

a De 24 a 30 individuos; para 60 días (MT × HT) sólo 14.

Los adultos F_1 obtenidos se cruzaron con individuos del sexo opuesto no irradiados, pero provenientes de pupas diapáusicas, y se realizaron las mismas observaciones hechas en los progenitores. Se encontró (tabla XI) que la fertilidad de la progenie F_1 proveniente de padres irradiados se vio más afectada que la proveniente de padres no irradiados, y que los progenitores en donde los machos fueron irradiados transmitieron mayor esterilidad a su progenie F_1 que las hembras irradiadas.

Al cruzarse individuos machos con hembras provenientes de padres irradiados en estadio de pupa diapáusica, se presentó completa esterilidad para los 30 y 60 días de almacenamiento a que estuvieron sometidos los padres, mientras que para los 15 días se tuvo una fertilidad de 2,65%.

En relación a la transferencia y/o recepción de espermatóforos, se vio que los individuos F_1 provenientes de padres irradiados mostraron menor capacidad para transferir o aceptar espermatóforos que los provenientes de padres no irradiados.

Cuando los progenitores tanto irradiados como sin tratar se almacenaron durante 15 días en estadio de pupa diapáusica, la progenie F_1 mostró mayor capacidad en la transmisión y/o recepción de espermatóforos en relación a los almacenados durante 30 y 60 días.

En relación a la longevidad de adultos F_1, cuyos padres fueron irradiados, parece que no hubo influencia en relación al almacenamiento, pero sí una reducción al comparar los tratamientos de radiación, siendo éste mayor cuando provienen de las cruzas de ambos padres irradiados (tabla XII).

CONCLUSION

En conclusión, de ambas series de experimentos se obtuvo una subesterilidad al irradiar pupas en diapausa con 2, 5 krad. Esta esterilidad,
como era de esperarse, fue más marcada en las cruzas de ambos sexos
irradiados. La edad de almacenamiento en las condiciones indicadas no tuvo
una influencia muy marcada sobre la fertilidad, pero las pupas de 2 a 3 días
de edad fueron las más susceptibles a la irradiación.

La longevidad de los adultos se reduce ligeramente conforme aumentan
el tiempo de almacenamiento de las pupas y si provienen de pupas irradiadas.
Lo mismo resulta, en términos generales, en relación a la transferencia
y/o aceptación de espermatóforos por machos y hembras respectivamente.

Al estudiar la generación F_1, la irradiación de los padres no parece
haber influido sobre el desarrollo de larvas, y pupas, pero los adultos
provenientes de cruzas con ambos padres irradiados presentaron más
anormalidades.

En relación a la fertilidad, los machos F_1 fueron más afectados que
las hembras F_1; la transferencia de espermatóforos por los machos o
aceptación por las hembras y la longevidad de adultos F_1 arrojó resultados
variables. Estudios parecidos se están continuando para obtener datos
más completos.

REFERENCES

[1] FLORES, G. R. , Comportamiento y fertilidad de adultos de Heliothis virescens Fabricius, al tratar pupas
     diapáusicas con radiaciones gamma, Programa de graduados, División de Ciencias Agropecuarias y
     Marítimas, Instituto Tecnológico de Monterrey, México (1971) (tesis sin publicar).
[2] SANCHEZ, I. , Efecto del almacenamiento en frío (5°C) y de radiaciones gamma (2, 5 krad) sobre pupas
     diapáusicas de Heliothis, Instituto Tecnológico de Monterrey, México (1972) (tesis sin publicar).
[3] ROACH, S. H. , ADKISSON, P. L. , Role of photoperiod and temperature in the induction for pupal
     diapause in the bollworm Heliothis zea, J. Insect Physiol. 16 (1970) 1591-97.
[4] PROSHOLD, F. E. , BARTELL, J. A. , Postembryonic growth and development of F_1 and F_2 tobacco budworms
     (Lepidoptera: Noctuidae) from partially sterile males, Can. Entomol. 104 (1972) 165-72.

INFLUENCE OF PUPAL AGE ON
THE RESPONSE OF THE ALMOND MOTH,
Cadra cautella Walker, TO DIFFERENT
DOSAGES OF GAMMA IRRADIATION*

B. AMOAKO-ATTA[†], G.J. PARTIDA
Department of Entomology,
Kansas State University,
Manhattan, Kans.,
United States of America

Abstract

INFLUENCE OF PUPAL AGE ON THE RESPONSE OF THE ALMOND MOTH, Cadra cautella Walker, TO DIFFERENT DOSAGES OF GAMMA IRRADIATION.

When 2, 4, 6 and 8 days old, almond moth pupae were irradiated at dosages of 0, 10, 20 and 30 krad. Irradiated 2-day-old pupae were the most susceptible (no adults emerged) followed by 4-day-old pupae. Mean percentage adult emergence from pupae 6 and 8 days old exposed to 10 or 20 krad was not affected; exposure to 30 krads significantly reduced emergence from 6-day-old pupae. Increasing the dosage from 20 to 30 krads increased the number of deformed adults emerging from 6-day-old pupae, but did not affect adults emerging from pupae irradiated when 8 days old. The effects on fecundity of ovipositing females indicated interaction of age, dose and sex. Irradiated females mated with normal males laid fewer eggs than normal females mated with males irradiated at the same dosage. Fecundity of adults emerging from pupae irradiated when 6 days old was more significantly reduced than was that of adults emerging from irradiated 8-day-old pupae. Developing pupae were more susceptible to radiation during earlier pupal ages, with a pronounced reduction in susceptibility between 4 and 6 days of age. Results indicated that effects on adults emerging from pupae irradiated with 20 krads when 6 days old and on those emerging from pupae exposed to 30 krads when 8 days old were not significantly different.

INTRODUCTION

Of the pests that infest dried cocoa beans in storage, Cadra cautella Walker is the most serious Lepidoptera species in Ghana, the world's leading producer of cocoa. C. cautella, called the almond moth in North America and the tropical warehouse moth in Ghana, is known as the fig moth in other parts of the world.

Rawnsley [1] described the limitations of insecticides for controlling the almond moth in stored cocoa in Ghana. Amuh [2] suggested using gamma radiation as an alternative measure for control, and currently the Ghana Atomic Energy Commission, with the assistance of the International Atomic Energy Agency, is studying that possibility. Part of the study is being conducted at Kansas State University.

Radiation studies on the almond moth date from 1952 [3-8]. Some of the studies determined some of the effects of radiation on the adult moth [5] and on different metamorphic stages [8]. However, detailed information is lacking on the response of pupae of different ages to gamma radiation.

* Work supported in part by the IAEA.
† Currently holding an IAEA Fellowship in the Dept. of Entomology, Kansas State University, Manhattan, Kans.

This study was conducted to determine how different radiation dosages affected pupae of different ages, adult emergence, adult deformities, female fecundity and sterility of moths.

MATERIALS AND METHODS

·   The experimental insects were obtained from the laboratory cultures maintained in the Stored-Product Insects Laboratory, Department of Entomology, Kansas State University. They were reared on a culture media of wheatbran, ground maize (corn meal) and glycerine in the ratio of 8:8:1 (wt/wt). Procedures used were similar to those described by Strong and co-workers [9]. Cultures were maintained in a room held at 27° ± 1°C and 65 ± 5% r.h. with 12 h of light and dark photoperiods. Last instar larvae ready to pupate were collected from 26-day-old cultures, sexed and reintroduced into modified pupation rolls 2 in. in diameter composed of corrugated cardboard (1 in. thick).

When 2, 4, 6 and 8 days old, pupae were irradiated at dosages of 0, 10, 20 and 30 krads in a ^{60}Co Gammacell 220. Four replications of 30 pupae each were used for each group and sex at each dosage level. The irradiated pupae were returned to the rearing room for adult emergence. The number of normal and deformed adults emerging were recorded.

The fertility and fecundity experiments were conducted on active adult moths that emerged from irradiated 6- and 8-day-old pupae. Irradiated females (I♀) were crossed with normal males (N♂), and irradiated males (I♂) were crossed with normal females (N♀). To determine oviposition, single pairs were mated in 2-oz glass vials with dry cotton wool as stoppers. All experiments were conducted in the rearing room. Mortality counts on the pairs were recorded during the oviposition period. Eggs were collected daily for 8 days, and the number of mated producing pairs recorded. Glass petri dishes, 5 cm diameter, were used to hold eggs incubated in the rearing room. The eggs were checked every 8 h until hatching was complete. Total number of eggs and percentage of egg hatch in each treatment were recorded.

Data were analysed using Duncan's multiple range test, at the 5% level, to evaluate the differences between means and to determine the radiation effect on the degree of sterility of mated pairs.

RESULTS

Effect of radiation on adult emergence and level of deformity

Irradiated 2-day-old pupae were the most susceptible; no adults emerged at any treatment level. From 4-day-old pupae exposed to 10 krads, 60 and 30% of the female and male adults respectively emerged; however, all adults died within 48 h after emergence. There was no emergence from pupae exposed to 20 or 30 krads when 4 days old.

The mean percentage adult emergence from pupae 6 and 8 days old exposed to 10 or 20 krads was not significantly affected (Table I). Exposing 6-day-old pupae to 30 krads did significantly reduce emergence.

TABLE I.   MEAN PERCENTAGE OF ADULTS EMERGING AND
PERCENTAGE OF THOSE DEFORMED FROM PUPAE IRRADIATED
WHEN 6 AND 8 DAYS OLD

Dose (krad)	Irradiated ♂♂ pupae		Irradiated ♀♀ pupae	
	Percentage adult emergence	Percentage deformed	Percentage adult emergence	Percentage deformed
6-day-old pupae				
0	93	2	91	1
10	86	2	94	0
20	87	13	84	16
30	38	90	63	83
8-day-old pupae				
0	95	0	89	0
10	87	0	85	0
20	89	0	83	0
30	74	10	90	2

TABLE II.   MEAN PERCENTAGE MORTALITY OF MALE AND FEMALE
ADULT MOTHS ON THE 2nd AND 6th DAY AFTER EMERGING FROM
PUPAE IRRADIATED WHEN 6 AND 8 DAYS OLD

Dose (krad)	Percentage adult mortality			
	2 days after emergence		6 days after emergence	
	6-day-old pupae	8-day-old pupae	6-day-old pupae	8-day old pupae
♂♂				
0	0	0	7	3
10	0	0	15	3
20	2	2	27	3
30	90	5	100	16
♀♀				
0	2	1	10	2
10	2	0	15	9
20	3	0	26	14
30	72	1	100	21

TABLE III. EFFECTS OF IRRADIATION DOSAGE ON FECUNDITY OF IRRADIATED FEMALES (I♀) CROSSED WITH NORMAL MALES (N♂), AND IRRADIATED MALES (I♂) CROSSED WITH NORMAL FEMALES (N♀)

Dosage (krad)	I♀ × N♂			I♂ × N♀		
	No. of females	Average no. of eggs/♀♀	Total no. of eggs laid	No. of females	Average no. of eggs/♀♀	Total no. of eggs laid
6-day-old pupae						
0	42	199	8358	57	200	11 400
10	32	70	2240	38	142	5396
20	21	37	770	16	52	832
30	3	23	69	6	32	192
8-day-old pupae						
0	50	181	9050	58	201	11 658
10	53	133	7049	34	168	5712
20	54	54	2946	35	115	4025
30	28	31	868	31	43	1333

For moths that emerged, radiation did not significantly affect length
of the mean pupal period, which for males was 9.7 days, significantly longer
than for females (9.2 days). The percentage of deformed adults emerging
from pupae irradiated when 6 and 8 days old indicated an age/dose interaction
but sex was not a factor. When the dosage for 6-day-old pupae was increased
from 20 to 30 krads, six times as many deformed adults emerged. Increasing
the dosage for 8-day-old pupae, however, did not cause a significant increase
in deformities in adults emerging, probably because pupae by that time had
almost completely developed.

Effect of radiation on adult mortality

As shown in Table II, adult moths emerging from pupae exposed when
6 or 8 days old to 0, 10 and 20 krads did not show a large increase in
mortality from the 2nd day to the 6th day after emergence. However, adults
emerging from pupae irradiated with 30 krads when 6 days old had a mean
percentage mortality of 72 and 90% for males and females respectively on
the 2nd day, and 100% mortality on the 6th day after emergence. Deaths
recorded for adults emerging from pupae exposed to 30 krads when 6 days
old coincided with the high percentage of adult deformities (Table I).

Effect of radiation on fecundity

The effects of irradiation on the egg-laying potential of irradiated moths
crossed with normal moths are summarized in Table III. Using analysis
of variance and testing for least significant difference, the results showed
that the irradiation effect on fecundity of ovipositing females reflected
interaction of age, dose and sex. Compared with oviposition of females
emerging from non-irradiated pupae, only one-third as many eggs were
oviposited by females emerging from pupae exposed to 10 krads when
6 days old and only one-fifth as many from females emerging from pupae
exposed to 20 krads at that age.

Fecundity of the adults emerging from pupae irradiated when 6 days
old was more significantly affected than was that of those emerging from
pupae irradiated when 8 days old. Results in Table III also show that the
oviposition potential of irradiated females was more significantly reduced
than that of normal females crossed with irradiated males, perhaps because
radiation affected the oogenesis of the developing female. Ovaries of
irradiated 6-day-old female pupae were in most cases smaller than normal.

Effect of radiation on moth fertility

Using mean percentage hatchability of eggs as a criterion for induced
sterility, it is obvious from data presented in Table IV that sterility in both
sexes increased when the radiation dose was increased, though the increase
in sterility was greater in females than in males. Dosages of 20 and 30 krads
induced greater sterility in adults emerging from pupae irradiated when
6 days old than from pupae irradiated when 8 days old.

Twenty krads was considered the dosage that induced sterility in
6-day-old pupae; 30 krads was the dosage required to induce the same level
of sterility in 8-day-old pupae.

TABLE IV.   HATCHABILITY OF EGGS LAID BY ADULTS EMERGING
FROM PUPAE IRRADIATED WHEN 6 AND 8 DAYS OLD

Dosage (krad)	I♀ × N♂		I♂ × N♀	
	No. of eggs	Percentage hatch	No. of eggs	Percentage hatch
	6-day-old pupae			
0	3590	86	3130	84
10	2200	38	3427	63
20	676	0.6	909	7
30	67	0	177	0.6
	8-day-old pupae			
0	2019	83	1898	87
10	3226	52	2956	64
20	2362	14	3089	40
30	851	0.2	1342	8

DISCUSSION AND CONCLUSION

Results (Table I) indicate that the age of the pupae at time of irradiation
influenced adult emergence.   The younger the pupae when irradiated the
more accentuated was the radiosensitivity.   Similar results were obtained
by Qureshi and co-workers [10] on Angoumois grain moth, Ouye and
co-workers [11] on pink bollworm, and Cogburn and co-workers [13] on
Angoumois grain moth and the Indian meal moth.

Between the 4th and 6th day, the almond moth pupae seemingly changed
in susceptibility to radiation.   Adults emerging from pupae exposed to
10 and 20 krads when 6 days old were least affected (Table I).   Adults
emerging from pupae exposed to 10 krads when 4 days old were seriously
affected; no adults emerged from pupae exposed to 20 krads at that age.
Ouye and co-workers [11] obtained similar results with the pink bollworm,
in which 5-day-old pupae marked the beginning of change in radiation
susceptibility.

This study's findings on adult moth mortality (Table II) are similar to
those of Ouye [12], who by irradiating different age groups of Lepidoptera
pupae with the same dosages showed that adults emerging from older pupae
had the lowest mortality.   As shown in Table III, the fecundity of irradiated
moths mated with normal moths of the opposite sex was reduced when the
radiation dose was increased.   The age of the pupae at the time of radiation
significantly affected fecundity.   The fecundity of irradiated females mated
with normal males was more significantly reduced than that of normal
females mated with irradiated males.   These results agree with those of
Cogburn and co-workers [13] and Ouye and co-workers [11].   Though
White and Hutt [14] observed significant reduction in fecundity of irradiated
female codling moth adults, they found that the reduction was not affected
by increase in dosage.

Calderon and Gonen [6] suggested that reduced female fecundity was caused by a smaller quantity of sperm being transferred by the male. Joubert [15] found that in Angoumois grain moth pupae the germarium of the ovariole ceased activity in the mid-pupal stage or just before that stage was reached, thus limiting the productivity of the female to the number of oocytes in the pupal ovariole. Ashrafi and co-workers [16] observed that when last instar larvae of P. interpunctella Hubner were irradiated, high radiation doses caused significant somatic and gonial injuries in the larval testes; he also found that the spermatogonia was most sensitive to radiation; spermatocytes and spermatids were in that order next most sensitive.

Results summarized in Table IV show that 20 krads gave 99 and 93% sterility in female and male moths respectively emerging from pupae irradiated when 6 days old, and that 30 krads was required to give the same level of sterility in moths emerging from pupae irradiated when 8 days old. These sterility levels for 6- and 8-day-old pupae did not cause any significant injuries (Table I) or deaths (Table II). Calderon and Gonen [6], Cogburn and co-workers [8] and Pendlebury and co-workers [4] found, as we did, that the female moths were sterilized at a comparatively lower radiation dose than were males.

Because of the lower radiation dosage required to induce sterility in 6- and 8-day-old almond moth pupae than in the adult moths, pupae instead of adults should be used in a sterile-male programme.

ACKNOWLEDGEMENTS

We wish to thank R. E. Faw, Head of the Nuclear Engineering Dept., Kansas State University, and M. J. McEwan, of the Nuclear Engineering Dept., for making available the ^{60}Co source. We appreciate the suggestions given by R. B. Mills and E. Horber of the Department of Entomology, and the assistance of L. Charlton who maintained the insect cultures. We also give special thanks to H. Marshall for typing the manuscript and J. Krchma who helped in the statistical analysis of the data.

REFERENCES

[1] RAWNSLEY, J., Ephestia cautella; Egg Laying after Knock-down by Synergized Pyrethrin Space Spray, Ghana Cocoa Marketing Publication No.6 (1958).

[2] AMUH, I.K.A., "Potentialities for application of the sterile-male technique to the control of the cocoa moth, Cadra cautella Walk.", Application of Induced Sterility for Control of Lepidopterous Populations (Proc. Panel Vienna, 1970), IAEA, Vienna (1971) 7.

[3] DENNIS, N.M., The effects of gamma-ray irradiation on certain species of stored product insects, J. Econ. Entomol. 54 (1961) 211.

[4] PENDLEBURY, J.B., JEFFRIES, D.J., BULL, J.O., "Some effects of gamma radiation on Rhizopertha dominica (F.), Cadra cautella (Wlk.), Plodia interpunctella (Hubn), and Lasioderma serricorne (F.)", The Entomology of Radiation Disinfestation of Grain (CORNWELL, P.B., Ed.), Pergamon Press, London (1966) 143.

[5] AHMED, M.S.H., AL-HAKKAK, Z., AL-SAQUR, A., "Exploratory studies on the possibility of integrated control of the fig moth, Ephestia cautella Walk.", Application of Induced Sterility for Control of Lepidopterous Populations (Proc. Panel Vienna, 1970), IAEA, Vienna (1971) 1.

[6] CALDERON, M., GONEN, M., Effects of gamma radiation on Ephestia cautella (Wlk.) (Lepidoptera, Phycitidae): 1. Effects on adults, J. Stored Prod. Res. 7 (1971) 85.

[7] GONEN, M., CALDERON, M., Effects of gamma radiation on Ephestia cautella (Wlk.) (Lepidoptera, Phycitidae): 2. Effects on progeny of irradiated males, J. Stored Prod. Res. 7 (1971) 91.

[8] COGBURN, R.R., TILTON, E.W., BROWER, J.H., Almond moth: gamma radiation effects on the life stages, J. Econ. Entomol. 66 (1973) 745.

[9] STRONG, R.G., PARTIDA, G.J., WARNER, D.N., Rearing stored product insects for laboratory studies: six species of moths, J. Econ. Entomol. 61 (1968) 1237.

[10] QURESHI, Z.A., WILBUR, D.A., MILLS, R.B., Sublethal gamma radiation effects on prepupae, pupae and adults of Angoumois grain moth, J. Econ. Entomol. 61 (1968) 1699.

[11] OUYE, M.T., GARCIA, R.S., MARTIN, D.F., Determination of the optimum sterilizing dosage for pink bollworms treated as pupae with gamma radiation, J. Econ. Entomol. 57 (1964) 387.

[12] OUYE, M.T., "Current Status of the co-ordinated research program on the use of sterile-insect technique against the rice stem borer", Application of Induced Sterility for Control of Lepidopterous Populations (Proc. Panel Vienna, 1970), IAEA, Vienna (1971) 113.

[13] COGBURN, R.R, TILTON, E.W., BURKHOLDER, W.E., Gross effects of gamma radiation on the Indian-meal moth and the Angoumois grain moth, J. Econ. Entomol. 59 (1966) 682.

[14] WHITE, L.D., HUTT, R.B., Effects of gamma irradiation on longevity and oviposition of the codling moth, J. Econ. Entomol. 63 (1970) 866.

[15] JOUBERT, P.C., The reproductive system of Sitotroga cerealella Olivier (Lepidoptera, Gelechiidae): 1. Development of the female reproductive system, S. Afr. J. Agric. Sci. 7 (1964) 65.

[16] ASHRAFI, S.H., BROWER, J.H., TILTON, E.W., Gamma radiation effects on testes and the mating success of the Indian meal moth, Plodia interpunctella (Hubn), Ann. Entomol. Soc. Am. 65 (1972) 1144.

DISCUSSION

G.W. RAHALKAR: What was the dose rate employed in your investigatio

B. AMOAKO-ATTA: The ^{60}Co Gammacell 220 supplied dose rates of 1.810 krad to 1.710 krad per minute during the experimental period.

G.W. RAHALKAR: Did you measure the temperature during irradiation?

B. AMOAKO-ATTA: If you mean the temperature within the radiation chamber, the answer is no.

I.A. MAYAS: You spoke of deformations of adults derived from pupae irradiated at four days. Could you tell us exactly what these deformations were? For example, did you observe deformation of the antennae or of the mouth parts?

B. AMOAKO-ATTA: Four criteria were used to establish deformities: (a) colour, shape and size of wings; (b) legs and antennal features — reduced number of segments or fused segments; (c) extension of the abdominal segments or protrusion of external genitalia from the last abdominal segment and (d) moths that could not completely emerge from pupal cases were also described as deformed. However, we did not look at the mouth parts since this insect is not an adult feeder.

R.R. BLUZAT: With reference to Table IV, have you examined the state of embryonic development of eggs which do not hatch and, if so, is the result a function of the sex of the irradiated parent?

B. AMOAKO-ATTA: Yes, our investigations indicated that embryonic abortions were higher in the eggs from irradiated female moths mated with normal males than in those from normal females mated with irradiated males.

HABITAT AS A FACTOR INDUCING DIVERSITY OF POPULATIONS IN THE CODLING MOTH AND OTHER ORCHARD PESTS, AND ITS RELEVANCE TO GENETIC CONTROL METHODS

B. NAGY, T. JERMY
Research Institute for Plant Protection,
Budapest, Hungary

Abstract

HABITAT AS A FACTOR INDUCING DIVERSITY OF POPULATIONS IN THE CODLING MOTH AND OTHER ORCHARD PESTS, AND ITS RELEVANCE TO GENETIC CONTROL METHODS.
Differences in behavioural and physiological characteristics of codling moth strains strongly indicate the genetic plasticity of the species. It is supposed that habitats differing substantially in environmental factors that are relevant to the codling moth and some other orchard pests would induce development of populations showing quite different behavioural and physiological characteristics. Having taken into consideration the basic differences in determining codling moth habitats in Hungary, it is suggested that the following types occur: (1) the "farm" ecotype of the commercial orchards, (2) the "semi-wild" ecotype of garden districts, and (3) the "wild" ecotype living on wild apple trees in woodland. Since laboratory rearing produces well-known alterations in the characteristics of the codling moth, such populations are regarded as the "domesticated" ecotype. These variations in the codling moth strains make the comparison of results obtained in various geographical regions very difficult, and necessitate that these factors must be taken into account in sterile-insect release experiments.
With the usual chemical control of the codling moth, Laspeyresia pomonella L., and other orchard pests, it was not necessary to investigate the question whether there are significant differences between populations of the same insect species living in different habitats. However, with the development of sterile-insect release (SIRM) techniques and other biological control methods using living control agents, the importance of this type of investigation becomes clear.

EVIDENCE OF VARIATIONS BETWEEN CODLING MOTH POPULATIONS IN NATURE

The best known variations between codling moth populations living in different geographical regions were shown by experiments carried out to determine the diapause response to photoperiods [1]. In northern regions full-grown larvae select overwintering sites at the base of tree trunks or in the soil, whereas in the south they over-winter on higher parts of the trunks [2].

Variations seemed to occur in the fecundity of Hungarian codling moth strains of different origin [3]. Variations were shown by Tadic [4] in the beginning of flight periods of various Yugoslavian codling moth populations. Bush found some differences by iso-enzyme analysis of different geographical strains even in the 24th laboratory generation; crossings, however, were successful between these strains [5].

These and similar differences between populations isolated by geographical distances are the natural consequence of the regional variations of ecological factors influencing the evolution of populations. However, it can be shown that codling moth populations colonizing different habitats in the same region may also be exposed to different selection pressures resulting in significant differences of physiological and behavioural characteristics.

The appearance of codling moth races resistant to insecticides is one of the well-known results of selection pressures represented by human influences [6-8]. Insecticide treatment, however, can also induce less visible changes. For example, Sluss et al.[1] have found that managed orchards in Washington State, United States of America, harbour codling moth populations that consist of approximately 90% multivoltine and 10% univoltine types of individuals, whereas in unmanaged orchards of the same region the ratio of the two types is almost the reverse: 5 and 95% respectively.

Management, that is the treated or untreated status of the orchards, proved to be responsible for different levels of natural control [9,10] and also for the composition of the injurious Lepidoptera complex [11] which also may represent variations in selection pressures.

MAIN TYPES OF HABITATS HARBOURING CODLING MOTH POPULATIONS

The original habitats of the codling moth in Eurasia were in all probability the forests in which wild apple trees, Malus silvestris L., Mill., are common. In central Hungary the infestation of wild apple trees in forests varies considerably. For instance, we found that in the vicinity of Budapest, in a swampy Alneto-Ulmetum type of forest, the fruits were free of codling moth infestation whereas in the drier Querceto-Carpinetum forests 8 to 23% of the fruits were damaged (August 1973). Since most of the Hungarian forests, representing 16% of the country's surface, are oak forests, the wild apple trees scattered throughout the woody areas allow the presence of extended, if not very abundant, codling moth populations that are not exposed to direct human influence. These populations can be regarded as the wild ecotype developing in all probability under stronger natural selection pressures than those in cultivated areas. Their abundance must be strongly regulated also by the intermittent fruiting of wild apple trees. We found that the fruits of such trees are sometimes totally destroyed by Anthonomus pomorum and winter-moths in the spring before the oviposition of the codling moth (Table I).

The wild ecotype has not been found on wild pear trees, Pyrus spp., which are infested not by the codling moth but by the pear moth, Laspeyresia pyrivora Dan., in the southwest of Hungary [12]. We succeeded in rearing a few specimens of the codling moth on Quercus acorns in the laboratory, but up to now we do not have any information on the natural infestation of acorns.

It can be supposed that the cultivated apple trees were originally infested in Euroasia by moths migrating from the forest into the orchards, since young apple orchards planted in the neighbourhood of forests soon become infested. This type of "radiation" effect of the forest community

[1] SLUSS, R.R., FERRO, D.N., BOGYO, T.P., "Studies of population structure and dynamics of the codling moth for the development of a model of the sterile-male release strategy", these Proceedings, IAEA-SM-186/40.

TABLE I.   RELATIVE IMPORTANCE OF SELECTION FACTORS
INFLUENCING CODLING MOTH ECOTYPES

Selection factors	Farm ecotype	Semi-wild ecotype	Wild ecotype	Domestic ecotype
Insecticides	+++	+	-	-
Host range	+	+++	- ?	-
Natural control	+	++	+++	(+)
Intermittence of cropping	+	++	+++	-

usually occurred also in the experimental orchard of the Research Institute
for Plant Protection near Budapest.   For instance, winter moths (Operophtera
brumata L. , Erannis defoliaria L. , etc.) were very abundant in 1972-74.
    On the other hand, the border zone of the forest may suffer from some
effects coming from the orchard.   The migration of codling moths from the
orchard into adjacent woodland has been shown from released and recaptured
adults [13, 14].   In such experiments, however, the role of a possible long-
range attraction by the sextraps cannot be excluded.
    The second, very extended, type of habitats harbouring probably most
of the codling moth populations is represented by small private gardens and
scattered fruit trees in vine-growing areas.   The importance of this garden
district can be demonstrated by the fact that 78% of all fruit trees in Hungary
are situated in small private gardens and vineyards in the neighbourhood of
built-up areas.
    The codling moth populations living in such garden districts can be
called a semi-wild ecotype since the selection pressure exerted by human
activities varies very much: some host trees or small apple orchards are
well managed whereas others are abandoned.   In this type of habitat not
only management is a selection factor of varying intensity but also the
presence of various other, mostly secondary, hosts of the codling moth
such as walnut, apricot, peach, pear and plum [13, 15, 16].   Sáringer
(to be published) has found different mortality ratios of codling moth larvae
reared on fruits of these plants in laboratory.   Therefore it can be supposed
that the basic differences between the ecological conditions of the forest and
the garden habitats result in behavioural differences of the wild and semi-
wild codling moth ecotypes.
    The total acreage of the commercial orchards is much smaller than that
of the garden districts, and the codling moth population density is very low
because of regular chemical treatments.   So the total codling moth population of the commercial orchards represents a considerably smaller biomass
than that inhabiting the garden districts.   Codling moth populations of such
areas are exposed to the strongest selection pressures in nature, and may
be regarded as the farm ecotype (Table I).

This intraspecific differentiation probably occurs also in other orchard pests ecologically similar to the codling moth, such as <u>Adoxophyes orana</u> Hbn., other tortricids, scales and spider mites. It can be supposed that the increasing industrialization of fruit production will further intensify the selection of the farm ecotype.

THE "DOMESTICATED ECOTYPE" OF THE CODLING MOTH

Field experiments carried out to develop the SIRM technique with the codling moth represent the release of populations which strongly differ from the natural ones, not only as a result of sterilization but also in several basic behavioural and physiological characteristics. The strong selection pressure present in all mass-rearing procedures alters significantly the populations reared. This has been found by several authors in experiments comparing laboratory moth strains with natural ones.

For example, Bulyginskaya and Sokolova [17] found that laboratory-reared males preferred mating with females of their own strain rather than with wild females of different origin. Wildbolz and Riggenbach [18] were able to select within eight generations a laboratory strain containing much less univoltine individuals than the original one. Jermy [19] found that larvae of a natural strain originating from Keszthely, Hungary, when reared on ripe apples, showed a higher percentage of diapause compared with larvae reared on green apples, whereas no such effect from the food quality could be found with the laboratory strain of Yakima, United States of America (Sáringer, to be published).

Laboratory strains very often proved to be more fertile than newly "domesticated" strains [17, 20].

Since rearing procedures used in different laboratories differ widely, various populations of the domesticated ecotype may also differ considerably. It would be a source of valuable information on the variability of the codling moth if an international comparison of behavioural and physiological characters of strains reared in several laboratories could be carried out as has been done by Bush on allozyme frequency.

CONCLUSIONS

Evidence collected during the last decades has proved the genetic plasticity of the codling moth. Genetic plasticity enables rapid changes to occur in physiological and behavioural characteristics when selection pressures vary. Such differences exist not only between strains of varying geographical origin but also between populations living in different habitats of the same region, and even more between natural and laboratory-reared strains. The inconsistencies often found in the results of experiments carried out by different authors is presumably caused in many cases by basic differences in the biological status of the moth populations used. It is therefore necessary to define thoroughly the insect material used in experiments.

These difficulties presumably also apply to other orchard pests that are similar to the codling moth, and particularly to experiments carried out with the SIRM technique and other biological control methods.

REFERENCES

[1] SHEL'DESHOVA, G.G., Ecological factors determining distribution of the codling moth, Laspeyresia pomonella L. (Lepidoptera, Tortricidae) in northern and southern hemispheres, Entomol. Obozr. 46 (1967) 583.

[2] KOT, J., Mozliwosci adaptacji gasienic owocowki jablkowki (Carpocapsa pomonella L.) do polnocnych warunkow zimowania, Ekol. Pol. Ser. B.4 (1958) 173.

[3] V. DESEÖ, K., Szaporodásbiológiai különbségek egy faj különbözö helyekröl származó populációi között, 21, Növényvédelmi Tudományos Értekezlet, Budapest 1 (1972) 53.

[4] TADIC, M., Jabucni smotavac (Carpocapsa pomonella L.), Posebna Izd. Hidroteh. Inst., Beogr. 4 (1957).

[5] ROBINSON, A.S., PROVERBS, M.D., Hybridization between geographical races of the codling moth (Lepidoptera: Olethreutidae), Can. Entomol. 105 (1973) 289.

[6] SMITH, L.C., DDT resistant codling moth: a report on the 1954/55 control trials, J. Dep. Agric. S. Aust. 55 (1955) 12.

[7] LESKI, R., GROMISZ, Z., KOTTER, A., The resistance of the codling moth (Carpocapsa pomonella L.) to metoxychlor in Lublin province, Zaklad Ochrony Roslin Inst. Sadownictwa, Ser. A 95 (1969) 313.

[8] SIKURA, N.M., Resistance of natural populations of Carpocapsa pomonella to DDT, Khimiya v Sel'skom Hozyaystve 3 (1972) 44.

[9] RUSS, K., On a remarkable occurrence of Beauveria bassiana (Bals.) Vuill. on Carpocapsa pomonella (L.), Pflanzenschutzberichte 31 (1964) 105.

[10] HAGLEY, E.A.C., The occurrence of fungal diseases of the codling moth in unsprayed apple orchards in Ontario, Proc. Entomol. Soc. Ont. 101 (1970) 45.

[11] NAGY, S., Comparative investigations of injurious Lepidoptera in cultivated and uncultivated orchards, Állat. Közl. 57 (1970) 93.

[12] PÁLFI, D., WIANDT, H., Occurrence and study of Laspeyresia pyrivora Dan. in the district of Zala, Növényvédelem 6 (1970) 251.

[13] NAGY, B., JERMY, T., On the host plants and distribution of the codling moth (Laspeyresia pomonella L.) in Hungary with special regard to the sterile release method, Acta Phytopathol. Acad. Sci. Hung. 7 (1972) 431.

[14] VOJNITS, A., Untersuchung der Vagilität und Dispersion des Apfelwicklers (Laspeyresia pomonella L.), Állat. Közl. 60 (1973) 161.

[15] SEPRÖS, I., TISZA, G., Gyümölcsmolyok, Mezögazd. Kiadó, Budapest (1970).

[16] V. DESEÖ, K., SÁRINGER, Gy., SEPRÖS, I., A szilvamoly (Grapholitha funebrana Tr.), Mezögazd. Kiadó, Budapest (1971).

[17] BULYGINSKAYA, M.A., SOKOLOVA, D.V., Comparison of laboratory and natural populations of the codling moth, N.A. Kholodov Memorial Lectures, 14 Apr. 1969 3 (1970) 3.

[18] WILDBOLZ, T.H., RIGGENBACH, W., Untersuchung über die Induktion und die Beendigung der Diapause bei Apfelwicklern aus der Zentral- und Ostschweiz, Mitt. Schweiz. Entomol. Ges. 42 (1969) 58.

[19] JERMY, T., Experiments on the factors governing diapause in the codling moth, Cydia pomonella L. (Lepidoptera, Tortricidae), Acta Phytopathol. Acad. Sci. Hung. 2 (1967) 49.

[20] JERMY, T., NAGY, B., "Genetic control studies of Carpocapsa pomonella (Linnaeus) in Hungary", Application of Induced Sterility for Control of Lepidopterous Populations" (Proc. Panel Vienna, 1970), IAEA, Vienna (1971) 91.

DISCUSSION

K. RUSS: Have you any evidence that the various ecotypes are really different, for example genetically?

B. NAGY: No, we have not yet done any experiments on possible genetic differences between these ecotypes.

T.P. BOGYO: Could you enlarge on the effect of nutrition on diapause. The results you mentioned in your paper for green apples versus ripe ones seem to be contrary to our findings.

B. NAGY: We, too, have had some inconsistent results in this connection. Jermy has established a nutritional effect (ripe-green apple) on the
diapause for the Hungarian strain, but Sáringer did not observe this in the
case of the Yakima strain.

T. P. BOGYO: I should like to draw your attention to the work done by
one of the students of Professor M. Barnes of Riverside, California, who
has studied the attack habits of various codling moth strains.

G. W. RAHALKAR: Do you think the various ecotypes you have mentioned
would markedly influence the effectiveness of the sterile-male release
method?

B. NAGY: Yes, I believe that they would, especially in the case of the
farm ecotype and the semi-wild ecotype. However, the main obstacles to
the use of the SIRM in orchard districts are associated with the mosaic
pattern of such habitats.

H. LEVINSON: Could you say how quickly a host preference can be
induced experimentally in the wild strain of codling moth?

B. NAGY: I am afraid not. However, data on preference induction in
other insect species have been published by various authors.

SOME INVESTIGATIONS ON THE
ECOLOGY OF THE CODLING MOTH,
Laspeyresia pomonella L., IN AUSTRIA
I. Generation dynamics; use of
pheromones, light traps and
corrugated cardboard rings

P. FISCHER-COLBRIE, K. RUSS, O. RUPF
Bundesanstalt für Pflanzenschutz,
Vienna, Austria

Abstract

SOME INVESTIGATIONS ON THE ECOLOGY OF THE CODLING MOTH, Laspeyresia pomonella L., IN AUSTRIA:
I. GENERATION DYNAMICS; USE OF PHEROMONES, LIGHT TRAPS AND CORRUGATED CARDBOARD RINGS.
　　Before practical tests for the application of the genetic control against codling moth can be initiated,
it is necessary to obtain a general view of the yearly changes of the generations of the moth. Investigations
on the generation dynamics of codling moth have been carried out since 1970 in various climatic regions
and at various altitudes in Austria. Procedures used to evaluate a possible second generation are reviewed.

1.　INTRODUCTION

　　Successful control of the codling moth, particularly by the sterile-
insect release method, requires a thorough knowledge of its ecology. Since
this pest produces either one or two generations yearly, depending on
several climatological parameters, our emphasis has been placed on the
investigation of these parameters in different regions and at various
altitudes in the Austrian apple-growing areas. These investigations were
initiated in 1970 and continued with improved methods in the following
years in about 120 apple orchards.

2.　INVESTIGATIONS ON THE NUMBER OF GENERATIONS IN AUSTRIA

　　It can be taken for granted that the number of codling moth generations
depend on the prevailing weather conditions, especially on the temperature
at various places. In addition it has been taken into account that the
photoperiodism influences by means of specific length of day the possible
number of generations in a certain area by inducing diapause [1-5]. In
general it can be stated that, under Austrian conditions (48°N), the induction
of diapause already begins about the middle of June. From then on, in an
increasing percentage, more and more caterpillars of the codling moth
enter into diapausal development, and by the end of July the population is
nearly 100% in diapause. It can therefore be concluded that the length of
day of 1 August ($16\frac{1}{2}$ hours) is finally the critical length of day for all
caterpillars to enter the diapause (Fig.1) [3].

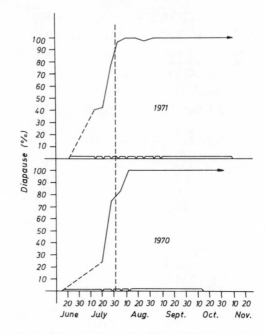

FIG. 1. Increasing percentage of diapausing larvae of codling moths in 1970 and 1971 at Kronberg, Austria.

2.1. Materials and methods

Initially we tried to estimate the number of generations by the flight activity of the moths, expressed by catches with a Robinson u.v. light trap. However, the disadvantages of this type of trap are well known. As soon as Codlemone became available in Austria in 1972, both trap systems were used to trace the flight peaks of this moth. The type of pheromone trap used was the IOBC standard trap (cylindrical trap, 90 mm ϕ, 155 mm length)

The investigations in 1973 in co-operation with the competent chambers of agriculture (Landeslandwirtschaftskammern) were based on about 40 light traps and 80 apple orchards equipped with 1-5 pheromone traps, depending on their size. Daily observations were made by observers, selected by the chambers of agriculture, who were obliged to fill in forms and send them to the Bundesanstalt für Pflanzenschutz every fourth day for analysis. Based on this information warnings against codling moth are issued by the Bundesanstalt through the radio, telephone and by postcards to subscribers.

In addition to the traps, cardboard rings were used to evaluate the generation dynamic of this moth. These corrugated cardboard rings were fixed to the trunks of the trees in May and were removed in the middle of October of the same year. The presence of the occurrence of a second generation was indicated by pupa integuments or the remains of a pupa skin in the strips.

This method, however, was not accurate enough always to reveal exact data on the occurrence of a second generation. Very often a pupa integument could not be found, although a second generation was to be expected from experience.

Since 1971, therefore, this method has been improved, at least in some important districts in our apple-growing areas, by fixing and removing the strips repeatedly during the season. In this case four corrugated cardboard strips were used, successively fixed and removed in the following system:

Strip No.	Date of fixing	Date of removing	Strips remained until
1	10 - 20 May	1 July	15 October
2	1 July	20 July	15 October
3	20 July	15 August	15 October
4	15 August	15 October	15 October

After the strips were removed (with the exception of No.4), they were placed in plastic boxes, which were hung up in the shady parts of the tree tops for further observations.

This method gave good results by using the pupa integument as well as the emerged moth as a parameter for a second generation of codling moth.

Apart from these direct investigations on generation dynamics, meteorological data were sampled continually in several apple-growing areas, which enabled us to interpret our findings on the number of generations. These meteorological centres had equipment to measure and record various temperatures of the air, stem and soil, relative humidity, duration of leaf wetness (a good tool to develop a forecasting system against apple scab), velocity and direction of wind and, last but not least, total radiation. An additional device was constructed to record the sum of the effective temperature.

3. RESULTS

3.1. Use of traps

Some examples of the flight activities of the codling moth, expressed by the number of moths trapped, are shown in Figs 2-5. Comparing both trap types it can be stated that pheromone traps are somewhat more efficient in the early season. In the later season the flight peaks are better expressed by the pheromone trap, which allows an easier timing of control measures against this pest.

3.2. Use of corrugated cardboard rings

Changing the cardboard rings three times in the season, and leaving them under natural conditions in special boxes in the shady parts of the crowns of the trees until harvest, proved to be the most accurate tool for determining the number of generations.

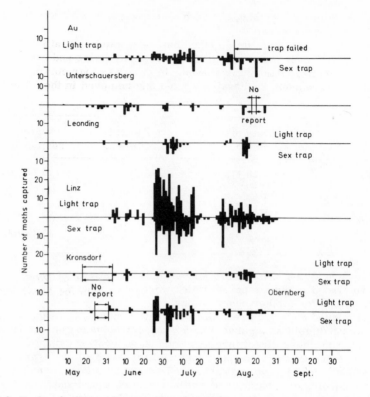

FIG. 2. Number of codling moths captured in various Austrian apple-growing areas using light and/or pheromone traps (Au, Unterschauerberg, Leonding, Linz, Kronsdorf, Obernberg).

The effectiveness of these cardboard rings was calculated by counting the total number of infested fruits and caught larvae. The average of caught larvae was 24.2% and ranged from 16.0% to 49.6%. A slight inverse correlat between the diameter of the crown of the trees and the percentage of caught larvae was found.

Generally a second generation of codling moth can be expected with an almost regular certainty up to 500 m above sea-level (yearly fluctuations up and down depending on varying climatological conditions were possible) (Figs 6-8). Such second generations can be extremely extensive locally. The highest degree of pupation was found in the first two cardboard strips, less in the third, with no pupation in the last one (Table I).

Investigations in higher altitudes, particularly in the Tirol, showed that codling moth may occur very well up to 1200 m (St. Ulrich, Ladis, Fig.6 A rather unexpected result was the occurrence of a second generation locally at 900 m above sea-level (Tulfes, Tirol) in the years 1970 and 1971 (Fig.6). These findings were surprising, as one would expect that at these altitudes in Austria a second generation would be prevented by the prevailing low temperature [6]. Meteorological data from Rinn, Tirol, from the years 1970 and 1971 (as Tulfes does not have a meteorological station, data were

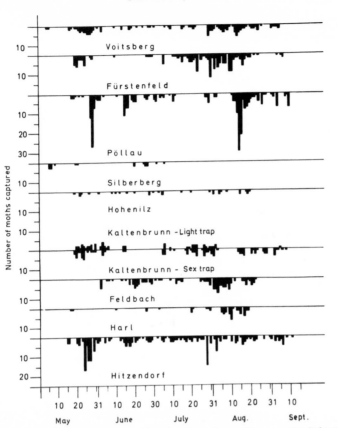

FIG. 3. Number of codling moths captured in various Austrian apple-growing areas using light and/or pheromone traps (Voitsberg, Fürstenfeld, Pöllau, Silberberg, Hohenilz, Kaltenbrunn, Feldbach, Harl, Hitzendorf).

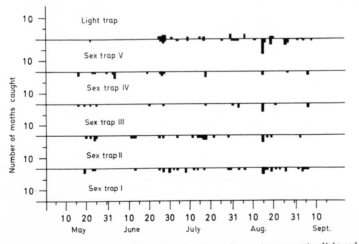

FIG. 4. Number of codling moths captured in various Austrian apple-growing areas using light and/or pheromone traps.

used from Rinn, 3 km from Tulfes), were compared with those from the
valley (Innsbruck, 570 m). Figure 9 shows that the effective temperature
sum of 525°C, which is necessary for development into a second generation [
was low in both years in Rinn (900 m) and high in Innsbruck (570 m). Total
radiation was nearly equal at both stations.

As the climatological factors do not appear to give a sufficient explanati
for this second generation at these altitudes, a possible adaptation of the
moth through more or less genetical changes was taken into consideration,
and was the subject of investigations during 1973.

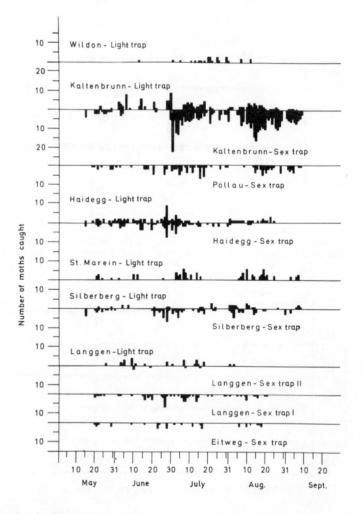

FIG. 5. Number of codling moths captured in various Austrian apple-growing areas using light and/or
pheromone traps (Wildon, Kaltenbrunn, Pöllau, Haidegg, St. Marein, Silberberg, Langgen, Eitweg).

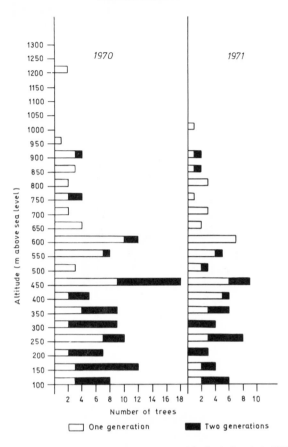

FIG. 6. Occurrence of a second generation at various altitudes in Austria in 1970 and 1971.

Diapausing larval material from four selected sampling locations in Austria (Tirol I = Tulfes 900 m; Tirol II = various sampling locations above 1000 m; Styria = various sampling locations between 300 and 560 m; and Vienna = up to 200 m), and a mixed collection from Czechoslovakia was stored at 5°C from 15 October until incubation at 24°C on 20 December.

The first interesting result was the fact that both Tirolean larval samples passed through diapause and pupated under laboratory conditions on an average 3 to 4 days earlier than larvae from the other samples (Fig.10). This means in any case better use of the seasonal developing possibilities, and may be one of the reasons for an occurrence of two generations at these altitudes.

Table II shows summarized results of a test with singly crossed males and females from the five locations mentioned above. There was no clear and significant indication for the occurrence of heavy genetical changes in all locations. However, an increase in fertility as well as fecundity within crossings of moths from higher altitudes (both Tirol locations and Czechoslovakia) with moths from lower altitudes (especially from Styria)

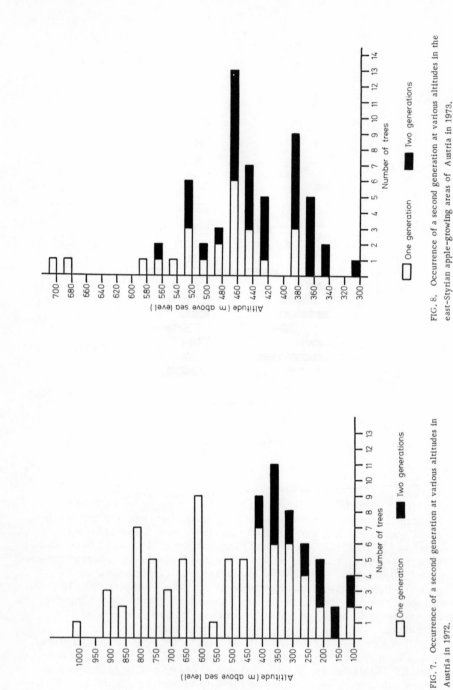

FIG. 8.   Occurrence of a second generation at various altitudes in the east-Styrian apple–growing areas of Austria in 1973.

FIG. 7.   Occurence of a second generation at various altitudes in Austria in 1972.

TABLE I. MEAN PERCENTAGE OF SECOND GENERATION FORMED IN FOUR SUCCESSIVE PERIODS AND AT VARYING ALTITUDES IN STYRIA

Periods [a]	Altitudes (m above sea level)													
	300	320	340	360	380	400	420	440	460	480	500	520	540	560
I	–	–	8.7	7.8	17.6	–	18.2	3.9	22.2	–	33.3	50.0	–	3.4
II	25.0	–	26.9	7.3	6.5	–	24.1	8.4	8.0	12.2	42.9	9.5	–	–
III	–	–	–	–	4.8	–	13.2	3.5	9.1	1.1	–	–	–	–
IV	–	–	–	–	–	–	–	–	–	–	–	–	–	–
Total	25.0	–	18.4	7.4	6.5	–	20.1	5.9	8.8	7.0	33.5	13.0	–	3.4

[a] Periods: I (10 May – 30 June) III (21 July – 15 August)
II (1 July – 20 July) IV (16 August – 15 October)

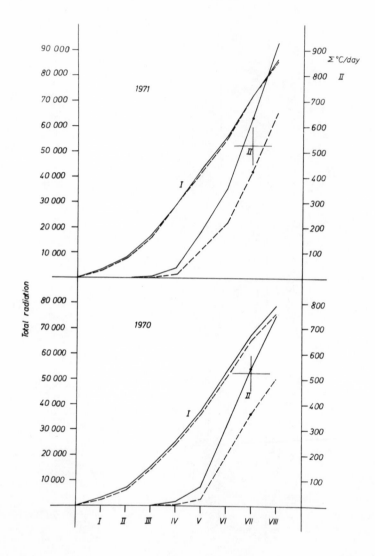

FIG. 9. Total radiation (I) and effective temperature (II) from Innsbruck (——, 570 m) and Rinn (-----, 900 m) in 1970 and 1971.

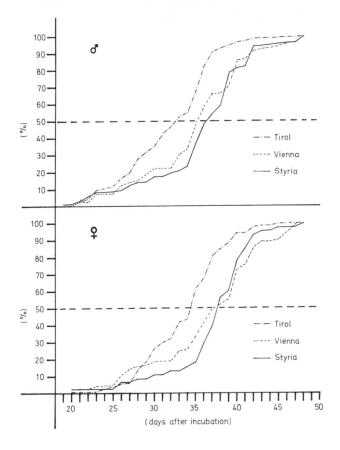

FIG. 10. Differences in the post-diapausal development of moths from various origins.

could point to certain differences in the genes. These possible differences
seen to be confirmed by a higher percentage of genetically fixed diapause in
crosses where moths from higher altitudes were involved (Table II).
Crossings using the F_1 generation of the original crossings show a similar
trend concerning fertility and fecundity as the parental generation (Table III).

A summary of observations on the dependence of fertility and fecundity
of the moths on the duration of the time to complete diapause is shown in
Table IV. In the course of the crossing studies, it was evident that moths
that passed through diapause later were significantly less fertile and fecund
compared with those that underwent diapause earlier. If this finding could
be confirmed by further investigations in the field, it would be very important
in selecting and establishing control measures against the codling moth.

TABLE II. CROSSING STUDIES: AVERAGE FERTILITY, FECUNDITY, MATING FREQUENCY AND PERCENTAGE DIAPAUSE OF F_1 LARVAE

Crossing symbols[a]	No. of replicates	No. of eggs per replicate	No. of larvae per replicate	Mating frequency (%)	Diapause F_1 larvae (%)
$T_I \times T_I$	20	5.3	4.1	25.0	27.6
$T_I \times T_{II}$	5	4.4	0.8	20.0	66.7
$T_I \times St$	16	15.9	10.8	25.0	15.7
$T_I \times V$	15	1.6	0.5	21.4	-
$T_I \times Cs$	10	16.3	9.8	20.0	14.5
$T_{II} \times T_I$	7	1.6	0.0	0.0	-
$T_{II} \times T_{II}$	3	13.7	0.0	0.0	-
$T_{II} \times St$	7	28.9	8.1	33.3	3.8
$T_{II} \times V$	6	0.0	0.0	16.7	-
$T_{II} \times Cs$	7	1.3	0.0	14.3	-
$St \times T_I$	16	8.4	1.3	37.5	0.0
$St \times T_{II}$	5	2.2	0.6	60.0	-
$St \times St$	16	9.2	6.4	31.3	0.0
$St \times V$	17	5.3	1.3	17.6	0.0
$St \times Cs$	10	6.7	4.1	10.0	40.0
$V \times T_I$	21	16.7	11.8	42.9	5.4
$V \times T_{II}$	6	9.8	0.0	0.0	-
$V \times St$	14	3.4	0.0	7.0	-
$V \times V$	17	3.7	2.4	17.6	0.0
$V \times Cs$	10	8.3	4.6	20.0	7.8
$Cs \times T_I$	10	4.5	2.4	40.0	61.1
$Cs \times T_{II}$	5	3.0	1.4	75.0	-
$Cs \times St$	10	12.7	9.9	30.0	0.0
$Cs \times V$	9	0.1	0.0	0.0	-
$Cs \times Cs$	8	15.5	0.0	16.7	-

[a] T_I = Tirol I (Tulfes 900 m).
T_{II} = Tirol II (different sampling locations above 1000 m).
St = Styria (sampling locations between 300 and 560 m).
V = Vienna (about 200 m).
Cs = Czechoslovakia (mixed collection).

TABLE III. CROSSING STUDIES: AVERAGE FERTILITY, FECUNDITY AND MATING FREQUENCY OF THE F_1 GENERATION

Crossing symbols [a]	No. of replicates	No. of eggs per replicate	No. of larvae per replicate	Mating frequency (%)
$T_I T_I \times T_I T_I$	6	15.5	11.0	50.0
$T_I St \times T_I St$	22	15.9	8.0	45.5
$T_I Cs \times T_I Cs$	13	0.5	0.4	15.4
$T_{II} St \times T_{II} St$	10	6.8	0.0	20.0
$StT_I \times StT_I$	7	0.4	0.0	0.0
StSt × StSt	12	4.9	3.5	41.7
$VT_I \times VT_I$	43	3.8	2.5	12.5
VCs × VCs	3	0.0	0.0	0.0
CsSt × CsSt	10	19.0	11.9	50.0

[a] T_I = Tirol I (Tulfes 900 m).
T_{II} = Tirol II (different sampling locations above 1000 m).
St = Styria (sampling locations between 300 and 560 m).
V = Vienna (about 200 m).
Cs = Czechoslovakia (mixed collection).

TABLE IV. OBSERVATIONS ON THE DEPENDENCE OF FERTILITY AND FECUNDITY OF CODLING MOTH ON THE DURATION OF THE TIME TO COMPLETE DIAPAUSE

Emergence period	No. of replicates	No. of eggs per replicate (%)	No. of larvae per replicate (%)	No. of matings per replicate
First half	122	72.3	84.3	0.26
Second half	112	27.7	15.7	0.23

REFERENCES

[1] DANILEVSKII, A.S., Photoperiodism and Seasonal Development of Insects, Oliver and Boyd, London (1965).
[2] BECK, S.D., Insect Photoperiodism, Academic Press, New York and London (1968).
[3] RUSS, K., Der Einfluss der Photoperiodizität auf die Biologie des Apfelwicklers (Carpocarpsa pomonella L.), Pflanzenschutzberichte 33 Sonderheft (1966) 27.
[4] SHEL'DESHOVA, G.G., Ecological factors determining distribution of codling moth, Laspeyresia pomonella L. (Lepidoptera, Tortricidae) in the northern and southern hemispheres, Entomol. Obozr. 46 (1967) 583 (English summary).
[5] WILDBOLZ, Th., RIGGENBACH, W., Untersuchungen über die Induktion und die Beendigung der Diapause bei Apfelwicklern aus der Zentral- und Ostschweiz, Mitt. Schweiz. Entomol. Ges. 42 (1969) 58.
[6] SCHNEIDER, F., Auftreten und Bekämpfung einiger Obstbaumschädlinge in Syrien, Z. PflKrankh. PflPath. PflSchutz 64 (1957) 613.
[7] RUSS, K., RUPF, O.,"Investigations on the generation dynamic of Laspeyresia pomonella (L.)", Proc. FAO Conference on Ecology in Relation to Plant Pest Control Rome, 1973, 47.

DISCUSSION

I.A. KANSU: Could you tell us more about your device for recording the effective temperature sum?

P. FISCHER-COLBRIE: I shall ask Mr. Russ, one of the co-authors, to answer this.

K. RUSS: We used an effective temperature recorder, constructed by Mr. Zyslovsky at the Bundesanstalt für Pflanzenschutz.

I.A. KANSU: Did you compare the sums of the effective temperatures and of the day-degrees obtained using the developmental threshold and the thermal constant of the hyperbolic formula B?

K. RUSS: We compared the effective temperature with the natural development in the field.

I.A. KANSU: Was the mating frequency of specimens of different origins significant or not?

P. FISCHER-COLBRIE: Unfortunately mating frequency was very low in crossings between moths of all origins, as is usual when diapausing larval material is brought into the laboratory from the field.

T.P. BOGYO: How do you distinguish a "second generation" insect from a late-emerging first-generation one?

P. FISCHER-COLBRIE: The use of cardboard rings prevents the capture of late-emerging first-generation insects. It would not be possible to distinguish between the two if we simply used ultra-violet or pheromone traps.

B. NAGY: The use of sex traps is becoming very extensive in codling moth research, but we do not know enough about the factors governing the attractivity of these traps. Have you performed any experiments in this direction?

P. FISCHER-COLBRIE: No.

I.A. MAYAS: You said that there was a difference in the times of commencement of codling moth capture by sex traps and light traps. Does the population level have any bearing on this difference?

P. FISCHER-COLBRIE: Yes, it does. In the early season, when the population is still low, the pheromone trap has a better chance of catching the moths because of its higher effectiveness. On the other hand, temperature could also play a certain role here, because the pheromone trap is able to catch moths during their flight peak at sunset, and the temperature in the early season might be just enough for flight activity at that time but too low during the night, when the u.v. light trap is working.

I.A. MAYAS: In Lebanon we have found that when the moths start to be attracted to the traps early in the season, there is a rapid increase in the number of males caught by the synthetic sex-lure trap; we have attributed this to protandry which is well-known in L. pomonella. Have you made the same observation?

P. FISCHER-COLBRIE: Yes.

G.R. RENS: How homogeneous were the samples you took from the different regions? Do you know whether genetic characteristics will change by selection, i.e. if strains are transferred to a different environment(altitude

P. FISCHER-COLBRIE: Samples from high altitudes were possibly not homogeneous, because the sampling points were scattered all over the Tirol. Our studies in 1974 are being concentrated in a few selected representative orchards. We have not investigated whether genetic characteristics are affected when strains are transferred to different environments but we are planning to make such a study.

INFLUENCE OF PARASITES AND PATHOGENES ON THE HIBERNATING POPULATION OF CODLING MOTH (Laspeyresia pomonella L.) IN AUSTRIA

K. RUSS, O. RUPF
Bundesanstalt für Pflanzenschutz,
Vienna, Austria

Abstract

INFLUENCE OF PARASITES AND PATHOGENES ON THE HIBERNATING POPULATION OF CODLING MOTH (Laspeyresia pomonella L.) IN AUSTRIA.
   During investigations covering a period of three years (1970-72), the following parasites and fungus diseases of hibernating codling moth larvae were identified:

Ichneumonidae:	Pristomerus vulnerator Grav.
	Trichomma enecator Rossi
	Ephialtes caudatus Ratz.
Braconidae:	Ascogaster quadridentatus Wesm.
	Microdus rufipes Nees.
Tachinae:	Elodia tragica Meig.
Perilampidae:	Perilampus tristis (hyperparasite)
Fungus diseases:	Beauveria bassiana (Bals.) Vuill.
	Paecilomyces farinosus (Dicks. ex Fr.) Brown et Smith
	Verticillium lecanii (Zimm.) Viegas
	Hirsutella sp.
	Tilachlidium sp.

The occurrence of different species at various altitudes and habitats is shown. In addition the fluctuation of the population density of the separate species of parasites as well as total parasitism were investigated.

INTRODUCTION

During our investigations on the ecology of codling moth covering several years, studies on the degree of parasitism and fungus diseases of hibernating codling moth larvae were carried out. These observations are closely connected with our efforts to integrate the sterile-male technique in pest control management against codling moth.

On the basis of repeated investigations over several years it was possible to obtain a survey of parasites and fungus diseases attacking hibernating larvae of codling moth in different regions of Austria.

Complete data from the years 1970, 1971 and 1972 are now available. A total of 3000-5000 hibernating larvae were sampled each year in corrugated-cardboard rings in various regions and at varying altitudes in Austria. The rings were mounted in the middle of May and dismantled in the middle of October in each year of investigation.

An average of about 120 localities per year in altitudes between 100 and 1200 m above sea-level were chosen for these investigations. After the caterpillars were taken into the laboratory, they were first investigated for their diseases and the degree of parasitism, and then either stored in a

cooling room (+ 5°C - + 7°C) or incubated and reared at + 25°C. This
procedure enabled hatched parasites and fungus diseases to be identified.
The work on fungus diseases and their identification was carried out by E.
Müller-Kögler, of the Biologische Bundesanstalt für Pflanzenschutz, Institut
für Biologische Schädlingsbekämpfung, Darmstadt, Federal Republic of
Germany. A detailed report on the type of pathogenes and their occurrence
will shortly be published.

RESULTS

From the rearing of the hibernating codling moth larvae the following
were found:

(a) Fungus diseases: Beauveria bassiana (Bals.) Vuill.
                        Paecilomyces farinosus (Dicks. ex. Fr.)
                                Brown et Smith
                        Verticillium lecanii (Zimm.) Viegas
                        Hirsutella sp.
                        Tilachlidium sp.

In addition to the fungus diseases mentioned above a number of other
fungi were found. However it seems that most of them might be sapro-
phytic, which would be understandable as part of the sampled larvae material
at the time of identification of fungi was already dead. Evidence of
Aspergillus sp. and Fusicladium sp. in particular may be mentioned. It is
also possible that some of these species may be really pathogenic.
Unfortunately more detailed investigations of this kind were not possible
because of lack of time.

(b) Parasites and hyperparasites:

    Ichneumonidae: Pristomerus vulnerator Grav.
                        Trichomma enecator Rossi
                        Ephialtes caudatus Ratz.
    Braconidae: Ascogaster quadridentatus Wesm.
                        Microdus rufipes Nees.
    Tachinae: Elodia tragica Meig.
    Perilampidae: Perilampus tristis Mayr. (hyperparasite)

The identification of the parasites was carried out by K.P. Carl, of the
Commonwealth Institute of Biological Control, Delemont, Switzerland.

Altitudes favoured by various parasites of hibernating codling moth larvae

A comparison of parasites from different altitudes showed the presence
of all parasites at all altitudes, but a cautious interpretation of the results
shows a trend towards favoured altitudes. Tables I-III provide the data.

(a) Ascogaster quadridentatus

Tables I-III show that Ascogaster was more or less frequent at all
altitudes. However, the percentage of its presence in codling moth larvae
at different altitudes fluctuated from year to year and from altitude to

TABLE I. PERCENTAGE OF PARASITISM BY DIFFERENT PARASITES AT VARIOUS ALTITUDES IN 1970

Altitude (m)	No. of larvae sampled	Percentage of parasitism by[a]						
		Asco.	Prist.	Trich.	Mic.	El.	Eph.	Peril.
Up to 300 m	2262	36.5	30.8	27.4	12.5	0.0	0.0	12.6
Up to 500 m	1446	12.4	1.6	3.4	53.1	21.4	0.0	13.1
Up to 700 m	988	30.9	13.5	5.7	0.0	0.0	0.0	0.0
Up to 900 m	449	20.0	54.0	63.4	34.3	78.5	0.0	74.2
Up to 1200 m	135	0.0	0.0	0.0	0.0	0.0	0.0	0.0

[a] Asco. = Ascogaster quadridentatus
Prist. = Pristomerus vulnerator
Trich. = Trichomma enecator
Mic. = Microdus rufipes

El. = Elodia tragica
Eph. = Ephialtes caudatus
Peril. = Perilampus tristis

TABLE II. PERCENTAGE OF PARASITISM BY DIFFERENT PARASITES AT VARIOUS ALTITUDES IN 1971

Altitude (m)	No. of larvae sampled	Percentage of parasitism by[a]						
		Asco.	Prist.	Trich.	Mic.	El.	Eph.	Peril.
Up to 300 m	2013	12.7	10.1	16.3	16.3	20.0	0.0	2.3
Up to 500 m	2009	4.3	7.5	13.9	16.8	0.0	0.0	6.53
Up to 700 m	621	6.0	12.8	11.2	70.7	80.0	0.0	21.4
Up to 900 m	64	9.7	31.1	10.9	0.0	0.0	0.0	52.2
Up to 1200 m	191	67.1	38.2	47.5	0.0	0.0	0.0	17.4

[a] Asco. = Ascogaster quadridentatus
Prist. = Pristomerus vulnerator
Trich. = Trichomma enecator
Mic. = Microdus rufipes

El. = Elodia tragica
Eph. = Ephialtes caudatus
Peril. = Perilampus tristis

TABLE III. PERCENTAGE OF PARASITISM BY DIFFERENT PARASITES
AT VARIOUS ALTITUDES IN 1972

Altitude (m)	No. of larvae sampled	Percentage of parasitism by [a]						
		Asco.	Prist.	Trich.	Mic.	El.	Eph.	Peril.
Up to 300 m	1390	27.1	26.9	55.0	20.4	0.0	29.1	47.0
Up to 500 m	932	20.8	13.7	14.6	14.0	100.0	0.0	17.9
Up to 700 m	584	24.9	29.7	24.6	60.1	0.0	70.8	35.2
Up to 900 m	402	27.1	29.5	5.6	5.2	0.0	0.0	18.5
Up to 1200 m	18	-	-	-	-	-	-	-

[a] Asco. = Ascogaster quadridentatus   El.   = Elodia tragica
   Prist. = Pristomerus vulnerator   Eph. = Ephialtes caudatus
   Trich. = Trichomma enecator   Peril. = Perilampus tristis
   Mic.   = Microdus rufipes

altitude. In general Ascogaster favours altitudes up to about 300 m although
in some years, especially in 1971, its occurrence was heavy above 900 m
(Tirol).

(b) Pristomerus vulnerator

Compared with Ascogaster, this species prefers higher altitudes.
During the investigations this species occurred most frequently above 700 m.
However, Pristomerus is present at all altitudes and plays a certain role
at lower altitudes.

(c) Trichomma enecator

Trichomma was found in 1970 and 1971 with greater frequency above
700 m, as well as above 900 m. However, in 1972 this species occurred
very frequently up to 300 m, although it was also frequently found at
altitudes up to 700 m.

(d) Microdus rufipes

The samples of this parasite species did not show a correlation with
any particular altitude. However it was striking that the larval material
was regularly more frequently infested between 500 m and 700 m.

(e) Ephialtes caudatus and Elodia tragica

Both these parasites were extremely rare in the sampled codling moth
material. Therefore it was not possible to check for a tendency to favour
certain altitudes. Ephialtes in particular was not present in the hibernating
material of the years 1970 and 1971. Elodia was found only rarely during
all three years of investigations.

(f)   Perilampus tristis (hyperparasite)

This hyperparasite species follows the same pattern as its main hosts
Ascogaster and Pristomerus.

Distribution of the various parasites in Austria

The distribution of the parasites doubtless coincides with the occurrence
of their hosts. There are also other factors to be considered; in particular
the influence of different culture and plant-protection methods on the host
plants of the codling moth is very important for the population density of the
parasites.

TABLE IV.   PERCENTAGE OF PARASITISM DURING 1970-72 IN
DIFFERENT AREAS OF AUSTRIA

Federal county	Percentage of parasitism		
	1970	1971	1972
Vienna	7.95	7.73	13.69
Lower Austria	6.73	8.03	15.97
Upper Austria	1.61	10.00	- a
Salzburg	0.88	0.34	- a
Tirol	4.63	18.07	7.98
Vorarlberg	3.28	0.00	5.43
Carinthia	-	2.20	- a
Styria	1.23	1.78	1.70
Burgenland	2.16	9.97	16.18

a Data are not representative because of lack of sufficient replications of samples.

TABLE V.   COMPOSITION OF PARASITES DURING 1970-72 IN
AUSTRIA

Year of investi-gation	Total No. of larvae sampled	Percentage composition of parasites[a]							
		Asco.	Prist.	Trich.	Mic.	El.	Eph.	Peril.	Total
1970	5280	47.8	25.1	11.0	4.9	1.2	0.0	9.8	3.0
1971	4898	40.7	19.7	27.4	2.8	0.4	0.0	8.6	8.2
1972	3226	17.4	21.0	36.2	9.2	0.2	0.5	15.1	11.6

a Asco. = Ascogaster quadridentatus   El.   = Elodia tragica
   Prist. = Pristomerus vulnerator   Eph.   = Ephialtes caudatus
   Trich. = Trichomma enecator   Peril. = Perilampus tristis
   Mic.   = Microdus rufipes

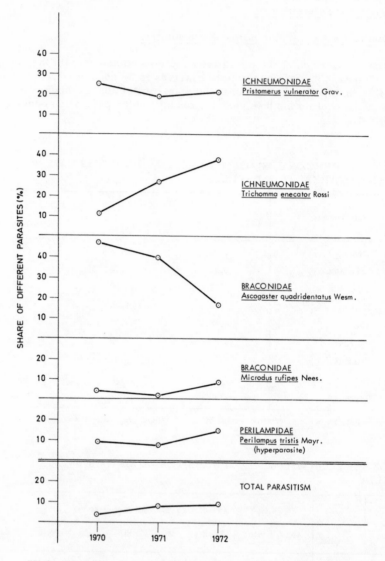

FIG.1. Composition of parasites of <u>Laspeyresia</u> <u>pomonella</u> L. in Austria (1970-72).

A comparison of the frequency of parasites occurring in different apple-growing areas of Austria (e. g. in the federal counties of Austria) showed different percentages of larval infestation (Table IV). During the years investigated Styria had an extraordinary low percentage of parasitized larvae. On the other hand, a higher parasitism was found in the Vienna area, Tirol and in Lower Austria in particular. The low percentage of parasitism in Styria, the most important apple-growing area in Austria, is certainly due to very intensive plant protection measures by which the parasites are controlled.

Composition of parasites during the years 1970-72 in Austria

Extreme fluctuations in the composition of the sampled parasites were noted from year to year. Table V and Fig. 1 summarize the results of these observations. The data show a slight decrease of Pristomerus from 1970 to 1971 and an increase in 1972. On the other hand, the percentage of Trichomma increased significantly from year to year.

The main predator of hibernating codling moth larvae in Austria, Ascogaster, decreased drastically during 1972. The other parasites showed a population trend similar to Pristomerus and Trichomma.

Total parasitism of codling moth larvae during 1970-72 in Austria

The total parasitism of hibernating codling moth larvae showed an increasing tendency during the years of investigation. This increase was mainly caused by the drastic increase in the population density of Trichomma enecator, one of the main parasites of hibernating codling moth larvae. As Table V shows, the average percentage of parasitized larvae was 3.0%, 8.2% and 11.6% in 1970, 1971 and 1972 respectively.

DISCUSSION

J. L. MONTY: Could the decrease in the abundance of Ascogaster be due to the activity of the hyperparasites?

K. RUSS: The decrease of Ascogaster is due to several factors, including Perilampus.

B. NAGY: Could you tell us how parasitism varies with habitat, for example, how does the degree of parasitism in a well-treated orchard compare with that in neglected trees?

K. RUSS: The amount of parasitism and fungus disease depends on whether and how pest control is applied in the target orchard. For example, in Styria, where there are numerous well-managed orchards, we did not find so much parasitism or disease.

B. NAGY: Did you find any evidence of the granulosis virus in codling moths sampled in the field?

K. RUSS: Yes, we sometimes found granulosis in field samples of codling moth larvae.

T. P. BOGYO: We have found that, even in unmanaged orchards, parasite and disease infestation is low when the codling moth population is low. Is this also your experience?

K. RUSS: Yes, especially in Styria.

SURVEY OF SYNTHETIC DIETS
FOR CODLING MOTHS

B. BUTT *
Division of Research and Laboratories,
International Atomic Energy Agency,
Vienna

Presented by D.A. Lindquist

Abstract

SURVEY OF SYNTHETIC DIETS FOR CODLING MOTHS.
　　Codling moth larval diets used at present can be classified as: (1) Wheat germ-sawdust; (2) Wheat germ-agar; (3) Soya-starch; (4) Bean; and (5) Calcium alginate gel. Pupae per kilogram of diet range from 76 to 600 and the range of ingredient costs is US $1.06 to 25.00 and for labour US $52 to 112.57 per 1000 pupae. Pupal weight ranges from 16 to 52.2 mg for females and 16 to 39.9 mg for males. Larval development time ranges from 10 to 25 days and rearing temperatures from 24° to 28°C. A survey of sources and costs of ingredients is included as well as a list of publications on codling moth rearing.

　　Codling moths (Laspeyresia pomonella L.) have been reared in laboratories on apples for many years [1]. There are problems in obtaining insecticide-free apples and storing them for one year. Since only thinning apples are normally used, one must plan the apple requirements one year in advance. A loss in storage may result in a reduction or cancellation of a programme. Because of these problems and the cost of buying, collecting and storing thinning apples, several synthetic diets have been developed for codling moth. Five such diets were compared with apples by Hathaway and co-workers [2]. The author has made a survey of synthetic codling moth diets and the results are reported here; Annex I is a list of persons responding to the survey.

　　The most widely used synthetic diets for mass-rearing codling moths are based on the wheat germ-sawdust diet developed by Brinton and co-workers [3] (Table I). This diet is cheap and easily prepared. In tray rearing, air is passed over the surface of the medium to prevent collection of moisture and the subsequent growth of harmful fungi.

　　Wheat germ-agar diets (Table II) have also been used to mass-produce codling moths [4]. The moisture problem in tray rearing is similar to that in sawdust-based diets. The wheat germ-agar diets are generally more expensive and more difficult to prepare; however, costs can be reduced. For an example, Wearing (personal communications) may be able to reduce the cost of producing 1000 adults from NZ$11.03 to 7.80 by excluding vitamin B_{12}.

---

* Present address: Entomology Research Division, US Department of Agriculture, Agricultural Research Service, 3706 West Nob Hill Boulevard, Yakima, Washington 98902, United States of America.

TABLE I. INGREDIENTS IN WHEAT GERM-SAWDUST DIETS REPORTED BY VARIOUS WORKERS

Ingredients	Grams/kg			
	Anderson	Brinton	Mani Charmillot	Wearing
Water	626.0	717.0	660.0	644.9
Sawdust	80.0	68.9	80.0	82.9
Whole wheat flour	120.0	98.6	120.0	121.1
Bran, wheat	21.0	18.0	20.0	
Sucrose	32.5	26.9	30.0	33.0
Casein	32.5	26.9	30.0	33.0
Wheat germ	10.8	9.0	16.0	11.0
Bran flakes				22.0
Cellulose				13.9
Paper pulp			15.0	
Wood pulp	16.3	12.4		
Cholesterol				0.2
Wesson's salts	7.6	6.2	6.8	7.6
Sorbic acid	1.1	0.9	0.9	0.9
Torula yeast	37.0			
Sunflower oil (pressed at low temperature)			4.0	
Ethyl alcohol				7.4
Linoleic acid				0.6
Nipagin-sodium				0.5
Aureomycine			0.088	
Chlortetracycline HCl	0.11	0.14		
Ascorbic acid	4.4	3.58	5.5	8.8
Citric acid	10.8	9.0	10.0	11.0
Choline chloride		1.0	1.0	1.0
Alpha-tocopherol			0.15	
Niacinamide		0.01	0.01	0.01
Calcium phanthenate		0.01	0.01	0.01
Riboflavin		0.005	0.005	0.005
Thiamine HCl		0.0025	0.0025	0.0025
Pyridoxin HCl		0.0025	0.0025	0.0025
Folic acid		0.0025	0.0025	0.0025
Vitamin B_{12} (0.1%)		0.002	0.02	0.002
Biotin (2%)		0.001	0.01	0.01

TABLE II. INGREDIENTS IN WHEAT GERM-AGAR DIETS REPORTED BY VARIOUS WORKERS

Grams/kg

Ingredients	Bulyginskaya	Falcon	Anderson	Howell	Pristavko and Yanishevskaya		
					No. 9	No. 16	No. 18
Water	To make 1 kg	837.8	847.0	861.0	To make 1 kg	To make 1 kg	To make 1 kg
Casein	40.0	34.1	34.5	27.1	34	34	32
Yeast	15.0				20	20	
Brewer's yeast							
Sucrose	40.0	25.9	26.4	27.1	25	32	24
Agar	15 - 20	24.3	18.6	10.7	24	24	23
Apple purée					500	100	80
Apple seeds					Three per container	10	
Linseed oil	5.0						
Cellulose (paper filler)							
Cellulose (powdered)					5	6	5
Alphacel		4.9	4.9	4.3			
Carboxy methyl cellulose							
Wheat germ	60 - 80	29.5	29.5	46.5	28	36	27
Ascorbic acid	8.0	8.2	4.2	3.2	8	7	8
Sorbic acid	0.9	7.7	0.65	1.0			

TABLE II. (cont.)

Ingredients	Grams/kg				Pristavko and Yanishevskaya		
	Bulyginskaya	Falcon	Anderson	Howell	No. 9	No. 16	No. 18
Potassium hydroxide		5.4	4.9				
Wesson's or mineral salts		9.7	9.8	7.7	9	12	9
Methyl-p-hydroxybenzoate		1.5	1.5		2	2	2
Ethyl alcohol					20	20	20
Glycine					2	1	
Cystine					2	1	
Cholesterol					2		
Chlortetracycline HCl 5.5%			0.14				
Chlorox				0.4			
Methyl parasept				1.1			
Formaldehyde		1.5		0.3			
Vitamin stock		9.7	9.8	9.7			
Choline chloride	0.005				1	1	1
Nicotinic acid	0.01				0.12	0.12	0.12
Calcium pantothenate	0.012				0.004	0.004	0.004
Pyridoxine hydrochloride	0.0018				0.006	0.006	0.006
Biotin	0.0005						
Citric acid	9.0						

Grams/kg

Ingredients	Bulyginskaya	Falcon	Anderson	Howell	Pristavko and Yanishevskaya		
					No. 9	No. 16	No. 18
Folic acid	0.001				0.02	0.02	0.02
Riboflavin					0.012	0.012	0.012
Inositol					0.1		0.1
Alpha-tocopherol					0.15	0.15	0.15
Cyanoco balamin					0.00005	0.00005	0.00005
Thiamine					0.012	0.012	0.012

TABLE III.   INGREDIENTS IN BEAN DIET REPORTED BY RUSS

Ingredients	Grams/1 kg
Water	706.2
Beans (soaked)	235.4
Brewer's yeast	35.3
Ascorbic acid	3.5
Methyl-p-hydroxybenzoate	2.2
Sorbic acid	1.1
Formaldehyde (40%)	2.2
Agar	14.1

TABLE IV.   INGREDIENTS IN SOYA-STARCH DIET REPORTED BY HOWELL

Ingredients	Grams/kg
Water	781.0
Starch	18.0
Soya	108.0
Wheat germ	36.0
Invert sugar	18.0
Mineral salts	1.2
Apple pomace	12.0
Methyl parasept	0.7
Sorbic acid	6.0
Vitamin mixture	6.0
Ascorbic acid	1.9
Formaldehyde	0.3
Propylene glycol	7.2
Cellulose fibre	1.8

TABLE V. INGREDIENTS IN CALCIUM ALGINATE GEL DIET
REPORTED BY NAVON

Ingredients	Grams/kg
Water	560.0
Sodium alginate	11.6
Autoclaved soy-beans	116.6
Vitamin-free casein	46.6
Sucrose	46.6
Torula yeast	46.6
Nipagin	0.9
Sorbic acid	0.4
Calcium phosphate (dibasic)	2.9
Ascorbic acid	0.4
Autoclaved peanut shells	153.2
Acetic acid (glacial)	14.2

S. Hatmosoewarno of this laboratory (personal communications) found
that a bean diet developed by Shorey and Hale [5] for noctuid moths was
satisfactory for codling moths, and this diet is being used for small-scale
rearing by Russ (personal communications) (Table III).

A diet with a soya-starch base was developed by Howell [6] (Table IV).
This diet appears to be very cheap, but detailed work has not been completed. A unique diet using a calcium alginate gel was developed by Navon
and Moore [7] (Table V).

Costs and productivity of various diets are presented in Table VI. All
information is not directly comparable because some data are for adults
rather than pupae. Also pupal weight varies with age. M.D. Proverbs, of
the Research Station, Canada Department of Agriculture, Summerland,
B.C., Canada (personal communications), has found that male pupal weights
drop from 29.6 to 26.5 mg in 8 days (167 mg adults) and female from 35.5
to 31.4 mg in 9 days (23.6 mg adults).

Annex II gives a list of mixers used by those answering the survey, and
Annex III a list of publications on codling moth rearing.

TABLE VI. COSTS AND PRODUCTIVITY OF VARIOUS SYNTHETIC CODLING MOTH DIETS[a]

Diet and reporter	Pupae/kg diet	Cost/1000 pupae		Pupal weight (mg)	Larval development (days)	Rearing temperature (°C)
		Ingredients	Labour			
Wheat germ-sawdust diet						
Anderson	81	$ 3.51 (Aust) / $ 5.23 (US)	$ 76.00 (Aust) / $112.57 (US)	-	29.5	26
Brinton	121-A	$ 1.58 (US)	$ 0.52 (US)	♀ 28.0 A / ♂ 18.5 A	32 - EA	28
Charmillot	100	7.31 (S.Fr.) / $ 2.12 (US)	-	♀ 42.41 ± 6.97N / ♂ 34.47 ± 4.79N	28	25.5 ± 0.5
Mani	150 - 200	3.65 - 4.80 (S.Fr.) / $ 1.06 - 1.39 (US)	15 - 20 (S.Fr.) / $ 4.35 - 5.80 (US)	♀ 40.4 N.D. / ♂ 32.2 N.D. / ♀ 32.2 D. / ♂ 27.6 D.	30 - A	27 ± 1
Wearing	77-A	$11.03 (NZ) / $15.36 (US)	1 man part-time	♀ 37.2 24.7-A / ♂ 28.9 16.8-A	31 - L-A	28
Wheat germ-agar diet						
Anderson	238	$ 4.84 (Aust) / $ 7.97 (US)	$67.00 (Aust) / $97.96 (US)	-	24.7	24
Bulyginskaya	230 - 250	-	-	♀ 21 - 23 A / ♂ 14 - 16 A	15 / 30 - 32 EA	25 ± 1
Falcon	150	$10.78 (US)	$21.00 (US)	16.6	10 - 14	25 ± 4
Hathaway	100 - 150	$24.00 (US)	$18.00 (US)	♀ 41.2 / ♂ 30.7	♀ 32.9 L-A / ♂ 32.6 L-A	27 ± 2
Howell	100 L	$ 2.33 (US)	$ 1.50 - 2.00 (US)	♀ 44.5 / ♂ 34.7	15 - 24 / Av. 18.9	28 ± 3

Diet and reporter	Pupae/kg diet	Cost/1000 pupae		Pupal weight (mg)	Larval development (days)	Rearing temperature (°C)
		Ingredients	Labour			
Soya-starch diet						
Howell		$ 1.35 (US)		♀ 52.2		
Bean diet						
Russ	206	328.40 (ASch) $ 15.24 (US)	35 hours	♀ 28.0 ♂ 23.8	25	25 ± 1
Calcium alginate gel diet						
Navon	250				16 ± 1	27 ± 2

a A - Adult.
EA - Egg to adult.
D - Diapause.
L - Larvae.
LA - Larvae to adult.
N - Apple.
ND - Non-diapause.

REFERENCES

[1] HAMILTON, D.W., HATHAWAY, D.O., "Codling moths", Insect Colonization and Mass Production (SMITH, C.N., Ed.), Academic Press, New York (1966) 339.

[2] HATHAWAY, D.O., CLIFT, A.E., BUTT, B.A., Development and fecundity of codling moths reared on artificial diets or immature apples, J. Econ. Entomol. 64 (1971) 1088.

[3] BRINTON, F.E., PROVERBS, M.D., CARTY, B.E., Artificial diet for mass production of the codling moth, Carpocapsa pomonella (Lepidoptera: Olethreutidae), Can. Entomol. 101 (1969) 577.

[4] HOWELL, J.F., CLIFT, A.E., Rearing codling moths on an artificial diet in trays, J.Econ. Entomol. 65 (1972) 888.

[5] SHOREY, H.H., HALE, R.R., Mass rearing of the larvae of nine noctuid species on a simple artificial medium, J. Econ. Entomol. 58 (1965) 522.

[6] HOWELL, J.F., Rearing the codling moth on a soya, wheat germ, starch medium, J. Econ. Entomol. 65 (1972) 636.

[7] NAVON, A., MOORE, I., Artificial rearing of the codling moth Carpocapsa pomonella on calcium alginate gel, Entomophaga 16 (1971) 381.

ANNEX 1

NAMES AND ADDRESSES OF CONTRIBUTORS TO SURVEY ON SYNTHETIC CODLING MOTH DIETS

1. Margot Anderson
    Division of Entomology
    P.O. Box 1700
    Canberra City
    A.C.T. 2601
    Australia

2. F.E. Brinton
    Research Station
    Canada Department of Agriculture
    Summerland
    British Columbia
    Canada

3. M.A. Bulyginskaya
    All-Union Science Institute
        of Plant Protection
    Leningrad
    USSR

4. P.J. Charmillot
    Federal Agricultural Research Station
    Château de Changins
    CH-1260 Nyon
    Switzerland

5. Louis A. Falcon
    Department of Entomological Sciences
    University of California
    Berkeley
    California
    USA

6. D.O. Hathaway
    ARS, USDA
    3706 West Nob Hill
    Yakima
    Washington 98902
    USA

7. J. Franklin Howell
    ARS, USDA
    3706 West Nob Hill
    Yakima
    Washington
    USA

8. E. Mani
    Swiss Federal Research Station for
        Arboriculture and Horticulture
    CH-8820 Wadenswil
    Switzerland

9. A. Navon
    Division of Entomology
    The Volcani Center of Agricultural Research
    P.O.B. 6
    Bet Dagan
    Israel

10. V.P. Pristavko
    Dept. of Biophysics
    Ukrainian Research Institute for
        Plant Protection
    Vasilkovskaia 33
    Kiev, 252127
    USSR

11. M. D. Proverbs
    Research Station
    Canada Dept. of Agriculture
    Summerland
    British Columbia
    Canada

12. K. Russ
    Institute for Plant Protection
    Trunnerstrasse 5
    A-1020 Vienna
    Austria

13. C.H. Wearing
    Entomology Division
    Department of Scientific and Industrial
        Research
    Private Bag, Auckland
    New Zealand

ANNEX 2

MIXERS USED TO PREPARE CODLING MOTH DIETS

Workers	Type of diet	Size of batch	Machine
Anderson	Wheat germ-sawdust	5.9 kg	Kenwood Peerless light industrial dough mixer
Anderson	Wheat germ-agar	0.914 kg	Kenwood Chef domestic vitamizer
Brinton	Wheat germ-sawdust	120.00 kg	Steam-jacketed kettle fitted with two sets of counter-rotating paddles (130 litres)
Charmillot	Wheat germ-sawdust	10.00 kg	Kenwood Peerless
Falcon	Wheat germ-agar	4 litre	Waring commercial blender, 4 litres
Hathaway	Wheat germ-agar	540 g	Waring commercial blender, 4 litres
Howell	Wheat germ-agar	160 litre	Eppenback homogenizer-mixer 3560 rev/min turbine speed; in steam jacketed kettle
Howell	Soya-wheat starch	8 litre	Waring commercial blender (4 litres) plus commercial type food mixer (28 litres)
Mani	Wheat germ-sawdust	45-50 kg	Steam jacketed kettle fitted with a planetary paddle "Artofex" RG-6 (60-801)
Navon	Calcium-alginate gel	860 g	Kenwood mixer/blender
Russ	Bean	2-4 kg	Waring commercial blender (4 litres)
Wearing	Wheat germ-sawdust	6538 g	Kenwood Major cake mixer

ANNEX 3

PUBLICATIONS RELATED TO CODLING MOTH REARING

BATISTE, W.C., OLSON, W.H., Codling moth mass production in controlled environment rearing units, J. Econ. Entomol. 66 (1973) 383, 66, 2.

BODE, W.M., The Codling Moth Laspeyresia pomonella (Lepidoptera: Olethreutidae): effects of an introduced granulosis virus on a field population and laboratory rearing on artificial diets, Diss. Abst. International B 32 (1971) 348.

BRINTON, F.E., PROVERBS, M.D., CARTY, B.E., Artificial diet for mass production of the codling moth, Carpocapsa pomonella (Lepidoptera: Olethreutidae), Can. Entomol. 101 (1969) 577.

CHAWLA, S.S., HOWELL, J.F., HARWOOD, R.F., Surface treatment to control fungi in wheat germ diets, J. Econ. Entomol. 60 (1967) 307.

COUTIN, R., Feeding of larvae of Laspeyresia pomonella L. (Lepidoptera: Tortricidae) on artificial media, Société de Biologie 146 (1952) 516-20 (in French).

DICKSON, R.C., BARNES, M.M., TURZAN, C.L., Continuous rearing of the codling moth, J. Econ. Entomol. 45 (1952) 65.

EDEL-MAN, N.M., Principles of developing nutritive media for the pests of reproductive organs of fruit crops (The codling moth as an example), Zool. Zh. 49 (1970) 1240 (in Russian).

EIDE, P.M., Oviposition cage for obtaining large quantities of codling moth eggs, US Dept. Agric. Bureau Entomol. Plant Quarantine E.J. 73 (1936).

FARBAR, M.D., FLINT, W.P., Rearing codling moth larvae throughout the year, J. Econ. Entomol. 23 (1930) 41.

FARRAR, M.D., McGOVRAN, E.R., Rearing technique, Trans. Ill. St. Acad. Sci. 28 (1935) 245.

HAMILTON, D.W., Rearing the codling moth (Carpocapsa pomonella) for use in laboratory testing, Proc. Entomol. Soc. Am. 12 (1957) 92.

HAMILTON, D.W., HATHAWAY, D.O., "Codling moths", Insect Colonization and Mass Production (SMITH, C.N., Ed.), Academic Press, New York (1966) 339.

HATHAWAY, D.O., Inexpensive cardboard trays for mass rearing codling moth, J. Econ. Entomol. 60 (1967) 888.

HATHAWAY, D.O., CLIFT, A.E., BUTT, B.A., Development and fecundity of codling moths reared on artificial diets or immature apples, J. Econ. Entomol. 64 (1971) 10880.

HATHAWAY, D.O., LYDIN, L.V., BUTT, B.A., MORTON, L.J., Monitoring mass rearing of the codling moth, J. Econ. Entomol. 66 (1973) 390.

HATHAWAY, D.O., SCHOENLEBER, L.G., LYDIN, L.V., Codling moths: plastic pellets or waxed paper as oviposition substrates, J. Econ. Entomol. 65 (1972) 756.

HERIOT, A.D., WADDEL, D.B., Some effects of nutrition on the development of the codling moth, Scient. Agric. 23 (1942) 172.

HOSTOUNSKY, Z., ABRAHAM, J., DVORACEK, J., A simple dosimeter of synthetic food for caterpillars, Acta Entomol. Bohemoslav 69 (1972) 69.

HOUSE, H.L., SINGH, P., BATSCH, W.W., Artificial Diets for Insects: A Compilation of References with Abstracts, Information Bull. No. 7, Res. Inst. Can. Dept. of Agric., P.O. Box 367, Belleville, Ont. (1971).

HOWELL, J.F., Paraffin film to control dehydration of an artificial rearing medium for codling moth, J. Econ. Entomol. 60 (1967) 289.

HOWELL, J.F., Problems involved in rearing codling moth on diet in trays, J. Econ. Entomol. 64 (1971) 631

HOWELL, J.F., Rearing the codling moth on an artificial diet, J. Econ. Entomol. 63 (1970) 148.

HOWELL, J.F., Rearing the codling moth on a soya, wheat germ, starch medium, J. Econ. Entomol. 65 (1972) 636.

HOWELL, J.F., Modification of the artificial diet for codling moth to improve larval acceptance and production of moths, J. Econ. Entomol. 65 (1972) 57.

HOWELL, J.F., CLIFT, A.E., Rearing codling moths on an artificial diet in trays, J. Econ. Entomol. 65 (1972) 888.

HUBER, J., BENZ, G., SCHMID, K., A method for semi-synthetic nutrition of the codling moth, Experientia 28 (1972) 1260.

HUTT, R.B., WHITE, L.D., SCHOENLEBER, L.G., SHORT, R.E., Automatic collection of mass reared codling moths by phototaxic response and a chilled environment, J. Econ. Entomol. 65 (1972) 1525.

MOFFITT, H.R., HATHAWAY, D.O., Effects of three systems of laboratory collection on the vigour of adult codling moths, J. Econ. Entomol. 66 (1973) 374.

NAVON, A., Development of an Economical Medium for Mass Rearing of the Codling Moth (Carpocapsa pomonella): Final Report of a Coordinated Programme of Insect Control by Irradiation, Volcani Inst. Agric. Res., Bet Dagan (1968).

NAVON, A., MOORE, I., Artificial rearing of the codling moth Carpocapsa pomonella on calcium alginate gels, Entomophaga 16 (1971) 381.

PETRUSHOVA, N.I., KOROBITSIN, V.G., DOMANSKY, V.N., SOKOLOVA, D.V., Continuous Rearing of the Codling Moth under Laboratory Conditions, Bull. of the Nikita Botanical Gardens No. 1 (1968) (in Russian).

PETRUSHOVA, N.I., PTITSYNA, N.V., SOKOLOVA, D.V., GRESS, P.Ya., DOMANSKY, V.N., Methods of Laboratory Rearing Laspeyresia pomonella L. on Fruit and Artificial Nutrient Media, Yalta (1971) (in Russian).

PRISTAVKO, V.P., Optimal control of insect rearing and radiation sterilisation, Abstract 14, International Congress of Entomology, Canberra (1972) 229.

PRISTAVKO, V.P., DEGTYAREV, B.G., Effect of temperature on the development and reproductive abilities of the codling moth Laspeyresia pomonella Lepidoptera: Tortricidae: Part I. Experimental data, Zool. Zh. 51 (1972) 517.

PRISTAVKO, V.P., DEGTYAREV, B.G., Effect of temperature on the development and reproductive abilities of the codling moth Laspeyresia Pomonella Lepidoptera: Tortricidae: Part II. Modelling and optimalization problems, Zool. Zh. 51 (1972) 994.

PRISTAVKO, V.P., YANISHEVSKAYA, L.V., Micro flora of the nutrient medium for breeding the codling moth and effectiveness of suppressing growth of microbes by some antiseptics, Biol. Nauki 15 7 (1972) 114.

PRISTAVKO, V.P., YANISHEVSKAYA, L.V., Methods in year-round rearing of codling moths on a synthetic nutrient medium, Zakhyst Roslyn 15 (1972) 22-6 (in Ukrainian).

PRISTAVKO, V.P., YANISHEVSKAYA, L.V., REZVATOVA, O.I., Some infectious diseases of the insectary reared codling moth Laspeyresia pomonella, Int. Colloq. Insect Pathol. Proc. 4 (1970) 262.

PROVERBS, M.D., LOGAN, D.M., A rotating oviposition cage for the codling moth Carpocapsa pomonella, Can. Entomol. 102 (1970) 42.

REDFERN, R.E., Concentrate media for rearing codling moth (Carpocapsa pomonella) (L.), and red-banded leafroller (Argyrotaenia velutinana Walker), Proc. Entomol. Soc. Am. 17 (1962) 126.

REDFERN, R.E., Concentrated medium for rearing the codling moth, J. Econ. Entomol. 57 (1964) 607.

ROCK, G.C., Aseptic rearing of the codling moth on synthetic diets: Ascorbic acid and fatty acid requirements, J. Econ. Entomol. 60 (1967) 1002.

ROCK, G.C., KING, K.W., Estimation by carcass analysis of the growth requirements of amino acids in the codling moth, Carpocapsa pomonella, Ann. Entomol. Soc. Am. 60 (1967) 1161.

SASANA, R.F., An improved oviposition cage for the codling moth, J. Econ. Entomol. 25 (1932) 140.

SENDER, C., Continuous rearing of the codling moth Carpocapsa (Laspeyresia) pomonella L. on a simple artificial diet, Ann. Zool. Ecol. Anim. 1 (1969) 321 (in French).

SENDER, C., Rearing of the codling moth on a new non-specific artificial medium, Ann. Zool. Ecol. Anim. 2 (1970) 93 (in French).

SHOMAKOV, E.M., "Difficulties and successes in the mass rearing of insects in the laboratory, and the possibility of autocidal control of some harmful species", Isotopes and Radiation in Entomology (Proc. Symp. Vienna,1967), IAEA, Vienna (1968) 219 (in Russian).

SINGH, P., Bibliography of Artificial Diets for Insects and Mites, Bull. 209, N.Z. Dept. of Scientific and Industrial Research (Entomol. Div.), Nelson, Wellington, N.Z. (1972).

SPEYER, W., Can the codling moth develop on leaves alone?, Land Forst 20 (1932) 183.

THERON, P.P., Studies on the Provision of Hosts for the Mass Rearing of Codling Moth Parasites, Union S. Africa Dept. Agr. Sci. Bull. 262 (1947).

VAN LEEUWEN, E.R., Increasing production of codling moth eggs in an oviposition chamber, J. Econ. Entomol. 40 (1947) 744.

YANISHEVSKAYA, L.V., Influence of some antiseptics on the diet sterility and development of the codling moth, Zakhyst Roslyn 10 (1964) 120.

DISCUSSION

R. GALUN: How do you account for the large variation in the prices of diets and labour at different laboratories?

D.A. LINDQUIST: It is probably associated with the different numbers of insects reared and the different uses to which they are put, i.e. mass-reared insects for release in the field or small-scale rearing for laboratory tests.

LARVAL DIETS FOR THE CODLING MOTH
(Laspeyresia pomonella L. , Lepidoptera: Olethreutidae)

S. HATMOSOEWARNO*, B. BUTT[†]
Division of Research and Laboratories,
International Atomic Energy Agency,
Vienna

Abstract

LARVAL DIETS FOR THE CODLING MOTH (Laspeyresia pomonella L., Lepidoptera: Olethreutidae).

Three diets for rearing codling moth (Laspeyresia pomonella L.) larvae were compared. Basically these diets were casein wheat-germ, bean and modified bean diets. Larvae were bred individually in small test tubes. Three successive generations were reared. Parameters used to measure success were larval survival, duration of larval stage, pupae obtained, adult emergence and egg production per female. Judging by the results and assuming data from individual rearing can be applied to large-scale rearing, all the diets would be satisfactory.

The sterile-male technique has been used experimentally at Yakima, Washington [1, 2], in integrated control of the codling moth (Laspeyresia pomonella L.). Since the sterile-male technique requires the release of large numbers of sterilized insects, mass rearing of the codling moth becomes important. Several authors [3-6] have reported on artificial diets for this insect.

The object of the present study was to compare three larval diets for codling moth rearing. Hathaway and co-workers [7] have compared several artificial diets with a diet of immature apples. However, these authors did not evaluate bean-based diets.

METHODS AND MATERIALS

Diet preparation

Three diets were evaluated: casein/wheat-germ [8], bean [9] and modified bean [10]. The formulae of the diets (Table I) are essentially the same as developed by the respective authors, except that local beans replaced the Pinto beans and brewer's yeast replaced torula yeast. Diet preparation was as described by the respective authors.

Rearing procedure

Larval rearing was conducted in 10 cm³ test tubes (6.0 × 1.5 cm). About 3 cm³ of diet was added to each; one newly hatched larva was placed in

---

* Present address: Yogyakarta Plantation Institute, Yogyakarta, Indonesia.
† Present address: Entomology Research Division, US Department of Agriculture, Agricultural Research Service, 3706 West Nob Hill Boulevard, Yakima, Washington 98902, United States of America.

TABLE I. FORMULAE OF CASEIN WHEAT GERM (CW), BEAN (B) AND MODIFIED BEAN (MB) DIETS FOR REARING CODLING MOTH LARVAE

Ingredients	Quantity (g or mlitre) of ingredient in indicated diets		
	CW	B	MB
Casein (vitamin-free)	49.9	0	0
Agar	28.5	20.0	0
Sucrose	57.6	0	0
Wheat germ	42.8	0	66.7
Yeast (brewer's)	42.8	50.0	41.7
H.W.G. Carageenan	0	0	15.3
Cholesterol	1.4	0	0
Alphacel	14.3	0	0
Wesson's salt	12.8	0	0
Beans (dry weight, soaked overnight)	0	333.3	91.7
Ascorbic acid	0	5.0	4.3
Sorbic acid	0	1.56	1.3
Nipagin	1.9	3.13	2.7
Aureomycin	0.0578	0	0
Sodium benzoate	2.6	0	0
Formaldehyde (40%)	0	3.13 mlitre	4.3 mlitre
Tetracyline HCl	0	0	0.0833
Vitamin mixture (Vanderzants)	12.8	0	5.0
Linseed oil (raw)	6.4 mlitre	0	0
Citric acid	11.6	0	0
KOH (4M)	3.5 mlitre	0	0
Water (de-mineralized)	1000.0 mlitre	1000.0 mlitre	1000.0 mlitre

TABLE II. COMPARISON OF CASEIN WHEAT GERM (CW), BEAN (B) AND MODIFIED BEAN (MB) DIETS FOR CODLING MOTH LARVAE REARING

225 larvae in individual vials per diet per generation

Diet	Average duration of larval stage (days)		Average yield pupae (%)	Average pupal weight (mg)		Sex ratio of pupae Males : Females	Average adult emergence (%)
	Males	Females		Males	Females		
Parent Generation							
CW	22.7	22.3	85	29.6	39.1	0.8	88.9
B	28.2	32.0	89	25.4	29.6	1.7	93.0
MB	26.0	28.5	82	31.5	37.7	0.8	79.2
F$_1$ Generation							
CW	27.0	27.0	65	29.4	43.2	1.8	85.4
B	29.4	33.4	82	26.1	32.1	0.6	80.9
MB	31.1	29.2	78	28.2	38.3	0.8	75.0
F$_2$ Generation							
CW	24.8	26.7	48	36.5	43.7	1.4	86.8
B	33.2	33.4	71	25.0	30.4	1.2	78.7
MB	28.1	30.0	86	27.8	32.9	0.9	84.2

TABLE III.   MATING PERCENTAGE AND EGG PRODUCTION OF CODLING
MOTHS REARED ON CASEIN WHEAT GERM (CW), BEAN (B) AND
MODIFIED BEAN (MB) DIETS[a]

Diet	Females mating[b] (%)	Egg production per female	Egg hatch (%)
	Parent Generation		
CW	46.7	36.8	57.9
B	40.0	42.1	53.1
MB	66.7	73.0	55.6
	F_1 Generation		
CW	60.0	73.3	60.7
B	60.0	62.4	79.0
MB	86.7	104.9	76.5
	F_2 Generation		
CW	70.0	45.7	79.2
B	80.0	69.2	65.9
MB	90.0	111.4	80.3

[a] Five pairs of moths per test.
[b] Based on the presence of spermatophores in the bursa copulatrix.

each container with the aid of a fine camel hair brush. After plugging with
sterile cotton, the vials were held at $26° \pm 1°C$, $59 \pm 1\%$ r.h. and a 17-hour
photoperiod.

Pupae were collected every other day and weighed on an electric balance
with a precision of 0.01 mg. After sexing by Peterson's [11] method, the
pupae were segregated in adult emergence boxes.

Fecundity test procedure

Five female and five male adults were paired in a plastic oviposition
cage (15.5 cm long, 9.5 cm diameter), lined with wax paper. A 3% sucrose
solution served as adult food. The wax paper was changed every other day
beginning when the first eggs were laid and continued until all females were
dead. The dead females were examined for presence of spermatophores.
The wax papers with eggs were surface sterilized for five minutes in 4%
aqueous formaldehyde followed by rinsing in water. After drying, eggs
were counted under a microscope. They were then incubated for 7 to 10
days for egg hatch data.

RESULTS AND DISCUSSION

Table II summarizes the data obtained on larval survival and development time, pupae obtained and adult emergence. Larval survival varied from 48-85% on casein wheat-germ diet, 71-89% on bean diet and 78-86% on modified bean diet.

The duration of the larval stage for both sexes in the F_1 generation was longer than in the P generation. The duration of larval development was consistently longer on the bean and modified bean diets than on the casein wheat-germ diet.

The percentage adult emergence varied from 75% to 93%. Most of the larvae reared on modified bean diet pupated inside the diet; removing the pupae during collection probably caused some induced adult emergence. The percentages of mated females, eggs per female and percentages of egg hatch are summarized in Table III. Judging by the presence of spermatophores in the bursa copulatrix, the proportion of females mating was consistently higher among those originating from the modified bean diet.

CONCLUSIONS

The data reported here indicate that rearing codling moth larvae on bean diet, which is cheaper and simpler than diets previously used, might be applied for large-scale breeding of this insect. Further investigations concerning mass rearing in trays should be conducted.

REFERENCES

[1] BUTT, B.A., HOWELL, J.F., MOFFITT, H.R., CLIFT, A.E., Suppression of populations of codling moths by integrated control (sanitation and insecticides) in preparation for sterile moth release, J. Econ. Entomol. 66 (1972) 411.

[2] BUTT, B.A., WHITE, L.D., MOFFITT, H.R., HATHAWAY, D.O., Integration of sanitation, insecticides, and sterile moth releases for suppression of populations of codling moths in the Wenas Valley of Washington, Environ. Entomol. 2 (1973) 208.

[3] BRINTON, F.E., PROVERBS, M.D., CARTY, B.E., Artificial diet for mass production of the codling moth Carpocapsa pomonella (Lepidoptera: Olethreutidae), Can. Entomol. 101 (1969) 577.

[4] HAMILTON, D.W., HATHAWAY, D.O., "Codling moths", Ch. 22, Insect Colonization and Mass Production (SMITH, C.N., Ed.), Academic Press, New York (1966).

[5] HOWELL, F.J., Rearing the codling moth on an artificial diet, J. Econ. Entomol. 63 (1970) 1148.

[6] HOWELL, F.J., Rearing the codling moth on a soya, wheat germ, starch medium, J. Econ. Entomol. 65 (1972) 636.

[7] HATHAWAY, D.O., CLIFT, A.E., BUTT, B.A., Development and fecundity of codling moths reared on artificial diets or immature apples, J. Econ. Entomol. 64 (1971) 1088.

[8] VANDERZANT, E.S., "Defined diets for phytophagous insects", Ch. 18, Insect Colonization and Mass Production (SMITH, C.N., Ed.), Academic Press, New York (1966).

[9] SHOREY, H.H., HALE, R.R., Mass rearing of the larvae of nine noctuid species on a simple artificial medium, J. Econ. Entomol. 58 (1965) 522.

[10] BURTON, R.L., Mass Rearing the Corn Earworm in the Laboratory, Rep. ARS, USDA 33-134 (1969).

[11] PETERSON, D.M., A quick method for sex determination of codling moth pupae, J. Econ. Entomol. 58 (1967) 576.

DISCUSSION

G.W. RAHALKAR: You have concluded that the bean diet is the most efficient, presumably because of the higher adult yield, but you are getting more females than males with this diet and this would certainly not be a good thing from the mass-rearing point of view.

S. HATMOSOEWARNO: Yes, I think that we shall have to carry out further experiments with more replicates.

G.W. RAHALKAR: In Table III, the fecundity and fertility for all three diets appear considerably lower than normal. Can you comment on this?

S. HATMOSOEWARNO: Unfortunately my study did not include a rearing experiment on the natural host, i.e. apples, so I am not certain of the normal fecundity and fertility of this insect.

E. BOLLER: What exactly is the diet ingredient H.W.G. Carageenan in Table I?

S. HATMOSOEWARNO: It is a thickening agent.

H. LEVINSON: Diets B and MB in Table I contain relatively large amounts of antimicrobial agents (sorbic acid, Nipagin (methyl-p-hydroxy benzoate), formaldehyde, and diet MB also contains tetracycline HCl). Is this burden to the developing codling moth larvae really essential?

S. HATMOSOEWARNO: The aim of my study was to gain experience in insect mass-rearing using a wide range of diets and I simply followed the diet formulae developed by the various authors cited.

A. HAISCH: The antimicrobial effect of sorbic acid depends essentially on the pH of the diet. What was the pH of your diets?

S. HATMOSOEWARNO: I did not measure the pH of the diets in my experiment.

B. NA'ISA: I understand that Mr. K. Russ has used the bean diet for small-scale rearing of codling moth larvae and I should be interested to know how his findings compare with the results reported by Mr. Hatmosoewarno.

P. FISCHER-COLBRIE: Perhaps I might answer that question. We have used the bean diet in our institute for experimental purposes. We are currently rearing the eighth generation and have not encountered any serious problems. Please refer to Table VI of Mr. Butt's paper[1] for details of costs and productivity.

J.E. SIMON F.: Mr. Hatmosoewarno, are you applying the experience you gained at Seibersdorf to your work on sugar-cane pests in Indonesia?

S. HATMOSOEWARNO: Yes, the experience I have gained is of great value for my work on the application of the sterile-male technique in general and the control of the white top sugar-cane borer in particular.

---

[1] BUTT, B., "Survey of synthetic diets for codling moths", these Proceedings, IAEA-SM-186/14.

GAMMA-INDUCED STERILITY
OF THE GREATER WAX MOTH,
Galleria mellonella L., Pyralidae, Lepidoptera

D.W. WALKER, H. SINGH*, K.P. MacKAY
Puerto Rico Nuclear Center,
University of Puerto Rico,
Mayaguez, Puerto Rico,
United States of America

Abstract

GAMMA-INDUCED STERILITY OF THE GREATER WAX MOTH, Galleria mellonella L., Pyralidae, Lepidoptera.

Adults were sterilized at 22 krad or less with cobalt gamma radiation. Males did not recover from sterilizing doses. Fractionated doses were more effective for female adults, sterilizing them at 6.6 krads, compared with 13.2 krads for a single dose. Fractionated and single doses were equally effective for male adults. Because the wax moth is a common pest of bees throughout the world its life history, habits and behaviour are well-known. It makes a good laboratory test animal since it is easily reared, it has a high reproductive rate, a short larval period, a rather long adult life span, and the sexes are easily distinguished in pupal and adult stages. The objects of this study were to determine: (1) the dose-range needed to sterilize male adults; (2) whether sterile males recovered virility; and (3) whether fractionated doses are more effective than a single dose.

METHODS

Moths were reared in one-gallon jars on a diet containing honey, glycerine, water, brewer's yeast, Pablum© baby food and vitamins [1]. Larvae from colony stock were placed directly on food in sterile jars, covered with a cloth, and held in the dark at $32° \pm 1°C$ and $70 \pm 5\%$ relative humidity.

Pupae were harvested and de-silked by immersion for 2 min in 5% sodium hypochlorite buffered with 5% sodium carbonate, water-rinsed and dried on filter paper [2]. Each pupa was sexed and maintained in a one-ounce jelly cup. Sex differences are described by Smith [3].

Two or more series of tests were made at each dose level, using four replicates of five individual pair matings in each replicate.

Moths were collected upon emerging. Those to be irradiated were treated within 24 hours in a 2000 Ci^{60}Co source using a dose rate of 1100 rad per minute. Moths were not aerated during exposure. After exposure the moths were individually paired and maintained in one-ounce jelly cups in the dark at 32°C. A one square centimetre wax envelope provided the oviposition site. Scotch tape attached to the inside of the cup trapped the larvae shortly after they hatched. If the larvae are not trapped and killed they will feed upon the unhatched eggs. Egg hatch was determined in two ways: by counting hatched eggs and by counting the larvae trapped in the tape.

---

* Present address: Knoxville College, Knoxville, Tenn., United States of America. The preliminary work reported here was done by H. Singh while on an Oak Ridge Associated Universities Grant to PRNC.

TABLE I. MALE STERILITY INDUCED BY A SINGLE EXPOSURE

Exposure (krad)	Fertile eggs laid per female	Eggs hatched	Percentage hatched
0	800	752	94.0
7.7	766	651	85.0
11.0	742	573	80.0
13.8	700	490	70.0
19.3	687	209	31.0
22.0	750	0	0
30.9	537	0	0
40.9	570	0	0
50.9	575	0	0

Usually courtship and mating took place immediately after pairing. Females began to lay eggs within three hours after mating. Each cluster contained 25 to 600 eggs. Eggs from normal adults developed to hatching in five days; embryonic development was slower in eggs from irradiated adults regardless of the sex irradiated.

RESULTS

Sterilizing dose to males

Males were fully sterilized at 22 krad and above (Table I). Less fertile eggs were laid at higher doses, particularly at 30.9 krads and above.

Male recovery

Males did not recover virility after radiation (Table II). There is a direct relationship between the number of eggs laid by the female and the age of the male, the second young female laying fewer eggs than the first young female in all instances. Three factors influenced female fertility: female age, age of male at time of mating, and dose received by male. Most of the infertile eggs were laid by old females.

Fractionated doses

In female adults the fractionated doses were more effective than single doses (Table III). The life span after exposure was the same from each type of exposure.

DISCUSSION

Nielsen and Lambremont [4] measured the effect of gamma-induced F_1 sterility in all life stages and discussed the use of the method for eradicating

TABLE II. TEST OF MALE RECOVERY FROM A SINGLE EXPOSURE [a]

Exposure (krad)		First female		Second female	
		Fertile eggs produced	Percentage hatched	Total fertile eggs produced	Percentage hatched
0	1st 5 days	579	95	500	85
	2nd 5 days	153	80	-	-
	Total	732	92	500	85
10	1st 5 days	454	82	850	80
	2nd 5 days	92	20	-	-
	Total	546	69.6	850	80
20	1st 5 days	453	10	450	0
	2nd 5 days	114	0	0	-
	Total	567	8	450	0
30	1st 5 days	623	0	737	0
	2nd 5 days	150	0	-	-
	Total	773	0	737	0
40	1st 5 days	520	0	725	0
	2nd 5 days	100	0	-	-
	Total	620	0	725	0
50	1st 5 days	575	0	500	0
	2nd 5 days	111	0	-	-
	Total	686	0	500	0

[a] After irradiation each male was mated with a 0 - 3-day-old virgin female. At the end of 5 days the male was mated with a second virgin female 0 - 3-days-old in a new cup. Egg counts were made from the first female in two time intervals, 0 - 5 days and 5 days until death. Egg counts from the second female were made after she died. All egg counts were made 10 days or more after females died.

or suppressing populations of this species. Fractionated exposures to cabbage looper pupae caused more inherited sterility than single exposures of equal size [5].

Sterilizing lepidoptera is difficult as very high doses are necessary to produce complete dominant lethality [6,7]. This has been attributed to the holokinetic nature of the lepidopteran chromosome [8].

Walker and Pedersen [9] suggested overflooding the natural population with partially sterile individuals to cause collapse of the natural population as a solution to the lepidopteran sterile release problem. This and other solutions are discussed by Smith and von Borstel [10]. The most serious

TABLE III. COMPARISON OF FRACTIONATED AND SINGLE EXPOSURE TO MALE AND FEMALE WAX MOTHS

Exposure (krad)	Fractionated [a]			Single		
	Average number of fertile eggs per female	Eggs hatched per female	Percentage eggs hatched	Average number of fertile eggs per female	Eggs hatched per female	Percentage eggs hatched
(a) Normal female; normal male						
0.0	650	624	96.0			
((b) Irradiated female; normal male						
1.1	489	445	91.0	409	204	49.9
2.2	426	394	92.5	413	153	37.0
3.3	536	357	66.5	371	68	18.3
6.6	287	0	0	567	67	12.0
13.2	380	0	0	542	0	0
19.8	358	0	0	601	0	0
(c) Irradiated male; normal female						
1.1	371	350	94.3	590	585	100
2.2	310	285	92.0	700	689	100
3.3	412	388	94.3	650	643	100
6.6	420	391	93.0	453	403	89
13.2	400	239	60.0	402	318	79
19.8	453	91	20.0	571	86	15

a Given in two equal doses, 24 ± 2 hours apart.

problem in the population collapse method is that additional reproduction will be taking place within the population if semi-sterile females are released. The natural solution to this dilemma is to overflood with partially sterile males in one isolated population and to overflood a second population (that is isolated from the first population) with fully sterile females. The high acute doses needed to sterilize females can be detrimental to mating competitiveness, whereas low doses have less effect on mating performance, and in some instances have actually increased the mating frequency (Walker, unpublished data).

We have shown that a fractionated dose to females produces more sterility than an equal acute dose. The increase in the net number of chromosome aberrations from fractionated doses to organisms with mono-kinetic chromosomes has been explained in terms of repair mechanisms [11-14]. Fractionated dose studies were made with Tribolium confusum [15] and the boll weevil [16]. The object of both these studies was to achieve sterility with minimal somatic damage so that the sterile adults would be effective when released. It was assumed that the amount of sterility induced would be equal regardless of whether the dose had been given as a single dose or as fractions. The effectiveness of total doses given in fractions that were less than the total acute dose needed for sterilizing males was not measured, and they were observing individuals that have monokinetic chromosomes. Toba and Kishaba [5] obtained more dominant lethals from fractionated doses to pink bollworm pupae.

Wolff [11] and Russell [14] explain the increase in net number of chromosome aberrations from fractionated doses in terms of repair mechanisms. Wolff cites evidence that restitutional restoration of breaks in Vicia is inhibited by prior exposure. Their conclusions are also supported by Tazima and Kondo [17]. It is confusing to interpret these hypotheses because of the effect of dose rate as pointed out by Tazima [18], and also the differential radio resistance due to the stage of development of the sex cells as pointed out by von Borstel and St.Amand [19] who worked with Habrobracon, and Sado [20] working with the silkworm. Kondo [12], also working with the silkworm, explains the phenomenon in terms of repair mechanisms that are intercellular and the interaction with certain multi-cellular organs. Whether this is specifically a function of protein repair phenomenon in chromosomes, as suggested by Kimball [13], is not known, but it provides a useful hypothesis of plausibility.

There are some data also to support this conclusion in holokinetic organisms [17,19-22]. Tazima [22] reported a higher mutation frequency in the genes responsible for eye colour in the silkworm larvae when doses were given as fractions to the embryos. He reported the highest mutation rate where there was 12 hours between the fractions. He interprets this to be an effect on the repair mechanism caused by the first exposure. Tazima makes a clear distinction between sensitivity of cells to killing by radiation and sensitivity of cells to mutation; he points out that the sensi-tivity to killing varies considerably according to the developmental stage of the cell, but mutation sensitivity does not vary much among the different types of gonia.

We speculated that the differences in response by males might be due to the fact that spermiogenesis and meiotic divisions of spermatogonia occur simultaneously whereas oogenesis does not. In the females this may be related to the effect on the trophocytes, nurse cells, surrounding and connected to each developing ovum. If this is so then we would expect that the younger trophocytes would be more sensitive than the mature ova. Therefore we would expect that the first group of eggs laid by treated females might hatch, but that later fertile eggs would not. We did not observe this. Where females were irradiated before mating there was no correlation between the time that the eggs were laid and the chance of egg hatch.

REFERENCES

[1] WATERHOUSE, D.F., Axenic culture of wax moth for digestion studies. Ann. N.Y. Acad. Sci. 77 (1959) 283.

[2] DUTKY, S.R., THOMPSON, J.V., CANTWELL, G.E., A technique for mass rearing the greater wax moth, Proc. Entomol. Soc. Wash. 64 (1962) 56.

[3] SMITH, T.L., "Breeding methods for Galleria mellonela", Culture Methods for Invertebrates, (LUTZ, E.F., WELCH, P.L., GALTSOFF, P.S., NEEDHAM, J.G., Eds), Dover, New York (1973) 349.

[4] NIELSEN, R.A., LAMBREMONT, E.N., J. Econ. Entomol., in press.

[5] TOBA, H.H., KISHABA, A.N., Cabbage loopers: pupae sterilized with fractionated doses of gamma irradiation, J. Environ. Entomol. 2 (1973) 118.

[6] LaCHANCE, L.E., SCHMIDT, C.H., BUSHLAND, R.C., "Radiation-induced sterilization", Pest Control: Biological, Physical and Selected Chemical Methods (KILGORE, W.W., DOUTT, R.L., Eds), Academic Press, New York (1967) 147.

[7] PROVERBS, M.D., Induced sterilization and control of insects, Annu. Rev. Entomol. 14 (1969) 81.

[8] BAUER, H., Die Kinetische Organisation der lepidopteran Chromosomen, Chromosoma 22 (1967) 101.

[9] WALKER, D., PEDERSEN, K., Population models for suppression of the sugarcane borer by inherited partial sterility, Ann. Entomol. Soc. Am. 62 (1969) 21.

[10] SMITH, R.H., VON BORSTEL, R.C., Genetic control of insect population, Am. Assoc. Adv. Sci. 178 (1972) 1164.

[11] WOLFF, S., Delay of chromosome rejoining in Vicia faba induced by irradiation, Nature (London) 173 (1954) 501.

[12] KONDO, S., "RBE of fast neutrons to gamma-rays for mutations in relation to repair mechanisms", Proc. Conf. Mechanisms of the Dose-Rate Effect of Radiation at the Genetic and Cellular Levels, Oso, 4-7 Nov. 1964; Supplement to the Japanese Journal of Genetics 40 (1965) 97.

[13] KIMBALL, R.F., "Repair and differential sensitivity to mutation induction: Summary and synthesis", Repair from Genetic Radiation Damage (SOBELS, F.H., Ed.), Macmillan, New York (1963) 427.

[14] RUSSELL, W.L., "The effect of radiation dose-rate and fractionation on mutation in mice", Repair from Genetic Radiation Damage (SOBELS, F.H., Ed.), Macmillan, New York (1963) 205.

[15] DUCOFF, H.S., VAUGHAN, A.P., CROSSLAND, J.L., Dose-fractionation and the sterilization of radiosensitive male confused flour beetles, J. Econ. Entomol. 64 (1971) 541.

[16] FLINT, H.M., BIBOW, W.R., LAHREN, C.K., Radiation studies with the boll weevil: lethal effects on larvae, pupae and adults; male sterility and dose-fractionation, J. Econ. Entomol. 59 (1966) 1249.

[17] TAZIMA, Y., KONDO, S., "Differential radiation-sensitivity of germ cells as a possible interpretation of sex differences in dose-rate dependence of induced mutation rates in the silkworm", Repair from Genetic Radiation Damage (SOBELS, F.H., Ed.), Macmillan, New York (1963) 237.

[18] TAZIMA, Y., "Mechanisms controlling two types of dose-rate dependence of radiation-induced mutation frequencies in silkworm gonia", Proc. Conf. Mechanisms of the Dose-Rate Effect of Radiation at the Genetic and Cellular Levels, Oso, 4-7 Nov. 1964; Jap. J. Genet. 40 Suppl. (1965) 68.

[19] VON BORSTEL, R.C., AMAND, W. St., "Stage sensitivity to X-radiation during meiosis and mitosis in the egg of the wasp Habrobracon", Repair from Genetic Radiation Damage (SOBELS, F.H., Ed.), Macmillan, New York (1963) 87.

[20] SADO, T., Spermatogenesis of the silkworm and its bearing on the radiation-induced sterility, Jap. J. Genet. 36 Suppl. (1961) 136.

[21] NAKANISHI, Y., IWASAKI, T., KATO, H., Cytological studies on the radiosensitivity of spermatogonia of the silkworm, Jap. J. Genet. 40 Suppl. (1965) 49.

[22] TAZIMA, Y., The Genetics of the Silkworm, Academic Press, New York (1964).

DISCUSSION

D.S. GROSCH: I suggest you study the components of reproductive failure of the wax moth females. If nuclear or chromosomal damage is not obvious, you may be dealing with cytosomal defects such as interference with yolk formation or its accumulative deposit.

D.W. WALKER: We have not studied yolk formation irregularities and I agree that this would be worth doing.

D.S. GROSCH: Your mention in the oral presentation of the importance of protein brings to mind this ingredient of the yolk. Furthermore, the correlation of defective eggs with age of female is consistent with what is known as senile decline in other insects, involving variations in the quantity and quality of the yolk.

D.W. WALKER: Thank you for the comment.

R. GALUN: Have you already performed any control experiments with sterile males and sterile females?

D.W. WALKER: Not with simultaneous release of both sexes. We are currently testing female overflooding in a field cage and the experiment will probably be completed this year. Last year we tested overflooding with males having inherited partial sterility. These tests were on a very small scale, but the results were promising; the population was zero by the time of the F_3 generation larval stage in each test cage compared with a three- to five-fold increase in number of each generation in the cages containing untreated insects.

W.J. NOORDINK: Did you consider using fast neutrons in your male sterilization programme?

D.W. WALKER: We tested the effects of sterilization by thermal neutrons on the sugar-cane borer some years ago, but our insects did not survive the treatment. However, I think that fast neutrons of known and specific energy levels should be tested. Klassen performed such experiments on the boll weevil about three years ago and, as I recall, he was able to induce considerable sterility, but the mortality rate was high.

M.S.H. AHMED: You report that in females the fractionated doses were more effective than single doses. Is there perhaps a relation between this and the lack of cross-overs in females?

D.W. WALKER: I do not think so. Niilo Virkki, the cytogeneticist at the University of Puerto Rico, has not seen cross-overs in any of the many descendant individuals (male or female) with inherited partial sterility that he has examined. It would appear that the lethal effect might be attributable to point mutations. Since I am working at much lower doses than North and others, I do not expect to find translocations, and apparently they are rare. However, we do have a semi-lethal effect that can be observed up to the F_7 generation.

A.M. FELDMANN: How would the release of fully sterile females cause a population collapse?

D.W. WALKER: There is only a very small number of normal males (and females too) in the habitat in relation to the number of females after release of the fully sterile females. Therefore we expect a very low mating frequency between normal females and normal males — one occurrence in 15. This would theoretically cause a reduction in the F_1 population to 1/15th of what it would have been if there had been no release. The female release ratio might have to be increased to more than 14 sterile females

to 1 normal wild female to eradicate the species. The future population
size will depend solely upon the relative mating occurrences between P
generation normal males and normal females, on the one hand, and normal
males and sterile females on the other.

THE SYNTHETIC SEX-PHEROMONE OF
Spodoptera littoralis Boisd. AND ITS USES
A field evaluation

D.G. CAMPION, B.W. BETTANY
Centre for Overseas Pest Research,
Overseas Development Administration,
London

BRENDA F. NESBITT, P.S. BEEVOR, R. LESTER,
R.G. POPPI
Tropical Products Institute,
Overseas Development Administration,
London,
United Kingdom

Abstract

THE SYNTHETIC SEX-PHEROMONE OF Spodoptera littoralis Boisd. AND ITS USES: A FIELD EVALUATION.
    The four components of the synthesized female sex-pheromone of S. littoralis are tetradecan-1-yl acetate
(I); cis-9-tetradecen-1-yl acetate (IIA); trans-11-tetradecen-1-yl acetate (IIB); and cis-9, trans-11-
tetradecadien-1-yl acetate (III). The synthetic pheromones dissolved in hexane were dispensed in small
polythene containers either singly or in combination over water-trough (WT) traps located in the potato-growing
region of south-east Cyprus. Only III was found necessary to attract male moths. A dose-response relationship
was established in that with increased loadings an increasing number of male moths were caught during the
20-day test periods. The persistence of attraction was also dependent on the loading of the pheromone within
the polythene dispensers. At a loading of 500 μg the material was attractive for approximately 20 days and at
a loading of 5000 μg attraction persisted for up to 40 days. IIA in combination with III in WT traps, or with
virgin females in vane traps, markedly reduced the catch of male moths. The level of inhibition was
proportional to the loading of IIA in the dispensers. The cis isomer of IIB also reduced the attractiveness of
female moths in vane traps, although it was a much less effective inhibitor than IIA. No marked attraction or
inhibition was shown for I or IIB, neither was any synergistic action demonstrated for these substances when in
combination with III. Their function is therefore at present unknown and may be related to close range mating
behaviour. A comparison of the catch of male moths in seven different traps baited with III indicated that the
water-trough trap was the most effective. The substitution of oil for water containing detergent had no adverse
effect on catch. Water-trough traps baited with III caught more males than a similar trap baited with virgin
females. Vane traps baited with virgin females caught more males than similar traps baited with synthetic
pheromone. The hourly catches of males in a WT trap between 1700 to 2300 h were consistently higher and
more uniform when compared with a similar trap baited with virgin females. Experiments using the confusion
technique were carried out. Two hundred polythene dispensers each containing 500 μg of III were evenly
distributed throughout a potato field of 2000 m^2. During the test period of 19 days virtually no males were
caught in a trap baited with virgin females located in the centre of the field. This was in contrast to catches
in the same field both before and after treatment and with catches in a similar trap on an untreated field
nearby. In another experiment using only 100 point sources of III at twice the loading, only a slight reduction
in catch compared with control catches was noted. The use of the synthetic sex-attractant for the control of
S. littoralis by mass-trapping or confusion techniques and in relation to a possible sterile-insect release
programme is discussed.

INTRODUCTION

Spodoptera littoralis Boisd. is a very important pest in the Middle East where the larval stage attacks a wide variety of crops. In Cyprus, where the present work was carried out, the insect attacks mainly irrigated autumn potatoes and lucerne. Strategies for the control of this insect in Cyprus and possibly elsewhere depend basically upon whether it regularly migrates from other neighbouring countries in great numbers, or whether infestations usually build up from small localized populations. During 1972, 50 traps baited with virgin females of S. littoralis were regularly distributed throughout Cyprus in an attempt to show whether or not such migrations occur by relating daily changes in the number and distribution of moths caught within the trap network to the meteorological situation [1]. Where there were significant wind runs to Cyprus from neighbouring countries along which insects could be carried, there was little evidence of their arrival in the daily distribution of catches within the trap network. In almost all cases where there were large increases in catch, they occurred within favourable areas such as lucerne or potato fields which were often geographically located at the opposite end of the wind run.

No diapause exists in S. littoralis, and it has been shown that very few insects manage to survive the cold winter of Cyprus (W. R. Ingram, private communication). In the absence of regular migrations from other countries, the mass-trapping of male moths in spring or early summer, supplemented by the release of sterile insects, may reduce the chances of population build-up. For mass-trapping to be possible on a large enough scale a synthetic attractant is essential. The female sex pheromone of S. littoralis was isolated and synthesized by Nesbitt and co-workers [2] who found that the female moth produced a complex of four pheromones: (I) tetradecan-1-yl acetate; (IIA) cis-9-tetradecen-1-yl acetate; (IIB) trans-11-tetradecen-1-yl acetate; and III cis-9, trans-11-tetradecadien-1-yl acetate. EAG responses to nanogram amounts of pheromones were in the order III > IIA ≫ IIB > I. This report presents the results of a field evaluation of the synthetic pheromone which was carried out in Cyprus during 1973. Trapping inhibition and pheromone dispensing techniques were similar to those used for the red bollworm Diparopsis castanea Hmps. in Malawi by Marks [3, 4]. Some of the results presented here have already been reported elsewhere [5].

MATERIALS AND METHODS

Relative attraction of the four components

The synthetic pheromones dissolved in hexane were transferred in the appropriate amounts to small lidded polythene bottles (3-mlitre capacity). The solvent was allowed to evaporate overnight with the lid open, after which the lid was closed. A small hot soldering iron was applied to the rim of the closed lid so that the plastic melted locally, and formed a watertight seal. The pheromones therefore had to pass through the polythene walls. The traps used in conjunction with the pheromone dispensers designed by Marks [3] consisted of a square trough (60 cm × 60 cm) made from galvanized metal over which was constructed a flat roof of the same material

The trough was partially filled with water containing a small quantity of detergent. A dispenser was hung from the centre of the undersurface of the roof approximately 5 cm above the surface of the water. The trough was supported by metal legs to a height of 30 cm above ground level. Moths attracted to the pheromone source passed through a gap of 7 cm between the roof and the rim of the trough and were drowned in the water.

For each test, four or five traps were spaced round the perimeter of potato fields located in the districts of Liopetri and Xylophagou in southeast Cyprus. The fields were approximately 40 000 m^2 in area.

As a control in each experiment to monitor the relative abundance of the local moth population, one vane-trap baited with virgin females was used following the procedures described by Campion [6]. To reduce positional bias on the catches of moths, each trap was moved one place anti-clockwise each day. Daily counts of catches were made and the water replenished in the traps. The usual exposure period for the pheromone source was 20 days. All the experiments were carried out during the period between August and October 1973.

Comparison of traps

A number of different pheromone traps, which had been reported to be effective for catching other Lepidoptera, were constructed. They were the mesh trap [7], the sticky disc and water trough trap [3, 4], the maze trap [8], the cup trap [9], the cylinder trap [6, 10] and the vane trap [6, 11]. One example of each kind of trap was set up in a fixed position at the perimeter of each of three lucerne fields, with the exception of the maze traps which were positioned in the middle of each field. Between each trap was a distance of at least 100 m. The fields were situated at the experimental farm of the Agricultural Research Institute, Nicosia. The traps were inspected daily and at the end of each 20-day period the standard pheromone dispensers containing synthetic pheromone were replaced with new ones. Each 20-day period constituted one replicate.

The relative effectiveness of the water trough and vane traps was compared using either synthetic pheromone or virgin females as lure. These experiments were also sited in a lucerne field of the experimental farm of the Agricultural Research Institute. The traps were positioned 150 m apart in a square formation. Each trap was moved one place in a clockwise direction each day. Daily inspections for catch were made and the females replaced every 3-4 days. As before, the lure in each replicate was exposed for 20 days. The standard pheromone dispensers were then replaced.

During the course of these experiments, hourly visits were made to the WT traps between 1700 and 2300 h for 13 nights and the catches at these times were noted.

Experiments to determine the effects on catch of substituting engine oil for water in trough traps baited with the synthetic pheromone were sited in the potato growing areas of southeast Cyprus following procedures described above. All the tests were carried out during the months of September to October 1973.

Inhibition tests

For each test, four or five vane traps were baited with virgin female moths following the procedure described by Campion [6]. The pheromone components to be tested were dispensed in the standard manner described above. The polythene dispensers were subsequently attached to the moth-holder of the vane trap. As a control, one vane trap in each test was baited with virgin females only. The traps were disposed evenly round the perimete of potato fields approximately 40 000 m² in area situated in the Liopetri district of southeast Cyprus. Each trap was inspected daily and the catch of male moths was noted for a total period of 20 days. To reduce the effects of position bias on catch, each trap was moved one position anti-clockwise each day. The tests were carried out during the months of August to November 1973.

Confusion experiments

Two plots were selected as treatment areas. Plot A was an isolated potato field of approximately 2000 m² at Avgorou with a control plot of similar size 150 m distant. Plot B of a similar size was the end segment of a much longer field at Xylophagou. The control plot in this case was the area at the other end of the field and separated from the treated area by 75 m In the plot A experiment 100 polythene dispensers each containing 1000 μg III (70% cis-trans) were evenly distributed throughout the plot by attaching them with string to the stems of the potato plants. In plot B, the same procedure was adopted except that 200 dispensers were used, each containing 500 μg III (70% cis-trans). A vane trap baited with virgin females was maintained in each of the treatment and control plots. Daily catches were recorded both before and during the test period. The treatment was terminated when the potato crop was harvested and most of the dispensers were buried. The traps were however maintained for a further 11-17 days in the harvested fields and the daily catches recorded.

RESULTS

The relative attractiveness of the four components of the pheromone complex

The results of exposing the four components of the pheromone complex either singly or in combination are shown in Table I. I, IIA and IIB alone at loadings of 500 μg, 100 μg and 100 μg respectively were not attractive to male moths, whereas III alone at a loading of 500 μg was highly attractive. Combinations of III at a loading of 500 μg with I or IIB either together or singly at loadings of 500 μg and 100 μg respectively neither greatly enhanced nor depressed the catches of male moths compared with III alone. The presence of IIA at a loading of 100 μg in any combination with III at a loading of 500 μg reduced the catch to almost zero.

A further series of experiments combining III at a fixed loading of 500 μg with various loadings of I and IIB were carried out to confirm whether or not these substances in any way modified the attractiveness of III. The results are shown in Table II. In all instances with I at loadings of 500 μg, 1000 μg and 5000 μg, and with IIB at 50 μg, 100 μg and 1000 μg, the total catch over the 20-day test period was slightly lower than that obtained with III exposed alone although the differences are probably not significant.

The effect on catch of varying the loading of III from 10-5000 μg are show in Table III. Between loadings of 10-500 μg of III a marked increase in catch was noted, whereas the increase between 500-5000 μg was relatively slight. When the loading was varied by increasing the number of polythene dispensers per trap, a greater persistence at the higher loadings was shown by the higher catches in the period from 20 to 40 days (Table IV).

To determine whether isomerically pure III is essential for good attraction or whether the more readily available mixed isomers (70% cis-trans with approximately equal amounts of cis-cis and trans-cis) would be adequate, the two materials were directly compared in the same plots.

The results are given in Table V and clearly show that the 100% cis-trans material was much more attractive. The cis-cis isomer of the diene at a loading of 500 μg was much less attractive than cis-trans at the same loading for the first 18 days of exposure; maximum catches were obtained between 19 to 25 days after exposure by which time the cis-trans diene was no longer attractive (Table VI).

Comparison of trap design

The catches obtained using the different traps baited with synthetic pheromone are shown in Table VII. The water trough traps consistently caught most males for the whole of the test period, although as the moth population declined the relative effectiveness of some of the smaller traps improved. Vane traps baited with synthetic pheromones (in contrast to the same traps baited with virgin female moths) were among the least effective trap devices.

The results of comparing the relative effectiveness of vane traps and water-trough (WT) traps baited with either synthetic pheromones or virgin females are shown in Table VIII. The WT trap baited with synthetic pheromone (III) overall caught most males, although the relative effectiveness of this trap system declined in the three successive replicates. Vane traps baited with virgin females were the next most effective trap system when based on the mean percentage of the total catch for the three replicates. When moths were most abundant however (experiment 1), the WT trap baited with females caught more moths than the vane trap baited with females. Vane traps baited with synthetic pheromone was the least effective trap combination.

The substitution of engine oil for water containing detergent in the trough traps had no adverse effects on catch and in some replicates the oil-based trap caught more male moths than the standard water trap (Table IX).

The total number of males arriving at hourly intervals from 1700 to 2300 h for the 13-day period at WT traps baited with III or virgin females, together with the mean figure for the remainder of the night period, is shown in Table X. The hourly catches of males in the trap baited with III were consistently higher compared with female-baited traps, and the rate of catch was more uniform from one hour to the next.

Inhibition experiments

The effect on the catch of male moths of each of the four components of the pheromone complex when in close proximity to virgin females in vane traps is shown in Table XI, where the totals for the 20-day test periods are presented. The presence of I or III at a loading of 500 μg was associated with an increase in catch when compared with the controls. The presence of IIB at a loading of 100 μg had no apparent effect. The presence of IIA

TABLE I. SUMMARY TABLE TO SHOW THE RELATIVE ATTRACTION OF THE FOUR COMPONENTS OF THE S. littoralis PHEROMONE COMPLEX WHEN EXPOSED SINGLY OR IN COMBINATION IN POLYTHENE DISPENSERS OVER WATER-TROUGH TRAPS. LOADINGS FOR III & I - 500 µg; FOR IIA AND IIB 100 µg. EXPS 1-7 WITH 70% cis-trans DIENE (III), EXP. 8 WITH 100% cis-trans DIENE (III). TOTAL CATCHES AFTER 20 DAYS GIVEN IN EACH INSTANCE

Pheromone formulation	III + IIA + IIB + I	III + IIB + I	III + IIA + I	III + I	III + IIA	III + IIB	III	IIA + IIB	IIA	IIB	I	Controls
Exp. No. 1		140	3			240						381
2	4	52			7							137
3		638					538			2		202
4		641							9		1	291
5		1356		542				4				810
6		-				310	413					311
7		283		402			476					424
8		518		472		275	476					365
Mean	4	518	3	472	7	275	476	4	9	2	1	365

TABLE II. EXPERIMENTS TO SHOW WHETHER THE PRESENCE OF IIB OR I AT VARIOUS RATIOS ENHANCES THE ATTRACTION OF III AT 500 μg TO MALE MOTHS OF S. littoralis WHEN EXPOSED IN WT TRAPS. CATCHES GIVEN ARE THE TOTALS FROM 20 DAYS' CONTINUOUS EXPOSURE

III[a] alone	+	III[a]I (500 μg)	III[a]I (1000 μg)	III[a]I (5000 μg)
1461		812	690	914
III[b] alone	+	III[a]IIB (50 μg)	III[a]IIB (100 μg)	III[a]IIB (1000 μg)
413		290	310	373

[a] diene 100% cis-trans.
[b] diene 70% cis-trans.

TABLE III. EXPERIMENTS TO SHOW THE RELATIVE ATTRACTIVE-NESS OF INCREASED LOADINGS OF III (100% cis-trans) WHEN EXPOSED IN WT TRAPS TO MALES OF S. littoralis. CATCHES GIVEN ARE THE TOTALS AFTER 20 DAYS CONTINUOUS EXPOSURE

	Loading of III (μg)						
	10	50	100	500	1000	2000	5000
A	29	143	160	635	-	-	-
B	-	-	449	720	453	497	-
C	-	-	-	156	151	235	325

TABLE IV. EXPERIMENTS TO SHOW THE RELATIVE ATTRACTIVENESS OF INCREASED LOADINGS OF III (70% cis-trans)+ WHEN EXPOSED IN WT TRAPS TO MALES OF S. littoralis, BY INCREASING THE NUMBER OF POLYTHENE DISPENSERS PER TRAP. EACH DISPENSER CONTAINED 500 μg III + 500 μg I

Number of dispensers	1 = 500 μg III	2 = 1000 μg III	4 = 2000 μg III	8 = 4000 μg III
20-day total	288	499	275	270
Additional catch after a further 20 days	27	101	155	253
Totals (40 days)	315	550	430	523

TABLE V. EXPERIMENTS TO COMPARE THE ATTRACTIVENESS OF
70% AND 100% cis-trans DIENE (III) WHEN EXPOSED OVER WATER-
TROUGH TRAPS. CATCHES OF MALE MOTHS ARE THE 20-DAY TOTALS.
ALL LOADINGS AT 500 μg

Exp.	100% cis-trans	70% cis-trans	Relative difference 100% : 70%
A	787	283	2.78
B	501	36	13.92
C	360	53	6.79

at a loading of 100 μg markedly reduced the catch of males by as much as
92% in comparison with the control catch; for 15 out of the 20 days of the
test period, no males were caught at all. The inhibitory effect of various
loadings of IIA ranging from 10-1000 μg is shown in Table XII.

The presence of III at a loading of 500 μg together with IIA at a loading
of 500 μg in no way reduced the inhibitory action of the latter (Table XIII).

The possible inhibitory effects on catch of other related monoenes is
shown in Table XIV. The presence of cis IIB at a loading of 100 μg reduced
the catch by 63% compared with the controls; IIA at the same loading
reduced the catch by 85%. IIB and trans IIA had no apparent effect at
loadings of 100 μg.

Confusion experiments

The results from plot A (100 $\times$ 1000 μg III) are shown in Table XV. Only
a slight reduction in catch occurred in the vane trap baited with virgin
females located in the centre of the test site compared with the catches in
the control plot trap. A marked reduction in catches of the trap located in
plot B (200 $\times$ 500 μg III) was apparent (Table XVI), and this was despite the
bias of a higher catch in the trap located in the treatment plot compared
with catches in the control plot trap before the pheromone was introduced.
During the 19-day test period only six moths were caught (compared with
88 in the control plot) whereas four out of the six were caught on the first
two nights, a period of time when the pheromone dispenser is known to be
least effective (unpublished data). For 15 out of the 19 nights, no catch was
recorded. Immediately after the potatoes were harvested by uprooting the
plants and when most of the pheromone dispensers were buried, the total
catches in the treatment and control plot traps were approximately equal after
a further 11 nights (Table XVI). Control pheromone dispensers remained
attractive in WT traps sited elsewhere in the district for the whole of the
test period.

Text continued on p. 606

TABLE VI. EXPERIMENT TO COMPARE THE PERSISTANCE OF ATTRACTION SHOWN BY THE cis-cis DIENE WITH THE cis-trans DIENE. BOTH AT LOADINGS OF 500 µg

Days	1	2	3	4	5	6	7	8	9	10	11	12	13	14	15	16	17	18	19	20	21	22	23	24	25	26	27	28	29	30	31	32	33	34	35
cc diene catches	13	28	9	19	31	6	8	0	18	10	7	2	3	12	3	10	6	7	48	58	19	64	26	26	20	8	20	6	20	2	1	1	1	0	0
ct diene catches	368	482	317	84	83	167	313	205	166	56	58	15	100	20	8	10	11	4	2	0	0	0	1	0	0	0	0	0	0	0	0	0	0	0	0

TABLE VII. CATCH[a] OF MALE MOTHS OF S. littoralis IN VARIOUS TRAPS BAITED WITH SYNTHETIC PHEROMONE (III - 500 μg)

Exp. No.	Date exp. started	Mesh	Sticky disc	Maze	Cup	Sticky cylinder	Water trough (WT)	Vane
1	23/9	34	21	378	50	27	511	23
2	3/10	-	3	19	46	21	-	-
3	4/10	-	9	53	21	27	54	-
4	23/10	12	14	150	10	8	172	8
5	23/10	-	2	7	25	3	30	-
6	26/10	-	2	9	12	4	28	-
7	14/11	-	0	0	2	5	7	-
8	14/11	-	1	1	2	7	0	-
9	16/11 [b]	-	6	2	2	6	8	-

[a] Totals in each instance after 20 days.
[b] Replicate lasted only 18 days.

TABLE VIII. TO COMPARE THE CATCH[a] OF MALE MOTHS IN WATER-TROUGH (WT) AND VANE TRAPS WHEN BAITED WITH EITHER VIRGIN FEMALES OF S. littoralis OR THE SYNTHETIC PHEROMONE (III)

Exp. No.	WT trap + females	Percentage catch of total	WT traps + synthetic pheromone	Percentage catch of total	Vane traps + females	Percentage catch of total	Vane traps + synthetic pheromones	Percentage catch of total
1	248	21.4	710	61.2	163	14.1	39	3.4
2	130	35.2	223	43.2	49	9.6	114	22.1
3	5	10.2	11	22.4	30	61.2	3	6.1
Totals	383		944		238		156	
Mean percentage catch		18.9		42.3		28.3		10.5

[a] Totals in each instance after 20 days.

TABLE IX. COMPARISON OF THE CATCH[a] OF MALE MOTHS IN TROUGH TRAPS CONTAINING EITHER WATER OR OIL WHEN BAITED WITH SYNTHETIC PHEROMONE (III) AT A LOADING OF 500 μg

Exp. No.	Trap + oil		Trap + water	
	A	B	A	B
1	3429	4096	2193	-
2	464	502	490	419

[a] Total catch after 20 days.

TABLE X. COMPARISON OF THE NUMBER OF MALE MOTHS ARRIVING AT HOURLY INTERVALS AT WATER-TROUGH (WT) TRAPS BAITED WITH EITHER VIRGIN FEMALES OF S. littoralis OR THE SYNTHETIC PHEROMONE III: TOTALS FROM 13 NIGHTS

	Hourly intervals						
	1700-1800	1800-1900	1900-2000	2000-2100	2100-2200	2200-2300	2300-04 (mean)
WT trap + synthetic pheromone (III)	0	81	55	46	49	65	43
WT trap + virgin females	1	5	9	8	25	36	23

TABLE XI. CATCHES OF MALE MOTHS IN VANE TRAPS BAITED WITH VIRGIN FEMALES IN THE PRESENCE OF THE FOUR COMPONENTS OF THE PHEROMONE COMPLEX; TOTAL CATCH AFTER 20 DAYS; I & III AT LOADINGS OF 500 µg; IIA & IIB AT LOADINGS OF 100 µg

Pheromone component	III	IIA	IIB	I	Controls[a]
Catch of males	277	11	116	585	143
Number of nights with no catch	3	15	4	1	3

[a] Virgin females only.

TABLE XII. INHIBITION OF THE CATCH[a] OF MALE MOTHS IN VANE TRAPS BAITED WITH VIRGIN FEMALES IN THE PRESENCE OF VARIOUS LOADINGS OF IIA

Loading	10 μg	100 μg	500 μg	1000 μg	Controls
No. of replicates	2	4	4	2	6
Mean catch	224	131	88	68	420
Percentage catch reduction	47	69	79	84	

[a] Catches are totals after 20 days.

TABLE XIII. INHIBITING EFFECT OF IIA ON THE CATCHING ABILITY OF VIRGIN FEMALES IN THE PRESENCE AND ABSENCE OF III

Synthetic pheromone components present	Totals after 20 nights	Percentage reduction	No. of nights with no catch
IIA (500 μg)	40	96	8
IIA (500 μg) + III (500 μg)	12	99	15
Controls (female moths only)	947		2

TABLE XIV. CATCHES OF MALE MOTHS IN VANE TRAPS BAITED WITH VIRGIN FEMALES IN THE PRESENCE OF VARIOUS MONOENES AT LOADINGS OF 100 μg

Monoene	Total after 16 nights	Total after 20 nights	Number of nights with no catch	Percentage catch reduction (16 nights)
IIA	22	183	9	85
trans IIA [a]	122	318	7	18
IIB	96	334	4	35
cis IIB [b]	55	351	5	63
controls	148	415		

[a] trans-9-tetradecen-1-yl acetate.
[b] cis-11-tetradecen-1-yl acetate.

TABLE XV.   CONFUSION EXPERIMENT TO SHOW THE EFFECT OF
DISTRIBUTING 100 × 1000 μg III (70% c̲ t̲) IN POLYTHENE DISPENSERS
THROUGHOUT A POTATO FIELD ON CATCHES OF MALE MOTHS IN A
VANE TRAP BAITED WITH VIRGIN FEMALES; TEST AND CONTROL
PLOTS 2000 m^2 (PLOT A)

Days	Pretreatment catches		During treatment catches		Post-treatment catches	
	Treated area	Control area	Treated area	Control area	Treated area	Control area
1	2	8	1	0	0	2
2	0	0	1	1	0	0
3	1	0	0	1	24	0
4	7	11	0	0	5	0
5	2	2	0	0	0	0
6			1	0	0	0
7			0	4	1	5
8			0	10	9	10
9			0	1	0	1
10			2	2	0	0
11			0	5	0	0
12			24	25	0	2
13			4	25	1	0
14			1	1	0	0
15			2	0	2	11
16			9	0	0	0
17			8	110	1	0
Totals	12	21	53	185	43	31
Mean	2.4	4.2	3.1	10.9	2.5	1.8

DISCUSSION

The diene (III) of the synthetic pheromone complex alone appears
sufficient for attraction when used in combination with WT traps.   No
attraction or inhibition of attraction was demonstrated for I or IIB nor any
synergistic action shown for these substances in the presence of III in
combination with WT traps.   Their function is therefore at present unknown
and may be related to close range mating behaviour.   The cis-cis isomer of
III was found to be only slightly attractive, and the increased catches after
18 days exposure in the field probably resulted from isomerization to the
more potent cis-trans isomer.

TABLE XVI. CONFUSION EXPERIMENT TO SHOW THE EFFECT OF
DISTRIBUTING 200×500 μg III (70% $\underline{c}\,\underline{t}$) IN POLYTHENE DISPENSERS
THROUGHOUT A POTATO FIELD ON CATCHES OF MALE MOTHS IN A
VANE TRAP BAITED WITH VIRGIN FEMALES; TEST AND CONTROL
PLOTS 2000 m^2 (PLOT B)

Days	Pretreatment catches		During treatment catches		Post-treatment catches	
	Treated area	Control area	Treated area	Control area	Treated area	Control area
1	0	1	1	1	7	0
2	0	0	3	4	5	7
3	1	0	0	4	5	0
4	3	0	0	0	0	0
5	8	5	0	0	0	0
6	2	1	0	0	0	0
7	0	0	0	4	0	0
8	3	0	0	19	0	0
9	27	5	0	3	0	0
10	6	2	0	9	0	0
11	28	8	1	27	53	75
12	184	53	0	4		
13			0	0		
14			0	2		
15			0	2		
16			1	2		
17			0	1		
18			0	4		
19			0	2		
Totals	262	75	6	88	70	82
Mean	21.8	6.3	0.3	4.6	6.4	7.5

Increased loadings of from 10 to 500 μg of III in the dispensers resulted
in a progressive increase in the total number of moths caught in WT traps
over the 20-day test period, but high loadings between 500 and 5000 μg
produced little further increase. At loadings of 500 μg, however, the daily
catch after 20 days was almost zero, whereas at loadings of 5000 μg daily
catches persisted for up to 40 days. Higher loadings in excess of 5000 μg
may therefore result in even greater persistence.

Of the seven trap designs tested, the WT trap consistently caught most moths, although for survey purposes the smaller cup trap containing water or the sticky-lined plastic cylinder might be adequate and certainly less expensive to produce. For the development of mass-trapping techniques the WT trap is to be preferred, and the substitution of oil for water may even improve catch efficiency as well as avoiding the constant need for replenishment due to water evaporation.

The 100% cis-trans III was also shown to be far more attractive than material containing approximately 70% cis-trans isomer mixed with other isomers. The purest possible material would therefore be required for any programme involving mass trapping.

It has been established for other Lepidoptera that geometrical isomers of the female sex pheromone strongly inhibit the response of male moths [12-14]. In S. littoralis the presence of IIA in combination with either III or virgin females or both together markedly reduces male catch and a similar situation was found for the red bollworm, Diparopsis castanea Hmps. [3, 4]. Such an inhibitor may prove of use as a pest control agent by utilizing an "inhibition" as opposed to a "confusion" technique. Compound IIA has been reported to be the sex pheromone of S. frugiperda [15] and one of the two components of the sex pheromone of S. eridania [16].

The confusion experiments using III were designed to show whether the principle was possible under field conditions in the potato-growing area of Cyprus. The experiments were not aimed at control since they were carried out towards the end of the growing season and the invasion of already mated females into the "treated" areas would be expected. However using 200 point sources in an area of 2000 m², disorientation of males within the plot for a minimum of 19 days was demonstrated.

CONCLUSIONS

The synthetic sex-pheromone dispensed in polythene containers has been shown to attract male moths of S. littoralis over relatively prolonged periods of time without need for replacement. Meanwhile improved formulations are to be expected. Techniques for the sterilization of this insect are known [17] and semi-artificial diets necessary for the development of adequate mass-rearing methods have already been established [18, 19]. Very low populations of S. littoralis occur in Cyprus for a large part of the year while no regular mass migration from elsewhere is expected. Not only therefore can control methods based on confusion/inhibition techniques be considered, but a pilot programme based on combination of mass-trapping with subsequent sterile insect release would seem most promising.

ACKNOWLEDGEMENTS

We thank J. Rosenberg and P. M. Symmons of the Centre for Overseas Pest Research and A. Krambias of the Cyprus Ministry of Agriculture for their assistance in the field. We also thank D. Kourris of the Agricultural Research Institute, Nicosia, for rearing the insects. We are indebted to Miss S.M. Green and R.F. Chapman for their advice and criticism.

This work formed part of a joint collaborative project between the Centre for Overseas Pest Research, Overseas Development Administration, and the Ministry of Agriculture and Natural Resources, Cyprus. We are therefore indebted to R.C. Michaelides, Director General of the Ministry of Agriculture and Natural Resources, for his help and encouragement. We also thank Th. Christou, Director of the Agricultural Research Institute, Nicosia, for providing work facilities and assisting us in many other ways.

REFERENCES

[1] CAMPION, D.G., McGINNIGLE, J.B., BETTANY, B.W., Pheromone traps in Cyprus to study possible migration and local movement of Spodoptera littoralis (unpublished).

[2] NESBITT, B.F., BEEVOR, P.S., COLE, R.A., LESTER, R., POPPI, R.G., Sex pheromones of two noctuid moths, Nature (London) New Biol. 244 (1973) 208.

[3] MARKS, R.J., Annual Rep. Agricultural Research Council of Malawi, 1972.

[4] MARKS, R.J., Annual Rep. Agricultural Research Council of Malawi, 1973.

[5] CAMPION, D.G., BETTANY, B.W., NESBITT, B.F., BEEVOR, P.S., LESTER, R., POPPI, R.G., Field studies of the female sex pheromone of the cotton leafworm Spodoptera littoralis (Boisd.) in Cyprus, Bull. Entomol. Res., 64 (1974) 89.

[6] CAMPION, D.G., Some Observations on the Use of Pheromone Traps as a Survey Tool for Spodoptera littoralis, Misc. Rep. Centre for Overseas Pest Research, No. 4 (1972).

[7] TOBA, H.H., KISHABA, A.N., WOLF, W.W., GIBSON, T., Spacing of screen traps baited with synthetic sex pheromone of the cabbage looper, J. Econ. Entomol. 63 (1970) 197.

[8] KILLINEN, R.G., OST, R.W., Pheromone-maze trap for cabbage looper moths, J. Econ. Entomol. 64 (1971) 310.

[9] BARIOLA, L.A., COWAN, C.B., Jr., HENDRICKS, D.E., KELLER, J.C., Efficacy of hexalure and light traps in attracting pink bollworm moths, J. Econ. Entomol. 64 (1971) 323.

[10] SNOW, J.W., Fall Armyworm: Use of Virgin Female Traps to detect Males and to determine Seasonal Distribution, Research Rep. US Dept. Agric., No. 110 (1969).

[11] TUNSTALL, J.P., Sex attractant studies in Diparopsis, Pest. Artic. & News Summ. (A) 11 (1965) 212.

[12] ROELOFS, W.L., COMEAU, A., Sex pheromone perception, Nature (London) 220 (1968) 600.

[13] ROELOFS, W.L., COMEAU, A., Sex pheromone specificity, taxonomic and evolutionary aspects in Lepidoptera, Science 165 (1969) 398.

[14] JACOBSON, M., Sex pheromone of the pink bollworm moth: biological masking by its geometrical isomer, Science 163 (1969) 190.

[15] SEKUL, A.A., SPARKS, A.N., Sex pheromone of the fall armyworm moth: isolation, identification and synthesis, J. Econ. Entomol. 60 (1967) 1270.

[16] JACOBSON, M., REDFERN, R.E., JONES, W.A., ALDRIDGE, M.H., Sex pheromones of the southern armyworm moth: isolation, identification and synthesis, Science 170 (1970) 542.

[17] TOPPOZADA, A., ABDALLAH, S., ELDERFRAWI, M.E., Chemosterilization of larvae and adults of the Egyptian cotton leafworm, Prodenia litura, by apholate, metepa and tepa, J. Econ. Entomol. 59 (1966) 1125.

[18] McKINLEY, D.J., The laboratory culture and biology of Spodoptera littoralis Boisduval, ALRC Occ. Rep. No. 18 (1971).

[19] MOORE, I., NAVON, A., A new gelling technique and improved calcium-alginate medium for rearing Spodoptera littoralis, J. Econ. Entomol. 66 (1973) 565.

CHAIRMEN OF SESSIONS

Session 1	L.E. LaCHANCE	Metabolism and Radiation Research Laboratory, Agricultural Research Service U.S.D.A., Fargo, N.Dak. 58102, United States of America
Session 2	B.G. LEVER	Plant Protection Ltd., I.C.I., Fernhurst, Haslemere, Surrey, United Kingdom
Session 3	K. RUSS	Bundesanstalt für Pflanzenschutz, 1020 Vienna, Austria
Session 4	E.F. BOLLER	Swiss Federal Research Station for Arboriculture, Viticulture and Horticulture, 8820 Wädenswil, Switzerland
Session 5	G.W. RAHALKAR	Biology and Agriculture Division, Bhabha Atomic Research Centre, Trombay, Bombay, India
Session 6	M.S.H. AHMED	Nuclear Research Institute Tuwaitha, Baghdad, Iraq
Session 7	L.C. MADUBUNYI	Department of Veterinary Pathology, University of Nigeria, Nsukka, Nigeria
Session 8	D. ENKERLIN S.	Instituto Tecnológico de Monterrey, Monterrey, Mexico

SECRETARIAT

Scientific
Secretary: D.A. LINDQUIST   Joint FAO/IAEA
                                      Division of Atomic Energy
                                      in Food and Agriculture,
                                      International Atomic
                                      Energy Agency, Vienna

Administrative
Secretary: CAROLINE DE MOL   Division of External Relations,
                VAN OTTERLOO   International Atomic
                                      Energy Agency, Vienna

Editor: S.M. FREEMAN   Division of Publications,
                                      International Atomic
                                      Energy Agency, Vienna

Records Officer: P.B. SMITH   Division of Languages
                                      and Policy-Making Organs,
                                      International Atomic
                                      Energy Agency, Vienna

LIST OF PARTICIPANTS

AUSTRIA

Altmann, H.
Studiengesellschaft für Atomenergie,
Lenaugasse 10, 1080 Vienna

Fischer-Colbrie, P.
Bundesanstalt für Pflanzenschutz,
Trunnerstr. 1-5, 1020 Vienna

Russ, K.
Bundesanstalt für Pflanzenschutz,
Trunnerstr. 1-5, 1020 Vienna

BELGIUM

Evens, F.M.
R.U.C.A. (University of Antwerp),
1, Middelheimlaan, 2020 Antwerp

BRAZIL

Wiendl, F.M.
Comissão Nacional de Energia Nuclear,
Departamento de Entomologia da "Escola Superior de
Agricultura Luiz de Aneiroz",
CNEN, 13400 Piracicaba S.P.

CANADA

Fitz-Earle, M.
Department of Biology,
Capilano College,
2055 Purcell Way,
North Vancouver, B.C.

CSSR

Bennett-Řežábová, B.
Institute of Entomology,
Czechoslovak Academy of Sciences,
Viničná 7, Prague 2

EGYPT, ARAB REPUBLIC OF

Wakid, A.-F.M.
Atomic Energy Establishment,
101 Kasr El-Eini St., Cairo

ETHIOPIA

Tesfa-Yohannes, T.-M.
Institute of Pathobiology,
Haile Selassie 1st University,
P.O. Box 1176, Addis-Ababa

FRANCE

Biémont, J.C.

Laboratoire d'écologie expérimentale,
Parc de Grandmont, 37000 Tours

Bluzat, R.R.

Université de Paris XI, 91405 Orsay

Chauvet, G.J.

E.I.D. (Entente interdépartementale) pour la
démoustication du littoral méditerranéen,
B.P. 6036, av. Paul Rimbaud,
34030 Montpellier

Cousserans, J.

E.I.D., B.P. 6036, av. Paul Rimbaud,
34030 Montpellier

Coz, J.F.

Office de la recherche scientifique et technique
Outre-Mer (ORSTOM),
24 rue Bayard, Paris 8e

Cuisance, D.

Institut d'élevage et de médecine vétérinaire des
pays tropicaux (IEMVT),
B.P. 286, Bobo-Dioulasso, Haute-Volta

GERMANY, FEDERAL REPUBLIC OF

Haisch, A.

Bayerische Landesanstalt für Bodenkultur und Pflanzenbau,
Menzingerstr. 54 (Willibaldstr. 40b),
8000 Munich 19

Hamann, H.J.

Gesellschaft für Strahlen- und Umweltforschung,
Institut für Strahlenbotanik,
Herrenhäuser Str. 2, 3000 Hanover

Hossain, M.M.

Institute for Radiation Technology,
Engesserstr. 20, 7500 Karlsruhe

Lerch, G.G.

c/o Leybold-Heraeus, Wilhelm-Rohn-Str.,
6450 Hanau

Levinson, H.

Max-Planck-Institut für Verhaltensphysiologie, MP IV,
83131 Seewiesen

Maier, W.A.

Institut für Medizin. Parasitologie,
5300 Bonn-Venusberg

Nogge, G.

Institut für angewandte Zoologie,
An der Immenburg 1, 5300 Bonn 1

Steffan, A.W.

Biologische Bundesanstalt für Land- und Forstwirtschaft,
Königin-Luise-Str. 19, D 1 Berlin 33

Wiesner, L.

Helmstedter Str. 44,
3000 Hanover

GHANA

Amoako-Atta, B.

Ghana Atomic Energy Commission,
P.O. Box 80, Legon

GREECE

Manoukas, A.G.

Department of Biology,
"Demokritos" Nuclear Research Centre,
Athens

HUNGARY

Nagy, B.

Research Institute for Plant Protection,
Herman O.-u. 15, 1525 Budapest

Szentesi, Á.

Research Institute for Plant Protection,
Herman O.-u. 15, 1525 Budapest

INDIA

Rahalkar, G.W.

Biology and Agriculture Division,
Bhabha Atomic Research Centre,
Modular Laboratories, Trombay,
Bombay 400085

INDONESIA

Hatmosoewarno, S.

Indonesian Atomic Energy Agency,
L.P.P. Jl. Sala 40A,
P.O. Box 6, Yogyakarta

IRAN

Navvab Gojrati, H.A.

Pahlavi University,
College of Agriculture, Shiraz

IRAQ

Ahmed, M.S.H.

Nuclear Research Institute Tuwaitha,
Baghdad

ISRAEL

Galun, Rachel

Israel Institute for Biological Research,
P.O. Box 19, Ness-Ziona

Spharim, Y.

Agricultural Research Organization,
Volcani Center, P.O. Box 6,
Bet Dagan

ITALY

Delrio, G.

Istituto de Entomologia Agraria,
Università di Padova,
Via Gradenigo 6, 35100 Padua

De Murtas, I.D.   CNEN, Laboratorio di Agricoltura Casaccia,
                                             Santa Maria di Galeria, Rome

Gasperi, G.   Istituto di Zoologia,
                                             Università di Pavia,
                                             Piazza Botta 9, 27100 Pavia

Girolami, V.   Istituto di Entomologia Agraria,
                                             Università di Padova,
                                             Via Gradenigo 6, 35100 Padua

Grigolo, A.   Istituto di Zoologia,
                                             Università di Pavia,
                                             Piazza Botta 9, 27100 Pavia

Sacchi, L.P.   Istituto di Zoologia,
                                             Università di Pavia,
                                             Piazza Botta 9, 27100 Pavia

KENYA

Rens, G.R.   National Agricultural Laboratories,
                                             P.O. Box 30028, Nairobi

LEBANON

Jalloul, A.M.   Department of Plant Protection,
                                             Ministry of Agriculture, Beyrut

Mayas, I.A.   Institut national de recherche agronomique,
                                             Fanar

MALAYSIA

Yunus, A.   Division of Agriculture,
                                             Ministry of Agriculture and Fisheries,
                                             Kuala Lumpur

MAURITIUS

Monty, J.L.   Agricultural Services, Reduit

MEXICO

Enkerlin S., D.   Instituto Tecnológico de Monterrey,
                                             Monterrey, N.L.

NETHERLANDS

Feldmann, A.   Association Euratom-ITAL,
                                             Keyenbergseweg 6, Wageningen

Heemert, C. van	Department of Genetics, Agricultural University, Generaal Foulkesweg 53, Wageningen
Leigh, B.	Department of Radiation Genetics and Chemical Mutagenesis, Wassenaarseweg 62, Leiden
Noordink, W.J.	Institute for Phytopathological Research, Postbus 42, Wageningen
Robinson, A.S.	Association Euratom-ITAL, Postbus 48, Wageningen
Ticheler, J.	Institute for Phytopathological Research, Postbus 42, Wageningen

NIGERIA

Amodu, A.A.	Nigerian Institute for Trypanosomiasis Research, P.M.B. 2077, Kaduna, N-C-S
Madubunyi, L.C.	Department of Veterinary Pathology, University of Nigeria, Nsukka, East Central State
Na'isa, B.	Tsetse and Trypanosomiasis Division, P.M.B. 2005, Kaduna

PAKISTAN

Qureshi, Z.A.	Atomic Energy Agricultural Research Centre, Tandojam Sind

PERU

Simon F., J.E.	Estación Experimental Agraria, La Molina, P.O. Box 2791, Lima

POLAND

Suski, Z.W.	Research Institute of Pomology, ul. Pomologiczna 18, Skierniewice

SPAIN

Ros, J.P.	Ministerio Agricultura, Instituto Nacional de Investigaciones Agronómicas, Avda. Puerta Hierro S/11, Madrid-3

SUDAN

El Amin, El Tigani M.   Agriculture Research Corporation,
                                          Entomology Section, Wad Medani

SWITZERLAND

Boller, E.F.   Swiss Federal Research Station for Arboriculture,
                                             Viticulture and Horticulture,
                                          8820 Wädenswil

THAILAND

Loaharanu, S.   Office of Atomic Energy for Peace,
                                          Srirubsook Rd., Bangkok 9

Minanandana, N.   Plant Protection Service Division,
                                          Department of Agricultural Extension,
                                          Bangkok

TURKEY

Kansu, I.A.   A.Ü. Ziraat Fakültesi,
                                          University of Ankara,
                                          Department of Entomology, Ankara

UGANDA

Opiyo, Elizabeth A.   East African Trypanosomiasis Research Organisation,
                                          P.O. Box 96, Tororo

UNION OF SOVIET SOCIALIST REPUBLICS

Andreev, S.V.   All Union Plant Protection Institute,
                                          Leningrad

Romantchenko, A.A.   All Union Plant Protection Institute,
                                          Insect Biology,
                                          Poltavskoe Chosse 7,
                                          Kichinev 31

UNITED KINGDOM

Bradbury, F.R.   University of Stirling,
                                          Department of Industrial Science,
                                          Stirling

Lever, B.G.   Plant Protection Ltd., I.C.I.,
                                          Fernhurst, Haslemere, Surrey

UNITED STATES OF AMERICA

Baranowski, R.M.

University of Florida,
Agricultural Research and Education Center,
18905 SW 280 St.,
Homestead, Fla. 33030

Bogyo, T.P.

Statistical Services,
Washington State University,
Pullmann, Wash. 99163

Cavin, G.E.

US Department of Agriculture,
Animal and Plant Health Inspection Service,
Federal Building, Rm. 646,
Hyattsville, Md. 20782

Grosch, D.S.

North Carolina State University,
3513 Gardner Hall,
Raleigh, N.C. 27607

LaChance, L.E.

Metabolism and Radiation Research Laboratory,
Agricultural Research Service U.S.D.A.,
Fargo, N. Dak. 58102

Walker, D.W.

Puerto Rico Nuclear Center,
University of PR/RUM,
Mayaguez, Puerto Rico 000708

Weidhaas, D.E.

US Department of Agriculture,
ARS, Insects Affecting Man, P.O. Box 14565,
Gainesville, Fla. 32604

ORGANIZATIONS

CEC (COMMISSION OF EUROPEAN COMMUNITIES)

Cavalloro, R.

Joint Nuclear Center of Euratom,
21020 Ispra (Varese), Italy

FAO (FOOD AND AGRICULTURE ORGANIZATION OF THE UNITED NATIONS)

Buyckx, E.J.

Via delle Terme di Caracalla,
00100 Rome, Italy

Finelle, P.R.

Div. de la production et de la santé animales,
Via delle Terme di Caracalla,
00100 Rome, Italy

Fletcher, B.S.

CSIRO, Division of Entomology,
Canberra ACT, Australia

ICIPE (INTERNATIONAL CENTRE OF INSECT PHYSIOLOGY AND ECOLOGY)

Ogah, F.

P.O. Box 30772,
Nairobi, Kenya

Syed, K.

P.O. Box 30772,
Nairobi, Kenya

WHO (WORLD HEALTH ORGANIZATION)

Pal, R.

1211 Geneva 27,
Switzerland

IAEA (INTERNATIONAL ATOMIC ENERGY AGENCY)

Bauer, B.

Joint FAO/IAEA Division of Atomic Energy
in Food and Agriculture,
IAEA, Kärntner Ring 11, P.O. Box 590,
1011 Vienna, Austria

Butt, B.

Division of Research and Laboratories,
IAEA, Kärntner Ring 11, P.O. Box 590,
1011 Vienna, Austria

Haucke, Margarethe

Division of Research and Laboratories,
IAEA, Kärntner Ring 11, P.O. Box 590,
1011 Vienna, Austria

Malekghassemi, B.

Division of Research and Laboratories,
IAEA, Kärntner Ring 11, P.O. Box 590,
1011 Vienna, Austria

Mourad, A.

Division of Research and Laboratories,
IAEA, Kärntner Ring 11, P.O. Box 590,
1011 Vienna, Austria

Nadel, D.J.

Division of Research and Laboratories,
IAEA, Kärntner Ring 11, P.O. Box 590,
1011 Vienna, Austria

Tateya, A.

Division of Research and Laboratories,
IAEA, Kärntner Ring 11, P.O. Box 590,
1011 Vienna, Austria

Wetzel, H.W.

Joint FAO/IAEA Division of Atomic Energy
in Food and Agriculture,
IAEA, Kärntner Ring 11, P.O. Box 590,
1011 Vienna, Austria

AUTHOR INDEX

(including participants in discussions)

The numbers underlined indicate the first page of a paper by an author.
Other numbers denote the page numbers of discussion comments.
Literature references are not indexed.

HOW TO ORDER IAEA PUBLICATIONS

Exclusive sales agents for IAEA publications, to whom all orders and inquiries should be addressed, have been appointed in the following countries:

UNITED KINGDOM	Her Majesty's Stationery Office, P.O. Box 569, London SE 1 9NH
UNITED STATES OF AMERICA	UNIPUB, Inc., P.O. Box 433, Murray Hill Station, New York, N.Y. 10016

In the following countries IAEA publications may be purchased from the sales agents or booksellers listed or through your major local booksellers. Payment can be made in local currency or with UNESCO coupons.

ARGENTINA	Comisión Nacional de Energía Atómica, Avenida del Libertador 8250, Buenos Aires
AUSTRALIA	Hunter Publications, 58 A Gipps Street, Collingwood, Victoria 3066
BELGIUM	Office International de Librairie, 30, avenue Marnix, B-1050 Brussels
CANADA	Information Canada, 171 Slater Street, Ottawa, Ont. K 1 A OS 9
C.S.S.R.	S.N.T.L., Spálená 51, CS-11000 Prague
	Alfa, Publishers, Hurbanovo námestie 6, CS-80000 Bratislava
FRANCE	Office International de Documentation et Librairie, 48, rue Gay-Lussac, F-75005 Paris
HUNGARY	Kultura, Hungarian Trading Company for Books and Newspapers, P.O. Box 149, H-1011 Budapest 62
INDIA	Oxford Book and Stationery Comp., 17, Park Street, Calcutta 16
ISRAEL	Heiliger and Co., 3, Nathan Strauss Str., Jerusalem
ITALY	Libreria Scientifica, Dott. de Biasio Lucio "aeiou", Via Meravigli 16, I-20123 Milan
JAPAN	Maruzen Company, Ltd., P.O.Box 5050, 100-31 Tokyo International
NETHERLANDS	Marinus Nijhoff N.V., Lange Voorhout 9-11, P.O. Box 269, The Hague
PAKISTAN	Mirza Book Agency, 65, The Mall, P.O.Box 729, Lahore-3
POLAND	Ars Polona, Centrala Handlu Zagranicznego, Krakowskie Przedmiescie 7, Warsaw
ROMANIA	Cartimex, 3-5 13 Decembrie Street, P.O.Box 134-135, Bucarest
SOUTH AFRICA	Van Schaik's Bookstore, P.O.Box 724, Pretoria
	Universitas Books (Pty) Ltd., P.O.Box 1557, Pretoria
SPAIN	Nautrónica, S.A., Pérez Ayuso 16, Madrid-2
SWEDEN	C.E. Fritzes Kungl. Hovbokhandel, Fredsgatan 2, S-10307 Stockholm
U.S.S.R.	Mezhdunarodnaya Kniga, Smolenskaya-Sennaya 32-34, Moscow G-200
YUGOSLAVIA	Jugoslovenska Knjiga, Terazije 27, YU-11000 Belgrade

Orders from countries where sales agents have not yet been appointed and requests for information should be addressed directly to:

Publishing Section,
International Atomic Energy Agency,
Kärntner Ring 11, P.O.Box 590, A-1011 Vienna, Austria

INTERNATIONAL
ATOMIC ENERGY AGENCY
VIENNA, 1975

PRICE: US $36.00
       Austrian Schillings 640,—
       (£15.50; F.Fr. 168; DM 88,—)

SUBJECT GROUP: I
Life Sciences/Entomology